Nonlinear Vibration and Dynamics of Smart Continuous Structures and Materials

Nonlinear Vibration and Dynamics of Smart Continuous Structures and Materials delves into intricate subjects concerning the analysis of nonlinear vibration issues in continuous structures. It covers general concepts and a history of nonlinear systems before evolving into kinetics and solution methods of continuous structures.

Exploring the implementation of new types of materials in various sectors of automobile, aerospace, and structural engineering, the book provides applicable information on the behaviors of smart structures. The book provides a set of mathematical formulations to solve nonlinear static and dynamic behaviors of smart continuous structures by applying principles of elasticity.

The book will interest academic researchers and graduate students studying structural engineering, mechanics of solids, and smart materials.

Nonlinear Vibration and Dynamics of Smart Continuous Structures and Materials

Majid Ghadiri

CRC Press
Taylor & Francis Group
Boca Raton London New York

CRC Press is an imprint of the
Taylor & Francis Group, an **informa** business

Designed cover image: Majid Ghadiri

First edition published 2025
by CRC Press
2385 NW Executive Center Drive, Suite 320, Boca Raton FL 33431

and by CRC Press
4 Park Square, Milton Park, Abingdon, Oxon, OX14 4RN

CRC Press is an imprint of Taylor & Francis Group, LLC

ISBN: 978-1-032-74747-7 (hbk)
ISBN: 978-1-032-74748-4 (pbk)
ISBN: 978-1-003-47069-4 (ebk)

DOI: 10.1201/9781003470694

Typeset in Times
by Apex CoVantage, LLC

Contents

About the Author

Majid Ghadiri is Associate Professor in the Department of Mechanical Engineering, Imam Khomeini International University (IKIU), Qazvin, Iran. His research interests include nonlinear vibrations, dynamics of micro and nano structures, applied mathematics, and fracture mechanics. Dr. Ghadiri has authored more than 100 high-quality, peer-reviewed research articles in his fields of interest. He has also edited and authored multiple books for well-known publishers. He is a distinguished reviewer whose expertise helps the editors of prestigious journals judge research articles.

Preface

The objective of this book is to delve into intricate subjects concerning the analysis of nonlinear vibration issues in continuous structures. The target audience for this material is graduate students seeking to expand their knowledge in this area. This document aims to present the fundamental principles of nonlinear vibrations in a clear and concise manner, both in theory and in practice. Numerous practical examples have been utilized to elaborate on nonlinear concepts comprehensively. The methodology employed for elucidating and articulating the subject matter is designed to familiarize the student and researcher with the primary and essential principles of linear vibrations. In essence, the approach involves a comprehensive coverage of these principles. Subsequently, the individual acquires knowledge of the fundamental principles underlying nonlinear vibrations, followed by an understanding of the techniques employed in the analysis and resolution of such issues. The researcher's cognitive abilities are significantly tested by the presentation of a practical and intricate example.

The literature reveals the lack of a comprehensive book concerning nonlinear vibration and dynamics of smart continuous structures. Because of the existence of such a lack in the literature, the proposed book will be arranged to analyze the nonlinear vibration and dynamics of smart continuous structures thoroughly. The proposed book, *Nonlinear Vibration and Dynamics of Smart Continuous Structures and Materials*, offers a wide range of application-based and practical considerations of state-of-the-art smart continuous structures. A fascinating exploration of the analytical and numerical solution procedures can be found in this comprehensive book, and each chapter provides and embraces detailed information of crucial characteristics of nonlinear vibration and dynamics of smart continuous structures. If you want to find a thorough all-in-one answer for all aspects of nonlinear vibration and dynamics of smart continuous structures, this book is highly recommended. In the proposed book, the first chapter will be dedicated to the principles of vibration. Chapter 2, "An Introduction to the Nonlinear Vibration", contains general concepts and a history of nonlinear systems. Qualitative analysis of nonlinear vibration will be discussed in Chapter 3. Solution methods of continuous structures will be presented in Chapter 4. In Chapter 5, nonlinear forced vibration will be discussed. Nonlocal systems and kinematics of the continuous structures will be explained in Chapter 6. An introduction to smart materials will be presented in Chapter 7. In Chapter 8, dynamics of nonlinear smart continuous structures—beams—will be discussed in detail, and in Chapter 9, dynamics of nonlinear smart continuous structures—plate—will be analyzed.

1 Principles of Vibrations

1.1 INTRODUCTION

This chapter starts the examination of oscillatory motion in a rather uncomplicated manner. Following a short historical overview, we analyze the significance of vibrations. Next, we enumerate the many stages employed in vibration analysis and provide concise explanations of the terminologies and principles that are involved. This chapter declines to provide a qualitative analysis of the subject.

1.2 HISTORICAL DEVELOPMENT OF VIBRATION

Presumably, upon the creation of the initial musical instruments such as whistles or drums, humans developed an interest in the concept of vibrations and started its research. The origins of wired musical instruments, like as the harp, may be traced to around 3,000 years ago through depictions found on ancient Egyptian artifacts. In that period, musicians and philosophers endeavored to investigate the principles governing sound and utilized these principles to enhance the functionality of musical instruments. The research of vibrational behavior has proven highly significant.

Over the course of time, researchers examined the vibrational characteristics of many systems and structures. Galileo was originally drawn to the Church of Pisa by the lamp's oscillating motion. In addition, Robert Hook (1703–1635) performed tests to ascertain the correlation between the length of a wire and the frequency of its vibrations. Indeed, Joseph Sauer (1716–1653) conducted these investigations and employed the term "acoustic" to refer to the study of phonology [1].

Joseph LaGrange (1813–1736) conducted an analysis of the oscillation of the vibrating wire, which was published in an essay by the Academy of Turin in 1759. In this analysis, he conceptualized the wire as a finite number of homogeneous mass particles positioned equidistantly from one another. In 1744, Euler and Bernoulli conducted the first examination of the vibrations of thick beams, which subsequently became known as the Euler-Bernoulli theory or thick beam theory.

Over the past few decades, there has been an increasing significance in the study of vibrations in intricate systems. In 1902, Fromm conducted a study on the behavior of torsion vibrations in the grasshopper shaft's design. Stefan Timoshenko (1872–1878) formulated a comprehensive theory on the oscillation of beams by studying rotation and shear deformation. This hypothesis is sometimes referred to as Timoshenko's bean theory or thick beam theory.

Mechanical vibrations may be categorized into several types, including free and forced vibrations, damped and undamped vibrations, linear and nonlinear vibrations, and regular and random vibrations. Free vibrations refer to the fluctuations of a system that occur without any external force acting on it, following the initial turbulence caused by either the initial displacement or velocity. On the other hand,

DOI: 10.1201/9781003470694-1

forced vibrations occur when a system varies under the influence of an external force. The term "undamped vibrations" refers to the movement of an oscillating system in which no energy is lost. Vibrations that occur when the energy of a system is lost due to any cause are referred to as damped. Linear vibrations refer to the movement of an oscillating system when all components, such as springs, masses, and dampers, exhibit linear behavior. On the other hand, if any of these components deviate from linear behavior, the movement of the system is referred to as nonlinear vibrations. Regular vibrations occur in an oscillating system when the amount of the effective excitation force is known at any given point, resulting in a predictable movement of the system. This type of excitation is referred to as regular excitation. Occasionally, the input force that stimulates a system is stochastic, and its magnitude cannot be ascertained at a certain instant. Random vibrations refer to vibrations that are created by unpredictable stimulus.

It is widely understood that the majority of dynamic phenomena, such as vibrations, are characterized by nonlinearity. Nonlinear systems exhibit behaviors that are not possible in linear systems. In 1892, Poincaré and Lyapunov pioneered the mathematical theory of nonlinear vibrations. In 1892, Poincaré devised the concept of perturbation to estimate the solution of mechanical problems. In 1920, Duffing and Van der Pol discovered the first solution to nonlinear vibrations [2]. Furthermore, the phenomenon of chaos has garnered the attention of scientists, such as Glieck [3] and Peitgen and Richter [4], as one of the nonlinear phenomena.

It is noteworthy that engineers in the 1950s developed the finite element approach, which enabled the analysis of intricate mechanical systems with several degrees of freedom [5]. The research of vibration is comprehensively described in Table 1.1.

TABLE 1.1
Providing a Concise Overview of the Historical Development of the Field of Vibrations

Year/Period	Event/Development
Ancient Times	Understanding of basic vibrations in musical instruments (e.g., strings and drums).
6th Century BC	Pythagoras studies vibrating strings and the relationship between length, tension, and pitch.
1st Century BC	Vitruvius, a Roman engineer, writes about the vibration of structures.
17th Century	Galileo Galilei investigates the oscillatory motion and pendulums.
1665	Robert Hooke discovers the law of elasticity (Hooke's Law) related to springs and vibrations.
1687	Isaac Newton's "Principia Mathematica" lays the foundation for classical mechanics and vibrations.
18th Century	Daniel Bernoulli and Leonhard Euler develop theories on the vibrations of beams and plates.
1738	Daniel Bernoulli publishes the principle of superposition in "Hydrodynamic".

TABLE 1.1 (*Continued*)
Providing a Concise Overview of the Historical Development of the Field of Vibrations

Year/Period	Event/Development
1746	Jean le Rond d'Alembert introduces the concept of damping in vibrating systems.
1822 Joseph	Fourier introduces Fourier series, allowing complex vibrations to be broken down into simpler components.
1842	Gustav Kirchhoff develops the theory of elastic vibrations.
1869	Lord Rayleigh publishes "The Theory of Sound", a fundamental text on acoustics and vibrations.
19th Century	Development of mathematical tools (e.g., differential equations) to analyze vibrations.
Early 20th Century	Advances in material science and engineering improve understanding of structural vibrations.
1940s	Development of vibration testing and analysis techniques during WWII for aircraft and machinery.
1950s	Emergence of electronic devices (e.g., accelerometers) for measuring vibrations.
1960s	Introduction of modal analysis and finite element methods (FEM) in vibration analysis.
1980s	Advances in computational methods and software for simulating and analyzing vibrations.
21st Century	Use of advanced sensors, machine learning, and real-time monitoring in vibration analysis.

1.3 SIGNIFICANCE OF STUDYING VIBRATION

Previously, scientists exerted much effort to comprehend the processes of nature and devise mathematical theories pertaining to vibrations. Currently, there is a growing focus on the study of vibrations and their practical applications. This is an endeavor to utilize the application of vibrations in the construction of machinery, foundations, buildings, engines, turbines, and control systems. The investigation of mechanical vibrations holds significance in the fields of engineering and science.

The significance of this matter stems from the following factors:

- **Enhancing the longevity of structures and equipment:** Monitoring and regulating vibrations can avert early and partial breakdown, hence, extending the operational lifespan of equipment.
- **Enhancing efficiency and performance:** Gaining a comprehensive understanding of vibrations and their management can contribute to the enhancement of machinery and equipment's efficiency and performance. This is because unmanageable vibrations have the potential to diminish efficiency and give rise to performance issues.

- **Safety:** The regulation and supervision of vibrations in structures and machinery can effectively mitigate accidents and potential threats to human life.
- **Analysis and design of structures:** In the design and analysis of structures and engineering systems, the recognition of vibrations plays an important role. This issue is vital to prevent resonance and resonance phenomena that can lead to the destruction of structures.
- **Sound and acoustics:** Mechanical vibrations are connected to the phenomena of sound and acoustics. Gaining insight into this correlation can aid in minimizing noise and enhancing sound fidelity across various settings.
- **Diagnosing and rectifying defects:** Assessing vibrations is a prevalent approach for diagnosing defects and issues in machinery and structures. Vibration analysis can assist in the early detection of flaws prior to the occurrence of significant failures.

Studying mechanical vibrations can generally result in enhanced design, heightened safety, decreased maintenance expenses, and optimized performance of systems and structures.

It is often observed that the engines of most automobiles are prone to fluctuations, leading to performance issues. For instance, a conventional diesel engine produces a significant amount of noise. In turbines, vibrations result in the failure of components.

Resonance occurs when the natural frequency of vibrations in a system matches the frequency of external stimulus. Hence, when engineering systems are designed, the ability of these systems to withstand vibrations is also considered (see Figure 1.1).

In contrast to the earlier described detrimental consequences, vibrations also possess beneficial industrial utilizations. For instance, vibrations find use in many devices such as conveyor, funnels, screens, washing machines, dental drills, timepieces, electrical massage devices, material oscillation testing, and even in the surface cleaning of mechanical parts (Figure 1.2).

1.4 THE FUNDAMENTAL PRINCIPLES OF VIBRATIONS

1.4.1 Vibration

Vibrational or oscillatory motion can be characterized in two distinct manners:

1. The process of continuous movement when potential energy and kinetic energy are alternately transformed.
2. A vibrating item is an elastic mass that exhibits oscillations.

Vibration or oscillation is the term used to describe any repetitive movement that occurs at regular time intervals. A pendulum is a weight suspended from a fixed point that swings back and forth under the force of gravity. Oscillation and elongated thread motion exemplify oscillatory motion. The theory of vibrations pertains to the examination of the oscillatory motion of objects and the forces that are influenced by it.

FIGURE 1.1 Nature of wind-induced vibration experienced by Tacoma Narrows Bridge before its failure. The bridge opened on July 1, 1940, and collapsed on November 7, 1940.

FIGURE 1.2 Screening machine using vibrations.

Oscillation systems can be categorized into many classifications:

1. Classification based on the mathematical representation:
   1.1: *Discrete systems* refer to situations when we encounter ordinary differential equations with the independent variable (time) constrained by a finite number of degrees of freedom.
   1.2: *Continuous systems* refer to systems having an infinite degree of freedom, characterized by the presence of partial or partial differential equations that involve both spatial and temporal variables.
2. Classification based on the system's behavior:
   2.1: Linear behavior refers to a situation where there is a direct and proportional relationship between two variables.
   2.2: Nonlinear phenomena.

Nonlinear behavior can be attributed to many factors:

Option A: Large displacements or rotations

Illustration 1.3 depicts the undamped oscillations of a simple pendulum.

The pendulum motion Equation (1.1) is displayed next.

$$
\begin{gathered}
-mg\sin\theta = ml\,\ddot{\theta} \\
\Rightarrow \ddot{\theta} + \frac{g}{l}\sin\theta = 0 \Rightarrow \ddot{\theta} + \frac{g}{l}\theta = 0
\end{gathered}
\tag{1.1}
$$

It is important to observe that when the sine of θ undergoes modest periodic displacements ($\sin\theta \simeq \theta$), the equation takes the form of a linear equation.

Illustration 1–4 depicts a substance that exhibits both linear and nonlinear characteristics and possesses elasticity.

a) Material with linear elastic behavior in the elastic range
b) Substance exhibiting nonlinear elastic behavior within the elastic range

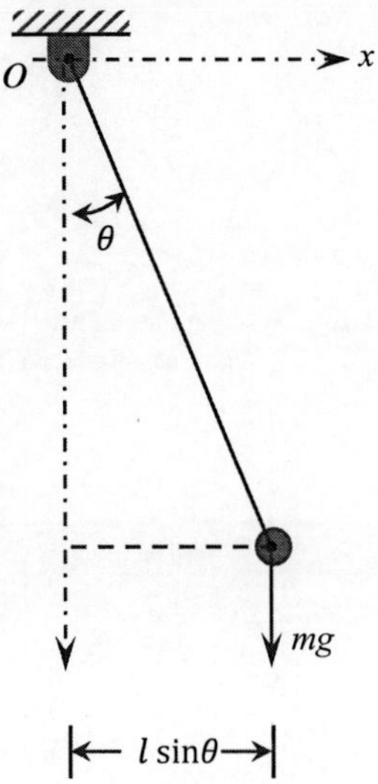

FIGURE 1.3 Display of a simple pendulum.

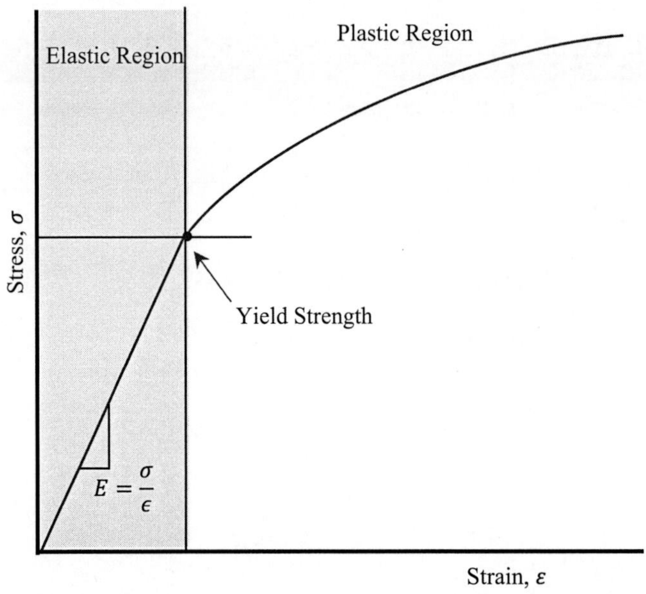

FIGURE 1.4 Display of the elastic behavior in both linear and nonlinear states.

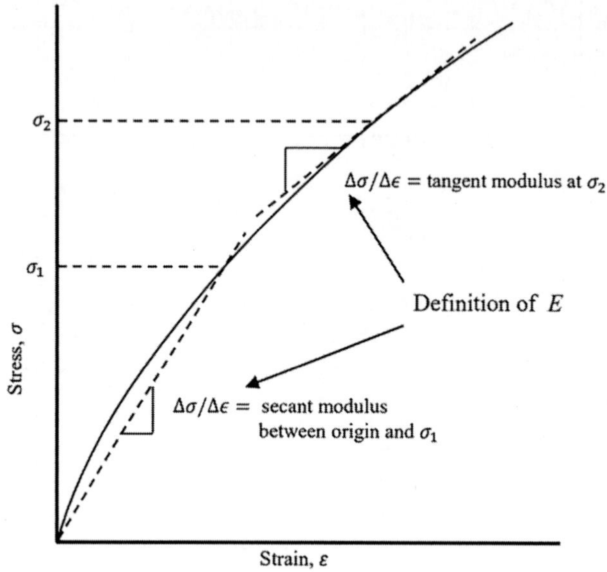

FIGURE 1.4 (Continued)

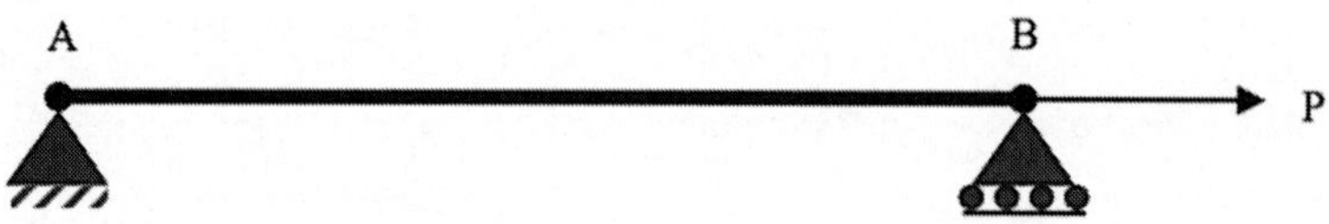

FIGURE 1.5 Display of a beam that has undergone transverse displacement due to axial force.

Illustration 1.5 depicts the impact of longitudinal stress on the beam with transverse displacement.

There exist two methods for representing oscillatory systems:

1. Utilizing public behavior as a basis for modeling:
   A) Free vibration: referring to oscillation occurring without any external force.
   B) Forced vibration: where oscillation is carried out in response to external stimulus and is categorized into three types: 1—harmonic, 2—periodic, 3—non-periodic (in general).
2. Energy consumption-based modeling:
   A. Without damper.
   B. With damper, which is further categorized into three types. The three categories are as follows: 1—viscose, 2—Coulomb, and 3—structural.

1.4.2 Components of a Vibrating System

An oscillating system typically comprises three main components: a potential energy storage device (such as an elastic spring or item), a kinetic energy storage device (such as a mass or inertia), and an energy loss system (such as a damper or frictional element). During oscillation, the system undergoes a conversion of potential energy into kinetic energy, and vice versa [6]. In the case of the damped system, there is a loss of energy in each oscillation cycle, as seen in Figure 1.6.

1.4.3 Degree of Freedom

The degree of freedom of a system is the minimum number of independent coordinates needed to determine the state of all system components at any one time. The system depicted in Figure 1.6 consists of a mass, a spring, and a damper, and it possesses a single degree of freedom. The movement of the mass-spring-damper system seen in Figure 1.6 may be described using coordinates x and y. Indeed, the x and y coordinates are interdependent. Alternatively, the symbol θ might denote the rotational movement (as seen in Figure 1.7).

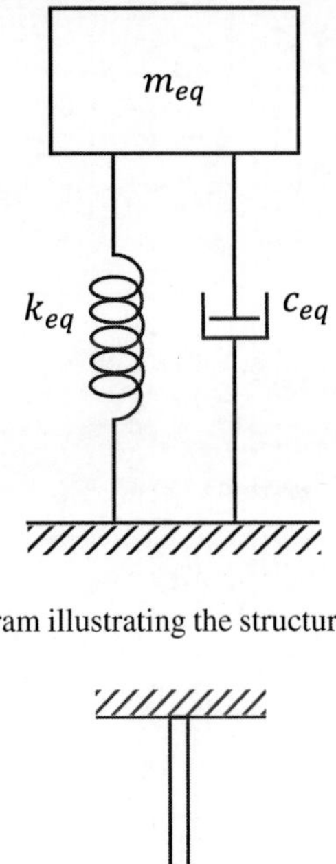

FIGURE 1.6 Display of a diagram illustrating the structure of the mass-spring-plain damper.

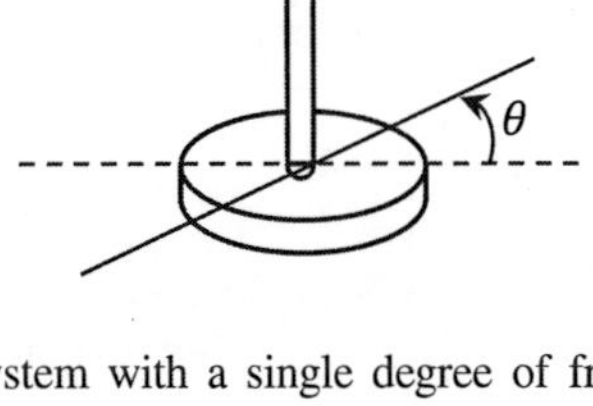

FIGURE 1.7 Display of a system with a single degree of freedom that exhibits a twisting motion.

FIGURE 1.8 Systems with two degrees of freedom.

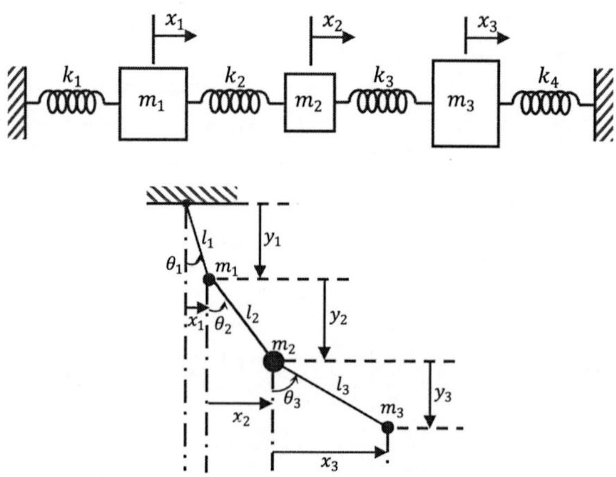

FIGURE 1.9 Systems that possess three degrees of freedom.

Figures 1.8 and 1.9 depict systems with two and three degrees of freedom, respectively.

1.5 REVIEWING MATHEMATICAL EQUATIONS IN MECHANICAL VIBRATIONS

1.5.1 The Differential Equation of Motion for a System with One Degree of Linear Freedom

Figure 1.10 presents the generic form of the differential equation of motion for a system with one degree of freedom, as described by Newton's second law.

$$m\ddot{x}(t) + c\dot{x}(t) + kx(t) = 0 \tag{1.2}$$

Symbols m,c and k represent the mass values, damping coefficient, and spring stiffness, respectively. The variable x represents the displacement from the equilibrium point.

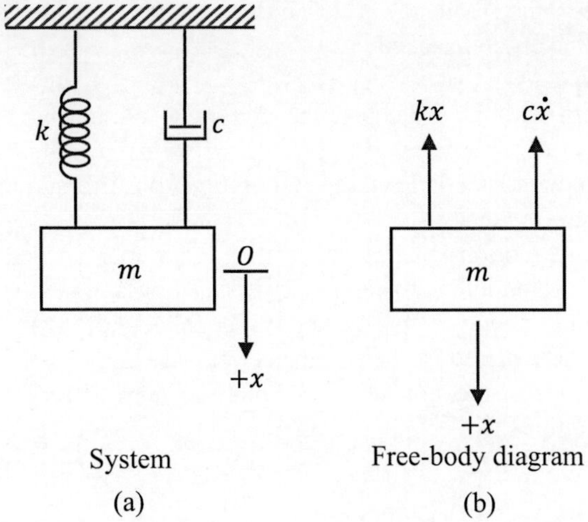

FIGURE 1.10 One-degree system of freedom with a damper: a) system and b) free-body diagram.

1.5.2 UNDAMPED FREE VIBRATION (C = 0)

Under the assumption of no energy loss in the oscillation process, the equation of motion may be reduced as follows:

$$m\ddot{x}(t) + kx(t) = 0 \tag{1.3}$$

The equation is homogeneous, and its solution is also homogeneous. The answer to this problem is represented in the following manner:

$$\begin{aligned} x(t) &= A\cos(\omega_n t - \varphi), \\ \omega_n &= \sqrt{\frac{k}{m}} \end{aligned} \tag{1.4}$$

The constants A and φ are derived from the initial conditions, and their values are unknown. To initiate movement according to the initial conditions, one must modify either the initial position or the first velocity, or a combination of both. The free vibration response may be determined by including the initial conditions into the equation of motion.

$$\begin{gathered} \begin{cases} x(0) = x_0 \\ \dot{x}(0) = V_0 \end{cases} \\ x(t) = x_0 \cos\omega_n t + \frac{V_0}{\omega_n}\sin\omega_n t \end{gathered} \tag{1.5}$$

From the analysis of the free vibration response, it is evident that the system would undergo harmonic oscillations indefinitely, maintaining a consistent amplitude and frequency ω_n, which represents the natural frequency of the system. The natural frequency refers to the frequency at which the system's free fluctuations occur. The following are some crucial aspects:

- In a system without a damper, the amplitude of the movement remains consistent.
- In a system without a damper, the system functions in a harmonious manner at its natural frequency.
- The system's motion is perpetual, with a continuous conversion of kinetic energy into potential energy and vice versa, extending indefinitely.
- The natural frequency of the system is independent of the initial conditions and solely determined by the parameters m and K.
- The amplitude of movement and the phase change differ according to the initial conditions.

1.6 HARMONIC MOVEMENT

Oscillatory movement may be classified into two types: regular repetition, as seen in pendulum movement, and erratic repetition, as observed in earthquakes. When a movement is repeated within the same time frame, it is referred to as periodic motion. Harmonic motion is the most basic form of periodic motion. In the Scotch yoke mechanism depicted in Figure 1.11, the mass m undergoes simple harmonic motion. Within this system, a crank undergoes rotational motion with a certain radius, denoted as A, around a fixed point known as Point O. Point P, located at the other end of the crank, is inserted into a slotted rod.

When the wheel spins at an angular velocity ω, the displacement x changes accordingly as follows:

$$x = A\sin\theta = A\sin\omega t \tag{1.6}$$

The velocity and acceleration of mass m are as follows:

$$\dot{x} = \frac{dx}{dt} = \omega A\cos\omega t \tag{1.7}$$

$$\ddot{x} = \frac{d^2x}{dt^2} = -\omega^2 A\sin\omega t = -\omega^2 x \tag{1.8}$$

This phenomenon is referred to as oscillatory motion, characterized by acceleration that is directly proportionate to the displacement. It is commonly known as simple harmonic motion. The vector $\overrightarrow{OP}$ (Figure 1.12) can be used to depict harmonic motion. This vector possesses a magnitude and undergoes rotational motion with a consistent angular velocity ω. The picture of the end of the vector $\vec{X} = \overrightarrow{OP}$ on the vertical and horizontal axes may be seen in Figure 1.12.

$$y = A\sin\omega t \tag{1.9}$$

$$x = A\cos\omega t \tag{1.10}$$

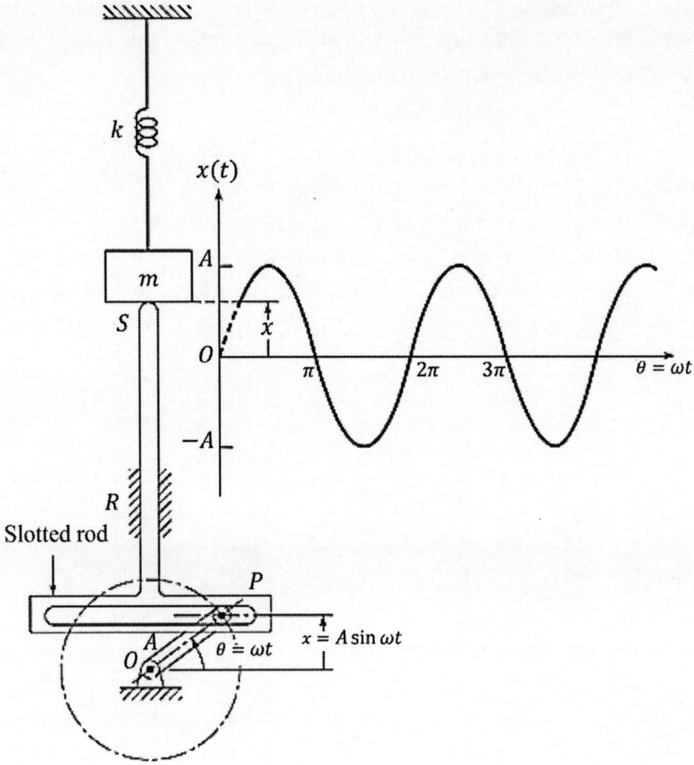

FIGURE 1.11 Scout yoke mechanism.

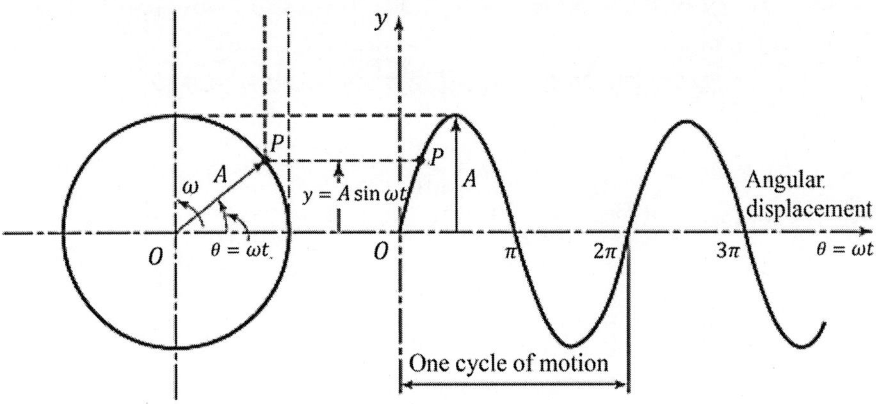

FIGURE 1.12 Harmonic motion as the vertical representation of the terminal point of a rotating vector.

Example 1–1: Calculate the natural frequency of the system depicted in Figure 1.13.

Response: Since the weight's horizontal location does not impact the system's natural frequency, we may conclude the following:

$$\sum M = I\ddot{\theta} \rightarrow ML^2\ddot{\theta} + ka^2\theta = 0 \rightarrow \omega_n = \frac{a}{L}\sqrt{\frac{k}{M}}$$

Example 1–2: Figure 1.14 depicts a cylinder with a mass of m and a moment of inertia of J_0. The cylinder is rolling without slipping and is confined by two linear springs with stiffness values of k_1 and k_2. Find the following: A) The natural frequency of vibration for the system. B) The optimal value of parameter a to maximize the natural frequency.

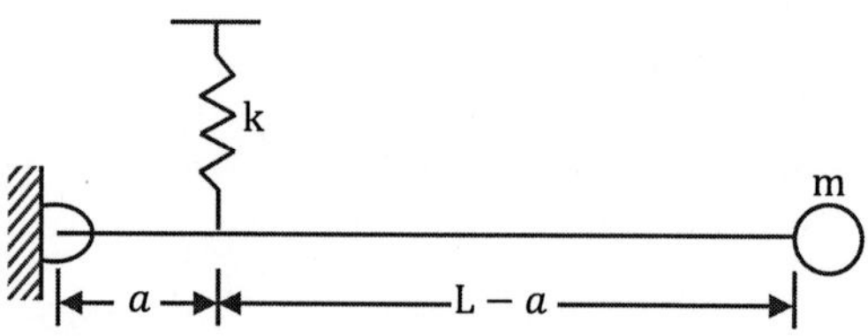

FIGURE 1.13 A rod with no mass that is accompanied by a concentrated mass at its head.

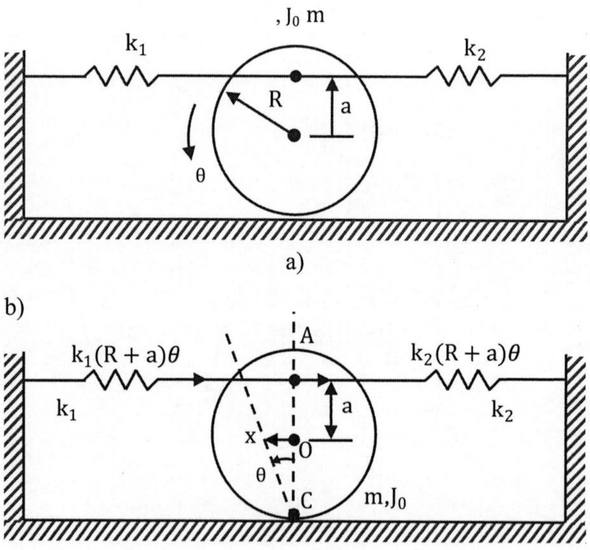

FIGURE 1.14 Roller cylinder is restrained by a spring a) before b) after.

Answer: A) The mass inertia for the cylinder may be expressed as follows:

$$J_0 = \frac{1}{2}mR^2, \qquad J_c = \frac{1}{2}mR^2 + mR^2$$

The equation of motion for the cylinder will be determined by a slight angular deviation, denoted as θ.

$$\Sigma M = J_c\ddot{\theta} \rightarrow J_c\ddot{\theta} + k_1(R+a)^2\theta + k_2(R+a)^2\theta = 0$$

$$\rightarrow \omega_n = \sqrt{\frac{(k_1+k_2)(R+a)^2}{J_c}} = \sqrt{\frac{(k_1+k_2)(R+a)^2}{\frac{3}{2}mR^2}}$$

B) We put $d\omega_n/da = 0$. The result is $a = R$.

1.7 FREE VIBRATION WITH VISCOUS DAMPER

Based on the general equation of motion (Equation (1.2)) and Figure 1.10, Equation (1.2) may be expressed in a more concise manner as follows:

$$\ddot{x}(t) + \frac{c}{m}\dot{x}(t) + \frac{k}{m}x(t) = 0 \tag{1.11}$$

Or

$$\ddot{x}(t) + 2\xi\omega_n\dot{x}(t) + \omega_n^2 x(t) = 0 \tag{1.12}$$

In Equation (1.12), ξ is a dimensionless quantity known as the damping ratio. The presence of a viscous damper in the system significantly influences the motion of the system due to the influence of ξ. The answer to the Equation (1.12) is as follows:

$$\begin{aligned} x(t) &= C_1e^{S_1t} + C_2e^{S_2t} \\ &= C_1e^{\left\{-\frac{c}{2m}+\sqrt{\left(\frac{c}{2m}\right)^2-\frac{k}{m}}\right\}t} + C_2e^{\left\{-\frac{c}{2m}-\sqrt{\left(\frac{c}{2m}\right)^2-\frac{k}{m}}\right\}t} \end{aligned} \tag{1.13}$$

The values of C_1 and C_2, which are unknown constants, can be calculated based on the initial conditions.

1.7.1 Critical Damping and Critical Ratio

The value of the damping constant, denoted as c, at which the square root term in Equation (1.13) equals zero, is referred to as the critical damping and is symbolized as C_c:

$$\left(\frac{C_c}{2m}\right)^2 - \frac{k}{m} = 0 \tag{1.14}$$

In other words:

$$C_c = 2m\sqrt{\frac{k}{m}} = 2\sqrt{km} = 2m\omega_n \tag{1.15}$$

The ratio between the constant damping and critical damping is referred to as the viscous damping ratio and is defined as follows:

$$\xi = \frac{c}{C_c} \tag{1.16}$$

Therefore, Equation (1.13) is expressed as follows:

$$x(t) = C_1 e^{\left\{-\xi+\sqrt{\xi^2-1}\right\}\omega_n t} + C_2 e^{\left\{-\xi-\sqrt{\xi^2-1}\right\}\omega_n t} \tag{1.17}$$

It has been noted that at $\xi = 0$, the vibration is undamped. Typically, there are three modes that may be taken into account for ξ damping:

1. System with a damping ratio less than 1 ($\xi < 1$). Given that ($\xi^2 - 1$) is a negative value, we may express it as follows:

$$\begin{aligned} x(t) &= C_1 e^{\left\{-\xi+i\sqrt{\xi^2-1}\right\}\omega_n t} + C_2 e^{\left\{-\xi-i\sqrt{\xi^2-1}\right\}\omega_n t} \\ &= e^{-\xi\omega_n t}\left\{C_1 e^{i\sqrt{1-\xi^2}\,\omega_n t} + C_2 e^{i\sqrt{1-\xi^2}\,\omega_n t}\right\} \end{aligned} \tag{1.18}$$

The variables $\phi_0, X_0, X, \phi, C_2, C_1$ are derived from the initial conditions. For the initial condition $x(t=0) = x_0$, and $\dot{x}(t=0) = \dot{x}_0$:

$$x(t) = e^{-\xi\omega_n t}\left\{x_0 \cos\left(\sqrt{1-\xi^2}\,\omega_n t\right) + \frac{\dot{x}_0 + \xi\omega_n x_0}{\sqrt{1-\xi^2}\,\omega_n}\sin\sqrt{1-\xi^2}\,\omega_n t\right\} \tag{1.19}$$

Also if:

$$C_1' = x_0, C_2' = \frac{\dot{x}_0 + \xi\omega_n x_0}{\sqrt{1-\xi^2}\,\omega_n} \tag{1.20}$$

We have:

$$X = \sqrt{(C_1)^2 + (C_2)^2} \tag{1.21}$$

$$\phi = \tan^{-1}(C_1/C_2) \tag{1.22}$$

Equation (1.19) represents the harmonic oscillation of damped system. The angular frequency of this movement, known as the frequency of damped vibrations, may be expressed as follows:

$$\omega_d = \sqrt{1-\xi^2}\,\omega_n \tag{1.23}$$

The frequency of damped vibrations is consistently lower than the natural frequency of the undamped vibration (ω_n). Figure 1.15 and 1.18 depict the graphs of Equations (1.19) and (1.23), respectively. It is important to mention that the underdamped vibration is the sole state in which movement exhibits oscillatory behavior [7].

2. Critically damped system ($\xi = 1$)
   For $\xi = 1$ in this situation, the answer to Equation (1.17) will be determined by the presence of dual roots in the characteristic equation.

$$x(t) = -(c_1 + c_2)e^{-\omega_n t} \tag{1.24}$$

By utilizing the initial conditions $x(t=0) = x_0$ and $\dot{x}(t=0) = \dot{x}_0$:

$$x(t) = \left[x_0 + \left(\dot{x}_0 + \omega_n x_0\right)t\right]e^{-\omega_n t} \tag{1.25}$$

Equation (1.25) denotes a non-periodic motion. According to this equation, when $t \to \infty$, $e^{-\omega_n t} \to 0$. As a result, the mentioned movement finally stops (Figure 1.19).

3. Overdamped system ($\xi > 1$)
   In this case, Equation (1.17) is given as follows:

$$x(t) = C_1 e^{\left\{-\xi + \sqrt{\xi^2 - 1}\right\}\omega_n t} + C_2 e^{\left\{-\xi - \sqrt{\xi^2 - 1}\right\}\omega_n t} \tag{1.26}$$

For the initial conditions $x(t=0) = x_0$, $\dot{x}(t=0) = \dot{x}_0$:

$$C_1 = \frac{x_0\omega_n\left(\xi + \sqrt{\xi^2 - 1}\right) + \dot{x}_0}{2\omega_n\sqrt{\xi^2 - 1}}$$
$$C_2 = \frac{-x_0\omega_n\left(\xi - \sqrt{\xi^2 - 1}\right) - \dot{x}_0}{2\omega_n\sqrt{\xi^2 - 1}} \tag{1.27}$$

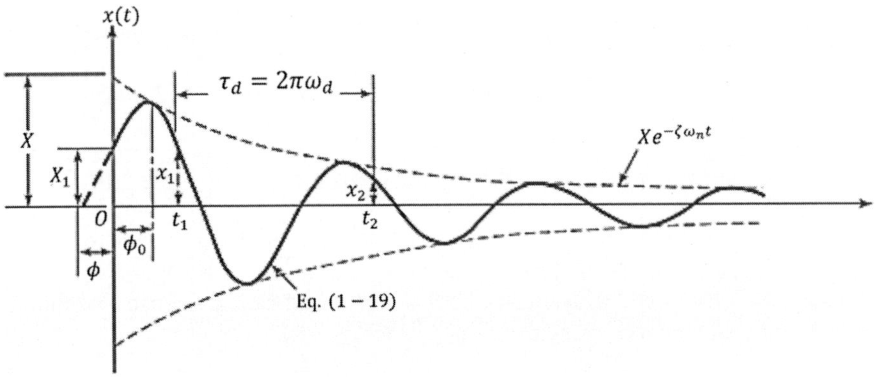

FIGURE 1.15 Vibration of the damped system.

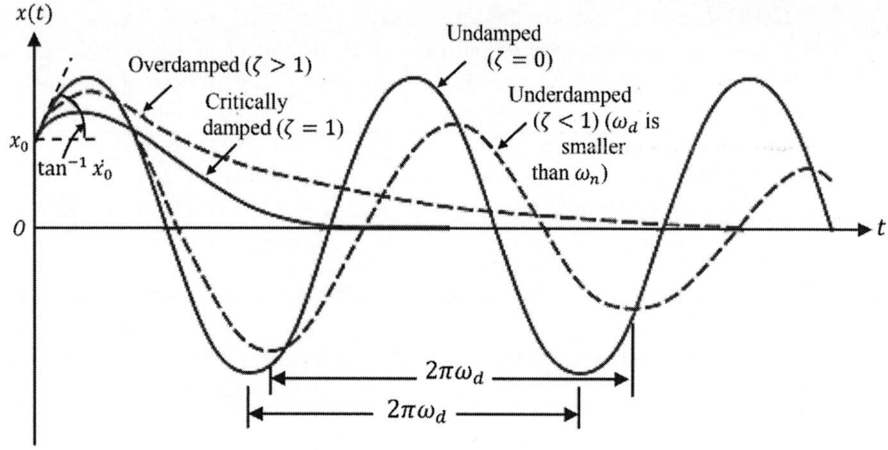

FIGURE 1.16 Comparison of motions based on different damping settings.

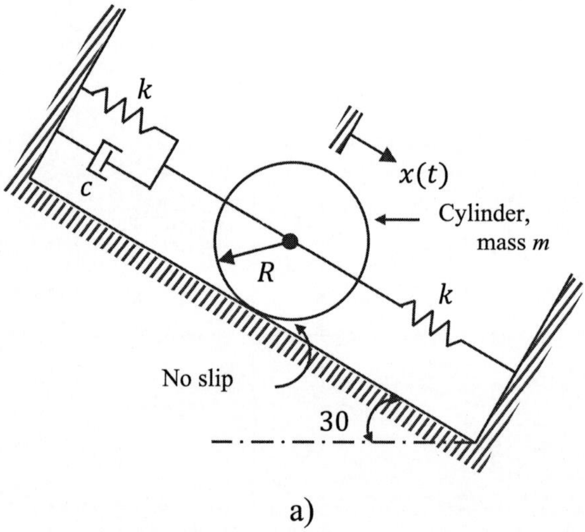

FIGURE 1.17 A rolling disk on an inclined surface that is impeded by a spring and damper.

Equation (1.26) demonstrates that regardless of the initial condition of the system, the motion remains continuous. As the roots are negative $\left(-\xi \pm \sqrt{\xi^2 - 1} < 0\right)$, the momentum experiences exponential decay over time (Figure 1.16).

Example 1–3: Determine the natural frequency and motion equation of the system Figure 1.17 (a).

Response: We have demonstrated the discrepancy of the cylinder in r relation to the state of static balance using the variable x (Figure 1.17 (b)):

$$\sum F = m\ddot{x} \rightarrow \frac{3}{2}m\ddot{x} + c\dot{x} + 2kx = 0 \rightarrow \omega_n = \sqrt{\frac{4k}{3m}}$$

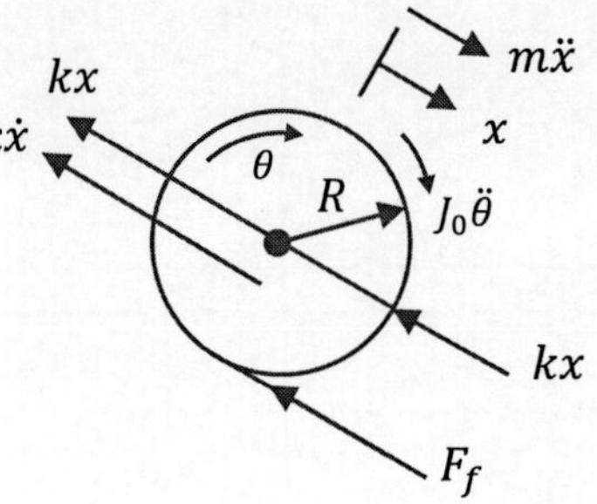

b)

FIGURE 1.17 (Continued)

Example 1–4: Derive the equation of motion for the oscillatory system shown in Figure 1.18 (a) and determine its natural frequency of vibration.

Response: We demonstrate the angular displacement of the rod in relation to the state of static balance using the symbol θ (as shown in Figure 1.18 (b)), and we obtain the following:

$$\sum M = J_o\ddot{\theta} \rightarrow -k\left(\frac{3}{4}l\theta\right)\left(\frac{3}{4}l\right) - c\left(\frac{l}{4}\right)\left(\frac{l}{4}\dot{\theta}\right) - 3k\left(\frac{l}{4}\right)\left(\frac{l}{4}\theta\right) = J_o\ddot{\theta}$$

$$J_o\ddot{\theta} + \frac{cl^2}{16}\dot{\theta} + \frac{3}{4}kl^2\theta = 0 \rightarrow \omega_n = \sqrt{\frac{4kl^2}{3J_o}}$$

In which:

$$J_o = \frac{7}{48}ml^2$$

1.8 FORCED HARMONIC OSCILLATIONS

1.8.1 Overview

Vibrations occur in a system when it experiences fluctuations due to external stimulus. The state of arousal can manifest in several forms, including harmonic, non-harmonic, periodic, non-periodic, or random patterns. The reaction of a system to harmonic excitation is referred to as its harmonic response. The reaction of a system to a sudden and sustained excitation that is abruptly applied is referred to as a transitory response.

This episode focuses on analyzing the reaction of systems that possess a single degree of freedom to harmonic excitation. If the excitation frequency matches the system's natural frequency, the system reaction will significantly increase. The occurrence of this condition, known as resonance, leads to system failure and should be actively prevented.

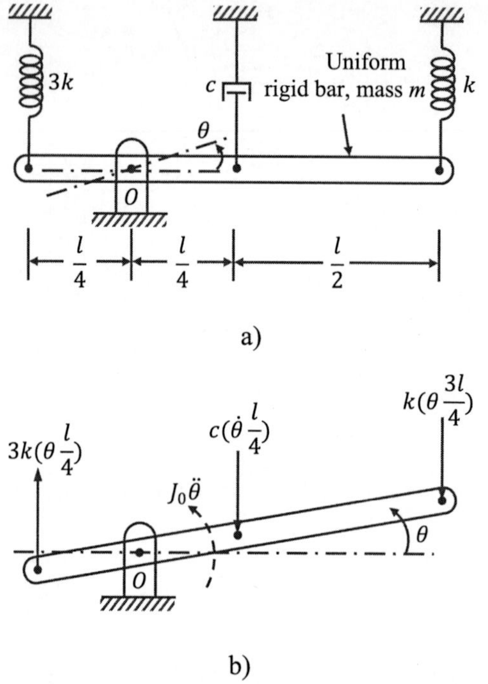

FIGURE 1.18 Rigid rod with a mass m that is constrained by a pin.

1.8.2 Motion Equation

The equation of motion for a mass-spring and damper system with the external force $F(t)$ applied at a specific location (as shown in Figure 1.19) may be expressed as follows:

$$m\ddot{x}(t)+c\dot{x}(t)+kx(t)=F(t) \tag{1.28}$$

The equation in question is heterogeneous, and its general solution is derived by adding the homogeneous solution $x_h(t)$ and the particular solution $x_p(t)$. The homogeneous component refers to the solution of the given problem.

$$m\ddot{x}(t)+c\dot{x}(t)+kx(t)=0 \tag{1.29}$$

This displays the oscillations of the system that occur without any external forces. As free vibrations decay with time, the general solution of Equation (1.28) reduces to its particular solution. The presence of external excitation ensures the continuity of motion. Figure 1.20 indicates that the homogenous response of $x_h(t)$ ceases to exist after time τ, leaving only the particular response of $x_p(t)$. Transient motion refers to the free vibrations that are attenuated due to dampening.

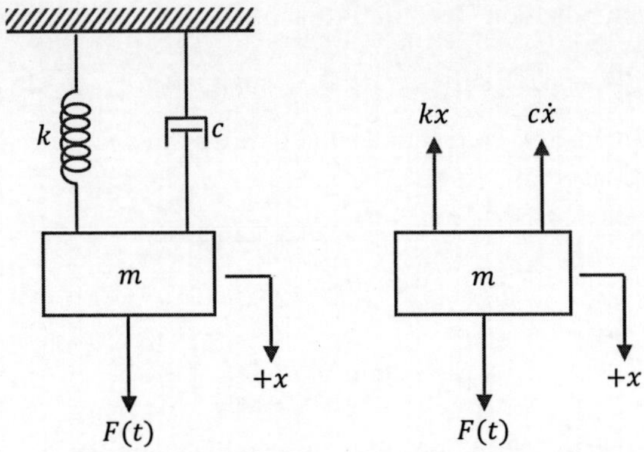

FIGURE 1.19 Mass-spring-damper system.

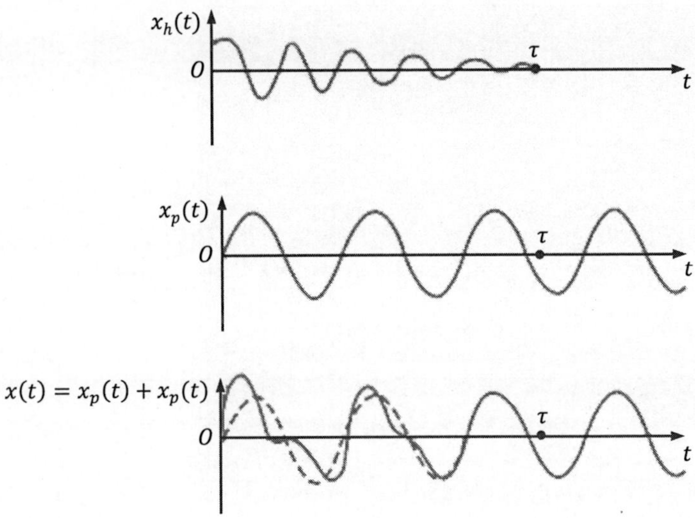

FIGURE 1.20 Homogeneous, particular, and combined solutions for Equation (1.28) in the underdamped system.

1.8.3 Harmonic Response of a Damped System

The equation of motion for a damped system, under the influence of the force $F(t) = F_0 \sin \omega t$, may be expressed as follows:

$$m\ddot{x}(t) + c\dot{x}(t) + kx(t) = F_0 \sin \omega t \tag{1.30}$$

The solution to Equation (1.30) is given by the particular solution.

$$x_p(t) = X \sin(\omega t - \phi) \tag{1.31}$$

By substituting the Equation (1.31) into Equation (1.30), the outcome is as follows:

$$X\left[\left(k-m\omega^{2}\right)\sin\left(\omega t-\phi\right)+c\omega\cos\left(\omega t-\phi\right)\right]=F_{0}\sin\omega t \tag{1.32}$$

By utilizing trigonometric connections and doing mathematical operations, we will obtain the following:

$$X=\frac{F_{0}}{\left[\left(k-m\omega^{2}\right)^{2}-c^{2}\omega^{2}\right]^{1/2}} \tag{1.33}$$

$$\phi=\tan^{-1}\left(\frac{c\omega}{k-m\omega^{2}}\right) \tag{1.34}$$

Now, let us establish the definitions of the following quantities:

$$\omega_{n}=\sqrt{\frac{k}{m}},\qquad c_{c}=2m\omega_{n}$$

$$\xi=\frac{c}{c_{c}}=\frac{c}{2m\omega_{n}}=\frac{c}{2\sqrt{mk}},\ \frac{c}{m}=2\xi\omega_{n}$$

$$\delta_{st}=\frac{F_{0}}{k}$$

$$r=\frac{\omega}{\omega_{n}}$$

So:

$$\frac{X}{\delta_{st}}=\frac{Xk}{F_{0}}=\frac{1}{\sqrt{\left[1-\left(\frac{\omega}{\omega_{n}}\right)^{2}\right]^{2}+\left[2\xi\frac{\omega}{\omega_{n}}\right]^{2}}}=\frac{1}{\sqrt{\left[1-r^{2}\right]^{2}+\left[2\xi r\right]^{2}}} \tag{1.35}$$

$$\phi=\tan^{-1}\left(\frac{2\xi\frac{\omega}{\omega_{n}}}{1-\left(\frac{\omega}{\omega_{n}}\right)^{2}}\right)=\tan^{-1}\left(\frac{2\xi r}{1-r^{2}}\right) \tag{1.36}$$

The quantity $M=\frac{X}{\delta_{st}}$ is called the magnification coefficient or amplitude ratio. The graphic in Figure 1.21 shows the variation of $\frac{X}{\delta_{st}}$ and φ with respect to the frequency ratio r.

Example 1–5: A beam with a uniform mass, denoted as m, is connected at Point O and supported by two springs and a damper, as seen in Figure 1.22. The terminal position of the P-point of the PQ spring experiences a sinusoidal displacement represented by the equation $x(t)=x_{0}\sin\omega t$. Determine the angular displacement required to reach the stable condition of the beam.

$$l=1m,\ k=1000\ N/m,\ c=500\ N-s/m,\ m=10\ kg,\ x_{0}=1\ cm,\ \omega=10\ rad/s$$

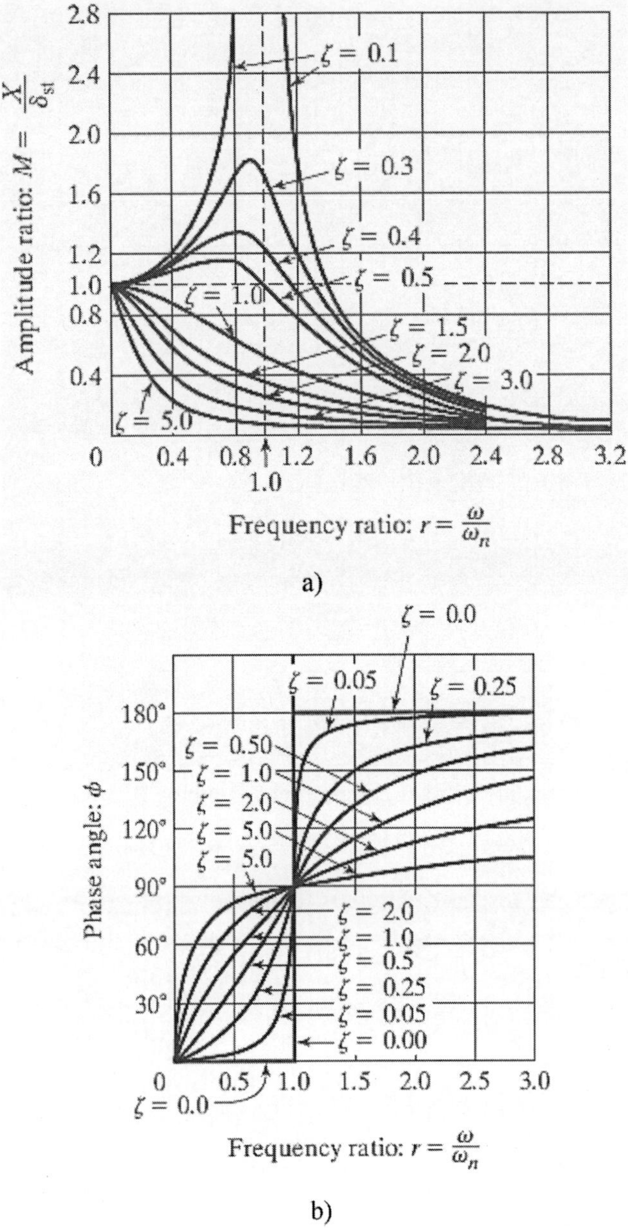

FIGURE 1.21 Changes in $\frac{X}{\delta_{st}}$ and φ by frequency ratio r.

The governing equation of the rod is as follows:

$$I_o\ddot{\theta} = -k\frac{l}{4}\theta\left(\frac{l}{4}\right) - c\frac{l}{4}\dot{\theta}\left(\frac{l}{4}\right) - k\left(\frac{3l}{4}\theta - x(t)\right)\frac{3l}{4}$$

$$\rightarrow I_o\ddot{\theta} + \frac{1}{16}cl^2\dot{\theta} + \frac{5}{8}kl^2 = \frac{3}{4}kl\ x(t) = \frac{3}{4}kl\ x_0 \sin \omega t$$

In which:

$$I_0 = \frac{1}{12}ml^2 + m\left(\frac{l}{4}\right)^2 = \frac{7}{48}ml^2 = \frac{7}{48}(10)(1^2) = 1.4583\ kg - m^2$$

Once the data is inserted into the equation, the equation of the system results in the following:

$$1.4583\,\ddot{\theta} + 32.25\,\dot{\theta} + 625.0\,\theta = 7.5 \sin 10t$$

Consequently:

$$m_{eq} = 1.4583, \quad c_{eq} = 31.25, \quad k_{eq} = 625.0, \quad M_0 = 7.5$$

The steady response of a rod will obtain as follows:

$$\theta(t) = \Theta \sin(\omega t - \phi)$$

In which:

$$\Theta = \frac{M_0}{\left[\left\{k_{eq} - m_{eq}(\omega)\right\}^2 + \left\{c_{eq}(\omega)\right\}^2\right]^{\frac{1}{2}}}, \qquad \phi = \tan^{-1}\left[\frac{c_{eq}(\omega)}{k_{eq} - m_{eq}(\omega)}\right]$$

The result is obtained after inputting numerical numbers.

$$\Theta = 0.01311\ rad, \qquad \phi = -0.5779\ rad$$

As a result, the stable response will be as follows:

$$\theta(t) = 0.01311\ \sin(10t - 0.5779)\ rad$$

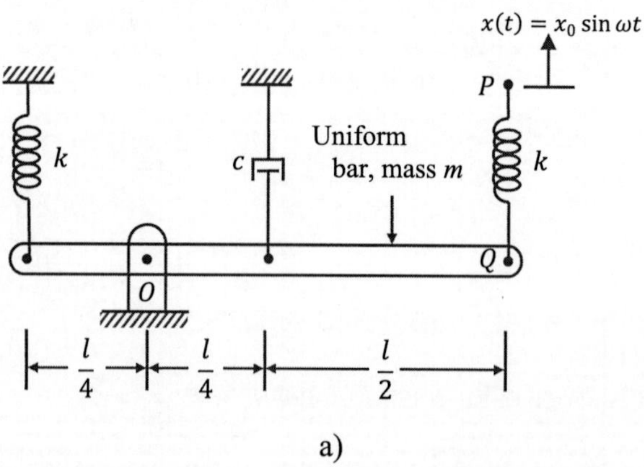

FIGURE 1.22 Uniform beam with mass m inhibited by a spring and a damper.

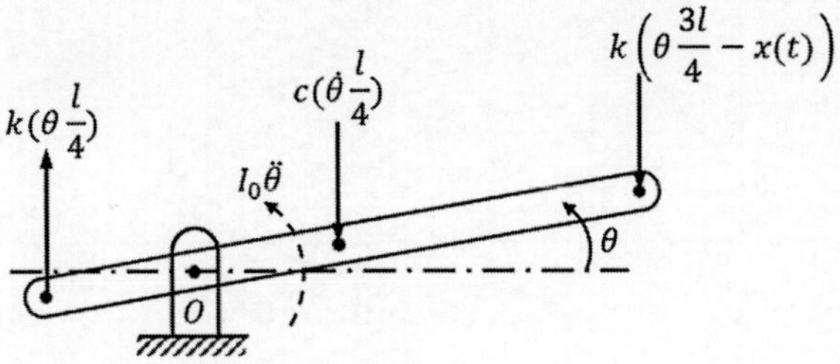

b)

FIGURE 1.22 (Continued)

1.9 A PRACTICAL ANALYSIS OF VIBRATIONS: FATIGUE FAILURE CAUSED BY VARIABLE LOADING

Fatigue is a potential consequence of vibration on a building or machine. Vibration control is crucial to prevent component failure caused by fatigue resulting from oscillating loads. This evaluation is crucial when the structure and/or machine undergoes resonance circumstances. In the occurrence of resonance, oscillation is linked to significant amplitudes. Consequently, the probability of parts experiencing failure or malfunction as a result of low cycle fatigue rises. In this part, we will provide a concise overview of the fundamental principles of fatigue and then elucidate the correlation between the phenomena of resonance and the subsequent malfunction or structural failure induced by fatigue.

1.9.1 Introduction to the Concept of Fatigue

During the tensile test of the material, the load is incrementally applied to provide sufficient time for complete strain to develop, resulting in the creation of the stress-strain graph. However, there are many instances in which tensions have altered or where they fluctuate between certain values (Figure 1.28).

σ_{min} = Minimum Stresses

σ_{max} = *Maximum* Stresses

σ_a = *Stress amplitude*

σ_a = *Average* Stresses

σ_r = *Stress range*

σ_s = *Static or Stable* Stresses

Frequently, machine components experience breakage as a result of repetitive or fluctuating pressures. However, thorough examinations indicate that the maximum stress levels generated were lower than the ultimate strength of the material and, in numerous instances, even lower than the yield strength. The salient characteristic of such failures is the frequent recurrence of tensions. This phenomenon is referred to as fatigue failure [8] due to this specific cause.

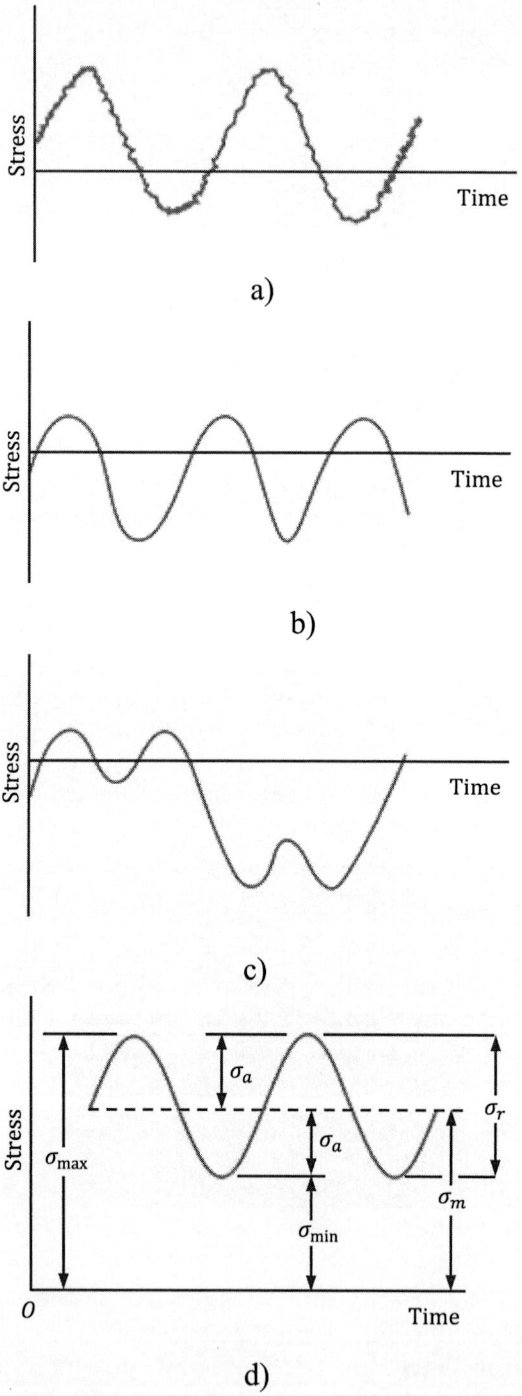

FIGURE 1.23 This figure illustrates many types of tension-time occurrences. The stress components depicted in Figure 1.23 are as follows.

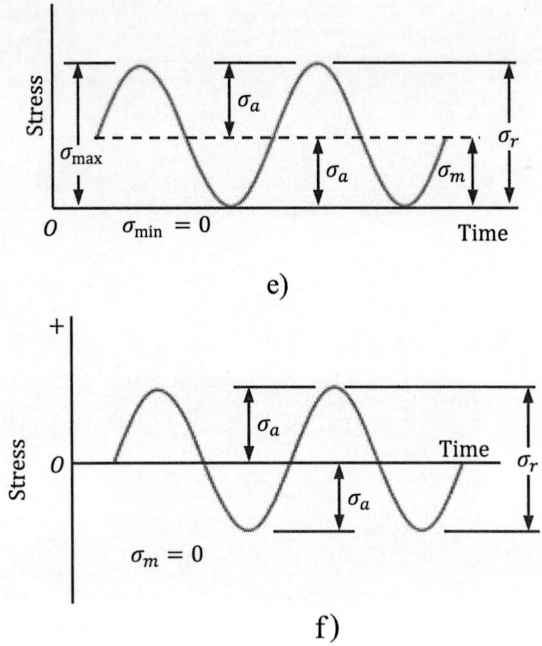

FIGURE 1.23 (Continued)

1.9.2 Relationships between Tension and Lifespan

In order to assess the durability of the material when subjected to fatigue loads, the samples are subjected to varying stresses, and the number of loading cycles required for the sample to fail is recorded. Subsequently, they represent the outcomes in the form of a graph known as S-N (Figure 1.24).

Basquin presented a mathematical equation to represent the S-N curve within the range of $N < (10)^6$. To enhance the fatigue resistance of the sample, it is advisable to subject it to a specific number of cycles.

$$S_f = a\,N^b \tag{1.37}$$

The constants a and b are defined as follows, where N is the number of cycles that result in failure:

$$a = \frac{(f\,S_{ut})^2}{S_e}, \quad b = -\frac{1}{3}\log\left(\frac{f\,S_{ut}}{S_e}\right) \tag{1.38}$$

The fatigue resistance coefficient, denoted as *f*, may be derived from the curve presented in Figure 1.25.

The objective of vibration analysis is to mitigate the occurrence of resonance phenomenon. In linear systems, when the frequency of the excitation matches the

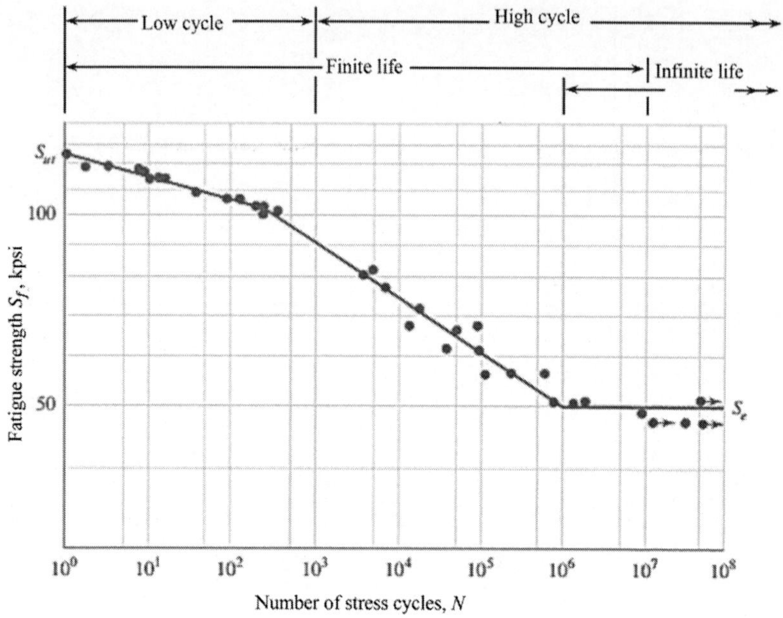

FIGURE 1.24 Diagram of the S-N drawn from the results of the fully reverse fatigue experiments for USN G41300, ($S_{ut} = 116\ kpsi$, max $S_{ut} = 125\ kpsi$)

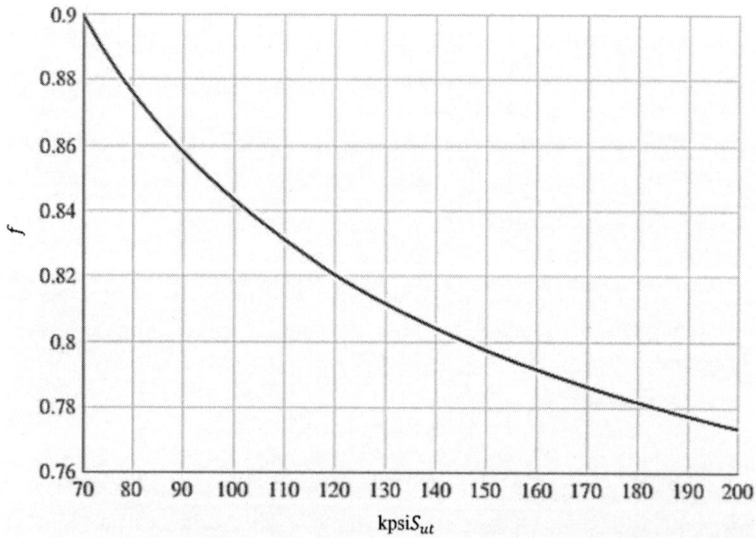

FIGURE 1.25 Fatigue resistance coefficient f, (S_{ut} in the number 10^3 cycles for $S_e = S_e' = 0.5\ S_{ut}$)

natural frequency, resonance occurs and results in a peculiar rise in the amplitude of oscillations. The occurrence of low cycle fatigue is observed when the device has prolonged operational circumstances. We will see a harmonic motion characterized by a significant oscillation range, resulting in the rapid deterioration of the machine components shortly after its initiation. Therefore, the examination of vibrations holds a prominent position in the field of engineering. In the context of machines functioning under near-escalation settings, the challenge of managing fluctuations to prolong equipment lifespan arises. This matter is addressed in control talks, coupled with the usage of energy absorbers.

REFERENCES

[1] Lindsay, R.B., The story of acoustics. *The Journal of the Acoustical Society of America*, 1966. **39**(4): p. 629–644.
[2] Nayfeh, A.H., *Perturbation methods*. 2008: John Wiley & Sons.
[3] Gleick, J., *Chaos: Making a new science*. 2008: Penguin.
[4] Peitgen, H.-O. and P.H. Richter, *The beauty of fractals: Images of complex dynamical systems*. 1986: Springer Science & Business Media.
[5] Rao, S.S., *The finite element method in engineering*. 2017: Butterworth-Heinemann.
[6] Rao, S.S., *Mechanical vibrations*. Sixth Edition. 2018: Pearson Education, United Kingdom.
[7] Caughey, T. and M. O'kelly, Effect of damping on the natural frequencies of linear dynamic systems. *The Journal of the Acoustical Society of America*, 1961. **33**(11): p. 1458–1461.
[8] Richard, G.B., *Shigley's mechanical engineering design*. 2019: McGraw-Hill Education.

2 An Introduction to the Nonlinear Vibration

2.1 INTRODUCTION

The previous chapter focused on the analysis of equations and linear systems. While we often analyze systems in a linear manner, in reality, these systems are nonlinear [1]. Nonlinear analysis should be employed if the range of motion is not limited. In order to accurately depict a physical system, it is typically necessary to employ a nonlinear model. This is because linear analysis of the system's whole behavior cannot effectively anticipate the events that arise in nonlinear systems [2]. In order to study nonlinear vibration issues, it is necessary to familiarize ourselves with the many phenomena and concepts associated with these systems. This chapter will cover many phenomena in nonlinear vibrations, as well as the topics that will be discussed in the subsequent chapters.

2.2 BRIEF REVIEW OF NONLINEAR OSCILLATIONS HISTORY

Nonlinear behavior manifests in numerous real-world occurrences. Consequently, scholars from various disciplines investigate nonlinear systems. The allure of nonlinear dynamics may elucidate why numerous scholars are drawn to the examination of nonlinear systems.

The study of dynamical systems originated in the mid-1600s with the introduction of differential equations by Newton. Poincaré played a pivotal contribution in the advancement of nonlinear dynamics throughout the late 1800s by utilizing qualitative analysis to study dynamical systems. His work was indeed a significant advancement in the field of nonlinear systems and led to the initial insight into chaos. Nonlinear oscillations are a fundamental aspect of dynamics. During the early 1900s, scholars were highly interested in studying nonlinear oscillations and their applications in the fields of physics and engineering. Over the course of these years, several influential scientists, including Van der Pol, Duffing, Cartwright, Levinson, Littlewood, Bogoliubov, Krylov, Levenson, Minorsky, Vitt, Andronov, Birkhoff, and Kolomogonov, have achieved a significant advancement in the field of nonlinear oscillatory systems. Table 2.1 succinctly outlines their contributions in the aforementioned field [3].

According to Table 2.1, Russian scientists made significant contributions to the study of nonlinear oscillatory systems in the early 1900s. The advancement of high-speed computers after the 1950s was a significant breakthrough in the study of nonlinear oscillations. The study of nonlinear oscillatory systems has led to significant advancements in the field, thanks to the use of newly developed computers. These advancements have provided both theoretical and practical insights into nonlinear systems [4].

DOI: 10.1201/9781003470694-2

TABLE 2.1
Important Contribution in the Field of Nonlinear Oscillatory Systems (Early 1900s)

Renowned Scientists	Reported Contributions in Nonlinear Oscillations
B. Van der Pol	Introduction of relaxation-oscillations
G. Duffing	Observation of cubic nonlinearity
M. Cartwright and J. E. Littlewood	Relaxation-oscillations and the topological approach for solving of nonlinear problems
N. N. Bogoliubov and N. M. Krylov	One of the first educational tools in the field of nonlinear mechanics and also developed the asymptotic methods in nonlinear mechanics
A. A. Andronov and A. A. Vitt	Self-excited oscillations and one of the first educational tools in the field of nonlinear oscillations
A. Kolmogorov	Nonlinear diffusion equation
G. D. Birkhoff	Dynamical systems with two degrees of freedom
N. Levinson	Transformation theory utilized for nonlinear equations
N. Minorsky	Parametric excitation
M. E. Levenson	Analyzed the Duffing equation
J. A. Shohat	Studied the Van der Pol equation

Between 1950 and 1955, Hayashi authored several papers and conducted research on subharmonic, forced oscillations, and the stability of nonlinear systems. In the 1950s and 1960s, a small number of researchers directed their attention towards studying nonlinear oscillations in the field of plasma physics. Over the course of these years, Crandall emerged as a highly engaged scientist in the area of nonlinear vibrations, making noteworthy advancements in random vibrations and the application of perturbation theory.

Notable scientists who have made significant advancements in the field of nonlinear oscillations, particularly in the area of structural mechanics, throughout the latter half of the 20th century include T. Yamamoto, Y. Ishida, A. H. Nayfeh, D. T. Mook, R. Rand, F. C. Moon, and E. H. Dowell, among others. Lorenz's discovery of chaos was a significant breakthrough in the field of nonlinear dynamics, with a profound impact on study in the area of nonlinear oscillatory systems.

Table 2.2 presents the primary achievements of the experts listed before in the field of nonlinear oscillatory systems [3].

The aforementioned experts formulated the fundamental principles underlying the science of nonlinear oscillations. After their initial studies, other researchers expanded the scope of the discipline to include not only large structures but also nonlinear oscillations in micro- and nano-systems. Currently, several research studies have been conducted on the interaction between fluids and structures, as well as the nonlinear modelling of their oscillatory systems. Chaotic vibration analysis of structures is a burgeoning research area in the study of nonlinear oscillatory systems.

Over the past 20 years, there has been a significant growth in the number of research initiatives focused on nonlinear oscillatory systems. Researchers such as

TABLE 2.2
Important Contributions in the Area of Nonlinear Oscillatory Systems (1950–2000)

Scientist	Major Contributions in Nonlinear Oscillations
A. H. Nayfeh	Enriching *field* of nonlinear vibrations with writing more than five highly cited books and 400 papers during 1970–2000
T. Yamamoto	He published more than 130 papers and a number of books in the *field* of nonlinear vibration with special focus on rotor dynamics
Y. Ishida	The main contribution of Ishida is nonlinear vibration, vibrations suppression and rotor dynamics
R. Rand	High influence in the *field* of nonlinear oscillations with publishing more than 150 papers about vibrations of beam, Van der Pol oscillators, parametric excitation, and chaos
F. C. Moon	He is one of the pioneers in the *field* of chaotic vibrations and explored chaos in structural systems
E. H. Dowell	One of the highly influential researchers in the *field* of nonlinear dynamics with special focus on fluid-structure interaction
M. P. Païdoussis	Païdoussis is one of the most influential scientists in the *field* of dynamical modelling of pipes and shells
R. A. Ibrahim	He is one of the pioneers in the *field* of fluid sloshing and nonlinear oscillations

Amabili, Chen, Kerschen, Bergman, Awrejcewicz, Mickens, Cveticanin, Inman, Balachandran, Esmailzadeh, and Spanos have made significant contributions to the advancement of novel works in the field of nonlinear oscillations. Table 2.3 displays a portion of their recent advancements in the field of nonlinear oscillatory systems [3].

2.3 NONLINEAR SOURCES

Nonlinear systems are characterized by the absence of the concept of superposition principles in their description. Nonlinearity is a common occurrence in structural mechanics, manifesting in many forms and conditions. These include nonlinearities associated with geometry, material properties, inertia, and friction.

Material nonlinearity refers to substances that have stress-strain relationships characterized by elastic-plastic behavior. Nonlinear geometry is designed to accommodate nonlinear interactions between displacement and strain. This particular form of nonlinearity is often addressed in study. The sources of this nonlinear behavior include the tension of the middle plate, significant curvature in structural parts, and substantial rotation. Inertial nonlinearities occur due to the concentration or dispersion of mass. These nonlinear types are expressed in the governing equations as the time derivatives of the displacements. Friction-induced nonlinearities of significant magnitude arise from dry friction, cohesion-slip, and residue. These nonlinearity phenomena are present in the governing differential equations [5, 6].

TABLE 2.3
Major Contributions in the Area of Nonlinear Oscillatory Systems (Last Two Decades)

Scientist	Contribution in Nonlinear Oscillations
M. Amabili	An influential researcher in the area of nonlinear oscillatory systems with focus on shell structures and fluid solid interaction
L. Q. Chen	An important figure in nonlinear vibration energy harvesting and vibration of axially moving continua
L. A. Bergman	Development of vibration-based energy harvesting and nonlinear structural dynamics
B. Balachandran	Analysis of nonlinear phenomena in beams, milling process, and different types of structures
R. E. Mickens	Development of new types of nonlinear differential equations arising in nonlinear oscillatory systems
A. F. Vakakis	Significant contribution in the field nonlinear normal modes and its applications in vibrations
P. D. Spanos	Analyzing of nonlinear and random vibrations is the main focus of his research
E. Esmailzadeh	Analyzing of nonlinear vibrations of beam- and plate-type structures

Another source of nonlinearity arises from boundary conditions that exhibit nonlinearity in terms of both equality and inequality. An instance of nonlinear inequalities may be observed in the interaction of elastic objects, where the relative displacement of contact sites along the contact route must not exceed or be equal to the initial distance between these places [7]. Table 2.4 provides a concise overview of the several origins of nonlinear mechanical vibrations.

2.4 EXAMPLES OF NONLINEAR VIBRATION

This section explores instances of nonlinear nature of the systems with a single degree of freedom, which are used to describe the nonlinear characteristics seen in physical systems.

We analyze a conservative system with one degree of freedom, where the governing equation is a simple nonlinear differential equation:

$$\ddot{x}(t) + f(x, \dot{x}) = 0 \tag{2.1}$$

2.4.1 Simple Pendulum

Examine the simple pendulum as depicted in Figure 2.1. The motion equation of this pendulum is given by the differential equation [8]:

$$ml^2\ddot{\theta}(t) + mgl\sin\theta = 0 \tag{2.2}$$

TABLE 2.4
Various Sources of Nonlinear Mechanical Vibrations

Source	Description
Material Nonlinearities	Arise from the inherent properties of materials where the stress-strain relationship is not linear, such as in rubber or biological tissues.
Geometric Nonlinearities	Occur when deformations are large enough that linear assumptions (like small angle approximations) no longer hold, e.g., in pendulums, beams, plates, and shells.
Boundary Condition Nonlinearities	Result from constraints that change with displacement, such as contact problems, or supports that exhibit nonlinear stiffness.
Friction and Damping Nonlinearities	Caused by nonlinear friction or damping mechanisms, such as Coulomb friction or damping that varies with velocity or displacement.
Electromechanical Coupling	Nonlinear behavior arising from the interaction between mechanical and electrical systems, as in piezoelectric materials and magnetoelastic effects.
Parametric Excitation	Generated by periodic variation of system parameters, such as in Mathieu's equation where stiffness varies with time.
Internal Resonances	Occur when energy is transferred between modes of vibration within a system, leading to complex motion and nonlinearity.
Rotational Effects	Nonlinear vibrations in rotating systems due to gyroscopic forces, centrifugal stiffening, or dynamic unbalance.
Impact and Collision	Nonlinear response from transient forces during impact or collision events, resulting in high-frequency content and large deformation.
Nonlinear Aeroelastic Effects	Arise from the interaction of aerodynamic forces with flexible structures, leading to phenomena like flutter or limit cycle oscillations.

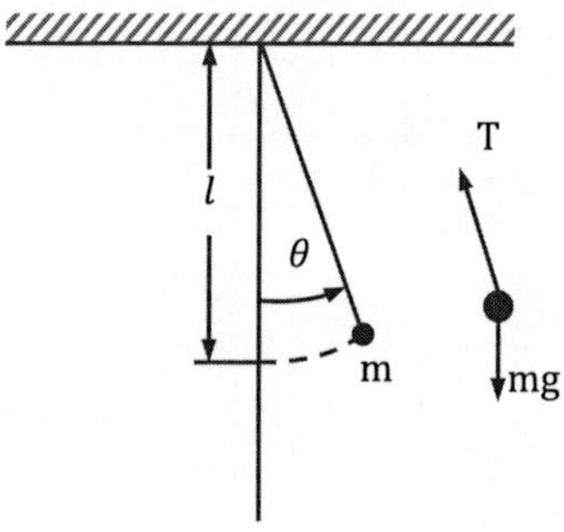

FIGURE 2.1 Simple pendulum.

Which can be rewritten as follows:

$$\ddot{\theta} + \frac{g}{l}\sin\theta = 0 \tag{2.3}$$

The foundation of a nonlinear differential equation is in the comparison of the Equation (2.1).

$$f(\theta) = \frac{g}{l}\sin\theta \tag{2.4}$$

It is a well-known fact that for small angles θ, the value of sin θ is about equal to θ ($\sin\theta = \theta$). Equation (2.2) is reduced into linear terms as a consequence.

$$\ddot{\theta} + \omega_0^2\theta = 0 \tag{2.5}$$

In which:

$$\omega_0 = \left(\frac{g}{l}\right)^{1/2} \tag{2.6}$$

The presence of significant angular momentum is responsible for the emergence of the nonlinear state in this particular example.

2.4.2 A Particle that is Confined and Connected by a Spring with a Nonlinear Behavior

For our second example, we will analyze the motion of an object with mass m on a horizontal plane without any friction. This object is coupled to a spring that does not follow a linear relationship (as seen in Figure 2.2). The differential equation governing the motion of the mass is given by the function $x(t)$ which describes its location [5].

$$\ddot{x}(t) + f(x) = 0 \tag{2.7}$$

The function $f(x)$ represents the force exerted by the spring on the mass. The linear spring equation, $f(x) = kx$, refers to a spring with a fixed value of k. The force exerted by a nonlinear spring is described by a deformed nonlinear function, as shown in Figure 2.2. The nonlinear component diminishes the force for a soft spring, but it amplifies the force for a hard spring. In this section, we consider the motion to occur in states where force is present and absent, along a curved path. We see that there is no loss, which results in the system being damped. In this particular instance, the presence of nonlinearity is mostly attributed to the behavior of material rather than to significant deformations.

Generally, the function $f(x)$ in Equation (2.7) can be expressed in three different ways:

1. $f(x) = kx$, representing a linear spring.
2. $f(x) = k\left(x + \frac{x^3}{3}\right)$, representing a hardening spring.

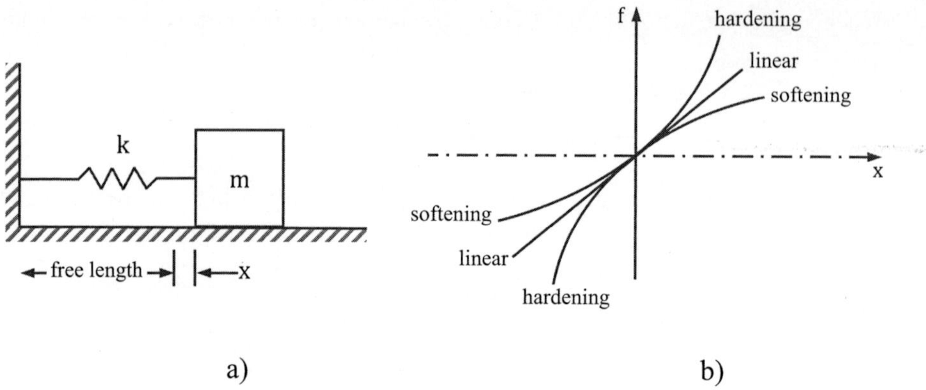

FIGURE 2.2 (a) System of mass and spring. (b) Specifications of spring.

3. $f(x) = k\left(x - \frac{x^3}{3}\right)$, indicating a softening spring.

2.4.3 Particle in a Central Force Field

As an illustration, we analyze the motion of a particle on a flat surface subjected to the effects of a central force field (Figure 2.3) [5].

The equation that governs the motion of the particle with mass m may be derived by expressing Newton's second law in the r-θ coordinate system:

$$m(\ddot{r} - r\dot{\theta}^2) + mF(r) = 0 \tag{2.8}$$

$$m\left(r\ddot{\theta} + 2\dot{r}\dot{\theta}\right) = 0 \tag{2.9}$$

If the field is gravitational, the variable m indicates the mass of the particle. Conversely, if the field is electric, m represents the electric charge of the particle. The Equation (2.9) can be expressed by means of integration as follows:

$$r^2\dot{\theta} = p \tag{2.10}$$

The variable p is a constant. This equation is a conservation angular momentum. Results obtained by eliminating $\dot{\theta}$ from the Equations (2.9) and (2.10) are as follows:

$$\ddot{r} - \frac{p^2}{r^3} + F(r) = 0 \tag{2.11}$$

By substituting the dependent variable r with its reciprocal $u = r^{-1}$ and replacing the independent variable t with θ, Equation (2.9) may be simplified. Ultimately, the derivatives are transformed in the following manner:

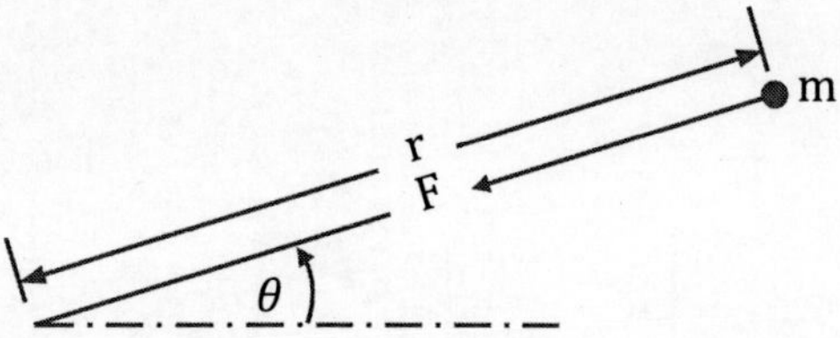

FIGURE 2.3 Particle in a central force field.

$$\dot{r} = \frac{dr}{dt} = \frac{dr}{d\theta}\dot{\theta} = -\frac{\dot{\theta}}{u^2}\frac{du}{d\theta} = -p\frac{du}{d\theta} \tag{2.12}$$

$$\ddot{r} = -p\frac{d^2u}{d\theta^2}\dot{\theta} = -p^2u^2\frac{d^2u}{d\theta^2} \tag{2.13}$$

Thus, Equation (2.11) becomes the following:

$$\frac{d^2u}{d\theta^2} + u - \frac{1}{p^2u^2}F\left(\frac{1}{u}\right) = 0 \tag{2.14}$$

The nonlinearity of the system in this instance arises from the effects of inertia and material characteristics.

If a particle with mass m is experiencing a gravitational force F, which is given by $F = G\frac{mm_0}{r^2}$, where m_0 is the mass of the fixed object, G is the universal gravitational constant, and r is the distance between the centers of the two masses (as shown in Figure 2.4), we can express this equation using the chosen variables as follows:

$$F = Gm\, m_0 u^2 \tag{2.15}$$

By inserting the aforementioned sentence into Equation (2.14), we will obtain the following:

$$-Gm_0u^2 = -h^2u^2\frac{d^2u}{d\theta^2} - \frac{1}{u}h^2u^4 \tag{2.16}$$

The result obtained following simplification as follows:

$$\frac{d^2u}{d\theta^2} + u = \frac{G\,m_0}{h^2} \tag{2.17}$$

The equation described by the second-order linear differential equation is nonhomogeneous. It is noted that the equation shows a linear relationship, when considering the force as the gravitational attraction [9].

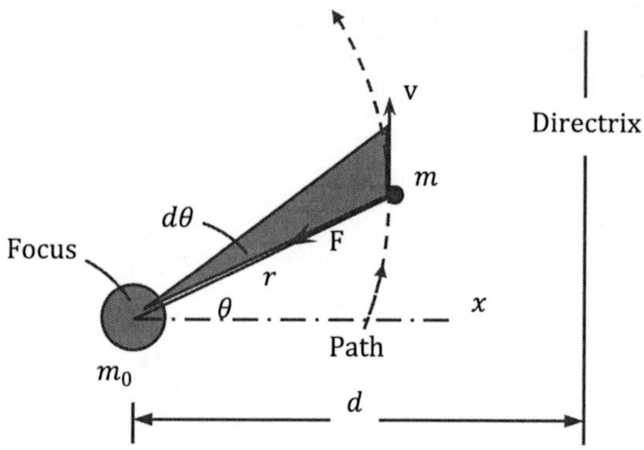

FIGURE 2.4 The particle m moving and under tensile force.

2.4.4 Mechanical Vibration with Dry Friction

The presence of nonlinearity is evident in the damping equation of the system (Figure 2.5). The system exhibits nonlinearity as a result of the dry friction between the mass m and the belt movement. This system involves two coefficients of friction: the static friction coefficient (μ_s) represents the force required to initiate movement of the item by dry friction, while the kinetic friction coefficient (μ_k) represents the force required to maintain the object's motion. The tangent component of the force is generated by the friction surface (F), which is the result of multiplying the friction coefficient by the force perpendicular to the surface [8].

The stages of the system movement, as depicted in Figure 2.5 (a), follow the velocity-force diagram illustrated in Figure 2.5 (b). Initially, the mass is positioned on the conveyor belt. The spring elongates as a result of the movement of mass m along the belt. As the spring is compressed, the spring force acting on the mass intensifies, surpassing the force of static friction. Consequently, the mass initiates sliding motion and swiftly moves towards the right. Consequently, the spring force is discharged until it is halted by the kinetic friction force. Subsequently, the spring starts the process of reaccumulating its potential energy. The variation of the damping force with the velocity of mass is seen in Figure 2.5 (b). The equation for mass movement may alternatively be expressed as follows:

$$m\ddot{x} + F(\dot{x}) + kx = 0 \tag{2.18}$$

The friction force F is a nonlinear function of $\dot{x}$, as shown in Figure 2.5 (b). When $\dot{x}$ is significantly big, the damping force is positive, resulting in the removal of energy from the system. When $\dot{x}$ has small values, the damping force acts in the opposite direction and transfers energy to the system. Despite the absence of external stimulus, this system exhibits oscillatory motion. The term used to describe these self-generated vibrations is mechanical vibrations.

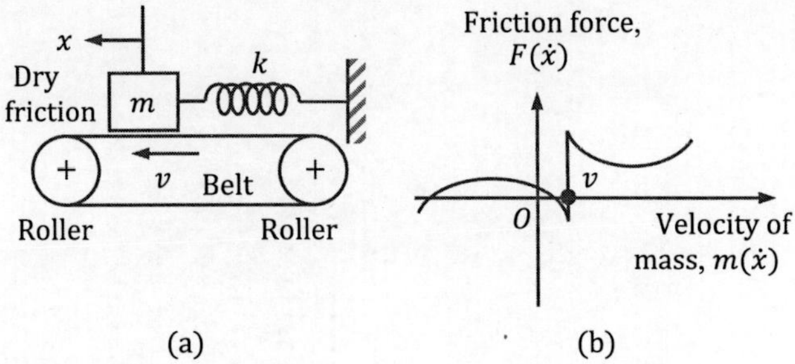

FIGURE 2.5 Dry friction damping.

2.4.5 Variable Mass System

Conventional systems assume that the mass and spring components remain constant during the solution process. However, in systems with changeable mass, a more comprehensive approach is required to accurately analyze the dynamics. A widely recognized example of a system that exhibits changing mass is a rocket. The fluctuating mass of a missile is a significant challenge in conducting a comprehensive and dynamic analysis since it introduces a nonlinear impact on the system.

Now contemplate the system including a mass that can vary, as seen in Figure 2.6. The mass of this system is dependent on x, in the case of big x shifts, and the motion equation is as follows:

$$\frac{d}{dt}(m\dot{x}) + kx = 0 \tag{2.19}$$

The aforementioned relationship is a nonlinear differential equation resulting from a changing mass.

2.4.6 Particle in a Spinning Circle

As an illustration, we analyze the movement of a mass (m) that is sliding without friction down the edge of a circle with a radius (R). This circle is also spinning at a constant angular velocity (Ω) while maintaining a distance equal to its horizontal diameter. The particle m is subject to the forces depicted in Figure 2.7 [5].

Based on the force diagram provided, while using Newton's second law in two distinct directions, we will have the following:

$$\begin{aligned} -mg + N\cos\theta &= mR\ddot{\theta}\sin\theta + mR\dot{\theta}^2\cos\theta \\ N\sin\theta &= mR\Omega^2\sin\theta - mR\ddot{\theta}\cos\theta + mR\dot{\theta}^2\sin\theta \end{aligned} \tag{2.20}$$

Now we can exclude N from the two aforementioned relationships:

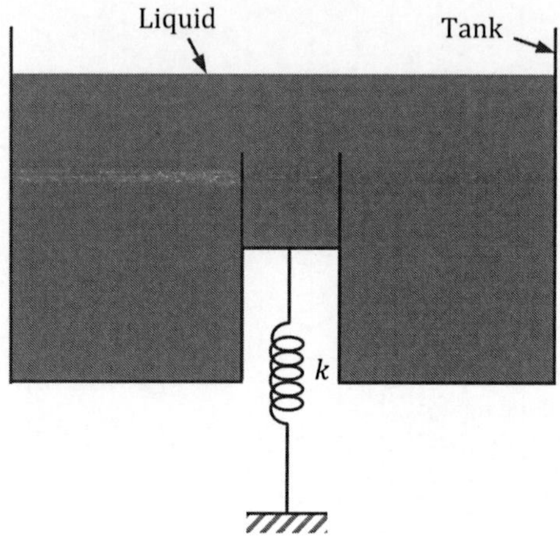

FIGURE 2.6 Variable mass system.

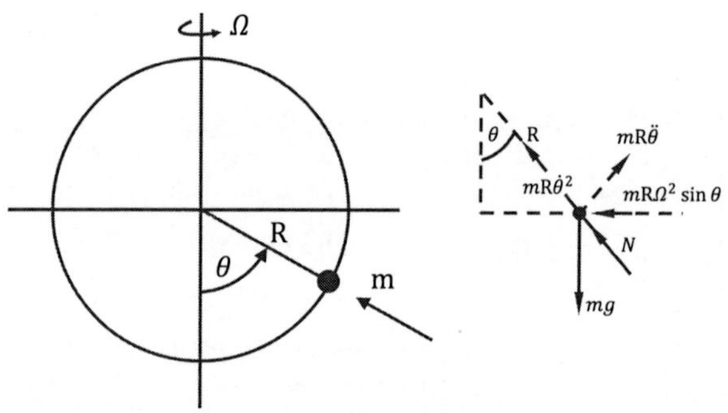

FIGURE 2.7 The moving particle on a soft rotary wire.

$$mR^2\ddot{\theta} = m\Omega^2 R^2 \sin\theta\cos\theta - mgR\sin\theta \tag{2.21}$$

In this particular example, the nonlinearity arises as a result of the combined effects of inertia and large deformation.

The governing differential Equation (2.21) can also be derived using the Lagrange technique. To express this, we shall arrange kinetic energy and potential energy in sequential order:

$$\mathrm{T} = \frac{1}{2} mR^2 \left(\dot{\theta}^2 + \Omega^2 \sin^2\theta\right) \tag{2.22}$$

$$\nu = -mgR\cos\theta \tag{2.23}$$

By formulating the Lagrange equation, we obtain the following:

$$\frac{d}{dt}\left(\frac{\partial T}{\partial \dot{\theta}}\right)-\frac{\partial T}{\partial \theta}=-\frac{\partial \nu}{\partial \theta} \tag{2.24}$$

Ultimately, by incorporating the energy relationships of the movement and potential into the Lagrange equation and conducting calculations, the equation of motion (2.21) is derived. Earlier, various nonlinear sources were discussed as an example. The subsequent discussion will explore the methods for analyzing nonlinear problems. Table 2.5 displays notable equations that describe nonlinear mechanical vibrations.

TABLE 2.5
Some Famous Nonlinear Mechanical Vibration Equations.

Equation Name	Equation	Description
Duffing Equation	$\ddot{x}+2\delta\dot{x}+\alpha x+\beta x^3=\gamma cos\,(\omega t)$	Describes the motion of a damped oscillator with a more complex restoring force.
Van der Pol Equation	$\ddot{x}-\mu(1-x^2)\dot{x}+x=0$	Models electrical circuits and biological systems; exhibits self-sustained oscillations.
Pendulum Equation	$\ddot{\theta}+\frac{g}{L}\sin(\theta)=0$	Governs the motion of a simple pendulum; nonlinear due to the $\sin(\theta)$ term.
Korteweg–de Vries (KdV) Equation	$u_t-6uu_x+u_{xxx}=0$	Describes waves on shallow water surfaces; famous for its soliton solutions.
Lorenz Equations	$\begin{bmatrix}\dot{x}=\sigma(y-x)\\ \dot{y}=x(\rho-z)-y\\ \dot{z}=xy-\beta z\end{bmatrix}$	Describes atmospheric convection; famous for chaotic solutions.
Mathieu Equation	$\frac{d^2x}{dt^2}+\left(a-2q\cos(2t)\right)x=0$	Describes parametrically excited systems; used in the study of vibrating systems with periodic coefficients.
Hill Equation	$\frac{d^2x}{dt^2}+f(t)x=0$	General form of a second-order linear differential equation with a periodic coefficient; arises in the stability analysis of dynamic systems.
Lotka-Volterra Equations	$\begin{Bmatrix}\dot{x}=\alpha x-\beta xy\\ \dot{y}=\delta xy-\gamma y\end{Bmatrix}$	Models predator-prey dynamics; nonlinear interaction terms between species populations.
Klein-Gordon Equation	$\Box\varnothing+m^2\varnothing+\lambda\varnothing^3=0$	Relativistic version of the wave equation with a nonlinear term; used in field theory.
Burgers' Equation	$u_t+uu_x=\nu u_{xx}$	Simplified model for turbulence and shock waves; nonlinear advection term.

(Continued)

TABLE 2.5 (*Continued*)
Some Famous Nonlinear Mechanical Vibration Equations.

Equation Name	Equation	Description
Rayleigh Equation	$\ddot{x}+\epsilon(\dot{x}^2-1)\dot{x}+x=0$	Models self-excited oscillations like those in a clarinet reed or steam engine governor.
Sine-Gordon Equation	$\varnothing_{tt}-\varnothing_{xx}+\sin(\varnothing)=0$	Appears in the study of crystal dislocations and nonlinear wave propagation.
Fermi-Pasta-Ulam-Tsingou (FPUT) Problem	$\ddot{x}_n=(x_{n+1}-x_n)-(x_n-x_{n-1})+\alpha[(x_{n+1}-x_n)^2-(x_n-x_{n-1})^2]$	Studies energy distribution in a nonlinear lattice; revealed insights into the foundation of chaos theory.
AW (Ablowitz-Weiss) Equation	$u_{tt}+u_{xx}+\alpha u^2+\beta u^3=0$	Nonlinear wave equation; appears in various physical contexts, including shallow water waves and nonlinear optics.

2.5 AN IMPORTANT POINT

When studying the behavior of nonlinear systems, we come across significant terms and phrases, such as primary resonance, secondary resonance, internal resonance, jumping phenomenon, parametric excitation, limit cycle, and saturation phenomenon. These terms are used to describe the behavior and phenomena observed in the analysis of nonlinear problems [4]. In the subsequent chapters, we shall elucidate several terminologies with meticulousness.

2.6 INTRODUCTION TO QUALITATIVE AND QUANTITATIVE ANALYSIS

There are two methodologies for examining the nonlinear systems [4]:

1. Qualitative analysis: The method of qualitative analysis, often employed in nonlinear dynamics, utilizes phase plate drawing concepts, such as stable equilibrium point, unstable equilibrium point, fixed point, separation point, saddle point, and others.
2. Quantitative analysis: The quantitative approach involves studying the behavior of the nonlinear system by analytical and semi-analytical approaches, such as perturbation, which is utilized in the analysis of nonlinear vibrations.

The subsequent chapters will thoroughly analyze the behavior of nonlinear systems using two distinct approaches: quantitative and qualitative.

REFERENCES

[1] Hayashi, C., *Nonlinear oscillations in physical systems*, Vol. 432. 2014: Princeton University Press.
[2] Sneddon, I., *WJ Cunningham, Introduction to nonlinear analysis*. 1959: McGraw-Hill, New York, 348 pp., 70s. *Proceedings of the Edinburgh Mathematical Society*, 1961. **12**(3): p. 167–167.
[3] Esmailzadeh, E., D. Younesian, and H. Askari, *Analytical methods in nonlinear oscillations*. 2018: Springer, Netherlands.
[4] Strogatz, S.H., *Nonlinear dynamics and chaos with student solutions manual: With applications to physics, biology, chemistry, and engineering*. 2018: CRC Press.
[5] Nayfeh, A.H. and D.T. Mook, *Nonlinear oscillations*. 2008: John Wiley & Sons.
[6] Nayfeh, A.H. and P.F. Pai, *Linear and nonlinear structural mechanics*. 2008: John Wiley & Sons.
[7] Nayfeh, A.H. and B. Balachandran, *Applied nonlinear dynamics: Analytical, computational, and experimental methods*. 2008: John Wiley & Sons.
[8] Rao, S.S., *Mechanical vibrations*, Fifth Edition. 2011: Pearson Ed Asia.
[9] Meriam, J.L., L. Glenn Kraige, and J.N. Bolton, *Engineering mechanics: Dynamics*. 2020: John Wiley & Sons.

3 Qualitative Analysis of Nonlinear Vibration

3.1 INTRODUCTION

Dynamics, despite its interdisciplinary nature, is fundamentally a field of physics. In the late 1800s, Poincaré introduced a novel approach to equations, prioritizing qualitative analysis above quantitative analysis. Instead of scrutinizing the precise locations of the planets at any given moment, he instead raised a different inquiry: Is the solar system inherently stable, and would the planets continue to travel indefinitely towards infinity? He devised a proficient geometric methodology to scrutinize such inquiries. The field of dynamics is commonly linked with nonlinear oscillators and their utilization in the domains of physics and engineering. Nonlinear oscillators are crucial in the advancement of several technologies, such as radio, radar, phase-looked loops, and lasers. Nonlinear oscillators have the capability to model novel mathematical techniques from a theoretical standpoint. Prior to delving into the realm of nonlinear systems, it is imperative to get a comprehensive understanding of the fundamental principles within this discipline. This chapter examines the fundamental principles involved in the qualitative analysis of nonlinear systems [1].

3.2 PHASE PLANE

The visual display phase screen can be defined as the distinctive features exhibited by certain forms of differential equations. Phase plane is utilized in applied mathematics, specifically in the domain of nonlinear systems analysis [1].

The differential equation regulating a nonlinear system is expressed in its generic form as follows:

$$\ddot{x} + f(x, \dot{x}, t) = 0 \tag{3.1}$$

If the time (t) does not appear explicitly in the differential Equation (3.1), the system is referred to as autonomous. Autonomous systems, by definition, lack any form of input. Put simply, their equations do not explicitly include the independent variable of time.

$$\ddot{x} + f(x, \dot{x}) = 0 \tag{3.2}$$

In order to represent the second-order differential equation in the state space, we shall define $x_1 = x$ and $\dot{x} = x_2$ as follows:

$$\begin{aligned} \dot{x}_1 &= x_2 \\ \dot{x}_2 &= -f(x_1, x_2) \end{aligned} \tag{3.3}$$

DOI: 10.1201/9781003470694-3

It is noted that the second-order differential Equation (3.2) can be transformed into two first-order differential Equations (3.3). Similarly, by representing the differential equation in the state space, a general n order differential equation can be converted into n first-order differential equations.

Alternatively, when and x_2 are plotted on a Cartesian coordinate system, the resulting graph is referred to as the phase plane. It represents the trajectory of x_1 and x_2, often known as the path line.

When x_2 is equal to zero and $f(x_1, x_2)$ is equal to zero, these points are referred to be singular points. This signifies the condition of balance or stability of the system. To determine fixed points, also known as fixed points, in a second-order differential equation, one only needs to substitute the differential Equation (3.2) into the state space Equation (3.3) and set it equal to zero:

$$\begin{aligned} x_2 &= \dot{x}_1 = 0 \\ \ddot{x} &= \dot{x}_2 = -f(x_1, x_2) = 0 \end{aligned} \tag{3.4}$$

The solutions derived from solving the aforementioned algebraic equation will result in the attainment of fixed points.

In order to determine the fixed points of the n -th order differential equation, we must first input the equation into the state space and then transform it into n -th order differential equations. This will provide the following results:

$$x = \begin{Bmatrix} x_1 \\ x_2 \\ \vdots \\ x_n \end{Bmatrix} \text{ and, } X = \begin{Bmatrix} X_1 \\ X_2 \\ \vdots \\ X_n \end{Bmatrix} \tag{3.5}$$

In summary:

$$\dot{x} = X(x) \tag{3.6}$$

To determine the singularity locations, it suffices to evaluate the function at $\dot{x} = 0$. Therefore, we will possess the following:

$$X_i(c) = 0 \qquad for \;\; i = 1, 2, \ldots, n \tag{3.7}$$

The symbol "c" denotes the collection of points that symbolize solitary points. Points that are not solitary are referred to as normal points. There are typically six types of fixed points, also known as singularity points:

- Stable node
- Unstable node
- Saddle point
- Stable focus

- Unstable focus
- Central point

These will be thoroughly examined during the conversation.

3.2.1 Phase Plane Analysis

Phase plane analysis is a graphical technique employed to examine the stability of second-order systems. According to this analysis:

1. Generating trajectories matching to various initial conditions, andsubsequently analyzing the qualitative qualities and attributes of the resulting curves.
2. We obtain valuable data on the system's stability and other patterns of motion.

The analysis was initially presented by mathematician Henry Poincaré in the 19th century AD [1]. The significance of this study is in its ability to examine second-order systems.

- The resulting response lines may be graphically shown as curves, which offer a straightforward means of comprehending the qualitative behavior of the system.
- The study of the behavior of nonlinear systems under various initial conditions may be conducted by employing analytical solutions for nonlinear equations.
- This technique is applicable not just to nonlinear weakening but also to systems exhibiting strong and rigid nonlinearity.
- Several pragmatic and utilitarian systems can be modeled as second-order systems for the purpose of analyzing the phase plane.

However, the phase plane analysis approach has a limitation in that it can only be used to second-order systems. The investigation of higher-order systems using this method becomes computationally and geometrically intricate.

3.2.2 Techniques for Generating Portrait on the Phase Plane

While the demand for phase drawing methods has diminished with the introduction of computers, the ability to create phase pictures remains valuable for regulating computer-generated outcomes. The techniques for applying a design onto the phase plane are as follows:

- Analytical method
- Isocline method
- Delta method
- Lienard method
- Pell method

3.2.2.1 Isocline Method to Ascertain the Trajectory of Motion in the Phase Plane

The term "isocline" is of Greek origin and signifies "same inclination". The isocline line approach involves determining the places on the x_1x_2 plane where the slope $\frac{dx_2}{dx_1}$ is constant. These lines and curves are referred to as the slope and are used to plot the course of the two-order systems. Indeed, by picking an arbitrary point as the initial point on the x_1x_2 plane, we may determine the trajectory by considering the multitude of lines present. Trajectories on the isoclines are tangent to the slopes of each identical isocline. Therefore, the isocline lines represent the set of points on the x_1x_2 plane that have the same slope $\frac{dx_2}{dx_1}$. Isocline lines can be either straight lanes or curved paths [2, 3].

By dividing Equation (3.3) by each other, we may deduce the following:

$$\frac{dx_2}{dx_1} = -\frac{f(x_1, x_2)}{x_2} = \phi(x_1, x_2) \tag{3.8}$$

If $x_2 = 0$ is equal to zero on the x_1 axis and $f(x_1, x_2) \neq 0$ is not equal to zero, then the slope of the trajectory, represented by the ratio of $\frac{dx_2}{dx_1}$, is infinite. Therefore, the trajectory becomes perpendicular to the x_1 axis.

Upon reevaluation, the slope of the trajectory line at Point x:

$$S(x) = \frac{dx_2}{dx_1} = \frac{f_2(x_1, x_2)}{f_1(x_1, x_2)} \tag{3.9}$$

An isocline is characterized by a constant slope α, where the function S(x) is equal to ± $(S(x) = \pm)$. All points on the curve $f_2(x_1, x_2) = \alpha f_1(x_1, x_2)$ have a tangent with a slope equal to α. Furthermore, it should be noted that the omission of time in this context implies that the replies $x_1(t)$ and $x_2(t)$cannot be immediately derived. Furthermore, it is only possible to examine qualitative aspects of behavior, such as stability or oscillatory response.

3.2.2.2 The Procedure for Generating a Phase Portrait Utilizing Isocline Method

1. Plot the α curve in the state space, also known as the phase plane, as shown in Figure 3.1.
2. Create short lines with an α slope. It is important to observe that the orientation of the line is determined by the positive or negative signs of f_1 and f_2 at that specific position.
3. Iterate the procedure for an enough quantity of α in order to populate the phase plane with a complete set of lines.

Example 3–1: Let's examine the mass and spring system illustrated in Figure 3.2, which is described by the following differential equation:

$$\ddot{x} + \omega^2 x = 0$$

Plot its trajectory in the phase plane.

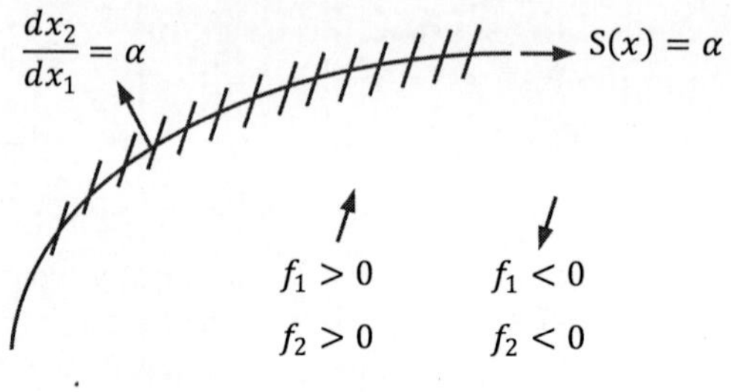

FIGURE 3.1 Schematic of the lines with positive slope.

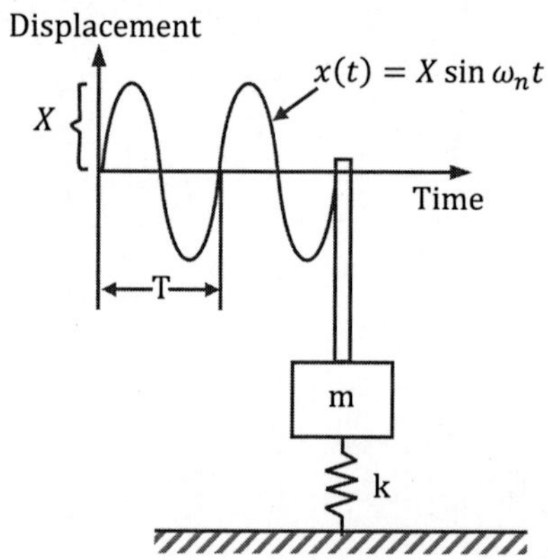

FIGURE 3.2 Amplitude diagram to time of a one-degree freedom system including mass and spring.

Response: The second-order differential equation is represented in the state space. Therefore, the following items will be included:

$$\begin{cases} \dot{x}_1 = x_2 \\ \dot{x}_2 = -\omega^2 x_1 \end{cases}$$

Next, we obtain the result by doing division on two equations:

$$x_2 dx_2 = -\omega^2 x_1 dx_1$$

Which can be written:

$$\frac{dx_2}{dx_1} = -\frac{\omega^2 x_1}{x_2} = \alpha \qquad \text{(a)}$$

Based on the aforementioned correlation, it may be expressed as follows:

$$x_2 = -\frac{\omega^2}{\alpha} x_1 \qquad \text{(b)}$$

The equation described earlier represents an equation with a slope of $-\frac{\omega^2}{\alpha}$, commonly referred to isoclines. By continuously varying α, it is possible to generate different lines.

Alternatively, it may be expressed based on the equation (a):

$$\frac{dx_2}{dx_1} = \alpha \tag{c}$$

Thus, based on the equations (b) and (c), it can be inferred that by choosing a certain value of α, a line can be constructed where all the lines of the trajectory have a same slope α.

Figure 3.3 demonstrates that joining lines with the same slope but different directions result in the creation of an oval curve along the line of those directions. If we desire to delineate the line of trajectory in an analytical manner, it may be expressed based on the equation (a) as follows:

$$x_2 dx_2 = -\omega^2 x_1 dx_1$$

By doing integration on both sides, we will obtain the following:

$$x_2^{\ 2} + \omega^2 x_1^{\ 2} = c^2$$

The equation earlier represents an oval and indicates that the locus of the garlic is the closed path of the oval. Figure 3.4 displays the plotted line of garlic for various values of c.

Example 3–2: Let's examine a pendulum without friction, as shown in Figure 3.5. The dynamic equations that control the state space are as follows:

$$\dot{x}_1 = x_2, \qquad \dot{x}_2 = -\sin x_1$$

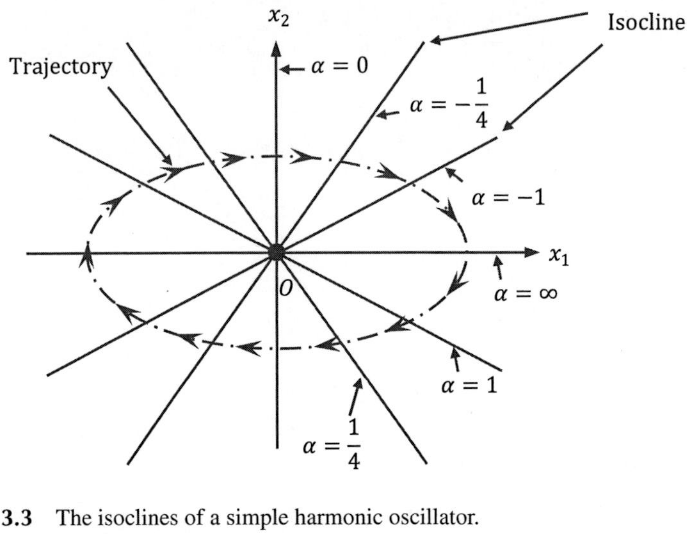

FIGURE 3.3 The isoclines of a simple harmonic oscillator.

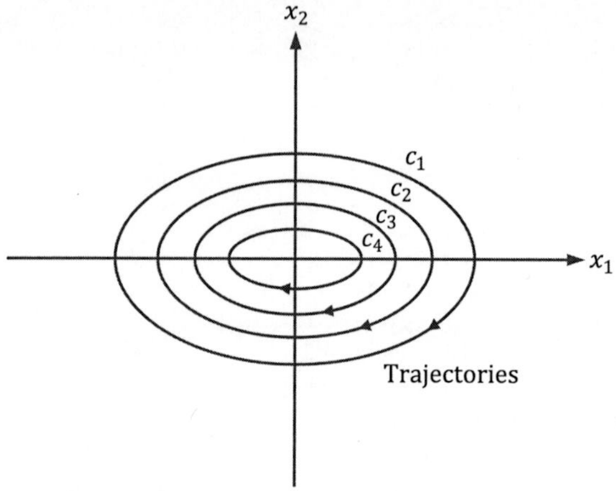

FIGURE 3.4 Drawing the trajectory for different values of c.

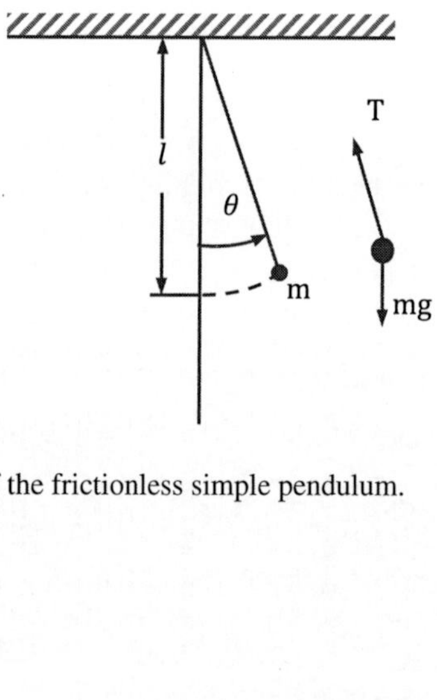

FIGURE 3.5 Schematic of the frictionless simple pendulum.

Which can be written:

$$S(x)=\frac{dx_2}{dx_1}=-\frac{\sin x_1}{x_2}=c$$

Consequently, the relationship of the lines will also be equivalent to the following:

$$x_2=-\frac{1}{c}\sin x_1$$

One can achieve the task of drawing a line by utilizing the aforementioned equation and seeking assistance from specific slopes on the isoclines. Given the initial point $x_0=\left(\frac{\pi}{2},0\right)$, the solution is depicted in Figure 3.6.

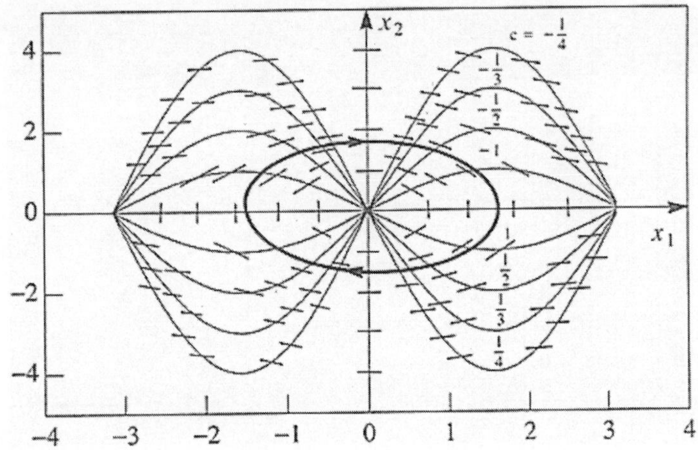

FIGURE 3.6 Geometric structure of the phase plane of the frictionless pendulum equation using the isoclines approach

Example 3–3: Let's examine a pendulum with friction, where its dynamic behavior may be described by the relationship between its state variables as follows:

$$\dot{x}_1 = x_2, \qquad \dot{x}_2 = -0.5\, x_2 - \sin x_1$$

Which can be written as follows:

$$S(x) = \frac{-0.5x_2 - \sin x_1}{x_2} = c$$

Response: The relationship between the lines may be expressed by the slope of the given relationship.

$$x_2 = -\frac{1}{0.5 + c} \sin x_1$$

By choosing the initial point $x_0 = \left(\frac{\pi}{2}, 0\right)$, it can be noticed that its trajectory is contracting and converging, like the behavior of a spring moving towards the origin. By varying the value of c, it is possible to design lines with varied slopes. Conversely, the value of the gradient on each line or curve, represented by $\frac{dx_2}{dx_1}$, is constant and equal to "c". This constant serves as a reliable reference for accurately drawing a line in the phase plane. Figure 3.7 depicts the procedure described earlier, which resulted in the creation of the trajectory.

The subsequent statement is a theorem that establishes the presence and uniformity principles utilized in qualitative analysis, which relies on the depiction of a sequential line.

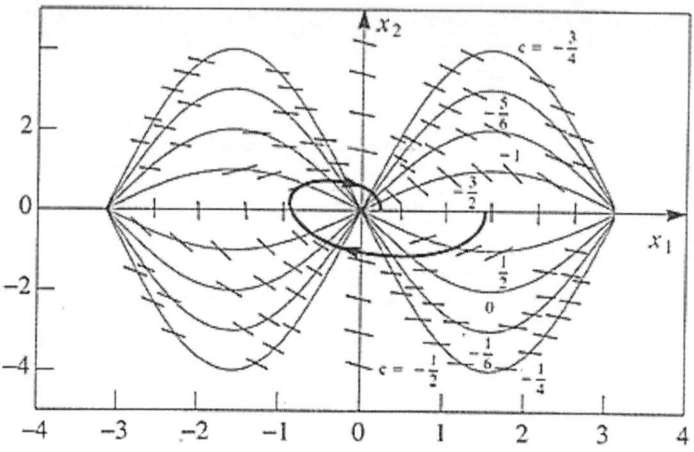

FIGURE 3.7 Geometric structure of the phase plane of the pendulum equation with friction, obtained using the method of isoclines.

3.3 THE INVESTIGATION OF EXISTENCE AND UNIQUENESS THEOREM

The theorem of existence and uniqueness implies that lines with distinct orientations do not intercept. If two lines of trajectory collide, there will be two solutions that originate from a same point, which coincides with the collision point, so contradicting the singularity aspect of the theorem. Put simply, a trajectory line is unable to travel in the other way.

Due to the nonintersecting nature of the trajectory lines, the fuzzy pictures consistently exhibit a tidy look in relation to them. Alternatively, they might deteriorate into a complex network of intersecting curves (Figure 3.8). Indeed, the presence of existence and uniqueness acts as a barrier to this occurrence.

In other words, the aforementioned can be said as follows:

Assuming that:

$$\dot{\vec{x}} = f(\vec{x}) \qquad \vec{x} \in R^2$$

If f is a smooth vector function and f', f are continuous, then there exists a solution for $\vec{x}(t)$, and it is unique for every initial condition. Thus, it can be concluded that the vector function f, being a uniform function, ensures that there is no interference based on the initial condition while sketching the trajectory.

3.4 STABLE SYSTEMS

Stable vibration systems are characterized by the preservation of energy inside the system, without any dissipation. The equation for these sorts of systems is as follows:

$$\ddot{x} + f(x) = 0 \tag{3.10}$$

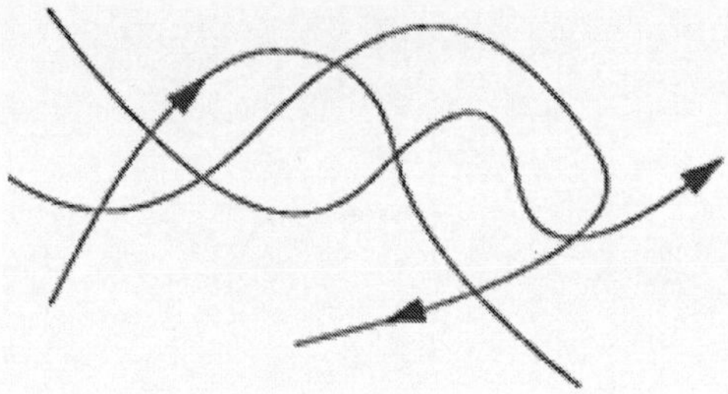

FIGURE 3.8 An illustration of trajectory lines.

Now, we substitute the variable $\ddot{x}$ with the value of $\dot{x}\dfrac{d\dot{x}}{dx}$, results in:

$$\dot{x}\frac{d\dot{x}}{dx}+f(x)=0 \tag{3.11}$$

Or

$$\dot{x}d\dot{x}+f(x)dx=0 \tag{3.12}$$

The generic form of equations for stable systems may be obtained by integration as follows:

$$\frac{\dot{x}^2}{2}+\int_0^x f(x)dx=E=constant \tag{3.13}$$

Based on the equation provided, it is noted that in stable systems, the total energy, which is the sum of kinetic energy and potential energy, remains constant. We will analyze the behavior of stable systems using graphical representations and phase planes.

Example 3–4: Examine the simple pendulum depicted in Figure 2.19.

Based on the preceding section, it can be readily demonstrated that the equation governing the motion of the pendulum is as follows:

$$\ddot{\theta}+\frac{g}{l}sin\theta=0$$

Or

$$\ddot{\theta}+\omega^2\sin\theta=0, \qquad \omega^2=\frac{g}{l} \tag{3.14}$$

Assuming $x_1=\theta$ and $x_2=\dot{\theta}$, we apply Equation (3.9) to the state variables, yielding the following outcome:

$$\begin{matrix}\dot{x}_1=x_2 \\ \dot{x}_2+\omega^2 sin\,x_1=0\end{matrix} \Rightarrow \begin{matrix}\dot{x}_1=x_2 \\ \dot{x}_2=-\omega^2 sin\,x_1\end{matrix}$$

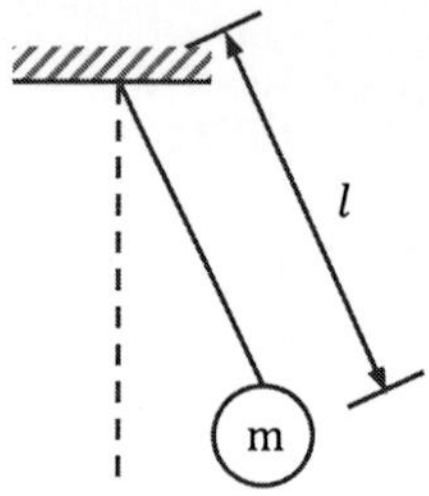

FIGURE 3.9 Simple pendulum with mass m.

Dividing earlier equations will result:

$$\frac{dx_2}{dx_1} = \omega^2 \frac{\sin x_1}{x_2} \tag{3.15}$$

We will multiply the sides.

$$x_2 dx_2 + \omega^2 \sin x_1 dx_1 = 0$$

Through the process of integration, it reaches a final determination:

$$\frac{1}{2} x_2^2 + \omega^2 \left(1 - \cos x_1\right) = E = constant \tag{3.16}$$

Regarding the Equation (3.16), E is a fixed quantity that is directly related to the overall energy of the system. An alternative method can also be used to establish the Equation (3.16). To fulfill this objective, it may be expressed as follows:

Kinetic energy $T = \frac{1}{2} m l^2 \dot{\theta}^2$

Potential energy $U = mgl\left(1 - \cos\theta\right)$

Thus, the total energy of the system will be equivalent to the following:

$$T + U = \frac{1}{2} m l^2 \dot{\theta}^2 + mgl\left(1 - \cos\theta\right)$$

The results are obtained by dividing the parties by ml^2.

$$\frac{1}{2}\dot{\theta}^2 + \frac{g}{l}\left(1 - \cos\theta\right) = \frac{Total\ energy}{ml^2} = E \tag{3.17}$$

Based on the Equation (3.13), the motion equation of a single degree of freedom ($\ddot{x} + f(x) = 0$) may be expressed in the generic form:

$$\frac{1}{2}\dot{x}^2 + F\left(x\right) = h$$

In a continuous system, the terms $\frac{1}{2}\dot{x}^2$ and $F(x)$ represent the kinetic energy and potential, respectively. These quantities are regarded as constant and are equal to h.

The subsequent are the classifications of fixed points that we meet in the qualitative examination of the behavior of dynamic systems. In order to achieve this objective, we examine the potential function $F(x)$ and the constant value h, as seen in Figure 3.10. The graph of Figure 3.10 may be generated by plotting the function $\frac{1}{2}\dot{x}^2 + F(x) = h$.

Figure 3.10 demonstrates a clear correspondence between the route of the line marked with (T_n) and the energy level (h_n). The points labeled as (S) represent the saddle points, whereas the one labeled as (C) represents the center point. Fixed points, also known as central points, correspond to extreme potential energy. Figure 3.10 shows a clear correlation between the saddle points and the highest energy level, whereas the centers (C) are associated with the lowest energy level.

Paths that intersect at the highest point in the saddle points $(\mathrm{T}_3 \text{ and } \mathrm{T}_5)$ (Figure 3.10) are referred to as separatrics. The particle's trajectory near Point (S) is considered unstable in the vicinity of saddle Point (S) because small displacements around Point (S) do not result in a closed path. The point that corresponds to the highest potential energy is referred to as the unstable fixed point. Conversely, in the central area of the neighborhood, the trajectories are not open, resulting in periodic answers. The particle's movement and its associated trajectory near the center are steady. The stable fixed point is the fixed point that corresponds to the minimal potential energy, as shown in Figure 3.11.

To gain a deeper comprehension of the notion of a stable and unstable fixed point, let us examine the pendulum depicted in Figure 3.12. The pendulum depicted in Figure 3.11 (a) exhibits a state of stable balance when it is positioned at its lowest point. By displacing the mass from its state of static equilibrium, it will undergo oscillations around its equilibrium position, which is located at the center (c). The trajectory of the mass in the phase plane will be curved. However, in the pendulum seen in Figure 3.11 (b), referred to as the reverse pendulum, the fixed point is located at its maximum height and belongs to the category of unstable fixed points. Therefore, even a small departure from the fixed point, known as the saddle point, will cause the mass to be displaced and prevent it from returning to its original position.

Analysis of the enclosed trajectory in Figure 3.10 reveals that the duration of rotation is dependent on the extent of movement. In nonlinear systems, the period of rotation or frequency is contingent upon the initial conditions. Typically, the trajectory on the left and right sides of the center is asymmetrical. As a result, the midpoint of the trajectory is shifted and moved away from the center of the stable equilibrium denoted by (C) by increasing the range of motion. This phenomenon is commonly referred to as a steady streaming or drift from the original trajectory. It is important to observe that when the amplitude of T_2 (Figure 3.10) increases, the trajectory, which encompasses the closed path of the oval, will move in a bigger oval shape. The center of this oval will be located distant from the center of static equilibrium.

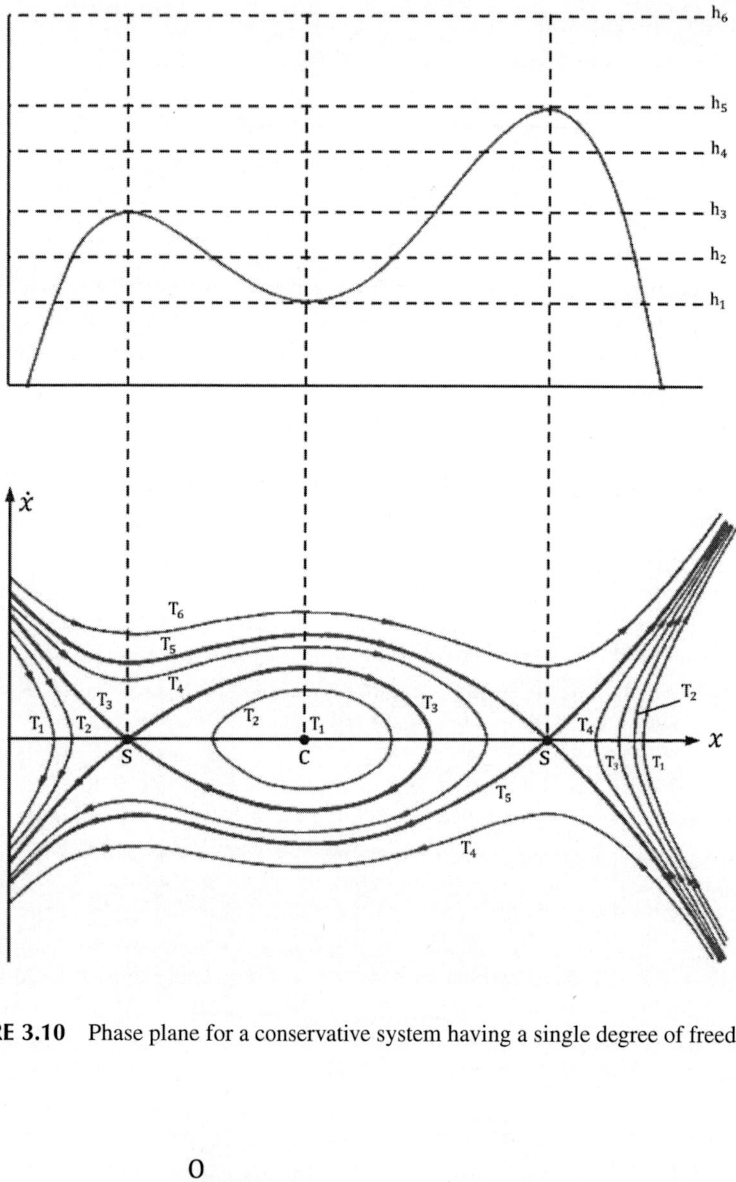

FIGURE 3.10 Phase plane for a conservative system having a single degree of freedom [4].

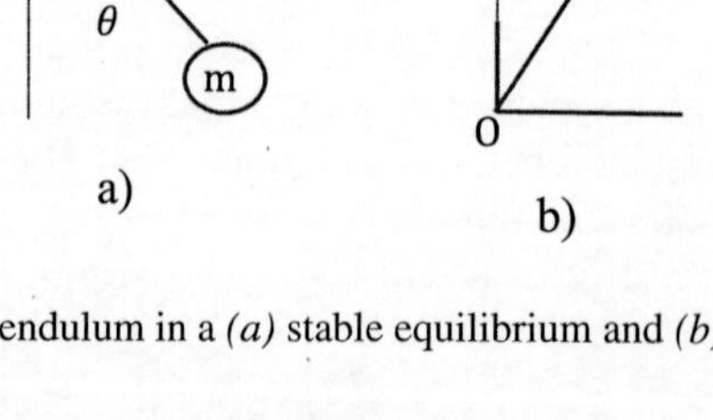

FIGURE 3.11 Simple pendulum in a *(a)* stable equilibrium and *(b)* unstable equilibrium.

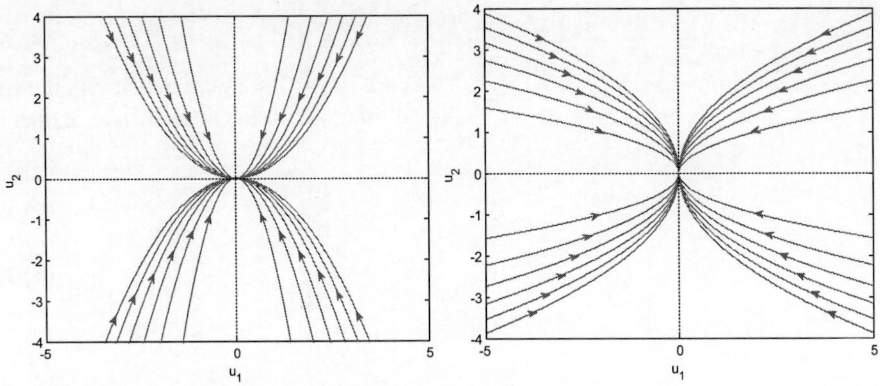

FIGURE 3.12 Stable node, (a) α > 1, (b) α < 1.

3.5 ANALYSIS OF THE QUALITATIVE BEHAVIOR OF SECOND-ORDER NONLINEAR DYNAMIC SYSTEMS USING THE LINEARIZATION APPROACH AROUND THE FIXED POINT

Prior to delving into the specifics of qualitative behavior analysis for second-order nonlinear dynamic systems using the linearization approach around the fixed point, it is crucial to take into account the following two key considerations [4]:

- The study of phase planes in nonlinear systems relies on the analysis of phase planes in linear systems, as nonlinear systems exhibit behavior that is almost identical to linear systems in the vicinity of the fixed point.
- Nonlinear systems have distinct features, such as multiple fixed points and limit cycles, which give rise to intricate patterns.

Initially, through the process of linearizing the differential equations that regulate a nonlinear system at the point of equilibrium, we want to conduct an analysis using a linear model. Given a second-order nonlinear differential equation in state space form, we possess the following:

$$\begin{aligned} \frac{dx_1}{dt} &= f_1(x_1, x_2) \\ \frac{dx_2}{dt} &= f_2(x_1, x_2) \end{aligned} \quad \Rightarrow \quad \frac{dx_2}{dx_1} = \frac{f_2(x_1, x_2)}{f_1(x_1, x_2)} \tag{3.18}$$

We have to get balance points:

$$\dot{x}_1 = \dot{x}_2 = 0 \qquad \Rightarrow \qquad f_1(x_1, x_2) = f_2(x_1, x_2) = 0 \tag{3.19}$$

$$f_1(x_{01}, x_{02}) = f_2(x_{01}, x_{02}) = 0 \tag{3.20}$$

The fixed point is located at (x_{01}, x_{02}). To simplify the issue without sacrificing its generic nature, we will suppose that the fixed point is located at the origin, namely, at (0,0). Therefore:

$$\begin{aligned} x = x_1 - x_{01} &\Rightarrow \frac{\partial x_1}{\partial t} = \frac{\partial x}{\partial t} \\ y = x_2 - x_{02} &\Rightarrow \frac{\partial x_2}{\partial t} = \frac{\partial x}{\partial t} \end{aligned} \tag{3.21}$$

The new fixed point is determined by the alteration of the aforementioned variable, resulting in the origin (0,0) remaining unchanged. Given two functions, f_1 and f_2, which are functions of two variables, their Maclaurin expansion around the fixed point (0,0) is as follows:

$$\begin{aligned} \dot{x} = f_1(x, y) = f_1(0,0) + a_{11}x + a_{12}y + \textit{Higher order sentences} \\ \dot{y} = f_2(x, y) = f_2(0,0) + a_{21}x + a_{22}y + \textit{Higher order sentences} \end{aligned} \tag{3.22}$$

In which:

$$a_{11} = \left.\frac{\partial f_1}{\partial x}\right|_{(0,0)}, \quad a_{12} = \left.\frac{\partial f_1}{\partial y}\right|_{(0,0)}, \quad a_{21} = \left.\frac{\partial f_2}{\partial x}\right|_{(0,0)}, \quad a_{22} = \left.\frac{\partial f_2}{\partial y}\right|_{(0,0)} \tag{3.23}$$

The utilization of linear expressions in the Maclaurin expansion and the omission of higher-order terms is justified by the insignificance of x and y fluctuations in the vicinity of (0,0). The Jacobian matrix is defined as the matrix resulting from the expansion coefficients based on the given Equation (3.23). In the vicinity of the fixed point, the trajectories of a nonlinear system have a behavior that closely resembles the linear trajectories of the system in that region. Presently, we possess the following:

$$\begin{pmatrix} \dot{x} \\ \dot{y} \end{pmatrix} = \begin{bmatrix} a_{11} & a_{12} \\ a_{21} & a_{22} \end{bmatrix} \begin{Bmatrix} x \\ y \end{Bmatrix} \tag{3.24}$$

Based on the response, we have the following format:

$$\begin{Bmatrix} x \\ y \end{Bmatrix} = \begin{Bmatrix} X \\ Y \end{Bmatrix} e^{\lambda t} \tag{3.25}$$

By substituting the given solution into the Equation (3.24), we obtain the following:

$$\lambda e^{\lambda t} \begin{Bmatrix} X \\ Y \end{Bmatrix} = \begin{bmatrix} a_{11} & a_{12} \\ a_{21} & a_{22} \end{bmatrix} \begin{Bmatrix} X \\ Y \end{Bmatrix} e^{\lambda t} \Rightarrow \begin{bmatrix} a_{11} - \lambda & a_{12} \\ a_{21} & a_{22} - \lambda \end{bmatrix} \begin{Bmatrix} X \\ Y \end{Bmatrix} e^{\lambda t} = 0$$

$$A = \begin{bmatrix} a_{11} - \lambda & a_{12} \\ a_{21} & a_{22} - \lambda \end{bmatrix}$$

We have a problem with getting eigenvalues:

$$\det A = 0 \Rightarrow \begin{bmatrix} a_{11} - \lambda & a_{12} \\ a_{21} & a_{22} - \lambda \end{bmatrix} = 0 \tag{3.26}$$

Therefore, the eigenvalues of the problem will be equal to the following:

$$\lambda_1, \lambda_2 = \frac{1}{2}(p \pm \Delta), \Delta = \sqrt{p^2 - 4q}, \quad p = a_{11} + a_{22}, \quad q = a_{11}a_{22} - a_{21}a_{12} \tag{3.27}$$

If $\begin{Bmatrix} X_1 \\ Y_1 \end{Bmatrix}$ and $\begin{Bmatrix} X_2 \\ Y_2 \end{Bmatrix}$ are eigenvectors associated with eigenvalues (λ_2, λ_1), respectively, and assuming they are $(\lambda_1 \neq 0)$ and $(\lambda_2 \neq 0)$ as well as $(\lambda_1 \neq \lambda_2)$, the answer can be written as follows: (where c_1 and c_2 are arbitrary constants).

$$\begin{Bmatrix} x \\ y \end{Bmatrix} = c_1 \begin{Bmatrix} X_1 \\ Y_1 \end{Bmatrix} e^{\lambda_1 t} + c_2 \begin{Bmatrix} X_2 \\ Y_2 \end{Bmatrix} e^{\lambda_2 t} \tag{3.28}$$

In the upcoming discussion, we will see that the qualitative dynamics of the system may be characterized by describing the distinct values. To solve the Equation (3.24), a straightforward approach involves decoupling and utilizing the T-transformation matrix in the following manner:

$$\begin{Bmatrix} x \\ y \end{Bmatrix} = \begin{bmatrix} X_1 & X_2 \\ Y_1 & Y_2 \end{bmatrix} \begin{Bmatrix} u_1 \\ u_2 \end{Bmatrix} = [T] \begin{Bmatrix} u_1 \\ u_2 \end{Bmatrix} \tag{3.29}$$

Matrix T represents the modal matrix, which contains eigenvectors. The generalized coordinates are denoted by $\begin{Bmatrix} u_1 \\ u_2 \end{Bmatrix}$. The governing differential equation may be expressed as follows:

$$\begin{bmatrix} X_1 & X_2 \\ Y_1 & Y_2 \end{bmatrix}^{-1} \begin{bmatrix} X_1 & X_2 \\ Y_1 & Y_2 \end{bmatrix} \begin{Bmatrix} \dot{u}_1 \\ \dot{u}_2 \end{Bmatrix} = \begin{bmatrix} X_1 & X_2 \\ Y_1 & Y_2 \end{bmatrix}^{-1} \begin{bmatrix} a_{11} & a_{12} \\ a_{21} & a_{22} \end{bmatrix} \begin{bmatrix} X_1 & X_2 \\ Y_1 & Y_2 \end{bmatrix} \begin{Bmatrix} u_1 \\ u_2 \end{Bmatrix} \tag{3.30}$$

Based on the theorem of linear algebra, specifically the algebra of matrices, the following is established:

$$\begin{bmatrix} X_1 & X_2 \\ Y_1 & Y_2 \end{bmatrix}^{-1} \begin{bmatrix} a_{11} & a_{12} \\ a_{21} & a_{22} \end{bmatrix} \begin{bmatrix} X_1 & X_2 \\ Y_1 & Y_2 \end{bmatrix} = \begin{bmatrix} \lambda_1 & 0 \\ 0 & \lambda_2 \end{bmatrix} \tag{3.31}$$

So we will have the following:

$$\begin{Bmatrix} \dot{u}_1 \\ \dot{u}_2 \end{Bmatrix} = \begin{bmatrix} \lambda_1 & 0 \\ 0 & \lambda_2 \end{bmatrix} \begin{Bmatrix} u_1 \\ u_2 \end{Bmatrix} \quad \Rightarrow \quad \begin{matrix} \dot{u}_1 = \lambda_1 u_1 \;\rightarrow\; u_1(t) = u_{10}\, e^{\lambda_1 t} \\ \dot{u}_2 = \lambda_2 u_2 \;\rightarrow\; u_2(t) = u_{20}\, e^{\lambda_2 t} \end{matrix} \tag{3.32}$$

Both u_{10} and u_{20} are constant. It is noted that two equations are distinct and may be readily solved using the aforementioned method. By eliminating the variable t from the aforementioned two equations, we obtain the following:

$$u_2(t) = u_{20} \left(\frac{u_1}{u_{10}} \right)^{\alpha}; \qquad \alpha = \frac{\lambda_2}{\lambda_1} \tag{3.33}$$

If $u_{10} = 0$, the half-trajectory corresponds to the axis u_2 (Figure 3.13).

Considering the aforementioned information, it can be concluded that by computing the eigenvalues of the coefficient matrix in the Equation (3.24), it becomes convenient to express the solution of the differential equation in terms of the Equation (3.32) rather than Equation (3.28). This approach facilitates the analysis of the qualitative characteristics of the equation. Based on the values of λ_1 and λ_2, the fixed points may be categorized as follows.

1. λ_1 and λ_2 are real and uneven values, i.e., $p^2 > 4q$.

   Next, given that λ_1 and λ_2 have the same or opposite sign, we specify the type of movement.

   a) if λ_1 and λ_2 have same symbols $(q > 0)$, the fixed point is called the node.

      a-1) if $\lambda_1, \lambda_2 < 0 \rightarrow p < 0$, as a result, the fixed point, a node, will be stable as Figure 3.12.

      a-2) if $\lambda_1, \lambda_2 > 0 \rightarrow p > 0$, as a result, the fixed point, a node, will be unstable as Figure 3.13.

      The behavior of the trajectory going through the origin is contingent upon the sign α. If α is a positive value (λ_1 and λ_2 have the same symbols), the origin is called a node or nodal point (as seen in Figure 3.12 and Figure 3.13). When the value of α is greater than 1 $(\alpha > 1)$, the trajectory intersects the axis (u_1) tangentially, as shown in Figure 3.12 a and Figure 3.13 a. Conversely, when α is less than 1 $(\alpha < 1)$, the trajectory intersects the axis (u_2) tangentially, as shown in Figure 3.14 b and Figure 3.13 b. Figure 3.12 depicts a stable node, indicating that both λ_1 and λ_2 are negative. Conversely, when λ_1 and λ_2 are positive, it signifies that as t grows, the point is diverging from the origin, which is referred to as an unstable node (Figure 3.13).

   b) If both λ_1 and λ_2 are real and equal, then the equation $p^2 = 4q$ holds. In this scenario, the trajectory will consist of straight lines that intersect the origin. If $\lambda < 0$, the fixed point (origin) will be stable, and if $\lambda < 0$, the fixed point will be unstable (Figures 3.14 a and b).

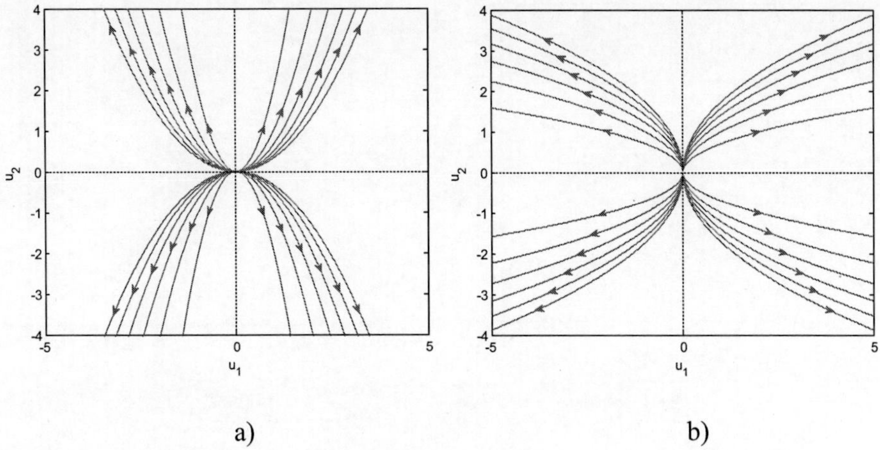

FIGURE 3.13 Unstable node, (a) α > 1, (b) α < 1.

For instance, we examine the scenario in which the roots possess the characteristics of being both real and equal. There are two potential modes in this case:

$$[B]=\begin{bmatrix}\lambda & 0\\ 0 & \lambda\end{bmatrix}$$

Or

$$[B]=\begin{bmatrix}\lambda & 1\\ 0 & \lambda\end{bmatrix}$$

We'll have [B] for the first case:

$$\dot{u}_1=\lambda u_1, \qquad \dot{u}_2=\lambda u_2 \tag{3.34}$$

So:

$$u_1=u_{10}\ exp(\lambda t), \qquad u_2=u_{20}\ exp(\lambda t), \qquad \frac{u_1}{u_2}=\frac{u_{10}}{u_{20}} \tag{3.35}$$

The point of origin is referred to as a node. The stability of the node's origin is determined by the sign of λ. If $\lambda<0$, the origin of the node is stable (Figure 3.14 a). Conversely, if $\lambda>0$, the origin of the node is unstable (Figure 3.14 b).

$$\lambda<0_1=\lambda_2<0 \qquad \lambda_1=\lambda_2>0$$

Regarding the second scenario, we will have [B]:

$$\dot{u}_1=\lambda u_1+u_2, \qquad \dot{u}_2=\lambda u_2 \tag{3.36}$$

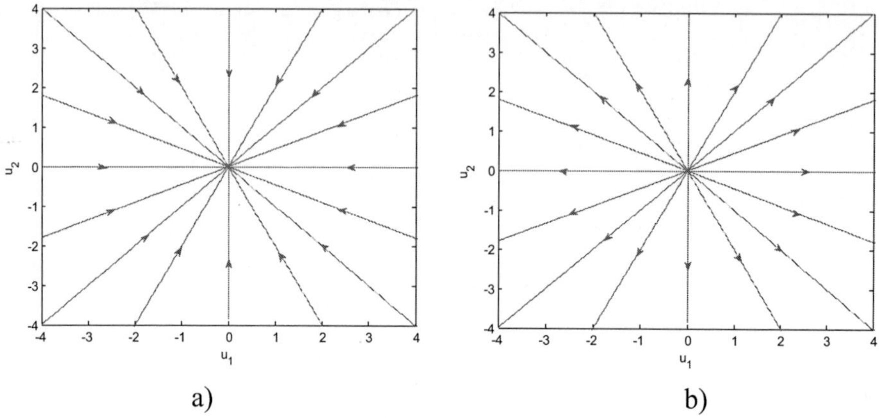

FIGURE 3.14 Stable node when the eigenvalues are equal and the Jacobin matrix is diagonal.

So:

$$u_1 = (u_{10} + u_{20}t)\ exp(\lambda t), \qquad u_2 = u_{20}\ exp(\lambda t) \tag{3.37}$$

Furthermore, in this particular scenario, the point of origin is referred to as a node. If $\lambda > 0$ (Figure 3.15 a), the origin of the node is stable. Conversely, if $\lambda < 0$ (Figure 3.15 b), the origin of the node is unstable.
The semicircle represents the point where $u_{20} = 0$ equals zero on the u_1 axis. Regarding the Equation (3.37), the following information is available:

$$\frac{u_2}{u_1} = \frac{u_{20}}{u_{10} + u_{20}t} \rightarrow \frac{1}{t} \quad as\ t \rightarrow \infty \tag{3.38}$$

Which is for $u_{20} \neq 0$ mode.
Based on the Equation (3.37), it is determined that u_2 is unable to undergo a change in sign. Consequently, every trajectory line must be positioned either in the top half or the bottom half of the plate. Furthermore, when t approaches infinity, the values of u_1 and u_2 converge, and the ratio u_2 / u_1 tends towards zero. Thus, all the lines of the trajectory pattern converge towards the origin in the lower half of the plane, exhibiting a horizontal slope and originating from the left side. All the lines of the trajectory converge towards the origin in the upper half of the plate, originating from the right side and with a gradient of zero (Figure 3.15).

2. If λ_1 and λ_2 are real but opposite to the sign $(q < 0)$ and the sign p is not important.
   In this scenario, one of the solutions converges towards the origin, while the other diverges towards infinity. In this particular situation, the origin represents the saddle point and corresponds to an unstable equilibrium.

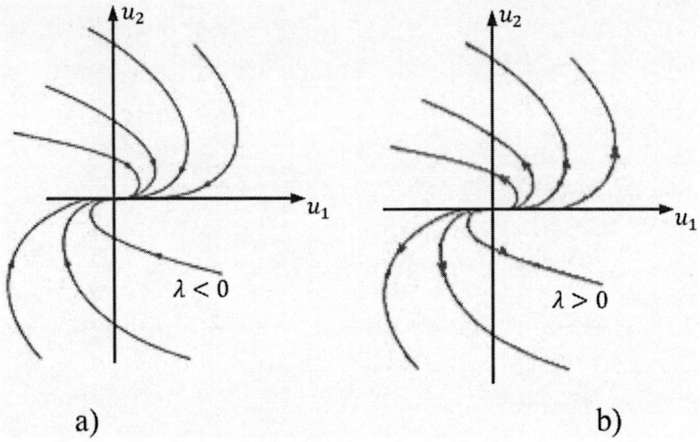

FIGURE 3.15 a) Stable node, b) unstable node.

Figure 3.16 displays the structure of the trajectory line. The axes u_1 and u_2 are unequivocally integral curves.

$$\lambda_2 < 0, \quad \lambda_1 > 0 \qquad \lambda_2 > 0, \quad \lambda_1 < 0$$

In mathematics, an integral curve is a parametric curve that provides a unique solution of an ordinary differential equation and/or system of equations. If the differential equation is represented as a vector field or slope field, then the associated integral curves are tangential to the field at every point.

If λ_1 and λ_2 are complex conjugate$(p^2 < 4q)$.

$$\lambda_1 = \theta_1 + i\theta_2 \rightarrow u_1(t) = u_{10}e^{\lambda_1 t} \rightarrow u_1(t) = \left(u_{10}e^{\theta_1 t}\right)e^{i\theta_2 t}$$
$$\lambda_2 = \theta_1 - i\theta_2 \rightarrow u_2(t) = u_{20}e^{\lambda_2 t} \rightarrow u_2(t) = \left(u_{20}e^{\theta_1 t}\right)e^{-i\theta_2 t}$$

The equations depicted earlier illustrate the motion of a logarithmic spiral. In this scenario, the point of equilibrium is referred to as the focus point or spiral point or focal point. If θ_1 is negative ($\theta_1 < 0$), both the asymptotic stable motion and the focus point will be stable. The expression $(p < 0)$ and $(q > 0)$ is equal to the condition shown in Figure 3.17.

If $\theta_1 > 0$, the focal point will be unstable. This is equivalent to $(p > 0)$ and $(q > 0)$ (Figure 3.18).

It is important to understand that the θ_2 sign just denotes the direction of rotation. If θ_2 is greater than zero ($\theta_2 > 0$), the motion is in the counterclockwise direction. Conversely, if θ_2 is less than zero ($\theta_2 < 0$), the motion is in the clockwise direction.

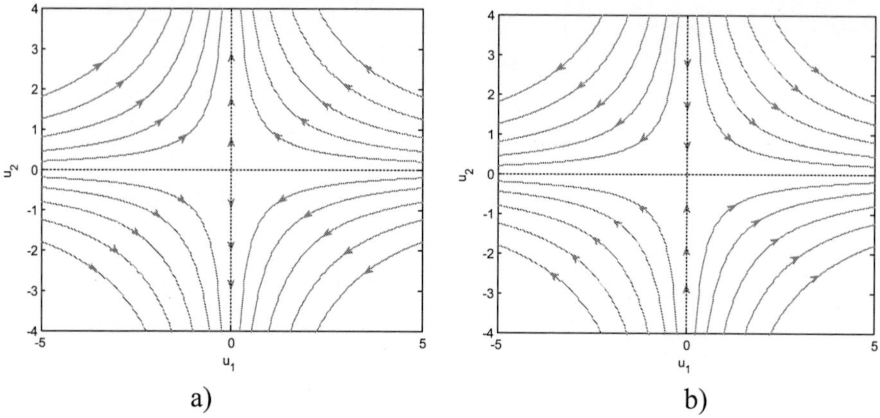

FIGURE 3.16 Saddle curve a) $\lambda_1 > 0$, $\lambda_2 < 0$, b) $\lambda_1 < 0$, $\lambda_2 > 0$.

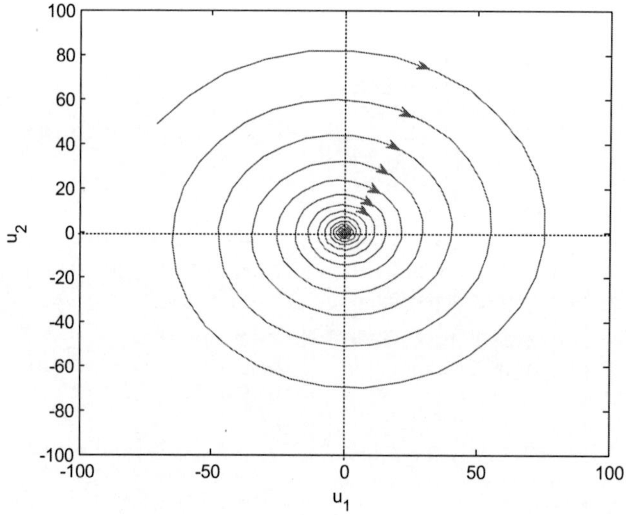

FIGURE 3.17 Stable focal point.

If θ_1 equals zero ($\theta_1 = 0$), meaning that p equals zero $(p = 0)$, then the mixed radial vector will have a constant value. Consequently, the phase plane is introduced as a circle with the center being the fixed point. There has been sporadic movement, resulting in stability. The fixed point in this instance is located in the center or vertex point (as seen in Figure 3.19).

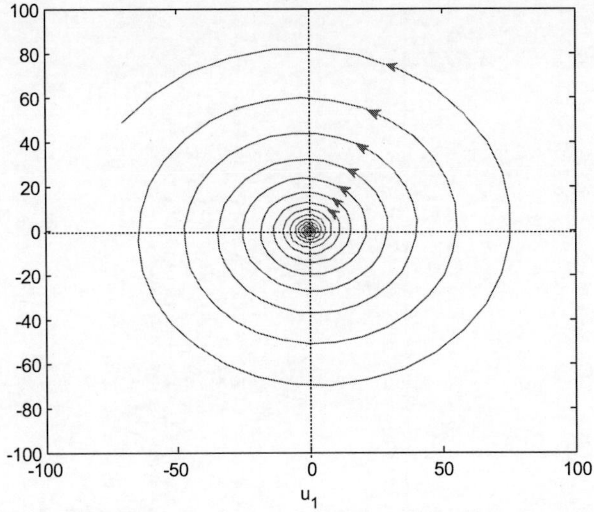

FIGURE 3.18 Unstable focal point.

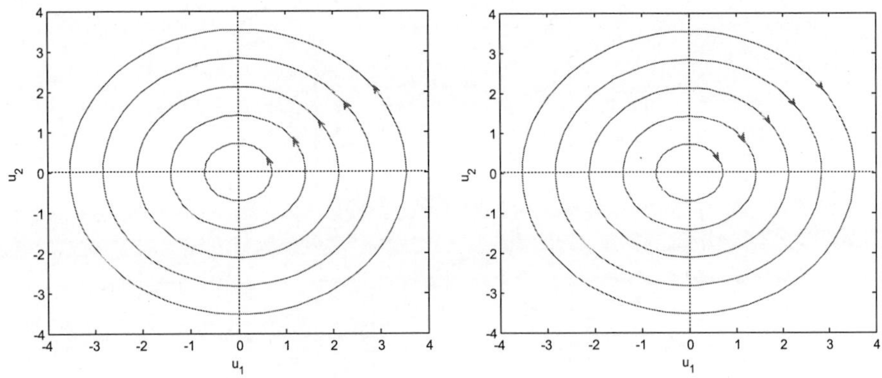

FIGURE 3.19 Vertex or center point.

3.6 REVIEW OF A FEW POINTS

1. The nature of the fixed point is determined by the polarity of the eigenvalues.
   An instability arises when the real component of the eigenvalues is positive, while the system remains stable when the real part of the eigenvalues is negative.
2. All the characteristics of linear systems are widely recognized.
3. The characteristics of nonlinear systems are only established within a limited region.

Figure 3.20 depicts the various states of the poles and the names of the fixed points associated with each condition.

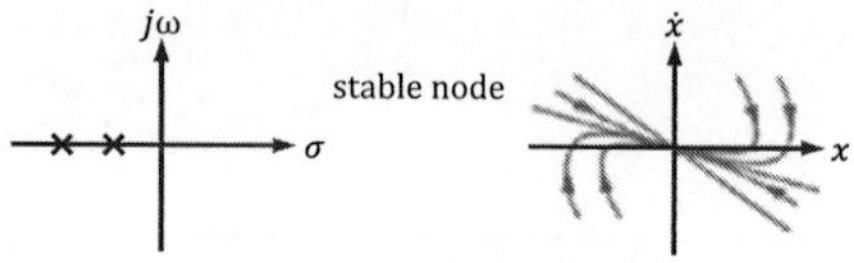

a) Stable node

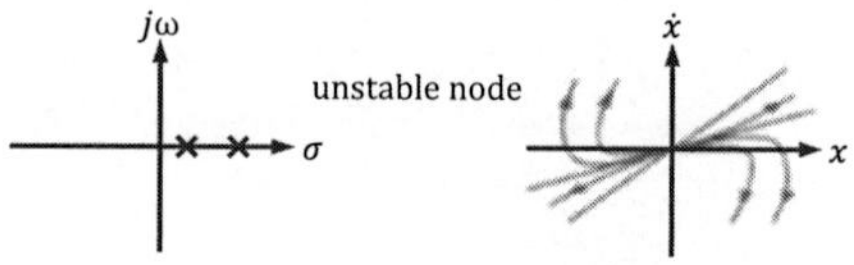

b) Unstable node

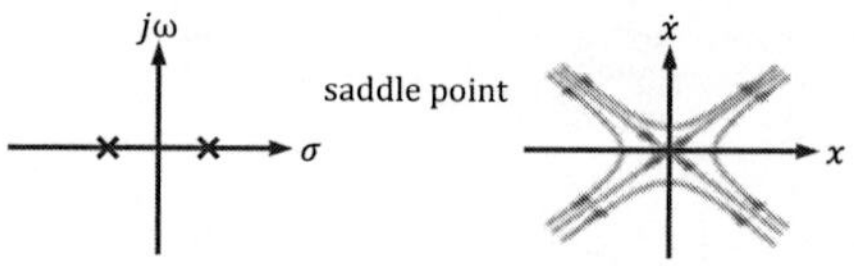

c) Saddle point

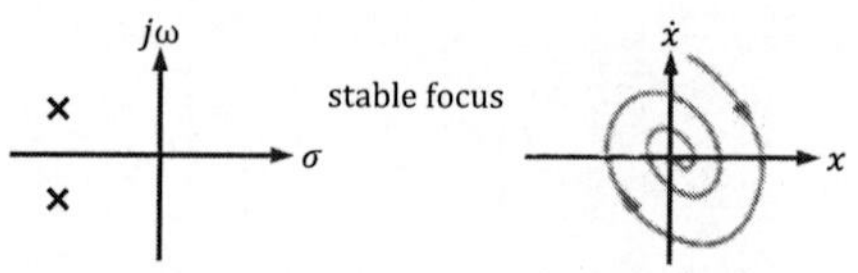

d) Stable focus

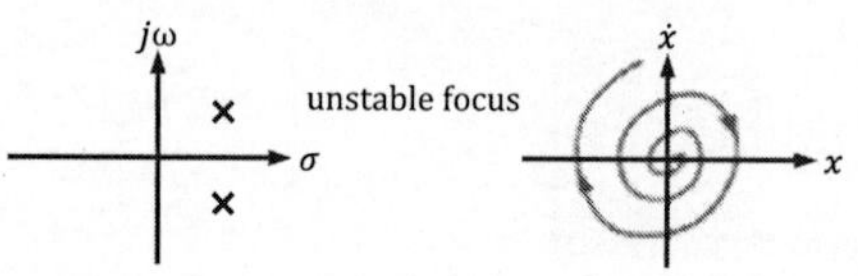

e) Unstable focus

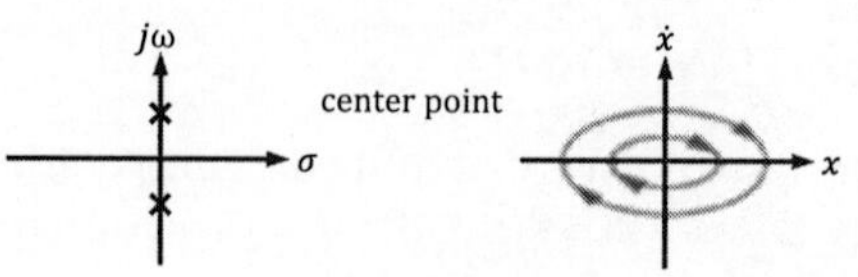

f) Center point

FIGURE 3.20 Different types of poles according to their eigenvalues and corresponding fixed points.

Example 3–5: Find the solution to the initial value issue given by the equations $\dot{x} = x + y$ and $\dot{y} = 4x - 2y$ using the initial conditions $(x_0, y_0) = (2, -3)$.

Answer: the corresponding matrix is equal to the following:

$$\begin{pmatrix} \dot{x} \\ \dot{y} \end{pmatrix} = \begin{bmatrix} 1 & 1 \\ 4 & -2 \end{bmatrix} \begin{pmatrix} x \\ y \end{pmatrix}$$

Firstly, we determine the unique values of matrix A. The equation can be described as a characteristic equation: $\lambda \hat{} 2 + \lambda - 6 = 0$. Therefore:

$$\lambda_1 = 2, \ \lambda_2 = 3$$

Next, we acquire distinct vectors $\boldsymbol{V} = (v_1, v_2)$ that correspond to certain values in such a way that:

$$\begin{bmatrix} 1-\lambda & 1 \\ 4 & -2-\lambda \end{bmatrix} \begin{pmatrix} v_1 \\ v_2 \end{pmatrix} = \begin{pmatrix} 0 \\ 0 \end{pmatrix}$$

Given $\lambda_1 = 2$, we may deduce that the matrix $\begin{bmatrix} -1 & 1 \\ 4 & -4 \end{bmatrix} \begin{pmatrix} v_1 \\ v_2 \end{pmatrix} = \begin{pmatrix} 0 \\ 0 \end{pmatrix}$ multiplied by the vector (v_1, v_2) equals the zero vector $(1,1)$ $(v_1, v_2) = (1,1)$. This equation has a trivial solution $(v_1, v_2) = (1,1)$ or any scalar multiple of it. (Certainly, every scalar multiple of a unique vector will always provide another unique vector, with an inclination towards selecting the most straightforward option, but all possibilities will be valid.) Similarly, when λ_2 is equal to –3, the equation takes on a eigen-value $\begin{bmatrix} -1 & 1 \\ 4 & -4 \end{bmatrix} \begin{pmatrix} v_1 \\ v_2 \end{pmatrix} = \begin{pmatrix} 0 \\ 0 \end{pmatrix}$. The equation $(v_1, v_2) = (1, -4)$ has a trivial solution $(v_1, v_2) = (1, -4)$. To summarize:

$$V_1 = \begin{pmatrix} 1 \\ 1 \end{pmatrix}, \ V_2 = \begin{pmatrix} 1 \\ -4 \end{pmatrix}$$

A generic response can be expressed as a linear mixture of specific replies. Thus, the answer will be applicable in all cases.

$$\boldsymbol{X}(t) = c_1 \begin{pmatrix} 1 \\ 1 \end{pmatrix} e^{2t} + c_2 \begin{pmatrix} 1 \\ -4 \end{pmatrix} e^{-3t}$$

To get the initial condition $(x_0, y_0) = (2, -3)$ in the final state, we compute the coefficients c_1 and c_2. At time $t = 0$, we will possess the following:

$$\begin{pmatrix} 2 \\ -3 \end{pmatrix} = c_1 \begin{pmatrix} 1 \\ 1 \end{pmatrix} + c_2 \begin{pmatrix} 1 \\ -4 \end{pmatrix},$$

Which is equivalent to the following algebraic equations:

$$\begin{aligned} 2 &= c_1 + c_2, \\ -3 &= c_1 - 4c_2. \end{aligned} \quad \rightarrow \; c_1 = 1, \; c_2 = 1$$

With a fixed date, the public response results:

$$\begin{aligned} x(t) &= e^{2t} + e^{-3t} \\ y(t) &= e^{2t} - 4e^{-3t} \end{aligned}$$

Which is solving the initial value problem.

Fortunately, it is not necessary to do all of these processes in order to create a phase picture of a linear system. Our sole requirement is to get knowledge of the distinctive values and their matching distinctive vectors.

Given that the system's eigenvalues are $\lambda_1 = 2$ and $\lambda_2 = 3$, let's draw phase shape. The initial unique answer experiences exponential growth, whereas the subsequent unique solution diminishes and eventually vanishes. This is the genesis of a zenith point. The manifold is defined by the line created by the vector $V_2 = (1,-4)$, which corresponds to the specific solution that is decreasing in magnitude. Likewise, there are a small number of unstable branches, such as the line formed by vector $V_1 = (1,1)$. Similar to the Zen dots, a single line of garlic moves towards several unstable branches, causing the need for $t \rightarrow \infty$ (as seen in Figure 3.21).

3.7 CLASSIFICATION OF FIXED POINTS

By now, you are likely fatigued from the numerous instances and eager for a straightforward category. Fortunately, this category exists. The associated equilibrium and stability point type may be visually shown in a straightforward diagram (Figure 3.22) [2].

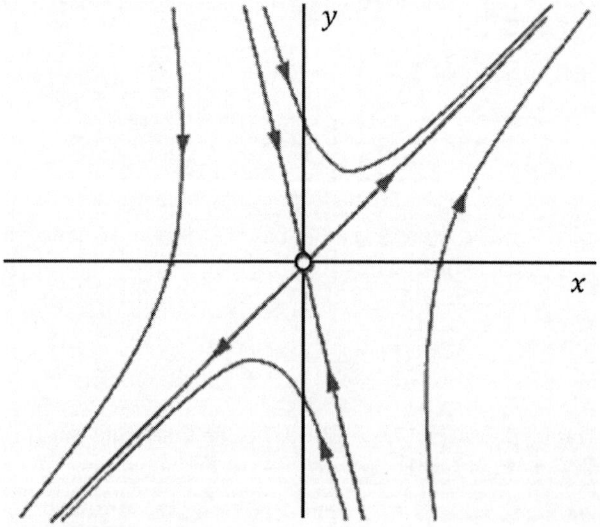

FIGURE 3.21 Phase plane image.

The second-order differential equation in the state space is defined based on the linear conditions surrounding the fixed point in the following manner:

$$\begin{Bmatrix} \dot{x} \\ \dot{y} \end{Bmatrix} = \begin{bmatrix} a_{11} & a_{12} \\ a_{21} & a_{22} \end{bmatrix} \begin{Bmatrix} x \\ y \end{Bmatrix} \tag{3.39}$$

Based on the response, we may observe the following structure:

$$\begin{Bmatrix} x \\ y \end{Bmatrix} = \begin{Bmatrix} X \\ Y \end{Bmatrix} e^{\lambda t} \tag{3.40}$$

By substituting the previous solution into the Equation (3.39), we obtain the following:

$$\lambda e^{\lambda t} \begin{Bmatrix} X \\ Y \end{Bmatrix} = \begin{bmatrix} a_{11} & a_{12} \\ a_{21} & a_{22} \end{bmatrix} \begin{Bmatrix} X \\ Y \end{Bmatrix} e^{\lambda t} \rightarrow \underbrace{\begin{bmatrix} a_{11} - \lambda & a_{12} \\ a_{21} & a_{22} - \lambda \end{bmatrix}}_{A} \begin{Bmatrix} X \\ Y \end{Bmatrix} e^{\lambda t} = 0$$

Therefore, we have a eigenvalues for solving the problem:

$$det\ A = 0 \rightarrow \begin{bmatrix} a_{11} - \lambda & a_{12} \\ a_{21} & a_{22} - \lambda \end{bmatrix} = 0 \rightarrow \quad \lambda_1, \lambda_2 = \frac{1}{2}\left(p \pm \sqrt{p^2 - 4q}\right)$$

$$p = a_{11} + a_{22}, \qquad q = a_{11}a_{22} - a_{21}a_{12}$$

In Figure 3.22, the p-axis, Trace, and the q-axis are the determinants of matrix A. The schematic information is derived from the subsequent relationships.

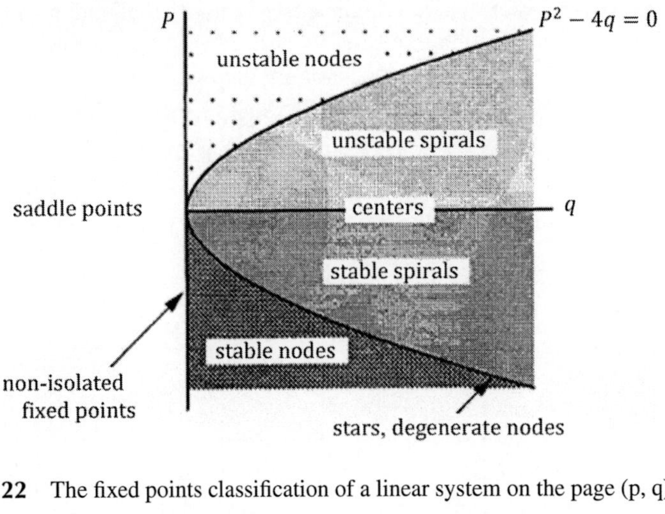

FIGURE 3.22 The fixed points classification of a linear system on the page (p, q).

$$\lambda_{1,2} = \frac{1}{2}\left(p \pm \sqrt{\Delta}\right), \Delta = p^2 - 4q \qquad q = \lambda_1\lambda_2, \qquad p = \lambda_1 + \lambda_2 \tag{3.41}$$

The second and third expressions in Equation (3.41) can be obtained by writing the characteristic equation in the form $(\lambda - \lambda_1)(\lambda - \lambda_2) = \lambda^2 - p\lambda + q = 0$.

According to Figure 3.22, the following is observed:

If q is < 0, the eigenvalues are real and have the opposite sign. So the fixed point is a saddle point.

If q is greater than zero ($q > 0$), the eigenvalues exhibit characteristics that resemble either sign (nodes) or complex conjugates of spirals and centers. It should be noted that the nodes exhibit a $p^2 - 4q > 0$ correlation, whereas the spirals adhere to a $p^2 - 4q < 0$ correlation. The equation $p^2 - 4q = 0$ represents the border separating nodes from spirals. This share includes both the star nodes and the degenerate nodes. p determines the stability of nodes and spirals. When the value of p is less than zero ($p < 0$), the real components of both eigenvalues are negative. Consequently, the fixed point is deemed stable nodes, and spirals that are unstable have a positive p value ($p > 0$). Stable centers are naturally positioned on the boundary line where $p = 0$, resulting in a combination of exceptional values.

If q equals zero ($q = 0$), then at least one of the eigenvalues is also zero. Therefore, center, a balancing point, does not exist in isolation. Under these circumstances, there are stationary points throughout the whole length of a line (as seen in Figure 3.23). Alternatively, if $A = 0$, a stationary points plate will be present.

Upon reexamination and with a more extensive analysis, it is evident that Figure 3.22 may be regarded as Figure 3.24. The primary components of the fixed points in this form are saddle points, nodes, and spirals. They are present in significant portions of spots on the plate (q, p). The centers and nodes of the star, as well as the equivalent and non-isolated fixed points, correspond to the boundary line curves on the plate defined by the coordinates (q, p). Among these borderline instances, centers have much more importance. They often arise in frictionless mechanical systems with steady energy.

It is emphasized that the parabolic curve with the equation $p^2 - 4q = 0$ in Figure 3.24 is a boundary line that consists of border nodes, star nodes, line $p = 0$ centers, and line $q = 0$ and is known as non-isolated fixed points. Among them, the

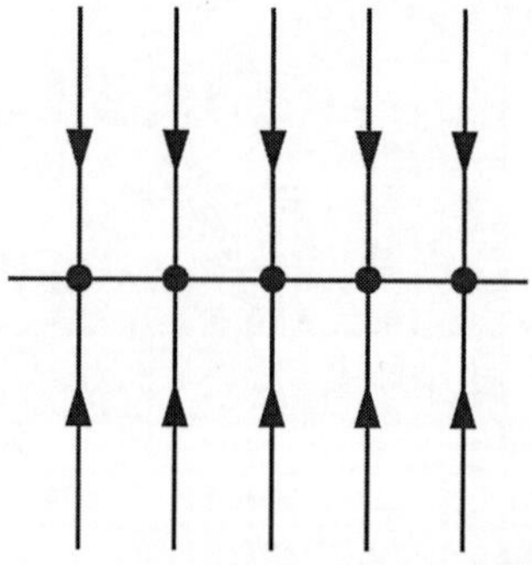

FIGURE 3.23 Phase plane corresponding to one of the eigenvalues of zero.

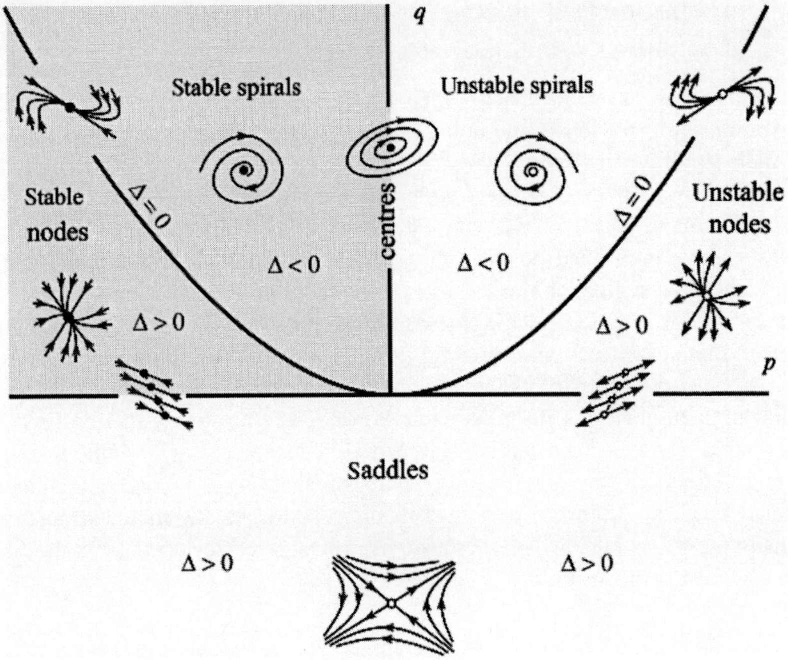

FIGURE 3.24 The fixed points classification for a linear system on the page (p, q).

centers have particular significance. Due to even the tiniest variation, the system may encounter either stable or unstable situations.

3.7.1 Tip

The trajectories of a nonlinear system are located in a neighborhood surrounding the fixed point, which is close to its linear trajectories.

If the equation originates from a linear state, a stable node, a stable focal point, or a saddle point, then in a limited vicinity of the fixed point, the trajectory of a nonlinear system will exhibit behavior similar to that of a stable node, a stable focal point, or a saddle point, respectively. If the linearized state equation originates from an unstable node, an unstable focal point, or a saddle point, then the behavior of the course line of a nonlinear system in a small neighborhood of the fixed point will resemble that of an unstable node, an unstable focal point, or a saddle point, respectively [4].

It is important to mention that the phrase mentioned earlier does not apply to boundary lines, such as central, equivalent, star nodes, and non-isolated fixed points. Undoubtedly, the center and non-isolated fixed points are quite sensitive within the boundary lines. By making a modest adjustment, such as including a damper, the central point of the system will tend to move towards either a stable or unstable spiral. This scenario is true for non-isolated fixed points. However, the stability criteria for equivalent and star nodes remain unchanged with minimum modifications. For example, unstable spirals transform into unstable nodes, yet their instability persists.

3.7.2 Hyperbolic Fixed Points, Topological Equivalence, and Structural Stability

In a second-order system, if both eigenvalues $Re(\lambda) \neq 0$ are not equal to zero, the fixed point is referred to as hyperbolic. The hyperbolic fixed points exhibit robustness and their stability type is unaltered by minor nonlinear perturbations. Conversely, fixed points that are not characterized by delusions are feeble and easily broken.

The Hartmann-Grobman theorem states that the local phase image closely approximates a hyperbolic fixed point, with a topological coefficient that aligns with the linearization phase image. The stability of the fixed point is specifically determined by linearization. Topological alignment refers to the process of achieving topological homeomorphism or mapping topology through continuous deformation and inversion. This involves drawing the image of the local phase onto another, ensuring that the line of the trajectories aligns with the line of other trajectories and the direction of time (path of vectors) is preserved. An illustrative instance of hyperbolic fixed points can be referred to as the saddle point, whereas for non-hyperbolic fixed point, as the center point. The Hartmann-Grobman theorem addresses the matter of stability of points in the phase plane when there is a little change or deviation from the equilibrium state, using eigenvalues.

3.8 QUALITATIVE BEHAVIOR OF *N*-ORDER NONLINEAR DYNAMIC SYSTEMS BY USING THE LINEARIZATION APPROACH AROUND THE FIXED POINT

Let's consider the differential equation that governs a system in state space with n dimensions. This equation is a first-order differential equation.

$$\dot{\vec{x}} = \vec{F}(\vec{x}) \qquad \vec{x} \varepsilon R^n, \tag{3.42}$$

Or

$$\dot{x}_i = F_i(x_1, x_2, x_3, \ldots, x_n) \qquad i = 1, 2, \ldots, n \tag{3.43}$$

F is a uniform function with a real value.

The function $\vec{F}(\vec{x})$ can generally exhibit nonlinearity. We need to compute the fixed points.

$$\dot{\vec{x}} = 0 \rightarrow \vec{F}(\vec{x}^*) = 0; \tag{3.44}$$

Given the desire to shift the fixed point to the origin, we may do this by modifying the variable as follows:

$$\begin{gathered} \vec{y}(t) = \vec{x}(t) - x^*; \\ \dot{\vec{y}}(t) = \dot{\vec{x}}(t); \end{gathered} \tag{3.45}$$

Assume that by substituting the variable change mentioned earlier into the Equation (3.43) and expanding the function $F_i(y_1, y_2, y_3, \ldots, y_n)$ around the origin, the resulting relationship is as follows:

$$\dot{\vec{y}}(t) = A\vec{y}(t); \tag{3.46}$$

Where the Jacobian matrix is defined as follows:

$$A = \begin{bmatrix} \frac{\partial F_1}{\partial y_1} & \frac{\partial F_1}{\partial y_2} & \cdots & \frac{\partial F_1}{\partial y_n} \\ \frac{\partial F_2}{\partial y_1} & \frac{\partial F_2}{\partial y_2} & \cdots & \frac{\partial F_2}{\partial y_n} \\ \vdots & & \ddots & \\ \frac{\partial F_n}{\partial y_1} & \frac{\partial F_n}{\partial y_2} & \cdots & \frac{\partial F_n}{\partial y_n} \end{bmatrix}_{y_i=0}; \qquad \vec{F} = \begin{bmatrix} F_1 \\ F_2 \\ \vdots \\ F_n \end{bmatrix} \tag{3.47}$$

Where the initial conditions are as follows:

$$\vec{y}(t) = \vec{y}_0 \qquad \text{at} \quad t = 0 \tag{3.48}$$

Let's define the vector $\vec{y}(t)$ as $\vec{y}(t) = T\vec{u}(t)$, where T is the modal matrix (a matrix holding eigenvectors) and u_i is are generalized coordinates. With this definition, the governing differential equation may be expressed as follows:

$$T\dot{\vec{u}}(t) = AT\vec{u}(t) \;\rightarrow\; \dot{\vec{u}}(t) = T^{-1}AT\vec{u}(t) \tag{3.49}$$

Where, $J = T^{-1}AT$ is referred to as the standard form of Jordan.

$$\dot{\vec{u}}(t) = J\vec{u}(t) \tag{3.50}$$

Based on the theorem of linear algebra, the following is known:

$$J = T^{-1}AT = \begin{bmatrix} \lambda_1 & \cdots & 0 \\ \vdots & \ddots & \vdots \\ 0 & \cdots & \lambda_n \end{bmatrix} \tag{3.51}$$

So the J matrix is a diagonal matrix of eigenvalues.

$$\begin{bmatrix} \dot{u}_1 \\ \dot{u}_2 \\ \vdots \\ \dot{u}_n \end{bmatrix} = \begin{bmatrix} \lambda_1 u_1 \\ \lambda_2 u_2 \\ \vdots \\ \lambda_n u_n \end{bmatrix} \rightarrow \vec{u}(t) = \Lambda(t)\vec{u}_0 \tag{3.52}$$

In which:

$$\Lambda(t)=\begin{bmatrix} e^{\lambda_1 t} & \cdots & 0 \\ \vdots & \ddots & \vdots \\ 0 & \cdots & e^{\lambda_n t} \end{bmatrix} \tag{3.53}$$

And $\vec{u}_0$ is obtained as follows according to the initial conditions of $\vec{y}_0$.

$$\vec{y}_0 = T\vec{u}_0 \rightarrow \quad \vec{u}_0 = T^{-1}\vec{y}_0 \quad \rightarrow \vec{u}(t) = \Lambda(t)T^{-1}\vec{y}_0 \tag{3.54}$$

As previously stated, the presence of the trajectories on the phase plane is indicative of second-order differential equations. Put simply, it is not feasible to depict an exceedingly intricate and unattainable curve by elevating the degree of the differential equation.

3.9 ANALYSIS OF NONLINEAR SYSTEMS USING A PHASE PLANE

When discussing phase plate analysis in nonlinear systems, it is crucial to take into account aspects related to limit cycles.

According to the Van der Pol oscillator system, besides the fixed points, the system also exhibits a closed curve. All the trajectories entering and exiting this curve will be inclined towards it. If the trajectory follows the curve, it will proceed along it and oscillate about the origin, eventually leaving the curve [4]. Figure 3.25 is a curve that exemplifies the occurrence of limit cycles in some nonlinear systems.

In the phase plane, a partial cycle is referred to as an independent closed curve. The closed phase route of this curve signifies the rotational character of its movement, while the discrete nature denotes the short duration of the cycle. Consequently, the nearby pathways either incline towards it or diverge from it. It is important to highlight that in the case of a mass or a satellite, the presence of several closed curves prevents these curves from being limit cycles. These cycles are interconnected. Limit cycles may be classified into three distinct types:

1. Stable limit cycles: have a characteristic where all trajectories surrounding them get increasingly directed towards the cycles as time progresses.
2. Unstable limit cycles: in these cycles, all the trajectories surrounding them are eliminated as time progresses.
3. Semi-stable limit cycles: some pathways around these cycles tend to approach them as time passes, while other paths move away from them.

Example 3–6: The system that follows is assumed:

$$\begin{cases} \dot{x}_1 = x_2 - x_1\left(x_1^2 + x_2^2 - 1\right) \\ \dot{x}_2 = -x_1 - x_2\left(x_1^2 + x_2^2 - 1\right) \end{cases} \tag{3.55}$$

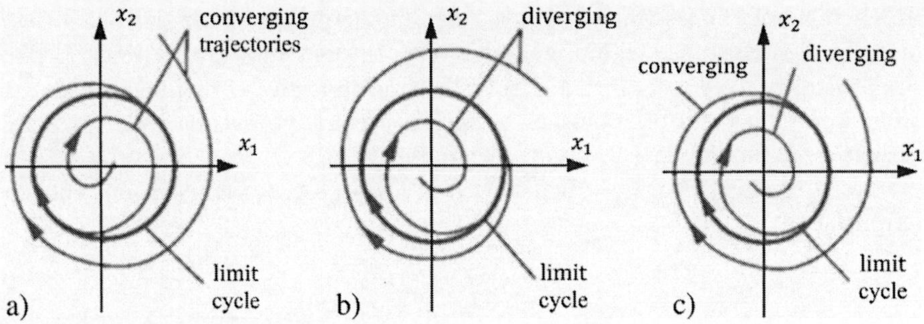

FIGURE 3.25 Types of limit cycles: a) stable, b) unstable, and c) semi-stable.

By establishing the polar coordinates using the following definition:

$$\begin{cases} r = \sqrt{x_1^2 + x_2^2} \\ \theta = \tan^{-1}\left(x_2 / x_1\right) \end{cases} \tag{3.56}$$

Now, derivative of Equation (3.56) based on time, we have the following:

$$\begin{aligned} \frac{dr}{dt} &= \frac{d\left(x_1^2 + x_2^2\right)^{1/2}}{dt} = x_1\dot{x}_1\left(x_1^2 + x_2^2\right)^{-1/2} + x_2\dot{x}_2\left(x_1^2 + x_2^2\right)^{-1/2} \\ &= x_1\left[x_2 - x_1\left(x_1^2 + x_2^2 - 1\right)\right]\left(x_1^2 + x_2^2\right)^{1/2} \\ &+ x_2\left[-x_1 - x_2\left(x_1^2 + x_2^2 - 1\right)\right]\left(x_1^2 + x_2^2\right)^{-1/2} \\ &= -\left(x_1^2 + x_2^2\right)\left(x_1^2 + x_2^2 - 1\right)\left(x_1^2 + x_2^2\right)^{-1/2} = -r\left(r^2 - 1\right) \end{aligned} \tag{3.57}$$

Conversely, in regard to θ, we have the following:

$$\frac{d\theta}{dt} = \frac{d\left(tan^{-1}\left(x_2 / x_1\right)\right)}{dt} = \frac{\dot{x}_2 x_1 - \dot{x}_1 x_2}{x_1^2\left(1 + \dfrac{x_2^2}{x_1^2}\right)} = \frac{-r^2}{r^2} = -1 \tag{3.58}$$

In the unit circle with a radius of 1 $(r = 1)$, the Equation (3.57) and (3.58) are represented as follows:

$$\frac{dr}{dt} = -r\left(r^2 - 1\right) = 0 \tag{3.59}$$

$$\frac{d\theta}{dt} = -1 \Rightarrow \theta = \theta_0 - t$$

If the value of r is less than 1 $(r < 1)$, then the derivative of r is positive. Consequently, the trajectories within the unit circle converge towards the circumference. If the value of r is greater than 1 $(r > 1)$, then the derivative of r will be negative. In other words, the trajectories outside the unit circle converge towards the circle. In this scenario, the limit cycle is stable, as seen in Figure 3.26.

Other examples later show that their limit cycles are, respectively, unstable and semi-stable:

$$\begin{cases} \dot{x}_1 = x_2 + x_1\left(x_1^2 + x_2^2 - 1\right) \\ \dot{x}_2 = -x_1 + x_2\left(x_1^2 + x_2^2 - 1\right) \end{cases}$$

The earlier relationships represent an unstable limit cycle, the diagram of which is shown in Figure 3.27.

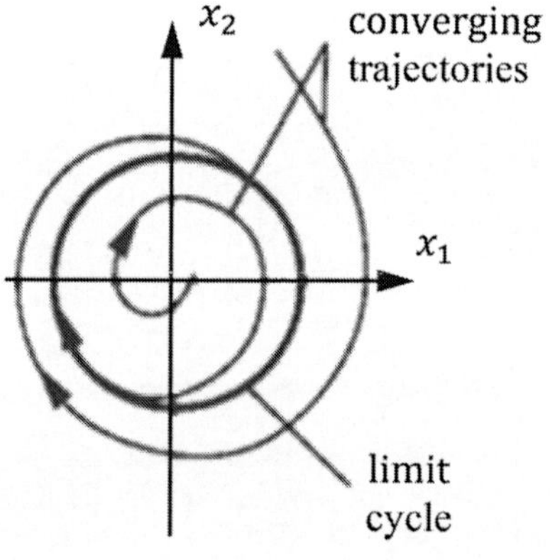

FIGURE 3.26 Stable limit cycle.

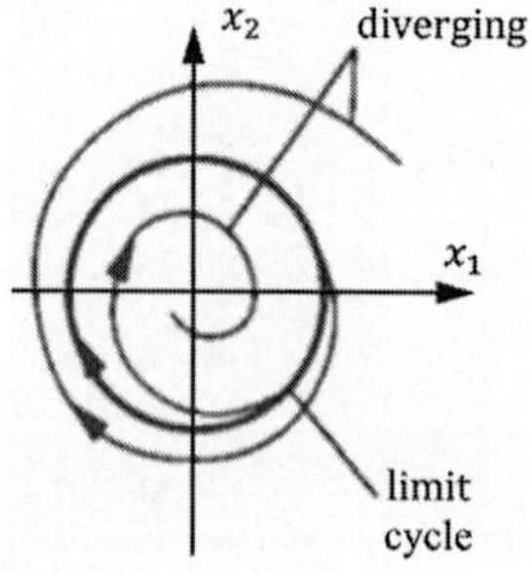

FIGURE 3.27 Unstable limit cycle.

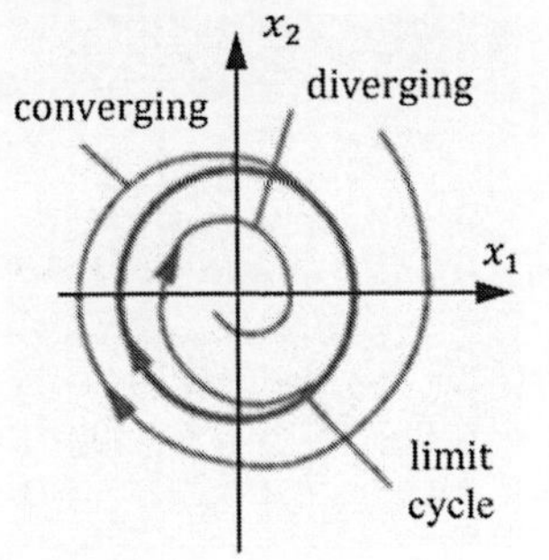

FIGURE 3.28 Semi-stable limit cycle.

The connections later depict a set of equations that pertain to a semi-stable limit cycle. The associated graph may be seen in Figure 3.28.

$$\begin{cases} \dot{x}_1 = x_2 - x_1\left(x_1^2 + x_2^2 - 1\right) \\ \dot{x}_2 = -x_1 - x_2\left(x_1^2 + x_2^2 - 1\right) \end{cases}$$

3.10 CASES OF THE EXISTENCE OF LIMIT CYCLES

In the discussion of the existence of limit cycles in nonlinear systems, three important theorems can be mentioned:

3.10.1 Poincaré Theorem

If an autonomous second-order system (without input) exhibits a limit cycle, then the number of limit cycles is equal to the sum of the number of stable fixed points and one ($N = S + 1$). Let N be the total count of nodes, centers, and centers enclosed by the limit cycle. S represents the count of saddle points encompassed by the limit cycle. The Poincaré theorem is sometimes referred to as the index theorem. An immediate consequence of this is that the limit cycle must encompass at least one fixed point (because the absence of a saddle point would leave us $N = 1$).

As we have seen, in previous examples, each limit cycle encloses at least one fixed point.

3.10.2 The Poincaré-Bendixon Theorem

If a path of the automated system of order two (3.56) remains in the finite area Ω, then one of the following three statements is true:

1. The path leads to a point of equilibrium.
2. The path leads to a stable limit cycle.
3. The path itself is a limit cycle.

$$\begin{cases} \dot{x}_1(t) = f_2(x_1, x_2) \\ \dot{x}_2(t) = f_2(x_1, x_2) \end{cases} \tag{3.60}$$

3.10.3 Bendixon's Theorem

In the context of an autonomous nonlinear system, the presence of a limit cycle in the Ω area of the phase plane is impossible if the Equation (3.61) remains unchanged and non-zero in that region.

$$\frac{\partial f_1}{\partial x_1} + \frac{\partial f_2}{\partial x_2} \tag{3.61}$$

Example 3–12: The following nonlinear system is assumed:

$$\begin{cases} \dot{x}_1 = g(x_2) + 4x_1 x_2^2 \\ \dot{x}_2 = h(x_2) + 4x_1^2 x_2 \end{cases} \tag{3.62}$$

According to Bendixon's theorem, it can be written as follows:

$$\frac{\partial f_1}{\partial x_1} + \frac{\partial f_2}{\partial x_2} = 4\left(x_2^2 + x_1^2\right) \tag{3.63}$$

Given that the equation mentioned earlier is positive at all points except the origin, it may be concluded that the system does not have a border phase plane at any location.

It is important to mention that the aforementioned situations are specifically applicable to second-order systems and do not apply to high-order systems. In higher-order systems, it is possible for complicated lateral behaviors to emerge, in addition to fixed points and limit cycles.

Example 3–7: Plot the amplitude of the changes in variable x with time t for the Van der Pol oscillator, given the initial conditions (–1,1) and (–3,3).

Answer: To plot a graph illustrating the variation of x with respect to t, we may utilize the Van der Pol oscillator based on the derived relationships from the preceding example.

$$\ddot{x} + \varepsilon\left(1 - x^2\right)\dot{x} + x = 0$$

$$\begin{cases} x = X_1 \\ \dot{x} = X_2 \end{cases} \rightarrow \begin{cases} \dot{X}_1 = X_2 \\ \dot{X}_2 = -\varepsilon X_2\left(1 - X_1^2\right) - X_1 \end{cases}$$

Therefore, by calculating the equations mentioned earlier, one may graph the amplitude of variations in the Van der Pol oscillator over time for various initial conditions as Figure 3.29.

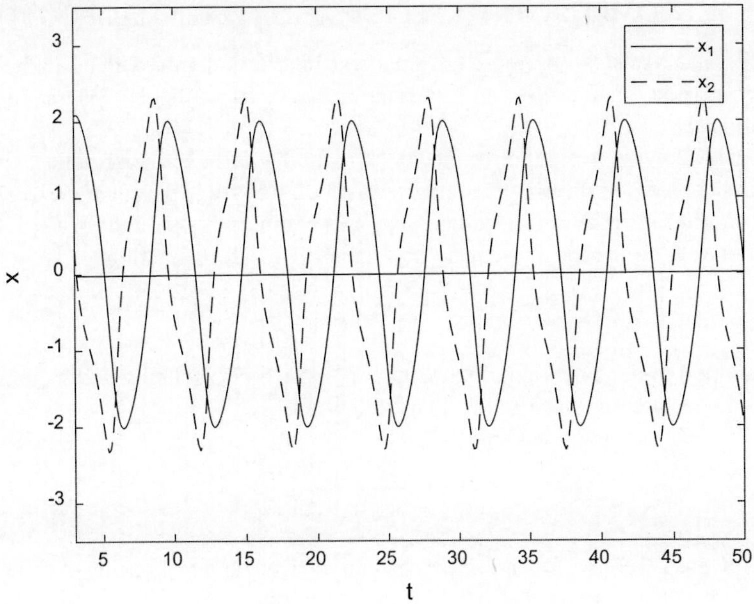

a) With preliminary conditions (–3,3)

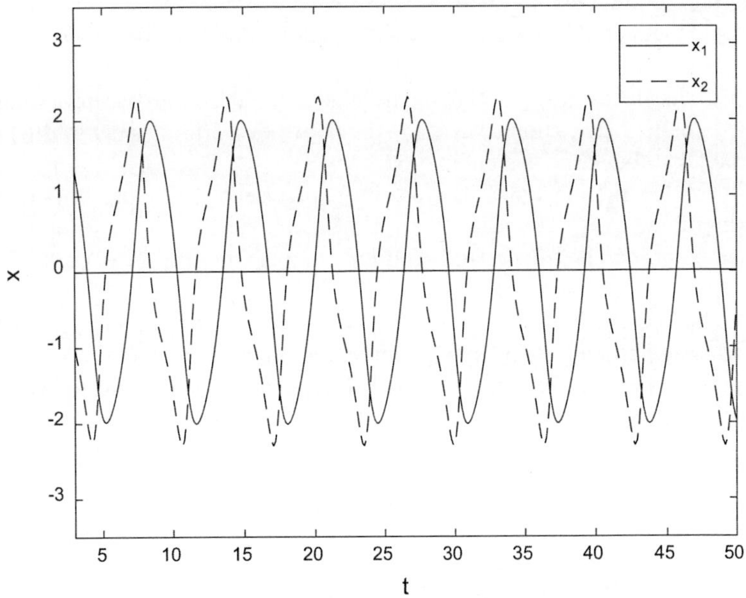

b) With preliminary conditions (–1,1)

FIGURE 3.29 Amplitude diagram of changes x by time t for the Van der Pol oscillator for different initial conditions (a) with initial conditions (–3,3) and (b) with initial conditions (–1,1).

3.11 A GEOMETRIC INTERPRETING A DIFFERENTIAL EQUATION

Graphical analysis can be more advantageous and less complex than mathematical formulas when qualitatively analyzing the behavior of nonlinear systems in certain scenarios [1].

In the following, we will propose a technique that utilizes vector fields to analyze certain behaviors of differential equations, which we will subsequently explain.

In order to introduce this methodology, we begin with a straightforward illustration. Let us examine the subsequent nonlinear differential equation:

$$\dot{x} = \sin(x) \tag{3.64}$$

The solution to this equation, obtained by the process of isolating and then integrating, can be expressed in closed form as follows:

$$t = \ln\left|\frac{csc\, x_0 + \cot x_0}{\csc x + \cot x}\right| \tag{3.65}$$

The previous response, albeit correct, had an intricate and perplexing explanation.

Now, the question at hand is whether it is feasible to respond to the subsequent question using the aforementioned answer:

Given the initial knowledge of $\dot{x}$ at time $t = 0$, it is feasible to ascertain the position of x at any specified time $t > 0$. However, what happens when t approaches infinity?

Contrary to the previous question, there exists a straightforward graphical analysis method that can provide a qualitative answer by depicting the vector field of the differential equation (60–3) in the form (Figure 3.30).

Prior to elucidating this approach, we shall initially establish the concept of fixed points.

Fixed points are positions where the derivative of x with respect to time is zero, indicating that the particle is not in motion.

At these specific locations, the velocity is precisely zero, indicating a complete absence of flow.

Fixed points can be categorized according to their stability as follows:

1. Stable fixed points, also known as attractors or sinks, are points in a system where a particle, when slightly moved, would travel back towards the fixed point. These locations are commonly referred to as "sinks" due to their ability to attract the flow.

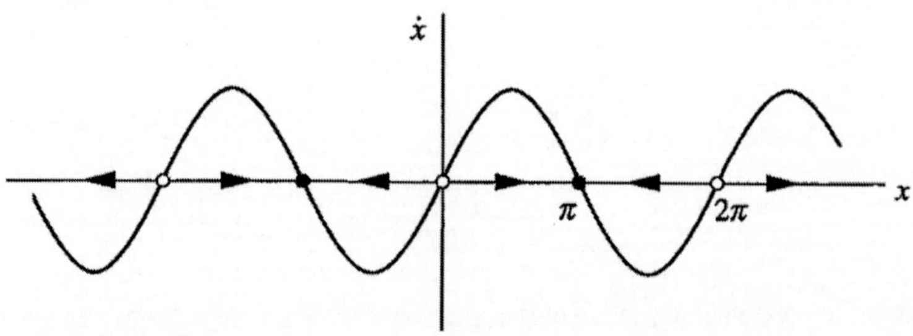

FIGURE 3.30 Displaying a vector field.

2. Unstable fixed points, often known as repellers or sources. In the case of an unstable fixed point, a particle will move away from the fixed point if it is even slightly displaced. These spots are sometimes referred to as "sources" due to their ability to reject the flow.

The concept of sinks and sources refers to the flow of substances or energy into or out of a system. A sink is a location or component that absorbs or removes substances or energy from the system, while a source is a location or component that adds or supplies substances or energy to the system.

Sinks are stable fixed sites towards which the flow is directed from both sides. Particles in close proximity to these locations will experience an attractive force and ultimately settle at the sink.

On the contrary, sources are unstable reference points from which the flow diverges. Particles in close proximity to these spots will experience repulsion and disperse away from the source.

Based on the definition provided earlier, we will proceed to explain the methodology.

In this context, t represents time, x represents the position of a hypothetical particle moving down a straight line, and $\dot{x}$ represents the velocity of that particle. The differential equation $\dot{x} = \sin(x)$ defines a vector field on the line, where it determines the velocity vector $\dot{x}$ at each x. In order to depict the vector field, it is advantageous to graph the derivative of x with respect to time ($\dot{x}$) against x and subsequently represent the velocity vector at each x by drawing arrows along the x-axis. The arrows indicate a rightward direction when the derivative of x is greater than zero, and a leftward direction when the derivative of x is less than zero.

Here's a more tangible perspective on the vector field: envision a continuous flow of fluid along the x-axis, where the velocity varies at different locations based on the rule $\dot{x} = \sin(x)$. As depicted in Figure 3.30, the direction of the flow is towards the right when the derivative of x is greater than zero ($\dot{x} > 0$) and towards the left when the derivative of x is less than zero ($\dot{x} < 0$). When the derivative of x with respect to time is equal to zero, there is no movement, and these points are referred to as fixed points. In Figure 3.30, there are two types of fixed points: stable fixed points, represented by solid black dots, and unstable fixed points, represented by open circles. Stable fixed points are typically referred to as attractors or sinks because the flow moves towards them, while unstable fixed points are known as repellers or sources.

3.12 THE THEORY OF THE BIFURCATION

In general, any qualitative change in the structure of the field vector or the same phase image (number of fixed points and/or stability of fixed points) in which fixed points are created, lost, or their type of stability changes is called bifurcation, and any places where these parametric changes occur are called bifurcation points [1].

The significance of bifurcation theory is in its capacity to identify the emergence of transient and unstable models inside a system as certain control parameters undergo changes. Take, for instance, the buckling of a beam. If a little weight is placed on the beam, as seen in Figure 3.31, the beam is capable of enduring the load and maintaining its upright posture. However, when the force is substantial, the vertical orientation of the beam becomes unsteady, perhaps causing the beam to buckle.

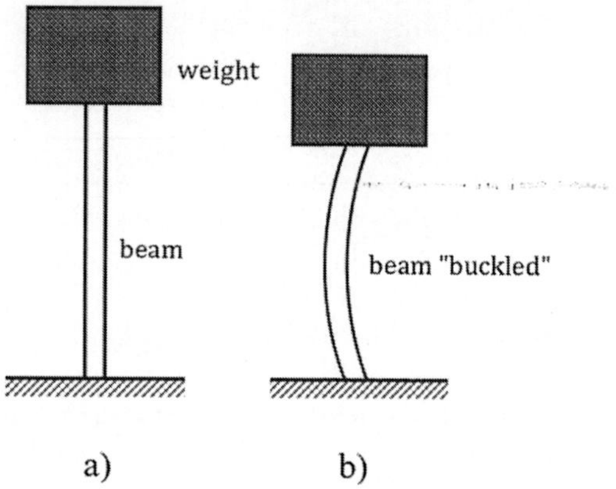

FIGURE 3.31 Beam a) in a stable state and b) in a buckled (unstable).

In this context, the weight acts as a control parameter that affects the upward movement of the beam from its vertical position. It serves as the dynamic variable x. Now, we will examine the many kinds of bifurcations.

3.12.1 Saddle-Node Bifurcation

A saddle-node bifurcation refers to a process where the formation and destruction of fixed points occur due to the alteration of a parameter. Indeed, within this bifurcation, one may see a transition in the system's condition from a state of stability to instability, or vice versa. Consequently, when a parameter undergoes a change, the two fixed points shift closer to one another, collide, and then vanish [1, 5–7].

An illustrative instance of a tangled obstruction in a system of primary order may be demonstrated as follows:

$$\dot{x} = r + x^2 \tag{3.66}$$

The parameter "r" can take on positive, negative, or zero values. When the value of r is negative, there will be two fixed points, one of which is stable and the other is unstable (as shown in Figure 3.32 (a)).

When r approaches zero from the bottom, the quota moves upward, and the two fixed points move towards each other. When $r = 0$, the fixed points at $x^* = 0$ become one at a semi-stable fixed point (Figure 3.32 (b)). This type of fixed point is very sensitive, and as soon as $r > 0$, it disappears, and in this case, there is no fixed point (Figure 3.32 (c)).

In this example, because the vector fields for $r > 0$ and $r < 0$ are qualitatively different, it is said that a bifurcation happened at $r = 0$.

3.12.2 Graphic Contracts

The most common way to describe a bifurcation is to reverse the axes of Figure 3.33 so that it is horizontally plotted in the role of an independent variable. It is logical

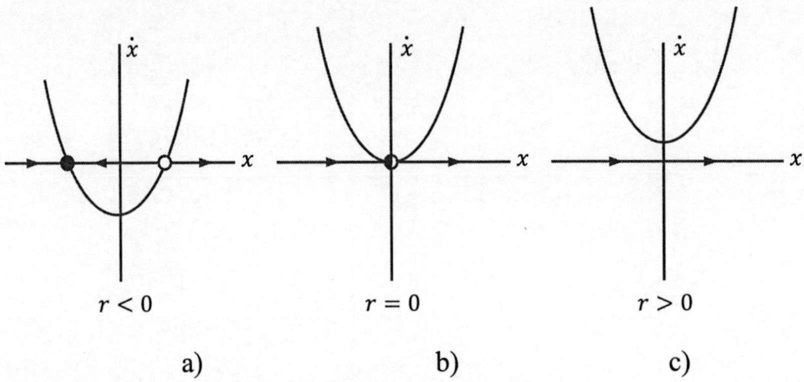

FIGURE 3.32 Fixed point in different situations.

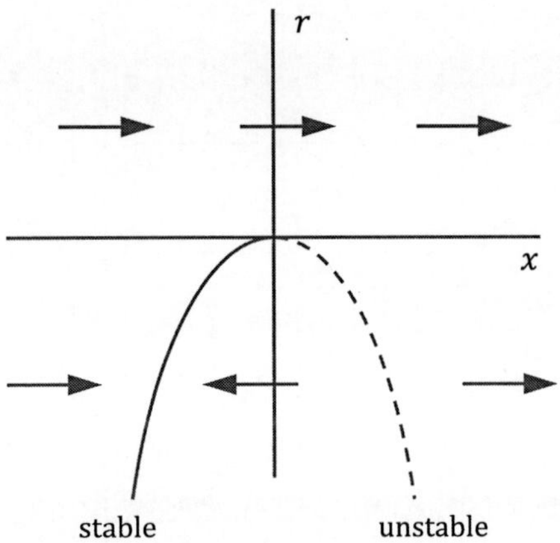

FIGURE 3.33 Saddle-node bifurcation (horizontal).

that r, playing the role of an independent variable, should be plotted horizontally (Figure 3.34). The disadvantage is that now, the x-axis has to be plotted vertically, which at first looks a little strange. Figure 3.34 is called a bifurcation diagram for a saddle-node bifurcation. The display of arrows, according to Figure 3.33, is sometimes and not always shown in some forms.

3.12.3 Transcritical Bifurcation

In some scientific issues, there are certain situations where there was a fixed point for all values of a parameter, and it does not go away. For example, in the logical equation and other simple models for the growth of a particular species type, regardless of the amount of its growth rate, there is a point of equilibrium in the zero population. Such a fixed point, however, may change its stability by changing the parameter, but there remains a fixed point. Transcritical bifurcation is a standard mechanism for

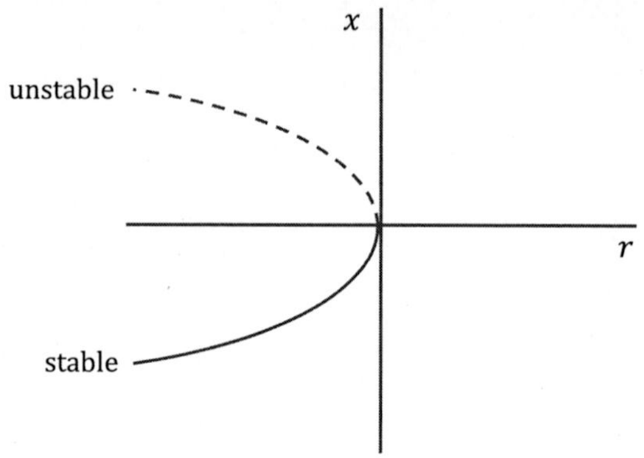

FIGURE 3.34 Saddle-node bifurcation (vertical).

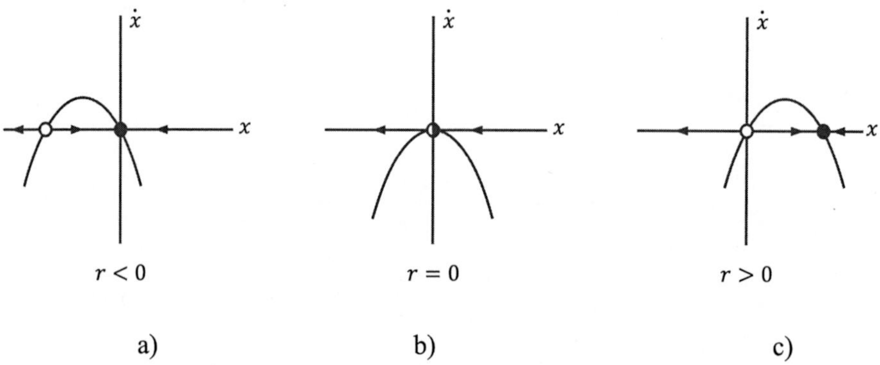

FIGURE 3.35 Fixed point behavior for different values of r.

such changes in sustainability. The conventional form for a transcritical bifurcation is as follows:

$$\dot{x} = rx - x^2 \tag{3.67}$$

The Equation (3.67) seems to be a logical equation, but now, we allow r and x to be able to have both negative and positive values. In Figure 3.35, we can see the field vector in exchange for the change r. Note that there is a fixed point in $x^* = 0$ for all r values. So that by changing the values of r, this fixed point does not disappear.

As can be seen in Figure 3.35, for $r < 0$, there is an unstable fixed point at $x^* = r$ and a stable fixed point at $x^* = 0$ (Figure 3.35 a). As r increases, the unstable fixed point approaches the origin and becomes one with it, i.e., when $r = 0$ (Figure 3.35 b). Finally, when $r > 0$, the origin has become unstable, and now, $x^* = r$ is stable (Figure 3.35 c). Here, a shift in sustainability is said to have taken place, between the two fixed points.

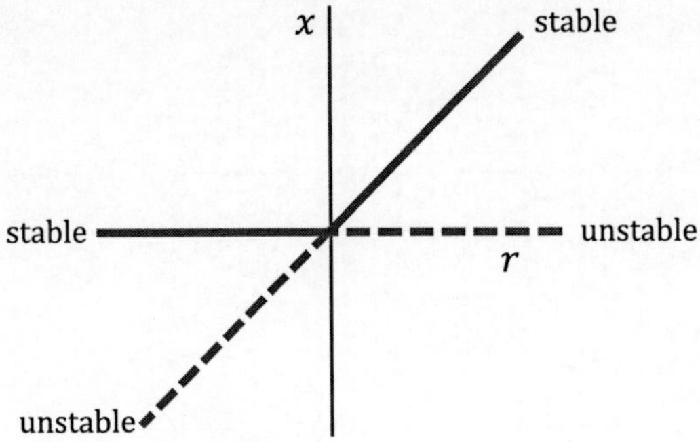

FIGURE 3.36 Transcritical bifurcation.

Pay attention to the difference between the saddle-node and transcritical bifurcations. In a critical state, two fixed points, after a bifurcation, do not disappear. But only their stability changes.

In Figure 3.36, the bifurcation diagram for the transcritical bifurcation is shown. So that the parameter r is considered an independent variable, and the fixed points $x^* = 0$ and $x^* = r$ are represented as dependent variables.

3.12.4 Pitchfork Bifurcation

Now, we are examining the third type of bifurcation, which is called pitchfork bifurcation. This bifurcation is common in physical issues that are symmetrical. In many issues, for example, there is a partial left-right symmetry. In such cases, fixed points tend to appear and disappear as symmetrical couples. In the example of the buckling beam (Figure 3.31) if the load is small, the beam is stable in the vertical position. In this case, there is a corresponding fixed point with zero deflection. But if the load exceeds the tolerance threshold, the beam may buckle left and/or right. In these conditions, the vertical position becomes unstable, and two new symmetrical fixed points, corresponding to the shape in the left or right arc, are born. There are two very different types of pitchfork bifurcation. The simplest variant of these bifurcations is called supercritical bifurcation, which will be discussed at the beginning [1, 5–7].

3.12.4.1 Supercritical Pitchfork Bifurcation

The conventional shape of the supercritical pitchfork bifurcation is as follows:

$$\dot{x} = rx - x^3 \tag{3.68}$$

Note that the Equation (3.68), under the variable $x \to -x$, is immutable. This means that if we paste x with –x and then remove the resulting negative sign on both sides of the equation, we return to the Equation (3.68). This immutability is the mathematical definition of left and right symmetry, which was mentioned earlier. Figure 3.37 illustrates the relation-related field vector (3.68) for different values of r.

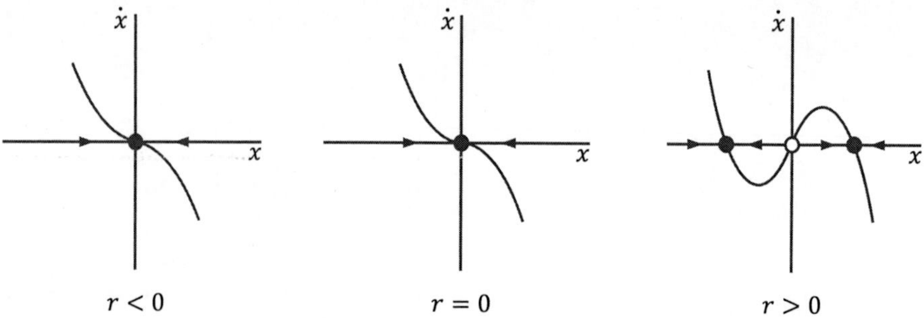

FIGURE 3.37 Field vector for different values of r.

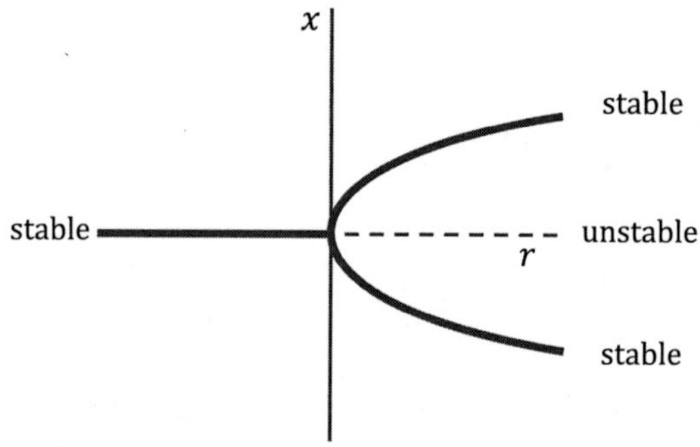

FIGURE 3.38 Supercritical pitchfork bifurcation diagram.

The reason for using the phrase "pitchfork" becomes more apparent when we draw the bifurcation diagram of Figure 3.38. In fact, the phrase "three-pronged fork" could have been a better phrase.

3.12.4.2 Subcritical Pitchfork Bifurcation

In the supercritical state, in which the relation $\dot{x} = rx - x^3$ was discussed, its third-order sentence is the system stabilizer such that it acts as a return force that pushes $x(t)$ back towards $x = 0$. Now instead of the third-order phrase, the unstable form is as follows:

$$\dot{x} = rx + x^3 \tag{3.69}$$

We will have a subcritical pitchfork bifurcation, the diagram of which is shown in Figure 3.39.

Compared to Figure 3.39, the fork is inverted. Non-zero fixed points $x^* = \pm\sqrt{-r}$ are unstable and exist only under the bifurcation $r < 0$ from which the subcritical expression is derived. More importantly, that, the origin is stable for $r < 0$, and unstable for $r > 0$, as it was in the supercritical state. But now, the instability for

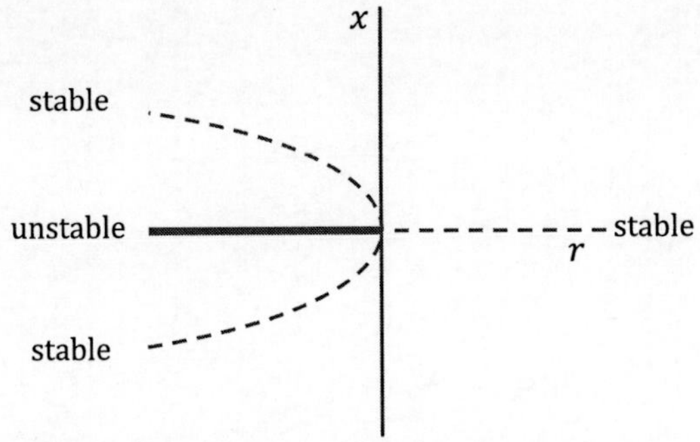

FIGURE 3.39 Diagram of the subcritical pitchfork bifurcation.

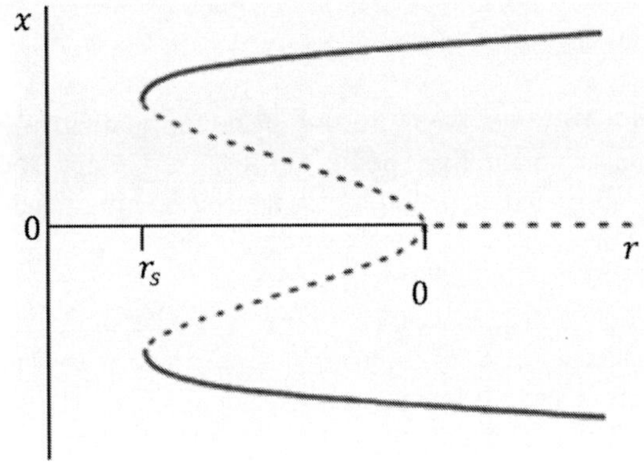

FIGURE 3.40 Symmetrical bifurcation relative to the x-axis.

$r > 0$, with the third-order expression, is not contradicted. In fact, the third-order sentence helps to move the line of trajectory towards infinity.

In real physical systems, the so-called explosive instability is in contrast to the sustainability effect of high order expressions. Suppose that the system is symmetrical relative to the x-axis (Figure 3.40). The first sustainability expression should be x^5. Thus, the standard example of a system with a subcritical pitchfork bifurcation is as follows:

$$\dot{x} = rx + x^3 - x^5 \tag{3.70}$$

Figure 3.40 illustrates a bifurcation diagram for the Equation (3.70). For small x values, the situation is the same as in Figure 3.39. The origin is stable for $r < 0$, and the branches bent backward make unstable fixed points, creating a bifurcation in $r = 0$. A new feature caused by the expression x^5 is that unstable branches return and become stable at $r = r_s$ (where $r_s < 0$). For all values of $r > r_s$, these stable branches of the large domain exist.

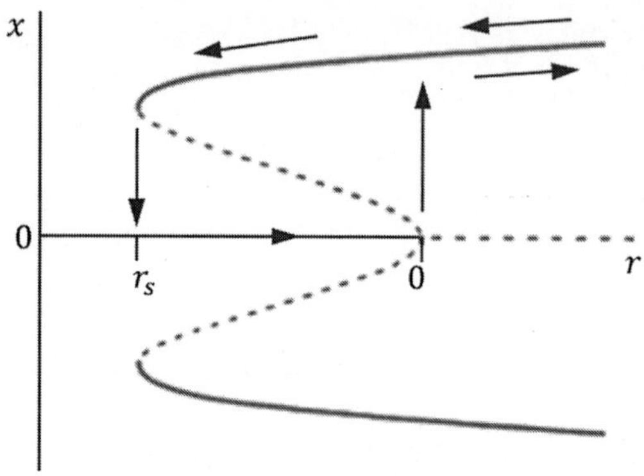

FIGURE 3.41 Phenomenon of jumping and hysteresis in the bifurcation.

Other points to be made about Figure 3.40 are as follows:

1. In the $r_s < r < 0$ range, there are two different qualitatively stable states, which are the origin and fixed points of the large domain. The initial condition x_0 determines which fixed point is approached in exchange for $t \to \infty$. One result is that the origin is stable in small disturbances, and this sustainability is a local one and not a general one.
2. Different stability modes make it possible to create jumping and hysteresis phenomena by changing r. Suppose the system is in state $x^* = 0$ and we increase the parameter r slowly (Figure 3.41). It remains stable until the value $r = 0$. At point $r = 0$, we see a jump to the large domain branch. As r increases further, the situation moves along the branch of the large amplitude. Even when r is reduced to less than zero, we need to reduce r to less than r_s to get back to origin from jump mode. This irreversible defect is known as hysteresis.
3. Bifurcation in r_s is a saddle-node bifurcation in which stable and unstable fixed points are born.

3.12.5 Technical Terms in the Bifurcation

Typically, in the theory of bifurcation, various other names are used. Supercritical bifurcation is sometimes called forward bifurcation and is dependent on continuity or second-order phase passage in statistical mechanics. Subcritical bifurcation is sometimes called upside-down or backward bifurcation. Subcritical bifurcation is linked to discontinuity and first-order phase transitions. In engineering texts, supercritical bifurcation is sometimes called soft and safe because non-zero fixed points are born on a small range. In contrast, subcritical bifurcation is difficult and dangerous because we see zero bounces to the large range.

REFERENCES

[1] Strogatz, S.H., *Nonlinear dynamics and chaos with student solutions manual: With applications to physics, biology, chemistry, and engineering*. 2018: CRC Press.
[2] Rand, R.H., *Lecture notes on nonlinear vibrations*. 2012: Cornell University.

[3] Knill, O., *Dynamical systems*. 2005: Harvard University.
[4] Nayfeh, A.H. and D.T. Mook, *Nonlinear oscillations*. 2008: John Wiley & Sons.
[5] Ghadiri, M. and S.H.S. Hosseini, Parametrically excited nonlinear dynamic instability of reinforced piezoelectric nanoplates. *The European Physical Journal Plus*, 2019. **134**(8): p. 413.
[6] Ghadiri, M. and S.H.S. Hosseini, Parametric excitation of pre-stressed graphene sheets under magnetic field: Nonlinear vibration and dynamic instability. *International Journal of Structural Stability and Dynamics*, 2019. **19**(11): p. 1950135.
[7] Hosseini, S.H.S. and M. Ghadiri, Nonlinear dynamics of fluid conveying double-walled nanotubes incorporating surface effect: A bifurcation analysis. *Applied Mathematical Modelling*, 2021. **92**: p. 594–611.

4 Solution Methods

4.1 INTRODUCTION

Nonlinear systems exhibit a broad spectrum of phenomena that are not observed in linear systems. The intricate nature of nonlinear systems results in intricate equations, which in turn makes it challenging to find precise solutions for these nonlinear equations [1]. This chapter provides a comprehensive quantitative analysis of methods used to solve nonlinear equations. When precise solutions are not available, the analysis of nonlinear systems is conducted using approximate and numerical solving methods [2].

This chapter is divided into two main sections. In the first sub-chapter, we analyze solutions in the time domain. One widely recognized and highly effective method is the perturbation method, which is further explored through various types of perturbation methods for solving nonlinear equations [3, 4].

In the second sub-chapter, we analyze solutions in the space domain. One of the methods that is widely used in the space domain is the solution of governing differential equations based on the weighted residual method [5, 6].

4.2 DIFFERENT WAYS OF SOLVING PROBLEMS USING THE PERTURBATION METHOD

1. Straight forward expansion method
2. Lindstedt-Poincaré method
3. Multiple time scales method
4. Averaging method
5. Harmonic balance method

Prior to delving into various ways of disruption, we will initially familiarize ourselves with the disruption method. It will become evident that the disruption method is an exceedingly effective tool for quantitatively analyzing nonlinear systems.

4.2.1 Introduction to Perturbation Method

Some engineering issues have a parameter that is tiny in magnitude. Typically, the differential equation, or more correctly, the function involved, is analytical. In the context of disturbance problems, it is common to come across a small parameter, denoted as σ. This parameter enables us to express the function $x(t)$, which solves the differential equation [7], as a series of powers that either increase or decrease in value.

$$x(\varepsilon,t) = x_0(t) + \varepsilon x_1(t) + \varepsilon^2 x_2(t) + \ldots \tag{4.1}$$

DOI: 10.1201/9781003470694-4

Next, we substitute the series (4.1) into the main equation and equate the coefficients, such as ε^k. Thus, we arrive at a system of equations derived from the solution of the variables $x_i(t)$. Regarding the convergence of the series (4.1), the Poincaré theorem demonstrates that this series converges in the vicinity of $\varepsilon = 0$. Therefore, using this approach, an accurate approximation may be obtained. We're going to look at the different types of the perturbation method.

4.2.2 Straightforward Expansion Method

Direct techniques are the fundamental and uncomplicated forms of the perturbation method. However, because to the requirement of several words to obtain the correct response, it typically yields limited success. For a comprehensive description of the technique, please refer to the citation [6].

4.2.3 Lindstedt-Poincaré Method

This approach is typically employed when seeking harmonic and quasi-harmonic solutions. Typically, a concise and effective response may be achieved with a limited amount of phrases. This approach will involve a limit cycle. To obtain further details, we shall employ the Lindstedt-Poincaré technique to solve the Duffing equation.

Examine the subsequent differential equation:

$$\frac{d^2x}{dt^2} + x + \varepsilon\alpha x^3 = 0, \qquad \varepsilon > 0 \tag{4.2}$$

The system in question is known as the Duffing oscillator, which is a model that incorporates nonlinear reinforcements such as springs.

The Lindstedt method is a straightforward technique used to analyze the relationship between the rotation period and scope in the Duffing Equation (4.2). The method involves expanding time to construct an approximate solution for the alternating domain-period relationship.

$$\tau = \omega t, \tag{4.3}$$

In which:

$$\omega = 1 + k_1\varepsilon + k_2\varepsilon^2 + \ldots \tag{4.4}$$

That unknown k_i coefficients are obtained in the solving process. We have a substitution connection between Equation (4.3) and Equation (4.2).

$$\omega^2 \frac{d^2x}{d\tau^2} + x + \varepsilon\alpha x^3 = 0 \tag{4.5}$$

Now, we expand the variable x as a series of powers in ε.

$$x(\tau) = x_0(\tau) + \varepsilon x_1(\tau) + \varepsilon^2 x_2(\tau) + \ldots \tag{4.6}$$

Given the expansion of the power series in Equation (4.3) and Equation (4.6), the results obtained using the Lindstedt method are only expected to be valid for small values. By substituting Equation (4.6) into Equation (4.5) and setting it equal to zero, we can determine the coefficients corresponding to ε^n.

$$\varepsilon^0 : \frac{d^2 x_0}{d\tau^2} + x_0 = 0 \tag{4.7}$$

$$\varepsilon^1 : \frac{d^2 x_1}{d\tau^2} + x_1 = -2k_1 \frac{d^2 x_0}{d\tau^2} - \alpha x_0^3 \tag{4.8}$$

$$\varepsilon^2 : \frac{d^2 x_2}{d\tau^2} + x_2 = -2k_1 \frac{d^2 x_0}{d\tau^2} - \left(2k_2 + k_1^2\right)\frac{d^2 x_0}{d\tau^2} - 3\alpha x_0^2 x_1 \tag{4.9}$$

The solution to Equation (4.8) is as follows:

$$x_0(\tau) = A\cos\tau \tag{4.10}$$

In this context, A represents the extent of movement, and we have selected the phase in a subjective manner. This is permissible due to the autonomous character of Equation (4.2), which lacks any dependence on the independent variable t. By substituting Equation (4.7) into Equation (4.8), we obtain the following:

$$\frac{d^2 x_1}{d\tau^2} + x_1 = 2Ak_1 \cos\tau - A^3\alpha\cos^3\tau \tag{4.11}$$

By simplifying the phrase $\cos^3\tau$, we have the following:

$$\frac{d^2 x_1}{d\tau^2} + x_1 = \left(2Ak_1 - \frac{3A^3\alpha}{4}\right)\cos\tau - \frac{A^3\alpha}{4}\cos 3\tau \tag{4.12}$$

To obtain a response that is periodic, we require the coefficients of the $\cos\tau$ on the right side of Equation (4.12) be zero. This crucial step is referred to as eliminating exacerbated or secular sentences. So:

$$2Ak_1 - \frac{3A^3\alpha}{4} = 0, \tag{4.13}$$

Which results:

$$k_1 = \frac{3}{8}\alpha A^2 \tag{4.14}$$

By achieving this outcome in Equation (4.3), we get the approximate correlation between amplitude and frequency as follows:

$$\omega = 1 + k_1\varepsilon + O\left(\varepsilon^2\right) = 1 + \frac{3}{8}\alpha A^2\varepsilon + O\left(\varepsilon^2\right) \tag{4.15}$$

The rotation period, denoted as T, may be expressed $T = 2\pi/\omega$.

$$T = \frac{2\pi}{\omega} = \frac{2\pi}{1+\frac{3}{8}\alpha A^2\varepsilon + O\left(\varepsilon^2\right)} = 2\pi\left(1-\frac{3}{8}\alpha A^2\varepsilon + O\left(\varepsilon^2\right)\right) \tag{4.16}$$

We can proceed to acquire approximations of higher order. By replacing Equation (4.14) with Equation (4.12), we may determine the solution for $x_1(\tau)$ as follows:

$$x_1(\tau) = \frac{A^3\alpha}{32}(\cos 3\tau - \cos\tau) \tag{4.17}$$

In this case, the amplitude of vibrations, denoted as A, is chosen such that in the complementary response (4.6), the initial displacement x_0 is equal to A. By solving Equation (4.17) for x_2, we may eliminate secular expressions and derive an expression for k_2. This process has the potential to continue endlessly.

To comprehend the dynamics of the Duffing Equation (4.2), we initiate the process by expressing it as a first-order system:

$$\frac{dx}{dt} = y, \qquad \frac{dy}{dt} = -x - \varepsilon\alpha x^3 \tag{4.18}$$

Equation (4.18) describes a trajectory on the $x-y$ phase plane, representing the movement of a point over time, given the initial condition $\left(x(0), y(0)\right)$. The trajectory of the integral curve passing through that location is defined by the following equation:

$$\frac{dy}{dx} = \frac{\frac{dy}{dt}}{\frac{dx}{dt}} = \frac{-x-\varepsilon\alpha x^3}{y} \tag{4.19}$$

Equation (4.19) represents a straightforward integral that yields:

$$\frac{y^2}{2} + \frac{x^2}{2} + \varepsilon\alpha\frac{x^2}{4} = constant \tag{4.20}$$

Equation (4.20) represents the fundamental concept of energy stability. Given that α is a positive value, Equation (4.20) exhibits a continuous arrangement of closed curves surrounding the origin. Each object exhibits a periodic motion described by Equation (4.2) that changes direction over time. Given that α is negative, any movements that start at the origin will be alternating, as depicted in Figure 4.1. In this particular scenario, Equation (3.14) exhibits two extra equilibrium points apart from the origin, namely, $x = \pm 1/\sqrt{-\varepsilon\alpha}$, $y = 0$. The integral curves passing through these locations delineate alternating movements between restricted and unconstrained growth, referred to as separatrix, so to speak (single: separatrix).

By performing numerical integration using Equation (4.2), we can observe that the period of oscillation in alternating motions is contingent upon the closed curves

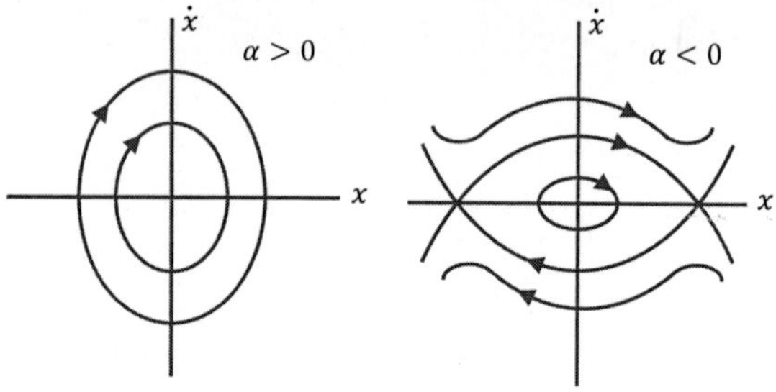

FIGURE 4.1 Phase plane for the Duffing equation.

present in the phase plane we were initially situated in. This phenomenon is a form of nonlinear vibration that is connected in the period based on the amplitude. In the upcoming episode, we will employ a perturbation technique to investigate it.

4.2.4 Multiple Time Scale Method

This approach is applicable to nearly any problem and, similar to the Poincaré method, provides a reasonably accurate solution using only a limited amount of phrases. The primary challenge associated with this approach is in the difficulty of converting it into a structured set of instructions in the form of programmed code. Put simply, it is extremely difficult to computerize.

To acquaint this technique, please examine the following differential equation:

$$\ddot{x}+2\varepsilon\dot{x}+x=0, \qquad \varepsilon \ll 1 \tag{4.21}$$

With intial conditions as follows:

$$x(0)=0, \qquad \dot{x}(0)=1 \tag{4.22}$$

We will employ the direct expansion technique to solve Equation (4.21), resulting in a non-uniform response. In order to address the issue, the concept of employing the multiple scale method will be proposed. By utilizing this approach, the challenges associated with the straightforward expansion method can be resolved efficiently, with minimal effort yielding positive outcomes. Hence, the overall response of the straightforward expansion method will be as follows:

$$x(\varepsilon,t)=x_0(t)+\varepsilon x_1(t)+\varepsilon^2 x_2(t)+\cdots \tag{4.23}$$

Let us evaluate the solution to the equation in the following manner:

$$x(\varepsilon,\, t)=x_0(t)+\varepsilon\, x_1(t)+O(\varepsilon^2) \tag{4.24}$$

By substituting Equation (4.24) into Equation (4.21), we obtain the following:

$$\ddot{x}_0+\varepsilon\,\ddot{x}_1+2\varepsilon(\dot{x}_0+\varepsilon\dot{x}_1)+x_0+\varepsilon x_1=0 \tag{4.25}$$

By substituting the initial condition (4.22) into the assumed answer (4.25), we obtain the following:

$$\begin{array}{l} x(0)=0 \rightarrow x_0(0)+\varepsilon\, x_1(0)=0 \\ \dot{x}(0)=1 \rightarrow \dot{x}_0(0)+\varepsilon\, \dot{x}_1(0)=1 \end{array} \Rightarrow \begin{cases} x_0(0)=x_1(0)=0 \\ \dot{x}_0(0)=1;\ \dot{x}_1(0)=0 \end{cases} \tag{4.26}$$

To obtain the result, we set the coefficients of ε^n to zero.

$$\begin{array}{l} \varepsilon^0 : \ddot{x}_0 + x_0 = 0 \rightarrow x_0(t) = c_1 \sin t + c_2 \cos t \overset{\text{Applying I.C.}}{\Rightarrow} x_0 = \sin t \\ \varepsilon^1 : \ddot{x}_1 + 2\dot{x}_0 + x_1 = 0 \rightarrow \ddot{x}_1 + x_1 = -2\cos t \end{array} \tag{4.27}$$

Based on the discussions on differential equations, it is established that the response $x_1(t)$ can be derived in the following manner:

$$x_1(t) = \overbrace{c_3 \sin t + c_4 \cos t}^{\text{Homogeneous response}} \ \overbrace{-t \sin t}^{\text{Particular response}} \tag{4.28}$$

By substituting the initial conditions stated in Equation (4.26) into Equation (4.28), we obtain the following:

$$\begin{array}{l} x_1(0)=0 \\ \dot{x}_1(0)=0 \end{array} \xrightarrow{c_3=c_4=0} x_1(t) = t \sin t \tag{4.29}$$

Therefore, the solution for $x(t)$ can be expressed as follows:

$$x(t) = x_0(t) + \varepsilon\, x_1(t) = \sin t - \varepsilon\, t \sin t \tag{4.30}$$

The expression ($\varepsilon\, t \sin t$) in equation (4.30) is referred to as secular, and it induces a non-uniform reaction in the system, as seen in Figure 4.2.

Figure 4.2 illustrates that for small time intervals, the answer obtained via the straightforward expansion method closely matches the expected response. However, for big-time intervals, the presence of variable t causes the response to deviate and demonstrate the extent of the increasing and growing process. Put simply, the system's response becomes faulty as the value of εt approaches 1 ($\varepsilon t \rightarrow 1$). According to the statement, the expansion fractures. Conversely, we are aware that the precise reaction of the system is as follows:

$$x(t) = \left(1-\varepsilon^2\right)^{\frac{1}{2}} e^{-\varepsilon t} \sin\left[\left(1-\varepsilon^2\right)^{\frac{1}{2}} t\right] \tag{4.31}$$

The amplitude of the answer can be expressed as follows:

$$A = \underbrace{(1-\varepsilon^2)^{\frac{1}{2}}}_{*} \underbrace{e^{-\varepsilon t}}_{**} = \underbrace{\left(1 - \frac{1}{2}\varepsilon^2 + \cdots\right)}_{\text{expansion of } *} \underbrace{(1 - \varepsilon t + \varepsilon^2 t^2 + \cdots)}_{\text{expansion of } **} \tag{4.32}$$

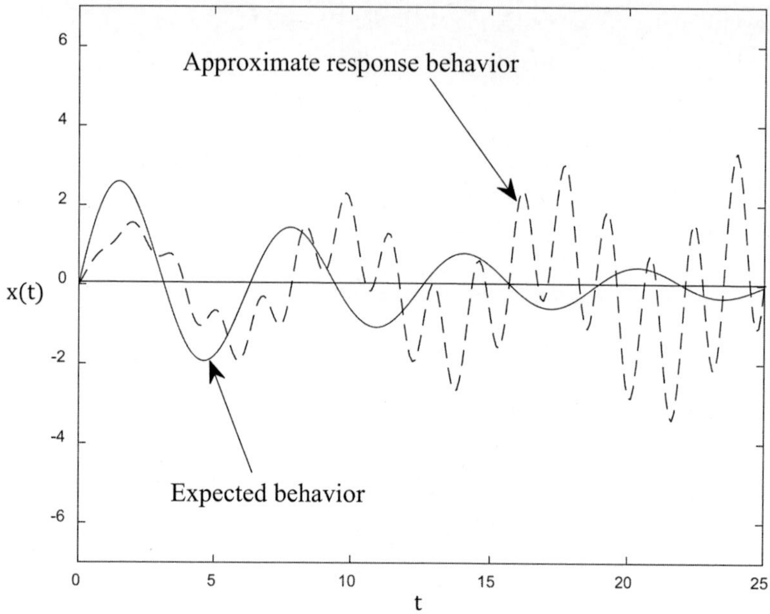

FIGURE 4.2 Non-uniform response to the system resulting from the straightforward expansion method.

The amplitude of answer A can be delineated as follows:

$$A = 1 - \varepsilon t + O(\varepsilon^2) \tag{4.33}$$

The answer in this instance will be as follows:

$$x(t) = (1 - \varepsilon t)\ \sin \omega t \tag{4.34}$$

In which:

$$\omega = 1 - \frac{1}{2}\varepsilon^2 \tag{4.35}$$

Upon comparing Equation (4.30) and Equation (4.34), it is evident that when ε is small, the precise solution and the solution derived through the straightforward technique are identical.

Subsequently, the problem of expansion interruption is addressed by employing the multiple time scale method. This method is utilized to resolve the non-uniform response that arises from the straightforward expansion method. Thus, the answer in the direct technique can be rectified in the following manner:

$$x(\varepsilon, t) \rightarrow x(T_0, T_1, T_2, \cdots) \tag{4.36}$$

The time scale T_n $(n = 0,1,2,\cdots)$ is defined as follows:

$$T_n = \varepsilon^n t \tag{4.37}$$

Typically, the following interpretation of time scales is employed:

$$T_0 = t \rightarrow \text{fast(sec)}$$

Which denotes swift fluctuations or consistent lengthy strides

$$T_1 = \varepsilon\, t \rightarrow \text{slow (min)} \tag{4.38}$$

It indicates a gradual transformation or incremental progress.

$$T_2 = \varepsilon^2 t \rightarrow \text{too slow (hr)}$$

What denotes subtle alterations or minute increments.

To rephrase the nonlinear differential Equation (4.21) and subsequently solve it using the method of multiple scales:

$$x(\varepsilon,t) \simeq x_0\left(T_0,T_1,T_2\right)+\varepsilon x_1\left(T_0,T_1,T_2\right)+\varepsilon^2 x_2\left(T_0,T_1,T_2\right)+\cdots \tag{4.39}$$

By assigning values to variables $\dot{x}$ and $\ddot{x}$, we obtain the following:

$$\begin{aligned} \dot{x} &= \frac{dx}{dt} = \frac{\partial x}{\partial T_0}\times\frac{dT_0}{dt}+\frac{\partial x}{\partial T_1}\times\frac{dT_1}{dt}+\frac{\partial x}{\partial T_2}\times\frac{dT_2}{dt} \\ \ddot{x} &= \frac{d^2x}{dt^2} = \frac{d}{dt}\left(\frac{dx}{dt}\right) = \frac{d}{dt}\left(\frac{\partial x}{\partial T_0}\times\frac{dT_0}{dt}+\frac{\partial x}{\partial T_1}\times\frac{dT_1}{dt}+\frac{\partial x}{\partial T_2}\times\frac{dT_2}{dt}\right) \end{aligned} \tag{4.40}$$

By establishing the D_n operator, the following may be deduced:

$$D_n = \frac{\partial}{\partial T_n} \tag{4.41}$$

In which:

$$\begin{aligned} \frac{d}{dt} &= D_0 + \varepsilon D_1 + \varepsilon^2 D_2 \\ \frac{d^2}{dt^2} &= D_0^2 + 2\varepsilon\, D_0 D_1 + \varepsilon^2\left(D_1^2 + 2D_0 D_2\right) \end{aligned} \tag{4.42}$$

The values of $\dot{x}$ and $\ddot{x}$ will be as follows:

$$\begin{aligned} \dot{x} &= D_0 x + \varepsilon D_1 x + \varepsilon^2 D_2 x \\ \ddot{x} &= D_0^2 x + 2\varepsilon\, D_0 D_1 x + \varepsilon^2\left(D_1^2 x + 2D_0 D_2 x\right) \end{aligned} \tag{4.43}$$

By substituting Equation (4.39) into Equation (4.43) and arranging the terms in ascending order of ε^n, we obtain the following:

$$\begin{aligned} \dot{x} &= D_0 x_0 + \varepsilon\left(D_1 x_0 + D_0 x_1\right) + \varepsilon^2\left(D_2 x_0 + D_1 x_1 + D_0 x_2\right) \\ \ddot{x} &= D_0^2 x_0 + \varepsilon\left(D_0^2 x_1 + 2D_0 D_1 x_0\right) + \varepsilon^2\left(D_0^2 x_2 + 2D_0 D_1 x_1 + D_1^2 x_0 + 2D_0 D_1 x_0\right) \end{aligned} \tag{4.44}$$

By substituting Equation (4.44) into Equation (4.21) and setting the coefficients corresponding to ε^n to zero, we obtain the following:

$$\begin{aligned}
&\varepsilon^0:\ D_0^2 x_0 + x_0 = 0 \rightarrow x_0(T_0,T_1,T_2) = a(T_1,T_2)\cos[T_0 + \phi(T_1,T_2)] \\
&= A(T_1,T_2)e^{iT_0} + \bar{A}(T_1,T_2)e^{-iT_0} = A(T_1,T_2)e^{iT_0} + CC \\
&\varepsilon^1:\ D_0^2 x_1 + x_1 = -2D_0 D_1 x_0 - 2D_0 x_0 \\
&\varepsilon^2:\ D_0^2 x_2 + x_2 = -2D_0 D_1 x_1 - 2D_0 D_2 x_0 - D_1^2 x_0 - 2D_1 x_0 - 2D_0 x_1
\end{aligned} \tag{4.45}$$

That CC is complex conjugate. The answer can be expressed in both triangle and polar formats, commonly employed to streamline the polar state. Therefore, it is feasible to write the following:

$$x(t) = Ae^{it} + \bar{A}e^{-it} \tag{4.46}$$

Given the equation $A = \frac{1}{2}ae^{i\phi}$, the following result may be obtained:

$$x(t) = \frac{1}{2}ae^{i\phi}e^{it} + \frac{1}{2}ae^{-i\phi}e^{-it} = \frac{1}{2}a\left[e^{i(t+\phi)} + e^{-i(t+\phi)}\right] = a\cos(t+\phi) \tag{4.47}$$

Now we have:

$$D_0^2 x_1 + x_1 = -2iAe^{iT_0} - 2iD_1Ae^{iT_0} + CC = \underbrace{-2i(D_1A + A)e^{iT_0}}_{\textit{secular term}} + CC \tag{4.48}$$

It is crucial to emphasize that the first differential equation, which involves x and its derivatives, has a solution consisting of the sum of the homogeneous response and the particular response. This combined solution is known as the general solution to the differential equation. Subsequent differential equations will have a homogeneous response that includes x_1, x_2, and so on. Furthermore, their derivatives are required. Put simply, in equations starting with x_1, only the particular response is considered.

Next, we must eliminate the secular terms in the resulting equation, which can be done by writing the following:

$$D_1A + A = 0 \rightarrow A(T_1,T_2) = B(T_2)e^{-T_1} \tag{4.49}$$

Given that we only consider the particular response as the answer in question, it can be concluded that the left-hand expression in Equation (4.48) is a secular expression and is set equal to zero, resulting in $(x_1 = 0)$. By substituting the solution of Equation (4.49) into the variable (x_0) in Equation (4.45), we obtain the following:

$$x_0(T_0,T_1,T_2) = B(T_2)e^{iT_0}e^{-T_1} + CC \tag{4.50}$$

To answer the third relationship, we will use Equation (4.45), which represents x_2:

$$D_0^2 x_2 + x_2 = -2D_0 D_2 x_0 - D_1^2 x_0 - 2D_1 x_0 \tag{4.51}$$

Since $(x_1 = 0)$, the remaining expressions are excluded. By substituting Equation (4.50) into Equation (4.51), we obtain the following:

$$D_0^2 x_2 + x_2 = \underbrace{\left(-2iD_2 B + B\right) e^{-T_1} e^{iT_0}}_{\text{secular term}} + CC \tag{4.52}$$

The response to the secular expression would be the following:

$$2D_2 B + iB = 0 \rightarrow B\left(T_2\right) = ce^{\frac{-i}{2}T_2} \tag{4.53}$$

In the absence of constraints on problem-solving, considering the equation $c = \frac{1}{2}ae^{i\phi}$, the following result will be obtained:

$$B\left(T_2\right) = \frac{1}{2}ae^{i\left(\phi - \frac{T_2}{2}\right)} \tag{4.54}$$

It is important to acknowledge that when we take into account the solution in a combined manner, all the constants are likewise combined. The answer (x_0) can now be expressed as follows:

$$x_0\left(T_0, T_1, T_2\right) = \frac{1}{2}ae^{i\left(\phi - \frac{T_2}{2}\right)} e^{iT_0} e^{-T_1} + CC \tag{4.55}$$

By substituting the values (T_0, T_1, T_2) into Equation (4.38) and carrying out mathematical calculations, we obtain the following:

$$x_0 = ae^{-\varepsilon t} \cos\left[t - \frac{1}{2}\varepsilon^2 t + \phi\right] \tag{4.56}$$

By substituting Equation (4.56) into Equation (4.39), the solution of Equation (4.21) using the multiple time scale method can be derived as follows:

$$x(t,\varepsilon) \simeq x_0 = ae^{-\varepsilon t} \cos\left[t - \frac{1}{2}\varepsilon^2 t + \phi\right] \tag{4.57}$$

In order to determine the precise solution of Equation (4.21) using the second-order differential equations with constant coefficients, the following expression will be obtained:

$$s^2 + 2\varepsilon s + 1 = 0 \ \rightarrow s = -\varepsilon \pm i\sqrt{1 - \varepsilon^2} \tag{4.58}$$

To streamline the precise solution to the equation, one must perform the following steps:

$$x\left(t\right) = ae^{-\varepsilon t} \cos\left[\left(1 - \frac{\varepsilon^2}{2}\right)t + \phi\right] \tag{4.59}$$

The final response utilized the following extension.

$$(1-\varepsilon^2)^{\frac{1}{2}} = 1-\frac{1}{2}\varepsilon^2+\cdots \tag{4.60}$$

Upon comparing the answers, Equation (4.57) and Equation (4.58), it becomes evident that the two solutions are the same, and the limitation of the small-time values are eliminated using the multiple time scales method. It is important to reiterate that when solving differential equations, if trigonometric functions are employed, they are considered to be constant throughout. In the complex number method, all the constants are complex, as noticed during the solution process.

It is important to note that assuming the initial conditions $x(0)=0$ and $\dot{x}(0)=1$, which resulted in the solution (4.31) using the straightforward expansion method, we can assume $\phi=\frac{\pi}{2}$ by applying it to the solution (4.59) and considering the expansion (4.60). This will yield the same solution (4.31) as obtained through the straightforward expansion method.

Now, we will attempt to solve the Duffing differential problem, which is a type of weak nonlinear differential equation. For this purpose, we initially solve the equation using a straightforward approach, which reveals the emergence of non-uniform response circumstances. Subsequently, we will finalize the answer using the multiple time scale method. The Duffing equation is widely recognized as the most renowned nonlinear differential equation, defined as follows:

$$\ddot{x}+x+\varepsilon x^3=0, \qquad \varepsilon \ll 1 \tag{4.61}$$

Equation (4.61) demonstrates that the presence of the nonlinear spring gives rise to the third-order nonlinear expression. Initially, we will attempt to solve the differential Equation (4.61) using analytical methods. Therefore, the following items will be available:

$$\ddot{x}=f(x) \tag{4.62}$$

And:

$$\begin{gathered}\ddot{x}\,dx=f(x)\,dx \rightarrow \int \dot{x}\,d\dot{x}=\int f(x)\,dx=h(x)\\ \frac{\dot{x}^2}{2}=h(x)\rightarrow \dot{x}=g(x)\rightarrow \int\frac{dx}{g(x)}=\int dt\end{gathered} \tag{4.63}$$

By acquiring knowledge about $g(x)$ and solving the integral mentioned earlier, one can determine $x(t)$. However, the analytical solution of the integral is quite intricate, necessitating the employment of numerical methods.

To solve the Duffing Equation (4.61) using the straightforward technique, we initially analyze the solution as follows:

$$x=x_0(t)+\varepsilon\, x_1(t)+\cdots \tag{4.64}$$

By following the instructions for solving the problem using the straightforward technique, we may substitute the answer (4.64) into the Duffing Equation (4.61) and include the appropriate coefficients ε^n. This will get the following result:

$$\begin{aligned} &\varepsilon^0:\ \ddot{x}_0 + x_0 = 0 \to x_0(t) = a\cos(t+\phi) \\ &\varepsilon^1:\ \ddot{x}_1 + x_1 + x_0^3 = 0 \to \ddot{x}_1 + x_1 = -a^3\cos^3(t+\phi) \end{aligned} \tag{4.65}$$

Applying the trigonometric relationship $\left(\cos^3 t = \dfrac{1}{4}\cos 3t + \dfrac{3}{4}\cos t\right)$, we obtain the following:

$$x_1(t) = \frac{1}{32}a^3\cos(3t+3\phi) - \frac{3}{8}ta^3\sin(t+\phi) \tag{4.66}$$

Since the statement $\left(\dfrac{3}{8}ta^3\sin(t+\phi)\right)$ is a secular expression and is multiplied by t, the final result of the equation obtained via the straightforward method is non-uniform. Now, employing the various time scale technique, we will once again solve Equation (4.61). Let us analyze the solution to Equation (4.61) in the following manner:

$$x(t,\varepsilon) = x_0(T_0,T_1) + \varepsilon x_1(T_0,T_1) + \cdots \tag{4.67}$$

By substituting Equation (4.67) into Equation (4.61) and subsequently isolating the coefficients ε^0 and ε^1 and setting them to zero, we obtain the following:

$$\begin{aligned} &\varepsilon^0:\ D_0^2 x_0 + x_0 = 0 \\ &\varepsilon^1:\ D_0^2 x_1 + x_1 = -2D_0D_1x_0 - x_0^3 \end{aligned} \tag{4.68}$$

For this instance, we will utilize trigonometric functions instead of mixed ones, which will naturally result in a little lengthier process. The solution to the first equation in Equation (4.68) is as follows:

$$x_0(t) = a(T_1)\cos\left[T_0 + \phi(T_1)\right] \tag{4.69}$$

By substituting the value of Equation (4.69) into Equation (4.68), we obtain the following:

$$\begin{aligned} D_0^2 x_1 + x_1 &= 2a'\sin(T_0+\phi) + 2a\phi'\cos(T_0+\phi) - a^3\cos^3(T_0+\phi) \\ &= \underbrace{2a'\sin(T_0+\phi) + \left(2a\phi' - \frac{3a^3}{4}\right)\cos(T_0+\phi)}_{\text{secular term}} - \underbrace{\frac{1}{4}a^3\cos(3T_0+3\phi)}_{\text{non-secular term}} \end{aligned} \tag{4.70}$$

In order to eliminate secular expressions, we employ the following procedure:

$$\begin{cases} a' = 0 \rightarrow a = a_0 = \text{Constant} \neq 0 \\ a\phi' = \dfrac{3}{8}a^3 \rightarrow \phi' = \dfrac{3}{8}a_0^2 \rightarrow \phi = \dfrac{3}{8}a_0^2 T_1 + \phi_0 \text{ , } \phi_0 = \text{Constant} \end{cases} \tag{4.71}$$

The equations derived in Equation (4.71) are referred to as frequency equations. The solution of Equation (4.61) can be expressed as follows:

$$x(t) = a_0 \cos\left[t + \frac{3}{8}a_0^2 \varepsilon t + \phi_0\right] + \varepsilon x_1 \tag{4.72}$$

By calculating the differential Equation (4.70) using the non-secular expression as the only non-homogeneous term, the general solution of the equation may be obtained. This will yield the expression $x_1\left(T_0,T_1\right)$.

$$x_1\left(T_0,T_1\right) = \frac{1}{32}a_0^3 \cos\left(3T_0 + 3\phi\right) \tag{4.73}$$

By substituting the value of Equation (4.73) into Equation (4.72), the final solution can be expressed as follows:

$$\begin{aligned} x(t) &= \overbrace{a_0 \cos\left[t + \frac{3}{8}a_0^2 \varepsilon t + \phi_0\right]}^{x_0} + \overbrace{\frac{1}{32}\varepsilon a_0^3 \cos\left(3t + 3\phi\right)}^{\varepsilon x_1} + O\left(\varepsilon^2 t\right) \\ &= a_0 \cos\left[t + \frac{3}{8}a_0^2 \varepsilon t + \phi_0\right] + \frac{1}{32}\varepsilon a_0^3 \cos\left[3t + \frac{9}{8}a_0^2 \varepsilon t + 3\phi_0\right] + O\left(\varepsilon^2 t\right) \end{aligned} \tag{4.74}$$

It is important to mention that in Equation (4.74), the error is represented as $O\left(\varepsilon^2 t\right)$. Typically, the error in the multiple time scales method is denoted as $O\left(\varepsilon^n t\right)$, which is also equivalent to $O\left(T_n\right)$.

Another instance involves solving the Duffing problem by making a little modification, employing the technique of Multiple time scale, and presenting the solution based on the algebra of mixed numbers. Examine the subsequent correlation:

$$\ddot{x} + \omega_0^2 x + \varepsilon \alpha x^3 = 0, \qquad \varepsilon \ll 1 \tag{4.75}$$

In the modified form of the Duffing Equation (4.75), it is important to mention that the α parameter is regarded as a coefficient for a nonlinear expression. This coefficient can be assigned positive or negative values based on certain conditions. The value ω_0^2 reflects the natural frequency of the linear differential equation. The solution to Equation (4.75) can be analyzed in the following manner:

$$x(t,\varepsilon) = x_0\left(T_0,T_1\right) + \varepsilon x_1\left(T_0,T_1\right) \tag{4.76}$$

By adding the Equation (4.76) into the Equation (4.75) and then separating the coefficients ε^0 and ε^1, we will have the following:

$$\begin{aligned} \varepsilon^0 &: D_0^2 x_0 + \omega_0^2 x_0 = 0 \\ \varepsilon^1 &: D_0^2 x_1 + \omega_0^2 x_1 = -2D_0 D_1 x_0 - \alpha x_0^3 \end{aligned} \tag{4.77}$$

Based on Equation (4.77), the x_0 response can be represented as follows:

$$x_0(T_0, T_1) = A(T_1)e^{i\omega_0 T_0} + CC \tag{4.78}$$

By substituting the value of Equation (4.78) into Equation (4.77), we obtain the following:

$$D_0^2 x_1 + \omega_0^2 x_1 = \left(-2i\omega_0 A' - 3\alpha A^2 \bar{A}\right)e^{i\omega_0 T_0} + CC + NST \tag{4.79}$$

The NST phrase in Equation (4.79) denotes the non-secular terms in equation. In order to exclude non-secular terms in Equation (4.79), we formulate it as follows:

$$2i\omega_0 A' + 3\alpha A^2 \bar{A} = 0 \tag{4.80}$$

The expression A' is $A' = \dfrac{dA}{dT_1}$. Given the response $\bar{A}$ and its complex-conjugate a, the following can be deduced:

$$\begin{cases} A = \dfrac{1}{2}a(T_1)e^{i\phi(T_1)} \\ \bar{A} = \dfrac{1}{2}a(T_1)e^{-i\phi(T_1)} \end{cases} \tag{4.81}$$

If A is partially heterogeneous, then both a and ϕ will be functions with real values. In respect to Equation (4.81), the coefficient of $\dfrac{1}{2}$ is used solely to establish proportionality in the final outcome. By substituting the values of Equation (4.81) into Equation (4.80), we arrive to the following assertion:

$$\begin{aligned} &i\omega_0 a' e^{i\phi} - a\omega_0 \phi' e^{i\phi} + \frac{3}{8}\alpha a^3 e^{i\phi} = 0 \rightarrow \left(i\omega_0 a' - a\omega_0 \phi' + \frac{3}{8}\alpha a^3\right)e^{i\phi} = 0 \\ &\xrightarrow{e^{i\phi} \neq 0} i\omega_0 a' - a\omega_0 \phi' + \frac{3}{8}\alpha a^3 = 0 \end{aligned} \tag{4.82}$$

Given that the final Equation (4.82) is a combination of several terms, it is necessary to arrange its real and imaginary forms in a manner that both have a value of zero:

$$\begin{cases} a' = 0 \rightarrow a = a_0 = \text{Constant} \neq 0 \\ a\phi' = \dfrac{3\alpha}{8\omega_0}a^3 \rightarrow \phi' = \dfrac{3\alpha}{8\omega_0}a_0^2 \rightarrow \phi = \dfrac{3\alpha}{8\omega_0}a_0^2 T_1 + \phi_0 \end{cases} \tag{4.83}$$

Hence, based on the outcomes derived from Equation (4.83), the response x_0 can be expressed in the following manner:

$$\begin{aligned} x_0\left(T_0,T_1\right) &= \frac{1}{2}ae^{i\phi}e^{i\omega_0 T_0} + CC = \frac{1}{2}a_0 e^{i\left(\omega_0 T_0 + \frac{3\alpha}{8\omega_0}a_0^2 T_1 + \phi_0\right)} + CC \\ &= a_0 \cos\left[\left(\omega_0 + \frac{3\alpha}{8\omega_0}a_0^2\varepsilon\right)t + \phi_0\right] \end{aligned} \tag{4.84}$$

Hence, the value of x, as determined by trigonometric functions, can be expressed in the following manner:

$$x(t) = a_0 \cos\left[\omega t + \phi_0\right] \tag{4.85}$$

In Equation (4.85), ω is defined as follows:

$$\omega = \omega_0\left(1 + \frac{3\alpha}{8\omega_0^2}a_0^2\varepsilon\right) \tag{4.86}$$

During the examination of the frequency response (4.86) from a physical standpoint, several observations can be made:

1. The frequency ω, which is not linear, depends on the domain a_0.
2. The expression $\left(\frac{3\alpha}{8\omega_0^2}a_0^2\varepsilon\right)$ represents a change in the linear frequency ω_0.
3. For $\alpha > 1$, the spring exhibits hardening behavior, while for $\alpha < 1$, it exhibits softening behavior, both of which have an impact on the frequency values.

4.2.5 Averaging Method

The method of averaging can be categorized into various techniques, such as Krylov-Bogoliubov method, the Krylov-Bogoliubov-Mitropolsky technique, the generalized method of averaging, averaging using focal variables, averaging using series and lie conversions, and averaging using Lagrangian. The subsequent approach employed is the as Krylov-Bogoliubov method. To obtain a more thorough analysis of the various methods of averaging, please consult the fifth chapter of Nayfeh's [7] book and the references [8–10].

To elucidate the Krylov-Bogoliubov approach, let us examine the generic structure of the subsequent weak, nonlinear second-order differential equation:

$$\ddot{u} + \omega_0^2 u = \varepsilon f\left(u,\dot{u}\right), \qquad \varepsilon \ll 1 \tag{4.87}$$

In Equation (4.87), $f\left(u,\dot{u}\right)$ is a nonlinear function. When the value of ε is equal to zero ($\varepsilon = 0$), the solution of Equation (4.87) can be expressed in the following manner:

$$u = a\cos(\omega_0 t + \beta) \tag{4.88}$$

Both a and β are constant variables that undergo gradual changes, in contrast to t, which experiences quick fluctuations. In order to achieve an approximate solution when $\varepsilon \neq 0$ and has a small value, Krylov and Bogoliubov made the assumption that the solution could still be expressed in terms of Equation (4.88) and be subject to the following condition, involving time-dependent variables a and β:

$$\dot{u} = -a\omega_0 \sin\left(\omega_0 t + \beta\right), \tag{4.89}$$

By taking the derivative of Equation (4.88) with respect to t, we obtain the following:

$$\dot{u} = -a\omega_0 \sin\left(\omega_0 t + \beta\right) + \dot{a}\cos\left(\omega_0 t + \beta\right) - a\dot{\beta}\sin\left(\omega_0 t + \beta\right), \tag{4.90}$$

So:

$$\dot{a}\cos(\omega_0 t + \beta) - a\dot{\beta}\sin(\omega_0 t + \beta) = 0 \tag{4.91}$$

The expression is regarded equivalent to zero because the variations in the variables a and β are significantly slower than the time variable. Specifically, $\dot{a} = O(\varepsilon)$ and $\dot{\beta} = O(\varepsilon)$ imply that they can be treated as almost constant in comparison. Once again, we shall do a derivation from Equation (4.89) with respect to t.

$$\ddot{u} = -a\omega_0^2 \cos\left(\omega_0 t + \beta\right) - \omega_0 \dot{a}\sin\left(\omega_0 t + \beta\right) - a\omega_0\dot{\beta}\cos\left(\omega_0 t + \beta\right) \tag{4.92}$$

By substituting Equation (4.92) into Equation (4.87) and utilizing the solution from Equation (4.88), the following are the results:

$$\begin{aligned} &\omega_0 \dot{a}\sin\left(\omega_0 t + \beta\right) + a\omega_0\dot{\beta}\cos\left(\omega_0 t + \beta\right) \\ &= -\varepsilon f\left[a\cos\left(\omega_0 t + \beta\right), -a\omega_0 \sin\left(\omega_0 t + \beta\right)\right] \end{aligned} \tag{4.93}$$

Equations (4.91) and (4.93) can now be employed to express $\dot{a}$ and $\dot{\beta}$:

$$\begin{cases} \dot{a} = -\dfrac{\varepsilon}{\omega_0} f\left[a\cos\left(\omega_0 t + \beta\right), -a\omega_0 \sin\left(\omega_0 t + \beta\right)\right]\sin\left(\omega_0 t + \beta\right) \\ \dot{\beta} = \dfrac{\varepsilon}{a\omega_0} f\left[a\cos\left(\omega_0 t + \beta\right), -a\omega_0 \sin\left(\omega_0 t + \beta\right)\right]\cos\left(\omega_0 t + \beta\right) \end{cases} \tag{4.94}$$

It is important to note that when we calculate the average of a slow variable such as $\dot{a}$ and $\dot{\beta}$ over a short period of time, these variables will converge to their mean value. This is the fundamental principle behind the averaging method. An illustrative instance is the process of calculating the mean value of a gradual variable, such as the alteration in human height, within a brief temporal interval, such as an hour. Indeed, the variable (height) and its average will be equivalent within the span of one hour.

By incorporating the parties involved in the initial Equation (4.94) into the time interval $(0, 2\pi)$, we will obtain the following:

$$\int_0^{2\pi} \dot{a}\, dt = \int_0^{2\pi} \varepsilon f\left[a\cos\left(\omega_0 t + \beta\right), -a\omega_0 \sin\left(\omega_0 t + \beta\right)\right]\sin\left(\omega_0 t + \beta\right) dt \tag{4.95}$$

When we integrate across a slow time scale, denoted by $\dot{a}$, with respect to a rapid variable, t, we assume that $\dot{a}$ is a constant value because there is no change in the time frame $(0, 2\pi)$. Hence, we can extract it from the integral and record it explicitly:

$$\dot{a} = \frac{\varepsilon}{2\pi\omega_0}\int_0^{2\pi} f\left[a\cos(\omega_0 t + \beta), -a\omega_0 \sin(\omega_0 t + \beta)\right]\sin(\omega_0 t + \beta)\, dt \quad (4.96)$$

By substituting the variable ϕ with $\omega_0 t + \beta$ ($\phi = \omega_0 t + \beta$), we obtain $d\phi = \omega_0 dt + \dot{\beta}\, dt$, and we can neglect $\dot{\beta}$ due to its small magnitude (as indicated by the order ε).

$$\dot{a} = \frac{\varepsilon}{2\pi\omega_0}\int_0^{2\pi} f\left[a\cos\phi, -a\omega_0 \sin\phi\right]\sin\phi\, d\phi \quad (4.97)$$

$$a\dot{\beta} = \frac{\varepsilon}{2\pi\omega_0}\int_0^{2\pi} f\left[a\cos\phi, -a\omega_0 \sin\phi\right]\cos\phi\, d\phi \quad (4.98)$$

The computable value can be derived from Equations (4.97) and (4.98) of domain a and phase β.

Example 4–1: Let's examine the subsequent Duffing equation:

$$\ddot{u} + \omega_0^2 u + \varepsilon\alpha u^3 = 0$$

Determine the frequency equations of the system using the method of averaging.

Answer: Considering that $f(u, \dot{u}) = \alpha u^3$ and taking into account the Equations (4.97) and (4.98), it is possible to express the following:

$$\begin{cases} u = a\cos\phi \\ \dot{u} = -a\omega_0 \sin\phi \end{cases}$$

By substituting "u" and "$\dot{u}$" into Equations (4.97) and (4.98), respectively, we obtain the following:

$$\dot{a} = \frac{\varepsilon\alpha}{2\pi\omega_0}\int_0^{2\pi} a^3\cos^3\phi \sin\phi\, d\phi = 0$$

$$a\dot{\beta} = \frac{\varepsilon\alpha}{2\pi\omega_0}\int_0^{2\pi} a^3\cos^3\phi \cos\phi\, d\phi = \frac{3\alpha\varepsilon}{8\omega_0}a^3 \Rightarrow \dot{\beta} = \frac{3\alpha\varepsilon}{8\omega_0}a^2$$

The system's frequency equations will ultimately be the following:

$$\dot{a} = 0 \Rightarrow a = a_0 = \text{Constant}$$

$$\dot{\beta} = \frac{3\alpha\varepsilon}{8\omega_0}a^2 \Rightarrow \beta = \frac{3\alpha}{8\omega_0}a_0{}^2 T_1 + \beta_0$$

The expected outcome for us, in the context of the Duffing equation, is that the domain "a" remains constant while the β-phase becomes a function of the normal frequency ω_0.

4.2.6 The Harmonic Balance Method

Harmonic balance is a technique employed to determine the stable state response of nonlinear differential equations, typically applied in the context of nonlinear electrical circuits. Harmonic balance is a frequency domain technique used to compute a stable state response, distinguishing it from other time domain methods for stable state analysis. The term "harmonic balance" originates from the technique that originated with Kirchhoff's current law in the frequency domain. When a sinusoidal signal is applied to the nonlinear component of a system, it generates harmonics at multiples of the fundamental frequency. This allows us to represent the system's response as a linear combination of sinusoidal functions. By ensuring that the current and voltage sinusoids are balanced, we may meet Kirchhoff's law. Additionally, this technique is frequently employed in the simulation of circuits containing nonlinear components and is also extensively utilized in feedback systems that exhibit limit cycles.

The harmonic balance method is utilized to achieve stable state responses when addressing nonlinear vibration problems. Indeed, this approach fails to provide us with the system's temporary answers. Furthermore, it should be noted that this strategy assumes that the system response is periodic.

To further elucidate the harmonic equilibrium method, let us once again examine the Duffing equation:

$$\ddot{u}+\omega_0^2 u+\alpha u^3=0 \tag{4.99}$$

Given that our solution is periodic, we express the answer to the problem as a harmonic series in the following manner:

$$\begin{aligned} u &= \sum_{m=0}^{M} A_m \cos\left(m\omega t + m\beta_0\right) \\ &= A_0 + A_1\cos\left(\omega t+\beta_0\right)+A_2\cos\left(2\omega t+2\beta_0\right)+\ldots \end{aligned} \tag{4.100}$$

Given the assumption that $\phi=\omega t+\beta_0$, the following equation holds:

$$u=A_0+A_1\cos\phi+A_2\cos 2\phi+\ldots \tag{4.101}$$

For the sake of facilitating calculations, we introduce the term "deviation sentence" or "drift" to refer to A_0. When there is a second-order nonlinear expression in the equation, this statement will indicate a value that is not zero. Because the Duffing equation does not contain a second-order nonlinear term, the equation can be simplified to $A_0=0$. The solution of Equation (4.99) can be succinctly described as follows:

$$u=A_1\cos\phi \tag{4.102}$$

By substituting the value of Equation (4.102) into Equation (4.99), we get the following:

$$-A_1\omega^2\cos\phi+\omega_0^2 A_1\cos\phi+\alpha A_1^3\cos^3\phi=0 \tag{4.103}$$

Using the trigonometric relationship of $\cos^3\phi = \frac{1}{4}\cos 3\phi + \frac{3}{4}\cos\phi$, we have the following:

$$-\left(\omega^2 - \omega_0^2\right)A_1 \cos\phi + \frac{1}{4}\alpha A_1^{\ 3}\cos 3\phi + \frac{3}{4}\alpha A_1^{\ 3}\cos\phi = 0 \tag{4.104}$$

Given that our approximation is limited to the term $\cos\phi$, we can omit the term $\frac{1}{4}\alpha A_1^{\ 3}\cos 3\phi$ in Equation (4.104). Next, the system's frequency response will be acquired in the following manner:

$$\cos\phi \neq 0 \Rightarrow -\left(\omega^2 - \omega_0^2\right)A_1 + \frac{3}{4}\alpha A_1^{\ 3} = 0 \overset{A_1 \neq 0}{\Rightarrow} \omega = \omega_0\left(1 + \frac{3\alpha}{4\omega_0^2}A_1^{\ 2}\right)^{\frac{1}{2}} \tag{4.105}$$

The solution to Equation (4.99) is as follows:

$$u = A_1 \cos\left(\omega t + \beta_0\right) \tag{4.106}$$

The values of A_1 and β_0 are derived from the provided initial conditions.

The natural frequency Equation (4.105) is determined by the A_1 domain resulting from the initial conditions, which is a characteristic of nonlinear systems.

Example 4–2: Let us examine the equation of a simple pendulum:

$$\ddot{u} + \sin u = 0$$

Or,

$$\ddot{u} + u - \frac{1}{6}u^3 = 0$$

Considering the system response as $u = A_1 \cos\phi$, we may conclude the following:

$$-A_1\omega^2\cos\phi + A_1\cos\phi - \frac{1}{6}A_1^{\ 3}\cos^3\phi = 0$$

By following the procedure outlined in the preceding section and taking into account that the solution solely comprises the terms $\cos\phi$, it can be expressed as follows:

$$\cos\phi \neq 0 \Rightarrow A_1\left[\left(1-\omega^2\right) - \frac{1}{8}A_1^{\ 2}\right] = 0 \overset{A_1 \neq 0}{\Rightarrow} \omega^2 = 1 - \frac{1}{8}A_1^{\ 2} \overset{A_1 \text{ is small}}{\Rightarrow} \omega = 1 - \frac{1}{16}A_1^{\ 2}$$

4.2.7 Examination of Nonlinear Vibrations in the Damper

For certain vibration problems with nonlinear damping, the paths of motion converge towards a closed curve. This closed curve represents a periodic solution of the system, regardless of whether the paths approach or move away from the origin.

This implies that any solution to the system, as the variable t approaches infinity ($t \to \infty$), is compelled towards an intermittent solution. The closed curve is referred to as the limit cycle.

To provide a more comprehensive clarification, we will analyze the Van der Pol equation [8] as follows:

$$\ddot{x} - \alpha\left(1 - x^2\right)\dot{x} + x = 0, \quad \alpha > 0 \tag{4.107}$$

The Van der Pol equation, introduced in 1922, describes a linear oscillator with nonlinear damping, as shown in Figure 4.3. Consult reference [8] for a comprehensive explanation of the process for deriving Equation (4.107) as illustrated in Figure 4.3. Let's examine the Van der Pol oscillator equation:

$$\ddot{x} + \omega^2 x + \varepsilon\left(x^2 - 1\right)\dot{x} = 0 \tag{4.108}$$

Later, we shall observe that solving this equation results in the emergence of a limit cycle, which is a characteristic of nonlinear systems. In order to solve the equation earlier, we will employ the Lindstedt-Poincaré approach. When the value of ε is equal to zero ($\varepsilon = 0$), we obtain a simple harmonic oscillator that exhibits a set of periodic responses, which are characterized by the parameter ω. We observe the proliferation of chaos as a recurring dominant pattern. Let's start by considering Equation (4.108). This relationship will hold true when ε is much smaller than 1 ($\varepsilon \ll 1$). Expanding turbulence to encompass a wider range of frequencies enables it to effectively accommodate nonlinear behavior through the introduction of the created time variable. For this particular situation, we compose the following:

$$\tau = \omega t \tag{4.109}$$

The frequency of the response, denoted by ω, is expressed as a series of capabilities based on the variable ε.

$$\omega = 1 + k_1\varepsilon + k_2\varepsilon^2 + \cdots \tag{4.110}$$

In Equation (4.110), $\omega_0 = 1$ represents the frequency of the basic harmonic oscillator. To determine the constant k_i ($i = 1, 2, \cdots$), further calculations are required. This

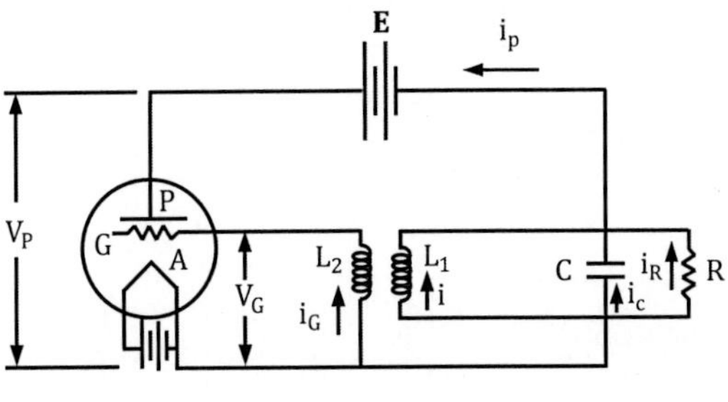

FIGURE 4.3 Flow circuit diagram for the Van der Pol oscillator.

incentive provides ample flexibility to ensure the elimination of secular expressions at every stage of the expansion of turbulence. The Van der Pol Equation (4.108) can be derived by employing variable change (4.109).

$$\omega^2 \frac{d^2x}{d\tau^2} + x + \varepsilon\left(x^2 - 1\right)\omega\frac{dx}{d\tau} = 0 \tag{4.111}$$

By substituting the series expansions for $x(\tau)$ and ω, we obtain the following:

$$\begin{aligned}&\left(1+2k_1\varepsilon+\left(k_1^2+2k_2\right)\varepsilon^2+O\left(\varepsilon^3\right)\right)\left[\frac{d^2}{d\tau^2}\left(x_0+\varepsilon x_1+\varepsilon^2 x_2+\cdots\right)\right]\\&+\left(x_0+\varepsilon x_1+\varepsilon^2 x_2+\cdots\right)+\varepsilon\left(x_0^2+2\varepsilon x_0 x_1+\varepsilon^2\left(x_1^2+2x_0x_1\right)+\cdots-1\right)\\&\left(1+\varepsilon k_1+\varepsilon^2 k_2+\cdots\right)\frac{d}{d\tau}\left(x_0+\varepsilon x_1+\varepsilon^2 x_2+\cdots\right)=0\end{aligned} \tag{4.112}$$

We have a category for each order of the associated coefficients ε that is equal to zero.

$$\begin{aligned}&O\left(\varepsilon^0\right):\ x_0'' + x_0 = 0\\&O\left(\varepsilon^1\right):\ x_1'' + x_1 = x_0'\left(1-x_0^2\right) - 2k_1x_0''\\&O\left(\varepsilon^2\right):\ x_2'' + x_2 = x_1'\left(1-x_0^2\right) - 2x_0x_1x_0' - 2k_1x_1''\\&\qquad -\left(2k_2+k_1^2\right)x_0'' + k_1\left(1-x_0^2\right)x_0'\end{aligned} \tag{4.113}$$

The equation $O(\varepsilon^0)$ represents the response of $x_0(\tau) = B_0 \cos\tau$. In order to eliminate secular terms in $O\left(\varepsilon^1\right)$, it is necessary to know the value of B_0. By employing the solution of $x_0(\tau)$ for $O\left(\varepsilon^1\right)$, the following result is obtained:

$$x_1'' + x_1 = -\beta_0 \sin\tau(1-B_0^2)\cos^2\tau + 2k_1B_0\cos\tau \tag{4.114}$$

The general solution of Equation (4.114) can be expressed as follows:

$$x_1(\tau) = \left[\frac{B_0^3}{4} - B_0\right]\left(-\frac{\tau}{2}\cos\tau\right) - \frac{B_0^3}{32}\sin 3\tau + 2k_1B_0\cos\tau + B_1\cos\tau + A_1\sin\tau \tag{4.115}$$

The constants A_1 and B_1 are arbitrary. It is important to observe that the function "$\tau\cos\tau$" produces an endless output when evaluated at t $t \to \infty$. Indeed, the solution can be written in the following order:

$$\begin{aligned}x(\tau) = B_0\cos\tau + &\left[\frac{B_0^3}{4} - B_0\right]\left(-\frac{\tau}{2}\cos\tau\right) - \frac{B_0^3}{32}\sin 3\tau\\&+2k_1B_0\cos\tau + \varepsilon B_1\cos\tau + \varepsilon A_1\sin\tau\end{aligned} \tag{4.116}$$

Thus, if τ is in order of $O(1/\varepsilon)$, and each phrase in the sequence is smaller than the one before it, this will disrupt the expansion of the disorder. The divergence of the response induced by the secular expression occurs when there is limitless and unbounded growth. Given our search for periodic solutions, it is necessary to exclude secular statements. If we select $B_0 = 2$, this secular phrase will be non-negative at the minimum at this level of approximation.

Following the resolution process, the final outcome will be denoted as $x_1(\tau)$:

$$x_1(\tau) = B_1 \cos\tau - \frac{1}{4}\sin 3\tau + \frac{3}{4}\sin\tau \tag{4.117}$$

The result is obtained by placing $x_0(\tau)$ and $x_1(\tau)$ on the right side of the relation $O(\varepsilon^2)$.

$$\begin{aligned} x_2'' + x_2 &= x_1'\left(1 - x_0^2\right) - 2x_0 x_1 x_0' - 2k_2 x_0'' \\ \Rightarrow x_2'' + x_2 &= \frac{1}{4}\cos\tau + 2B_1 \sin\tau - \frac{3}{2}\cos 3\tau \\ &\quad + 3B_1 \sin 3\tau + \frac{5}{4}\cos 5\tau + 4k_2 \cos\tau \\ &= \left(4k_2 + \frac{1}{4}\right)\cos\tau + 2B_1 \sin\tau - \frac{3}{2}\cos 3\tau + 3B_1 \sin 3\tau + \frac{5}{4}\cos 5\tau \end{aligned} \tag{4.118}$$

By choosing $k_2 = -\frac{1}{16}$, and $B_1 = 0$, the secular expressions are excluded. The frequency $\left(O(\varepsilon^2)\right)$ will now be represented as follows:

$$\omega = 1 - \frac{\varepsilon^2}{16} \tag{4.119}$$

Equation (4.119) demonstrates that the presence of a nonlinear expression leads to a decrease in the frequency of oscillation, resulting in an increase in the period. The magnitude of the oscillation is determined by the following equation:

$$\begin{aligned} x(\tau) = x_0(\tau) + \varepsilon x_1(\tau) + \varepsilon^2 x_2(\tau) &= 2\cos\tau - \varepsilon \frac{[\sin 3\tau - 3\sin\tau]}{4} \\ &\quad - \varepsilon^2 \frac{[5\cos 5\tau - 18\cos 3\tau + 12\cos\tau]}{96} \end{aligned} \tag{4.120}$$

Figure 4.4 showcases a captivating aspect of the response. Irrespective of their starting conditions, all trajectory lines are eventually approached by a closed curve called the limit cycle. This curve depicts a non-harmonic oscillation of the stable state. This phenomenon is exclusive to specific nonlinear vibration problems and does not occur in any other concerns. If the starting point is located within the limit cycle, the resulting response manifests as outwardly expanding spiral arcs. Conversely, if the starting point is located beyond the limit cycle, the resulting behavior will be inward

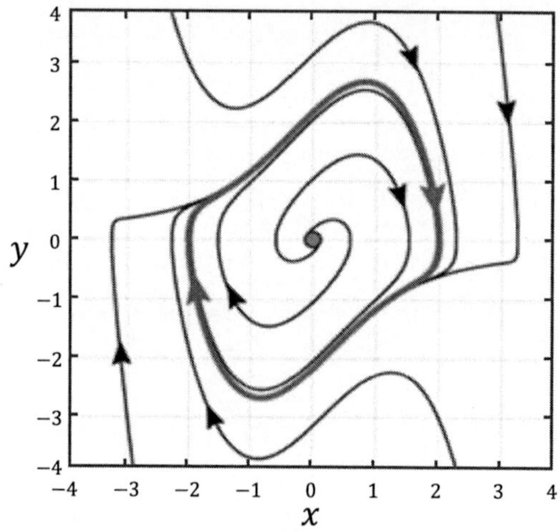

FIGURE 4.4 Schematic of trajectories and limit cycle for the Van der Pol oscillator equation.

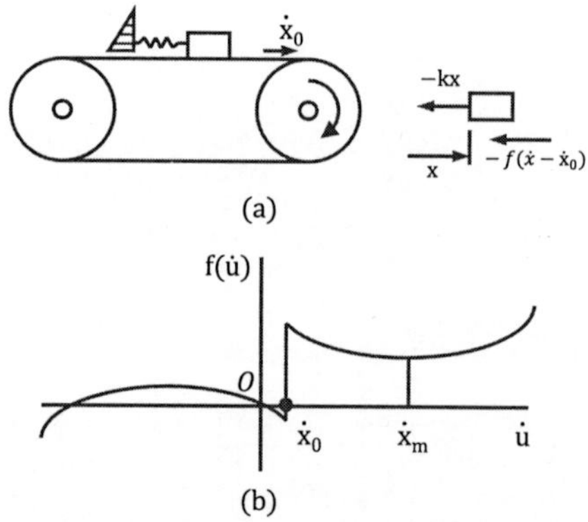

FIGURE 4.5 Mechanical system to perform self-regulating oscillations.

spiraling curves. As previously mentioned, the limit cycle in both stages results in a specific closed curve. A notable attribute of the limit cycle is that the greatest value of x, irrespective of the value of α as determined by Equation (4.107), consistently approximates 2. The value $B_0 = 2$ after solving the Van der Pol oscillator equation, which is the range of values for the answer template (4.120).

Allow me to provide an additional illustration of a mechanical system that exhibits negative damping. Examine Figure 4.5, which depicts a block with mass m placed on a rough belt that is traveling at a constant speed of $\dot{x}_0$. A mass m is connected to a spring that is attached to a rigid object. Let x represent the displacement of the block from the equilibrium position of the spring [8, 11].

$$m\ddot{x} + kx - f\left(\dot{x} - \dot{x}_0\right) = 0 \tag{4.121}$$

The variable "f" represents the friction force acting between the block and the belt in Coulomb's experiment. Let us now add the new variable u, which is defined as follows:

$$u = x - k^{-1} f\left(-\dot{x}_0\right) \tag{4.122}$$

By variable change of (4.122), Equation (4.121) will be as follows:

$$\ddot{u} + \omega_0^2 u + F(\dot{u}) = 0 \tag{4.123}$$

Where $\omega_0^2 = k/m$ and,

$$F(\dot{u}) = m^{-1}\left[f\left(-\dot{x}_0\right) - f\left(\dot{u} - \dot{x}_0\right)\right] \tag{4.124}$$

If the value of $\dot{x}_0$ is not excessively big, the function $F(\dot{u})$ will exhibit a level of curvature similar to that depicted in the diagram of Figure 4.5. It should be noted that if $\dot{x}_0 < \left|\dot{x}_m\right|$, then the slope of $F(\dot{u})$ at the origin is negative. If the value of $\dot{x}_0$ at the origin is quite big, then the derivative of $F(\dot{u})$ at the origin will be positive. Therefore, negative damping only happens when the value of $\dot{x}_0 < \left|\dot{x}_m\right|$.

It is important to observe that dry friction can be employed to demonstrate negative damping in various other mechanical systems. Rayleigh had employed the same reasoning to elucidate the creation of vibrations in the violin wire resulting from the action of drawing the bow across its width. Dry friction induces self-excited oscillations in a pendulum attached to a rotating shaft, resulting in the shaft exhibiting an unsteady motion within a loosely supported bearing. Dry friction can be employed to elucidate the collision between the brake shoes and train wheels during braking.

Example 4–3: It is preferable to address the issue of the Van der Pol oscillator by employing the technique of multiple time scales and illustrating its phase diagram to depict the occurrence of the limit cycle.

Response: Please examine the subsequent connection:

$$\ddot{x} - \epsilon\left(x^2 - 1\right)\dot{x} + x = 0; \quad x(0) = a_0, \quad \dot{x}(0) = 0,$$

By substituting the solution as a series expansion $\left(x_0 + \epsilon x_1 + \epsilon^2 x_2 + \ldots\right)$ into the given relation, we obtain the following:

$$\begin{aligned}
&\left[\frac{\partial^2}{\partial T_0^2} + 2\epsilon \frac{\partial^2}{\partial T_1 \partial T_2} + \epsilon^2\left(\frac{\partial^2}{\partial T_1^2} + \frac{2\partial^2}{\partial T_2 \partial T_0}\right)\ldots\right]\left(x_0 + \epsilon x_1 + \epsilon^2 x_2 + \ldots\right) \\
&+\epsilon\left\{\left[\left(x_0^2 + 2\epsilon x_0 x_1 + \epsilon^2\left(x_1^2 + 2\epsilon x_0 x_2\right)\right) - 1\right]\left(\frac{\partial}{\partial T_0} + \epsilon \frac{\partial}{\partial T_1} + \epsilon^2 \frac{\partial}{\partial T_2}\ldots\right)\left(x_0 + \epsilon x_1 + \epsilon^2 x_2 + \ldots\right)\right\} \\
&+\left(x_0 + \epsilon x_1 + \epsilon^2 x_2 + \ldots\right) = 0
\end{aligned}$$

Through re-separation in each order ϵ, the following relationships are established:

$$O(\epsilon^0):\ Lx_0 = 0$$

$$O(\epsilon^1):\ Lx_1 = -2\frac{\partial^2 x_0}{\partial T_1 \partial T_0} - \left(x_0^2 - 1\right)\frac{\partial x_0}{\partial T_0}$$

$$O(\epsilon^2):\ Lx_2 = -2\frac{\partial^2 x_0}{\partial T_1 \partial T_0} - \left(\frac{\partial^2}{\partial T_1^2} + \frac{2\partial^2}{\partial T_2 \partial T_0}\right)x_0 - \left(x_0^2 - 1\right)\frac{\partial x_0}{\partial T_1}$$
$$-\left(x_0^2 - 1\right)\frac{\partial x_1}{\partial T_0} - 2x_0 x_1 \frac{x_0}{\partial T_0}$$

In which:

$$Lx_i = \left[\frac{\partial^2}{\partial T_0^2} + 1\right]x_i$$

The solution to the problem $\left(O\left(\epsilon^0\right)\right)$ will be as follows:

$$x_0 = A\left(T_1, T_2\right)e^{iT_0} + \bar{A}\left(T_1, T_2\right)e^{-iT_0}$$

By substituting this expression for x_0 into x_1, we obtain the following:

$$O\left(\epsilon^1\right):\ Lx_1 = -2i\left(\frac{\partial A}{\partial T_1}e^{iT_0} - \frac{\partial \bar{A}}{\partial T_1}e^{-iT_0}\right)$$
$$-i\left\{\left(|A|^2 A - A\right)e^{iT_0} - \left(|A|^2 \bar{A} - \bar{A}\right)e^{-iT_0} + A^3 e^{3iT_0} - \bar{A}^3 e^{-3iT_0}\right\}.$$

In order to avoid secular phrases with a time complexity of $O\left(\epsilon^1\right)$, the condition $2\frac{\partial A}{\partial T_1} = A - |A|^2 A$ is imperative to establish. To explicitly solve for A by substituting $A = \frac{R\left(T_1\right)e^{i\theta\left(T_1\right)}}{2}$, we derive distinct equations for the amplitude and phase A.

$$\frac{\partial R}{\partial T_1} = \frac{R}{2} - \frac{R^3}{8}$$

$$\frac{\partial \theta}{\partial T_1} = 0,$$

It is indicated that $\theta\left(T_1\right)$ remains constant, and we choose $\theta\left(T_1\right) = 0$ as well.

$$R\left(T_1\right) = \frac{2R(0)e^{\frac{T_1}{2}}}{\sqrt{\left(e^{T_1} - 1\right)R(0)^2 + 4}}$$

The given initial conditions $x(0) = a_0$ and $\dot{x}(0) = 0$ indicate that $R(0) = a_0$. Therefore:

$$x_0 = \frac{a_0 e^{\epsilon t/2} e^{it}}{\sqrt{\left(e^{\epsilon t} - 1\right) a_0^2 + 4}} + CC = \frac{2}{\sqrt{1 + \left(4 / a_0^2 - 1\right) e^{-\epsilon t}}} \cos t$$

It has been observed that while considering the first sentence as an extension response, the oscillator limit cycle of the Van der Pol remains stable. The initial conditions of this response are exponentially absorbed, resulting in a somewhat poor approximation for the limit cycle. If we advance the procedure once more, we will regain the frequency correction of the Lindstedt-Poincaré approach. This matter can be investigated as a practical exercise.

We will create a phase plane of the Van der Pol oscillator to visualize the limit cycle behavior.

$$\ddot{x} + \varepsilon\left(1 - x^2\right)\dot{x} + x = 0$$

$$\begin{cases} x = X_1 \\ \dot{x} = X_2 \end{cases} \rightarrow \begin{cases} \dot{X}_1 = X_2 \\ \dot{X}_2 = -\varepsilon X_2\left(1 - X_1^2\right) - X_1 \end{cases}$$

Figure 4.6 displays the phase plane of the Van der Pol oscillator, illustrating the complete cycle of the stable limit. Consult Refs. [8, 12] for a more comprehensive analysis and comprehension of the image produced by the phase plane.

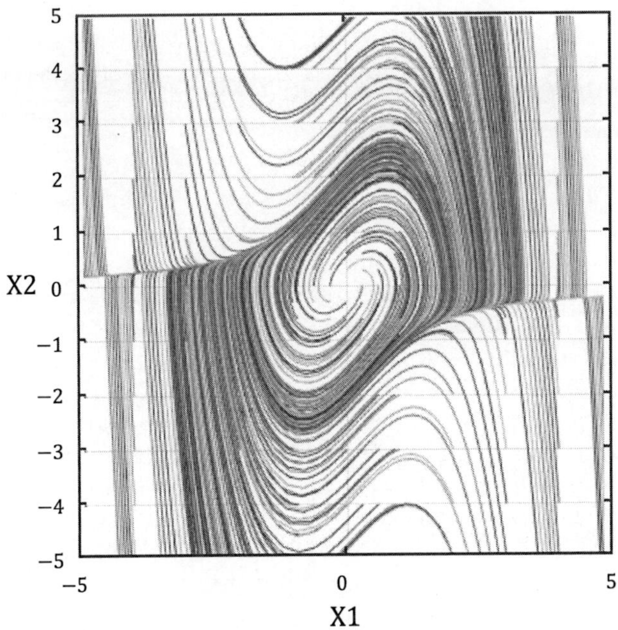

FIGURE 4.6 Phase plane of the Van der Pol oscillator.

4.3 GENERALIZE DIFFERENTIAL QUADRATURE METHOD

4.3.1 INTRODUCTION

In the second section of this chapter, we study the finite element method. The finite element method (FEM) is a versatile technique utilized in engineering to solve a broad spectrum of equations, typically expressed as partial differential equations. Over the past few decades, various solution techniques have been introduced to address the governing equations of diverse problems. The physical formulation of a problem can be transformed using three methods: the direct method, the variation method (also known as Rayleigh-Ritz), and the weighted residual method. This sub-chapter will explore the various methods for solving the problem such as GDQ and perturbation methods. Analytical methods in solving engineering problems are always considered due to their high accuracy and are the first option in solving a problem. But in many cases, these methods face limitations, including the complexity of the governing equations, or the geometry of the problem, or the existence of discontinuity in the problem-solving range. This issue has led to the emergence and growth of numerical methods so that nowadays, various numerical methods have been presented to solve engineering problems, among which we can refer to the finite difference method, the boundary element method, the differential transform method, the Ritz method, Galerkin Method, and so on.

4.3.2 HISTORY

The DQ approach, introduced by Belman in the 1970s, offers a superior alternative to finite difference, finite element, and boundary element techniques for solving initial value and boundary value problems in the fields of physics and engineering. This method had an edge over earlier strategies in terms of its faster calculations and utilization of fewer nodes. The DQ approach estimates the partial derivatives of a function at a specific location by calculating a linear combination of the function's values at all points in the domain. This is in contrast to prior methods like finite difference. This assumption was made since there was an improvement in accuracy and a notable decrease in the amount of calculations required. Nevertheless, this approach was hindered by constraints such as challenges in dividing the domain into discrete sections and accurately determining the weight coefficients of each domain point. The GDQ technique was introduced in the early 1990s by Shu and Richards as an adaptation of the DQ method [13]. It was developed to address the limitations of the DQ approach in solving two-dimensional Navier-Stokes equations. Since then, the GDQ method has been widely applied in solving elasticity issues for sheets and shells. Calculating the weighted coefficients of the derivatives in this method is more straightforward than the DQ method and can accommodate a greater number of points without being restricted by point selection limitations. This approach approximates the derivative at each place by combining the nodal values at other points. Consequently, by utilizing this approach, the equations that control the problem are transformed into algebraic equations, which can either be linear or nonlinear depending on the nature of the problem. This method possesses the capability to solve a wide range of equations, encompassing equations with variable coefficients, nonlinear equations, and eigenvalue equations. The GDQ approach has been employed in numerous studies to address vibration issues and stress analysis of rectangular sheets and cylindrical shells.

4.3.3 Principles

The general principle in the method of DQ is to divide the problem-solving interval into a network of points and solve the differential equation in this discrete network of points. In this method, the goal is to estimate the derivatives of the function of different orders in each of the problem-solving points according to the value of the function in all points. In order to achieve this goal, consider a number of points like $(x_1, f_1), (2, f_2), \ldots, (x_N, f_N)$. To fit a unique polynomial for these points, Lagrange provided the following relation [14]:

$$f(x) = L_1(x)f_1 + L_2(x)f_2 + \ldots + L_N(x)f_N = \sum_{j=1}^{N} L_j(x)f_j \tag{4.125}$$

In which $L_j(x)$ are the Lagrange coefficients that:

$$L_j(x) = \begin{cases} 1 & x = x_j \\ 0 & x \neq x_j \end{cases} \tag{4.126}$$

In other words, the following relationship is established at point $x = x_j$:

$$L_j = 1 \; L_1 = L_2 = L_3 = \ldots = L_{j-1} = L_{j+1} = \ldots = L_{N-1} = L_N = 0 \tag{4.127}$$

In order to calculate the derivatives of the function with r times of derivation from both sides, Equation (4.125) can be written as follows [15, 16]:

$$\frac{d^r f}{dx^r} = \sum_{j=1}^{N} \frac{d^r L_j(x)}{dx^r} f_j \tag{4.128}$$

Therefore, to calculate the derivative of the function, it is enough to calculate the derivative of the Lagrange coefficients. Many functions can be defined that satisfy Equation (4.128), but the simplest function that can be considered is a polynomial function that has the following general form:

$$L_j(x) = \frac{(x - x_1)(x - x_2)\ldots(x - x_{j-1})(x - x_{j+1})\ldots(x - x_N)}{(x_j - x_1)(x_j - x_2)\ldots(x_j - x_{j-1})(x_j - x_{j+1})\ldots(x_j - x_N)} = \frac{\prod_{\substack{k=1 \\ k \neq j}}^{N}(x - x_k)}{\prod_{\substack{k=1 \\ k \neq j}}^{N}(x_j - x_k)} \tag{4.129}$$

In this relation, Π is the operator of successive multiplication of several expressions. For ease of derivation, we can express the Equation (4.129) using the mathematical operator Ln and its properties as follows:

$$\mathrm{Ln}\left[L_j(x)\right] = \sum_{\substack{k=1 \\ k \neq j}}^{N} Ln(x - x_k) - \sum_{\substack{k=1 \\ k \neq j}}^{N} Ln(x_j - x_k) \tag{4.130}$$

By deriving both sides of the equation, Equation (4.130) can be written as follows:

$$\frac{1}{L_j(x)}\frac{dL_j(x)}{dx}=\sum_{\substack{k=1\\k\neq j}}^{N}\frac{1}{x-x_k} \tag{4.131}$$

And according to the Equation (4.129), the following relationship can be expressed:

$$\frac{dL_j(x)}{dx}=\frac{\prod_{\substack{k=1\\k\neq j}}^{N}\left(x-x_k\right)}{\prod_{\substack{k=1\\k\neq j}}^{N}\left(x_j-x_k\right)}\sum_{\substack{k=1\\k\neq j}}^{N}\frac{1}{x-x_k} \tag{4.132}$$

The Equation (4.128) can be rewritten as follows for the value of the first derivative (r = 1) at an arbitrary point such as $x=x_i$.

$$\left.\frac{df}{dx}\right|_{x=x_i}=\sum_{j=1}^{N}\left.\frac{dL_j(x)}{dx}\right|_{x=x_i}f_j=\sum_{j=1}^{N}A_{ij}^{(1)}f_j \tag{4.133}$$

In this regard, according to the Equation (4.132), the following relation can be presented.

$$A_{ij}^{(1)}=\frac{\prod_{\substack{k=1\\k\neq j}}^{N}\left(x_i-x_k\right)}{\prod_{\substack{k=1\\k\neq j}}^{N}\left(x_j-x_k\right)}\sum_{\substack{k=1\\k\neq j}}^{N}\frac{1}{x_i-x_k} \tag{4.134}$$

The Equation (4.133) can be expressed in the form of the following matrix.

$$\left\{\frac{df}{dx}\right\}_{N*1}=\left[A^{(1)}\right]_{N*N}\left\{f\right\}_{N*1} \tag{4.135}$$

It should be noted that the superscript (1) in the relationship $\left[A^{(1)}\right]$, indicates the estimation of the first derivative of the function.

According to Equation (4.133), it can be concluded that in expression $A_{ij}^{(1)}$, the first subscript indicates the point where the value of the first derivative is estimated, and the second subscript indicates the point where this value is the corresponding coefficient in the estimation of the first derivative.

By expanding Equation (4.134), it can be shown that all sentences will be equal to zero, except for one sentence that will be ambiguously $\frac{0}{0}$, which can be shown by disambiguating as:

$$A_{ij}^{(1)}=f(x)=\begin{cases}\dfrac{\prod_{\substack{k=1\\k\neq j}}^{N}\left(x_i-x_k\right)}{\prod_{\substack{k=1\\k\neq j}}^{N}\left(x_j-x_k\right)}, i,j=1,2,3,\ldots,N: i\neq j\\[2ex] \displaystyle\sum_{\substack{k=1\\k\neq j}}^{N}\frac{1}{x_i-x_k}, i,j=1,2,3,\ldots,N\end{cases} \tag{4.136}$$

To estimate the second derivative, consider the following equation:

$$\left\{\frac{d^2 f}{dx^2}\right\}=\left[A^{(1)}\right]\left\{\frac{df}{dx}\right\} \tag{4.137}$$

which is expressed as follows using the Equation (4.135)

$$\left\{\frac{d^2 f}{dx^2}\right\}=\left[A^{(1)}\right]\left[A^{(1)}\right]\{f\} \tag{4.138}$$

As a result, the value of the second derivative can be estimated as follows:

$$\left\{\frac{d^2 f}{dx^2}\right\}=\left[A^{(2)}\right]\{f\} \tag{4.139}$$

in which,

$$\left[A^{(2)}\right]=\left[A^{(1)}\right]\left[A^{(1)}\right] \tag{4.140}$$

As a result, in general, the following relationship can be expressed to estimate the derivative r of a function.

$$\left\{\frac{d^r f}{dr}\right\}=\left[A^{(r)}\right]\{f\} \tag{4.141}$$

In which,

$$\left[A^{(r)}\right]=\left[A^{(1)}\right]\left[A^{(r-1)}\right] \mathrm{r}=2,3,\ldots \tag{4.142}$$

It should be noted that with the aim of creating ease in notation in this booklet, the following convention is used to show the corresponding matrixes for different derivatives.

$$A=A^{(1)}, B=A^{(2)}, C=A^{(3)}, D=A^{(4)},\ldots \tag{4.143}$$

As a result, the following relationship can be expressed:

$$\left\{\frac{df}{dx}\right\}=[A]\{f\},\left\{\frac{d^2 f}{dx^2}\right\}=[B]\{f\},\left\{\frac{d^3 f}{dx^3}\right\}=[C]\{f\},\left\{\frac{d^4 f}{dx^4}\right\}=[D]\{f\},\ldots \tag{4.144}$$

in which,

$$[B]=[A][A],[C]=[A][B],[D]=[A][C],\ldots \tag{4.145}$$

It should be noted that matrixes $[A],[B],[C],\ldots$ are known as weight coefficients matrixes. Considering that the derivatives of the function are considered only in certain points of the problem solving domain, it is obvious that with the increase of the number of points considered in the problem-solving domain, the accuracy of the obtained answers increases, and finally, in a certain number of convergence points, it will converge. In addition to the number of points that are considered, how the points are distributed in the domain of problem-solving is also very important. The simplest explanation that comes to mind in the first step is the uniform distribution of points along the interval. If this type of explanation is used, the convergence of the problem will be created for a very high number of points because with the increase in the order of the derivative, the approximation error will grow at a high rate compared to the increase in the order of the derivative. The best type of point distribution that has been considered so far is the cosine explanation known as Chebyshev-Gauss-Lobatto distribution. The characteristic of this type of distribution is that the density of points considered in this type of distribution is higher near the border points than the middle points. This distribution is calculated for the interval $[a,b]$ from the following equation.

$$x_i = a+\frac{b-a}{2}\left\{1-\cos\left[\frac{(i-1)\pi}{N-1}\right]\right\} \quad i=1,2,3,\ldots, N \tag{4.146}$$

Example: Using the DQM, we solve the given differential equation under the corresponding boundary conditions.

$$f''+2f'-4f=4x(2-x) \tag{4.147}$$

$$\text{B.C}:\ f(0)=f(1)=0 \tag{4.148}$$

Using Equation (4.144), Equation (4.147) can be expressed in the form of the following matrix:

$$[B]\{f\}+2[A]\{f\}-4\{f\}=\{q\} \tag{4.149}$$

In which:

$$q_i = 4x_i\left(2-x_i\right) \tag{4.150}$$

The Equation (4.149) can be expressed as follows:

$$[K]\{f\}=\{q\} \tag{4.151}$$

In which:

$$[K]=[B]+2[A]-4I \tag{4.152}$$

And in this relation, I is identity matrix of order N. The given boundary conditions for the vector f at the beginning and end points of the interval is equal to zero $(f_1 = f_N = 0)$.

As a result, in order to satisfy the boundary conditions, the first and last row of the vectors $\{f\}$ and $\{q\}$, as well as the first and last row and columns of the matrix $[K]$, can be removed, and the Equation (4.151) can be expressed as follows:

$$[K^*]\{f^*\} = \{q^*\} \tag{4.153}$$

In which:

$$[K^*] = \begin{bmatrix} K_{22} & \cdots & K_{2(N-1)} \\ \vdots & \ddots & \vdots \\ K_{(N-1)2} & \cdots & K_{(N-1)(N-1)} \end{bmatrix}, \{f^*\} = \begin{Bmatrix} f_2 \\ \vdots \\ f_{N-1} \end{Bmatrix}, \{q^*\} = \begin{Bmatrix} q_2 \\ \vdots \\ q_{N-1} \end{Bmatrix} \tag{4.154}$$

To solve the system of equations obtained in Equation (4.153), we can write the following:

$$\{f^*\} = [K^*]^{-1}\{q^*\} \tag{4.155}$$

And as a result, the vector of unknowns $\{f\}$ according to the boundary conditions of the problem will be obtained as follows:

$$\{f\} = \begin{Bmatrix} 0 \\ \{f^*\} \\ 0 \end{Bmatrix} \tag{4.156}$$

It can be shown that the exact solution of Equation (4.148) is as follows:

$$f_{exact} = x^2 - x \tag{4.157}$$

In Figure 4.7, the numerical solution of Equation (4.147) is drawn by the method of square differences for different values of N, along with the exact solution of this equation. These figures show well the convergence and high accuracy of the DQM.

4.4 WEIGHTED RESIDUAL METHOD

4.4.1 Introduction

The finite element method is a versatile approach used to solve various equations in the realm of engineering, particularly those that manifest as partial differential equations. Over the past few decades, numerous techniques for resolving the equations that govern various issues have been suggested. There are three methods for converting the physical formulation of the problem: the direct method, the variational method (also known as Rayleigh-Ritz), and the weighted residual method [17]. The challenge of determining the eigenvalue of Rayleigh and Rayleigh-Ritz can be solved using methods that rely on the preservation of the Rayleigh residual

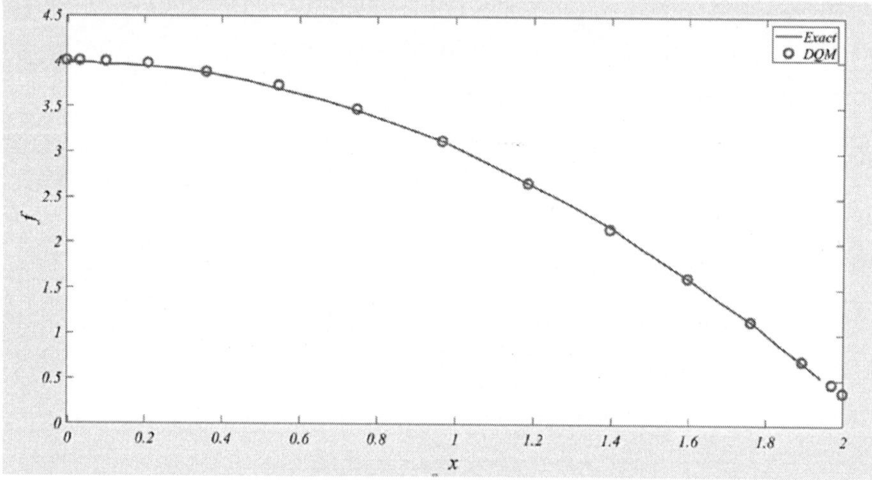

FIGURE 4.7 Comparison of the exact and numerical solution.

method. These methods can be classified as variation methods due to the association of Rayleigh residual with variation methods. Weighted residual approaches, which belong to a distinct category, are employed for solving vibrational problems. Galerkin method, collocation method, subdomain collocation, and the least squares method are all classified as weighted residual methods. Weighted residual methods directly address the governing differential equation and the boundary conditions of a problem [18]. There are two distinct approaches to utilizing the weighted residual method. The first approach involves partitioning the geometry of the issue into numerous small elements and employing the governing differential equation in conjunction with the required shape function for each of these constituents. It is unnecessary to establish boundary requirements for these components. Ultimately, the integration of the components results in the attainment of a fully formed geometry. This technique is employed in the widely utilized finite element method. However, in the second approach, a test function is used to make an educated approximation for the entire object's response. This chapter will examine the fundamentals of the second weighted residual approach, specifically focusing on the utilization of the test function.

4.4.2 Principles

In general, the weighted residual approach is a highly powerful method for obtaining approximation answers to ordinary differential equations or partial differential equations [18].

Let us examine the given partial differential equation:

$$\Omega A(u) - f = 0 \tag{4.158}$$

In domain:

$$(\tilde{x}) = (x_1, x_2, x_3, x_4 ......, x_n) \tag{4.159}$$

Which A is differential operator. The given boundary condition is as follows:

$$G(u) = g \quad @\, \Gamma \tag{4.160}$$

The differential operator G is used to represent the boundary condition.

Many engineering issues stated as ordinary differential equations and partial differential equation can be solved by the following approximation:

$$\begin{gathered} U_N(\tilde{x}) = \sum_{i=1}^{n} C_i\, \phi_i(\tilde{x}) \\ (\tilde{x}) = (x_1, x_2, x_3, \ldots, x_n) \end{gathered} \tag{4.161}$$

Let $i = 1,2,\ldots,n$ and let $\phi_i(x)$ represent test functions that must satisfy the following two conditions:

1. The boundary conditions have been implemented.
2. Match the physics of the problem.

By substituting $\mathrm{U}(\tilde{x})$ in Equations (7.158) and (7.160), we obtain the following:

$$\begin{gathered} A(U) - f \neq 0 \\ G(U) - g \neq 0 \end{gathered} \tag{4.162}$$

The R_s and R_b residues for the given differential equation within the problem's domain and subject to the boundary conditions can be expressed as follows:

$$\begin{gathered} R_s = A(U) - f \\ R_b = G(U) - g \end{gathered} \tag{4.163}$$

If the precise solution of the differential Equation (7.158) is available, the residues R_s and R_b will be equal to zero. Nevertheless, in numerous real-world scenarios, the precise solution is unattainable, resulting in non-zero values for R_s and R_b. To minimize the residues mentioned earlier, we can set the integral to zero by equating the weighted integrals of the residues to zero, with the appropriate determination of the coefficients W_i and W_j. Consequently, it can be expressed as follows:

$$\begin{gathered} \int_{\Omega} W_i R_s \, d\Omega = 0 \\ \int_{\Gamma} W_j R_b \, d\Omega = 0 \end{gathered} \tag{4.164}$$

Let W_i and W_j, where $(i, j = 1,2,\ldots,n)$ represent a collection of weight functions for R_s and R_b residues, respectively. If the selected form's functions meet the boundary criteria, the value of R_b becomes zero and the equation earlier is transformed accordingly:

$$\int_{\Omega} W_i R_s d\Omega = 0 \tag{4.165}$$

When expressing the remainder of R_s in the general case using R, the outcome is as follows:

$$\int_{\Omega} W_i R d\Omega = 0 \tag{4.166}$$

The expression earlier reflects the weight residual approach, which is commonly employed in numerical formulations like the finite element method. By substituting the Equation (7.163) into the Equation (7.165), we will obtain the following:

$$\int_{\Omega} W_i \left[A(U) - f\right] d\Omega = 0 \tag{4.167}$$

The equation $U_N(\tilde{x}) = \sum_{i=1}^{n} C_i \phi_i(x)$ represents a set of equations for $i = 1, 2, ..., n$ that can be expressed as follows:

$$\begin{cases} \int_{\Omega} W_1 \left[A\left(\sum_{i=1}^{n} C_i \phi_i(x) \right) - f \right] d\Omega = 0 \\ \int_{\Omega} W_2 \left[A\left(\sum_{i=1}^{n} C_i \phi_i(x) \right) - f \right] d\Omega = 0 \\ \quad \cdot \\ \quad \cdot \\ \quad \cdot \\ \int_{\Omega} W_n \left[A\left(\sum_{i=1}^{n} C_i \phi_i(x) \right) - f \right] d\Omega = 0 \end{cases} \tag{4.168}$$

The set of equations mentioned earlier represents the n equation, which is used to find the unknown n coefficients c_i.

The selection of a weight function in residual weight methods has a substantial influence on the method's performance. The primary distinction among various weight techniques lies in the selection of weight functions.

The remaining weight methods that were previously stated are as follows:

1. The Ritz method
2. Point collocation method
3. Sub-point collocation method
4. Least squares method
5. Galerkin method

4.4.2.1 Ritz Method

In this method, we have $w_i = 1$, so:

$$\int_{\Omega} W_i R d\Omega = \int_{\Omega} R d\Omega = 0 \tag{4.169}$$

4.4.2.2 Point Collocation Method

In this method, we have $W_i = \delta(x - x_i)$ or $W = \delta(x - x_1, x - x_2, x - x_3, \dots, x - x_n)$, so:

$$\int_{\Omega} W_i R d\Omega = \int_{\Omega} \delta(x - x_i) R d\Omega = 0 \tag{4.170}$$

4.4.2.3 Sub-point Collocation Method

In this way, we have the following:

$$W_i(x) = \begin{cases} 1 & \text{for } x \text{ in } \Omega_i \\ 0 & \text{for } x \text{ is not in } \Omega_i \end{cases} \tag{4.171}$$

4.4.2.4 Least Squares Method

The initial step of this approach involves the definition of a function in the following manner:

$$J(c_i) = \int_{\Omega} R.R d\Omega \tag{4.172}$$

We're going to have a derivative of J:

$$\frac{\partial J}{\partial c_i} = \int_{\Omega} \frac{\partial (R.R)}{\partial c_i} d\Omega \tag{4.173}$$

In accordance with the principle of the weighted residual method, the weight function in the least squares method is chosen as follows:

$$W_i = \frac{\partial \text{R}}{\partial c_i} \tag{4.174}$$

We'll have a result:

$$\int_{\Omega} W_i R d\Omega = \int_{\Omega} \frac{\partial \text{R}}{\partial c_i} R d\Omega = 0 \tag{4.175}$$

4.4.2.5 Galerkin Method

In this procedure, we will have $W_i = \phi_i(x)$ as a result:

$$\int_{\Omega} W_i R d\Omega = \int_{\Omega} \phi_i R d\Omega = 0 \tag{4.176}$$

Example 4–1: To effectively disperse the temperature in a one-dimensional blade, find the differential equation of heat transmission. The blade has a length of $L = 1$.

$$\frac{d^2T}{dx^2} + 1000x^2 = 0, \quad T(0) = 0, \quad T(L) = 0$$

The precise solution is given by the equation $T(x) = \frac{250}{3}x(1-x^3)$.

Solution: Initially, we hypothesize the test function.

$$\frac{d^2T}{dx^2} + 1000x^2 = 0, \quad T(0) = 0, \quad T(L) = 0$$

When $i = 1$, we successfully solve the issue and obtain a result.

$$T_1(x) = c_1 x(1-x^2) \Rightarrow R = \mathrm{R}_1 = \frac{d^2T_1}{dx^2} + 1000x^2 = c_1(-6\,x) + 1000\,x^2$$

Ritz method:

$$\int_\Omega W_1 R d\Omega = \int_0^l 1 \times (-6c_1 x + 1000\,x^2)dx = 0 \Rightarrow c_1 = \frac{1000}{9}$$

$$T_1(x) = \frac{1000}{9}x(1-x^2)$$

Point collocation method:

The approximate solution can be expressed by choosing the point $x = \frac{L}{2} = 0.5$:

$$\int_\Omega W_1 R d\Omega = \int_0^1 \delta\left(x - \frac{1}{2}\right)\left(-6c_1 x + 1000\,x^2\right)dx = 0 \Rightarrow c_1 = \frac{250}{3}$$

$$T_1(x) = \frac{250}{3}x(1-x^2)$$

Least squares method:

$$W_i = \frac{\partial \mathrm{R}}{\partial c_i} = -6x \Rightarrow \int_0^1 (-6x)\left(-6c_1 x + 1000x^2\right)dx = 0 \Rightarrow c_1 = 125$$

$$T_1(x) = 125\,x(1-x^2) = \frac{250}{2}x(1-x^2)$$

Galerkin method:

$$\begin{cases} T_1(x) = c_1 \Phi_1(x) \\ \Phi_1(x) = x(1-x^2) \Rightarrow \int_0^1 x(1-x^2)\left(-6c_1 x + 1000\,x^2\right)dx = 0 \Rightarrow c_1 = \dfrac{625}{6} \\ W_1 = \phi_1(x) \end{cases}$$

$$T_1(x) = \frac{625}{6}x(1-x^2)$$

Figure 4.8 displays the comparison of the obtained responses.

Example 4–2: Let's examine a basic component of an element structural system. As depicted in Figure 4.9, this element is a solid with high formability, characterized by one dimension that is significantly greater than the other two dimensions. The force is applied just in the x-direction, which aligns with the member's direction. The axial displacement, denoted as u, is determined by solving the equilibrium equation that governs this system.

$$EA\frac{d^2u}{dx^2} + b(x) = 0$$

The letter E represents the young module, A is the cross-sectional area, and $b(x)$ represents the external axial load applied along the length of the element. Given the boundary conditions $u(x=0)=0$ and $u(x=L)=0$, where L represents the length of the element and assuming $L=1$, $b(x)=12x^2$, $A=1$, $E=1$. Determine the axial displacement of this structural component.

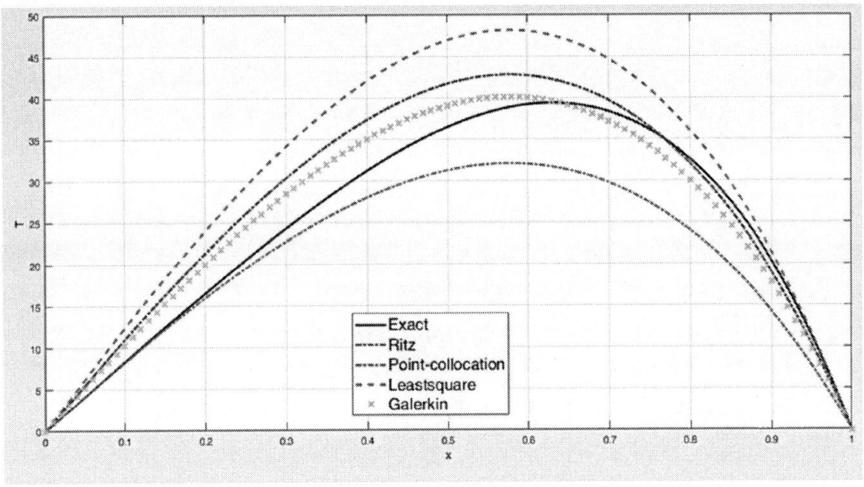

FIGURE 4.8 Temperature distribution response along the blade for precise method compared to different methods.

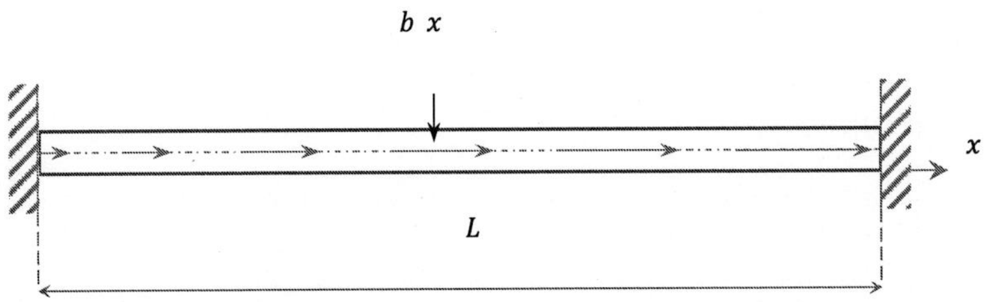

FIGURE 4.9 Simple structure under a distributed longitudinal load.

Response: The precise solution is $u_{exact} = -x^4 + x$, given that the estimated solution is as follows:

$$u_i(x) = c_i x^{i-1} x(x-L)$$

The unknown coefficients are represented by c_i, and the test functions are denoted by $\varnothing_i(x) = x^{i-1}x(x-L)$. By substituting the relation $u(x)$ into the differential equation of the truss member, the resulting function is as follows:

$$R(x) = \frac{d^2u(x)}{dx^2} + b(x)$$

When $i = 1$, the estimated solution for $u(x)$ can be expressed as follows:

$$u_1(x) = c_1 x(x-L)$$

The residue corresponding to it is denoted as $R(x)$ when it is placed in the given relation:

$$R = R_1(x) = \frac{d^2u_1(x)}{dx^2} + b(x) \Rightarrow R = R_1(x) = 2c_1 + 12x^2$$

When $i = 2$, the approximate answer can be expressed as follows:

$$u_2(x) = c_1 x(x-L) + c_2 x^2(x-L)$$

The residue corresponding to it is denoted as $R(x)$ through its placement in the relation:

$$R = R_2(x) = \frac{d^2u_2(x)}{dx^2} + b(x) \Rightarrow R = R_2(x) = 2c_1 + c_2(6x-2) + 12x^2$$

The Ritz method:

Given the assumption that $i = 1$, the following is true:

$$\int_\Omega W_1 R d\Omega = \int_\Omega R d\Omega = 0 \Rightarrow \int_0^1 \left(2c_1 + 12x^2\right) dx = 2c_1 + 4 = 0$$
$$\Rightarrow c_1 = -2$$

And

$$u_1(x) = -2x(x-1)$$

The Ritz approach is employed just for $i = 1$, whereas for $i = 2$ and higher, the resulting equations will be the same.

Point collocation method:

Given the assumption that $i = 1$ and the selection of the location $x = \frac{L}{2} = 0.5$, the approximate solution can be expressed as follows:

$$\int_\Omega W_1 R d\Omega = \int_0^1 \delta(x-0.5) R d\Omega = 0 \Rightarrow \int_0^1 \delta(x-0.5)\left(2c_1 + 12x^2\right) dx = 0 \Rightarrow c_1 = -\frac{3}{2}$$

Where:

$$u_1(x) = -\frac{3}{2}x(x-1)$$

Given that $i = 2$ and choosing two points on the truss member, we may determine that $x = \frac{L}{3} = \frac{1}{3}$ and $x = \frac{2L}{3} = \frac{2}{3}$.

$$\int_{\Omega} W_i R d\Omega = 0 \Rightarrow \int_0^1 \delta\left(x - \frac{1}{3}\right)\left(2c_1 + c_2(6x-2) + 12x^2\right)dx = 0$$

Also:

$$\int_0^1 \delta\left(x - \frac{2}{3}\right)\left(2c_1 + c_2(6x-2) + 12x^2\right)dx = 0$$

The values of c_1 and c_2 are determined by solving the integrals mentioned earlier. Specifically, $c_1 = -\frac{2}{3}$ and $c_2 = -2$. Therefore, the following can be concluded:

$$u_2(x) = -\frac{2}{3}x(x-1) - 2x^2(x-1)$$

Sub-point collocation method:
We assume that $i = 1$.

$$\int_{\Omega} W_1 R d\Omega = \int_{\Omega} R d\Omega = 0 \Rightarrow \int_0^1 \left(2c_1 + 12x^2\right)dx = 2c_1 + 4 = 0 \Rightarrow c_1 = -2$$

So:

$$u_1(x) = -2x(x-1)$$

Given that $i = 2$ and splitting the domain into two subdomains as described, we obtain the following:

$$\begin{cases} \Omega_1 : \mathrm{W}_1 = 1, & 0 \le x \le 0.5 \\ \Omega_2 : \mathrm{W}_2 = 1, & 0.5 \le x \le 1 \end{cases}$$

$$\int_{\Omega} W_i R d\Omega = 0 \Rightarrow \int_0^{0.5} \left(2c_1 + c_2(6x-2) + 12x^2\right)dx = 0$$

$$\int_{0.5}^1 \left(2c_1 + c_2(6x-2) + 12x^2\right)dx = 0$$

By evaluating the integrals mentioned before, we get that $c_1 = -1$ and $c_2 = -2$. Therefore:

$$u_2(x) = -x(x-1) - 2x^2(x-1)$$

Least squares method:

Assuming that the $i = 1$:

$$W_1 = \frac{\partial R}{\partial c_1} = 2$$

$$\int_\Omega W_1 R d\Omega = \int_\Omega \frac{\partial R}{\partial c_1} R d\Omega = \int_0^1 2\left(2c_1 + 12x^2\right)dx = 0 \Rightarrow c_1 = -2$$

$$u_1(x) = -2x(x-1)$$

Given the assumption that $i = 2$, we can conclude:

$$\begin{cases} W_1 = \dfrac{\partial R}{\partial c_1} = 2 \\ W_2 = \dfrac{\partial R}{\partial c_2} = 6x - 2 \end{cases}$$

$$\int_\Omega W_i R d\Omega = 0 \Rightarrow \int_\Omega \frac{\partial R}{\partial c_1} R d\Omega = 0, \quad \int_\Omega \frac{\partial R}{\partial c_2} R d\Omega = 0$$

$$\Rightarrow \int_0^1 2\left(2c_1 + c_2(6x-2) + 12x^2\right)dx = 0$$

$$\int_0^1 (6\mathrm{x} - 2)\left(2c_1 + c_2(6x-2) + 12x^2\right)dx = 0$$

By evaluating the integrals mentioned before, we get that $c_1 = -1$ and $c_2 = -2$. Therefore:

$$u_2(x) = -x(x-1) - 2x^2(x-1)$$

Galerkin method:

Considering the function $\mathrm{u}_i(x) = c_i x^{i-1} x(x-L)$ and the function $U_N(\tilde{x}) = c_i \phi_i(x)$, where $\varnothing_i = x^{i-1}x(x-L)$, we may conclude that for $i = 1$:

$$\varnothing_1 = x(x-L)$$

$$\int_\Omega W_1 R d\Omega = \int_\Omega \phi_1 R d\Omega = 0 \Rightarrow \int_0^1 x(x-L)\left(2c_1 + 12x^2\right)d\Omega = 0$$

The value of $c_1 = -1.8$ is determined by calculating the integral mentioned earlier.

$$u_1(x) = -1.8x(x-1)$$

For $i = 2$

$$\varnothing_2 = x^2\left(x - L\right)$$

$$\int_{\Omega} W_i R d\Omega = \int_{\Omega} \phi_i R d\Omega = 0$$

$$\Rightarrow \int_0^1 x\left(x-L\right)\left(2c_1 + c_2\left(6x-2\right) + 12x^2\right) dx = 0$$

$$\int_0^1 x^2\left(x-L\right)\left(2c_1 + c_2\left(6x-2\right) + 12x^2\right) = 0$$

By evaluating the integrals mentioned earlier, we see that $c_1 = -0.8$ and $c_2 = -2$. Therefore: $u_2(x) = -0.8x(x-1) - 2x^2(x-1)$

The answers to the problem are compared in mode $i = 1$ in Figure 4.11 and in mode $i = 2$ in Figure 4.10.

Based on the diagrams, the Galerkin response is more accurate compared to the precise answer due to its reliance on the minimum energy approach. The Galerkin response is consistent with the Rayleigh-Ritz variation method due to its shared reliance on energy minimization.

Example 4–3: Take into account the beam that is simply supported, with the torque M_0 applied at both ends.

The differential equation that governs the behavior of this beam, along with its corresponding boundary conditions, can be described as follows:

$$EI\frac{d^2y}{dx^2} - M_0 = 0 \qquad x \in [0, L]$$

$$y(0) = 0, \qquad y(L) = 0$$

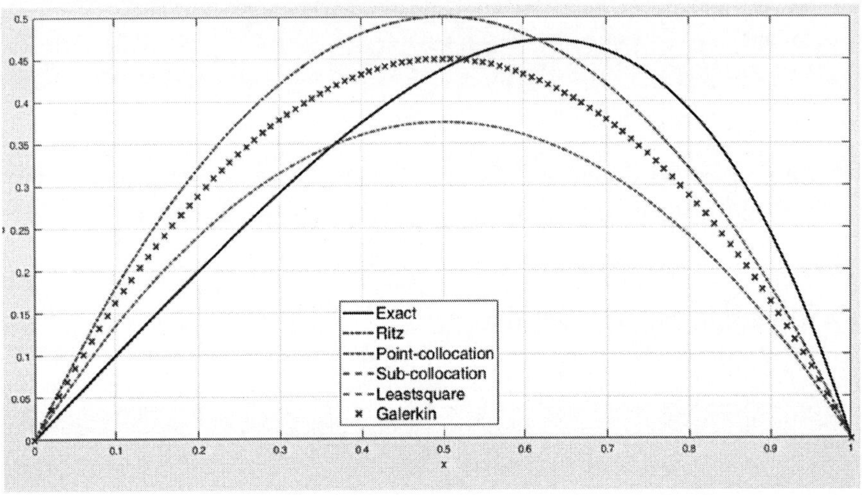

FIGURE 4.10 Diagram of the bar longitudinal displacement, comparing the results of numerical methods with exact response in state $i = 1$.

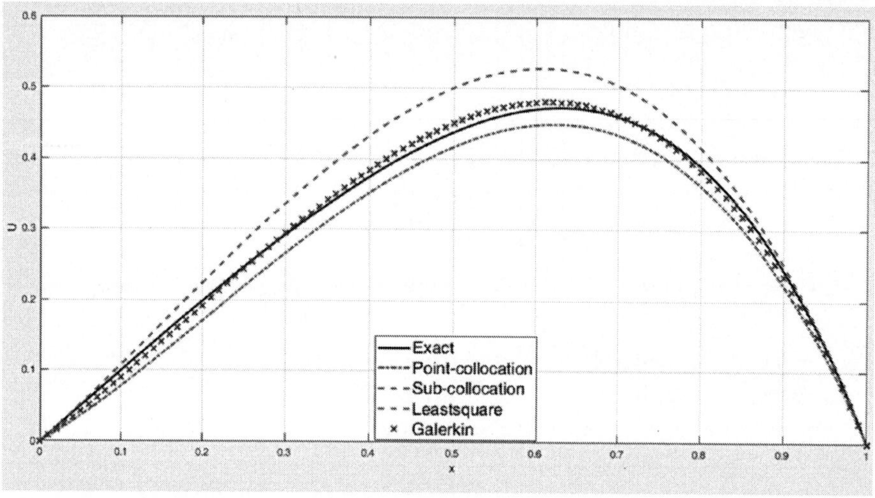

FIGURE 4.11 Diagram of the bar longitudinal displacement, comparing the results of numerical methods with exact response in state $i = 2$.

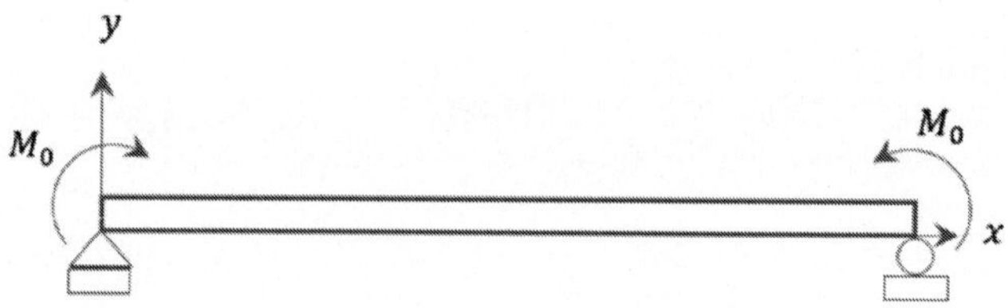

FIGURE 4.12 Beam with simple supports exposed to bending torques.

The precise solution to the differential equation governing the motion of the beam is as follows:

$$y(x) = -\frac{M_0}{2EI}x(L-x)$$

Using the Galerkin method to solve the differential equation of the beam is preferable.

Solution: We hypothesize the test function in the following manner:

$$u(x) = Asin(Bx)$$

Where A and B are fixed coefficients.

The boundary conditions are incorporated into the test function, as seen in Figure 4.12.

$$u(x) = Asin\left(\frac{\pi x}{L}\right) = C_1\varnothing_1$$

$$\varnothing_1 = sin\left(\frac{\pi x}{L}\right)$$

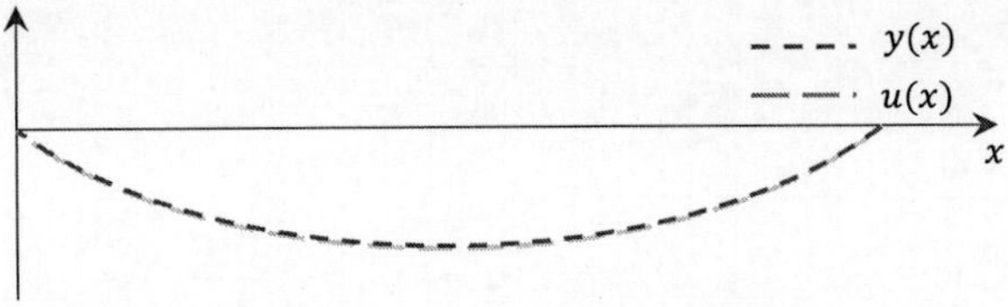

FIGURE 4.13 Comparison of the beam deflection obtained by Galerkin method with exact response.

$$R = EI\frac{d^2u}{dx^2} - M_0 = -EI\frac{A\pi^2}{L^2}\sin\left(\frac{\pi x}{L}\right) - M_0$$

$$\int_\Omega \phi_1 R d\Omega = 0 \Rightarrow \int_0^L sin\left(\frac{\pi x}{L}\right)\left(-EI\frac{A\pi^2}{L^2}\sin\left(\frac{\pi x}{L}\right) - M_0\right)dx = 0$$

$$\left(\frac{EI\pi^2}{2L}\right)A + \frac{2M_0L}{\pi} = 0 \Rightarrow A = -\frac{4M_0L^2}{\pi^3 EI}$$

$$u(x) = -\frac{4M_0L^2}{\pi^3 EI}\sin\left(\frac{\pi x}{L}\right)$$

Figure 4.13 shows the beam's bending obtained from the Galerkin method and the exact response.

It is important to highlight that the Galerkin approach is not exclusively restricted to the components method. The Galerkin approach predates the development of the finite element method. The primary distinction between the Galerkin technique and the finite element method lies in the fact that, unlike the Galerkin method, the finite element method does not specify the approximation function across the full physical domain. The approximation function is specifically designed for individual elements inside the physical domain. The Galerkin method is commonly employed in the finite element method standard to derive element equations [18, 19].

The admissible functions corresponding to the various boundary conditions of the plates are displayed. It is worth mentioning that as the beam can be viewed as a one-dimensional object, the authorized functions described in Ref. [20] can also be utilized for the beams.

The frequency ω is determined by the values of K and N obtained from Ref. [20] for $\nu = 0.25$, according to the following equation:

$$\omega^2 = \frac{\pi^4 D}{a^4 \rho}\frac{K}{N} \tag{4.177}$$

By employing the weighted residual method described earlier, problems involving partial derivatives, such as those related to location and time variables, can be transformed into ordinary differential equations that just depend on the time variable. These equations can then be solved using semi-analytical methods. The particular

TABLE 4.1
Admissible Functions

Boundary Condition	Mode Shape	N	K
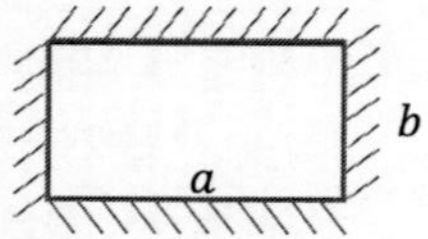	$\left(\cos\frac{2\pi x}{a}-1\right)\left(\cos\frac{2\pi y}{b}-1\right)$	2.25	$12+8\left(\frac{a}{b}\right)^2+12\left(\frac{a}{b}\right)^4$
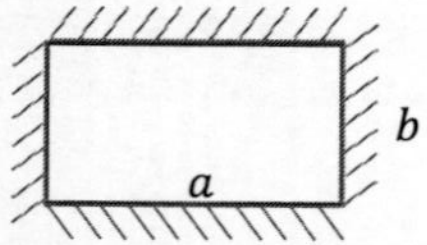	$\left(\cos\frac{3\pi x}{2a}-\cos\frac{\pi x}{2a}\right)\left(\cos\frac{2\pi y}{b}-1\right)$	1.50	$3.85+5\left(\frac{a}{b}\right)^2+8\left(\frac{a}{b}\right)^4$
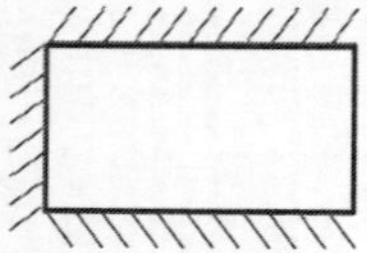	$\left(1-\cos\frac{\pi x}{2a}\right)\left(\cos\frac{2\pi y}{b}-1\right)$	0.340	$0.0468+0.340\left(\frac{a}{b}\right)^2+1.814\left(\frac{a}{b}\right)^4$
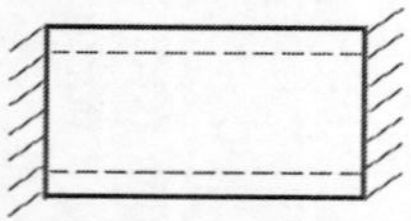	$\left(\cos\frac{2\pi x}{a}-1\right)\sin\frac{y}{b}$	0.75	$4+2\left(\frac{a}{b}\right)^2+0.75\left(\frac{a}{b}\right)^4$
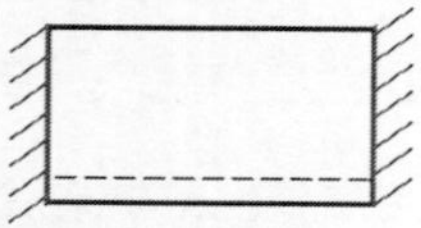	$\left(\cos\frac{2\pi x}{a}-1\right)\frac{y}{b}$	0.50	$2.67+0.304\left(\frac{a}{b}\right)^2$
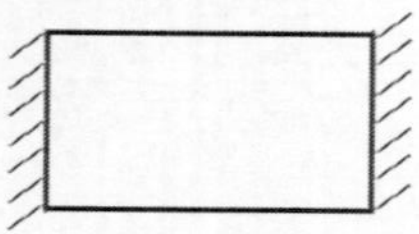	$\cos\frac{2\pi x}{a}-1$	1.50	8

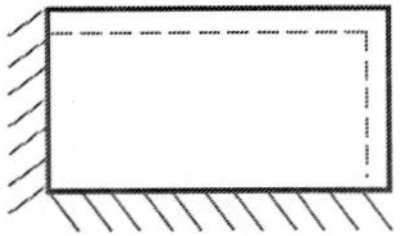	$\left(\cos\frac{3\pi x}{2a}-\cos\frac{\pi x}{2a}\right)\left(\cos\frac{3\pi y}{2b}-\cos\frac{\pi y}{2b}\right)$	1.00	$2.56+3.12\left(\frac{a}{b}\right)^2+2.56\left(\frac{a}{b}\right)^4$
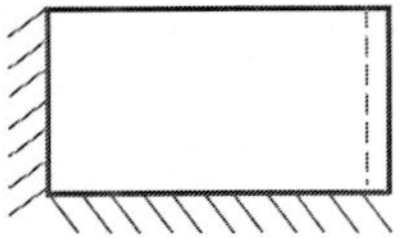	$\left(\cos\frac{3\pi x}{2a}-\cos\frac{\pi x}{2a}\right)\left(1-\cos\frac{\pi y}{2b}\right)$	0.227	$0.581+0.213\left(\frac{a}{b}\right)^2+0.031\left(\frac{a}{b}\right)^4$
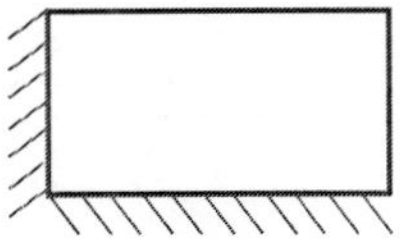	$\left(1-\cos\frac{\pi x}{2a}\right)\left(1-\cos\frac{\pi y}{2b}\right)$	0.0514	$0.0071+0.024\left(\frac{a}{b}\right)^2+0.0071\left(\frac{a}{b}\right)^4$
	$\left(\cos\frac{3\pi x}{2a}-\cos\frac{\pi x}{2a}\right)\sin\frac{\pi y}{b}$	0.50	$1.28+1.25\left(\frac{a}{b}\right)^2+0.50\left(\frac{a}{b}\right)^4$
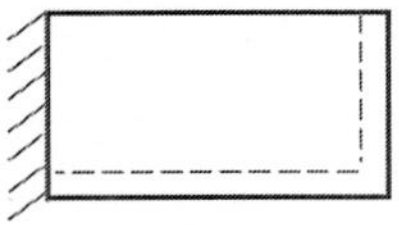	$\left(\cos\frac{3\pi x}{2a}-\cos\frac{\pi x}{2a}\right)\frac{y}{b}$	0.333	$0.853+0.190\left(\frac{a}{b}\right)^2$
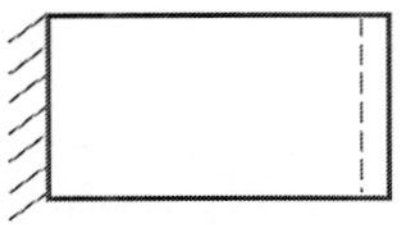	$\cos\frac{3\pi x}{2a}-\cos\frac{\pi x}{2a}$	1.00	2.56

(Continued)

TABLE 4.1 (*Continued*)
Admissible Functions

Boundary Condition	Mode Shape	N	K
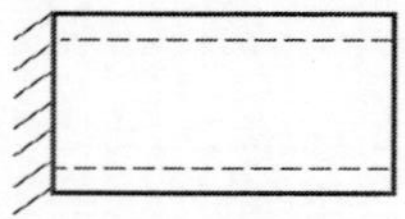	$\left(1-\cos\frac{\pi x}{2a}\right)\frac{\pi^2}{b^2}\sin\frac{\pi y}{b}$	0.1134	$0.0156+0.0852\left(\frac{a}{b}\right)^2+0.1134\left(\frac{a}{b}\right)^4$
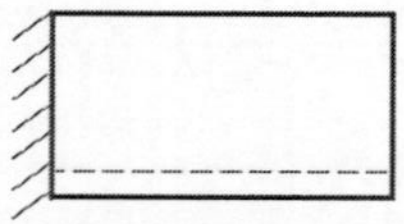	$\left(1-\cos\frac{\pi x}{2a}\right)\frac{y}{b}$	0.0756	$0.0104+0.0190\left(\frac{a}{b}\right)^2$
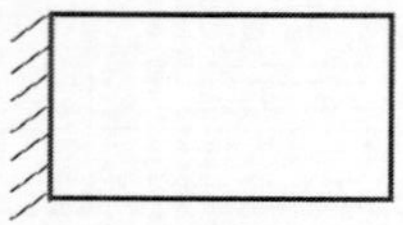	$1-\cos\frac{\pi x}{2a}$	0.2268	0.0313
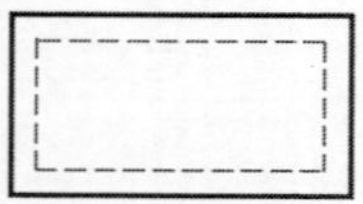	$\sin\frac{\pi x}{a}\sin\frac{\pi y}{b}$	0.25	$0.25+0.50\left(\frac{a}{b}\right)^2+0.25\left(\frac{a}{b}\right)^4$
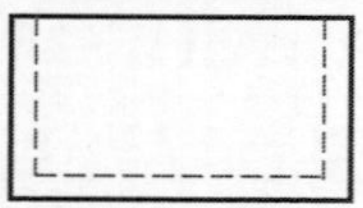	$\left(\sin\frac{\pi x}{a}\right)\frac{y}{b}$	0.1667	$0.1667+0.0760\left(\frac{a}{b}\right)^2$
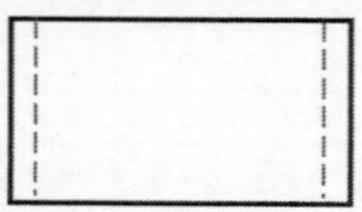	$\sin\frac{\pi x}{a}$	0.50	0.50

solution is determined based on the type of boundary condition and simplified utilizing the deliberate circumstances of the problem. This results in the partial differential equation being transformed into a linear or nonlinear ordinary differential equation with time variable [21]. The section on perturbation theory provided an explanation of the technique for solving ordinary linear and nonlinear differential equations [22].

REFERENCES

[1] Ghadiri, M. and M. Kazemi, Nonlinear vibration analysis of a cable carrying moving mass-spring-damper. *International Journal of Structural Stability and Dynamics*, 2018. **18**(02): p. 1850030.

[2] Ghadiri, M. and M. Safi, Nonlinear vibration analysis of functionally graded nanobeam using homotopy perturbation method. *Advances in Applied Mathematics and Mechanics*, 2017. **9**(1): p. 144–156.

[3] Ghadiri, M., A. Rajabpour, and A. Akbarshahi, Non-linear vibration and resonance analysis of graphene sheet subjected to moving load on a visco-Pasternak foundation under thermo-magnetic-mechanical loads: An analytical and simulation study. *Measurement*, 2018. **124**: p. 103–119.

[4] Noroozi, M. and M. Ghadiri, Nonlinear vibration and stability analysis of a size-dependent viscoelastic cantilever nanobeam with axial excitation. *Proceedings of the Institution of Mechanical Engineers, Part C: Journal of Mechanical Engineering Science*, 2021. **235**(18): p. 3624–3640.

[5] Ghadiri, M. and N. Shafiei, Nonlinear bending vibration of a rotating nanobeam based on nonlocal Eringen's theory using differential quadrature method. *Microsystem Technologies*, 2016. **22**: p. 2853–2867.

[6] Shafiei, N., M. Kazemi, M. Safi, and M. Ghadiri, Nonlinear vibration of axially functionally graded non-uniform nanobeams. *International Journal of Engineering Science*, 2016. **106**: p. 77–94.

[7] Nayfeh, A.H., *Perturbation methods*. 2008: John Wiley & Sons.

[8] Nayfeh, A.H. and D.T. Mook, *Nonlinear oscillations*. 2008: John Wiley & Sons.

[9] Hannsgen, K.B., Nonlinear ordinary differential equations (DW Jordan and P. Smith). *SIAM Review*, 1979. **21**(2): p. 264.

[10] Nayfeh, A.H. and P.F. Pai, *Linear and nonlinear structural mechanics*. 2008: John Wiley & Sons.

[11] Rao, S.S., *Mechanical vibrations*, Sixth Edition. 2018: Pearson Education, United Kingdom.

[12] Strogatz, S.H., *Nonlinear dynamics and chaos with student solutions manual: With applications to physics, biology, chemistry, and engineering*. 2018: CRC Press.

[13] Shu, C. and B.E. Richards, Application of generalized differential quadrature to solve two-dimensional incompressible Navier-Stokes equations. *International Journal for Numerical Methods in Fluids*, 1992. **15**(7): p. 791–798.

[14] Civalek, Ö., Application of differential quadrature (DQ) and harmonic differential quadrature (HDQ) for buckling analysis of thin isotropic plates and elastic columns. *Engineering Structures*, 2004. **26**(2): p. 171–186.

[15] Ghadiri, M., N. Shafiei, and H. Alavi, Thermo-mechanical vibration of orthotropic cantilever and propped cantilever nanoplate using generalized differential quadrature method. *Mechanics of Advanced Materials and Structures*, 2017. **24**(8): p. 636–646.

[16] Shafiei, N., M. Ghadiri, H. Makvandi, and S.A. Hosseini, Vibration analysis of Nano-Rotor's Blade applying Eringen nonlocal elasticity and generalized differential quadrature method. *Applied Mathematical Modelling*, 2017. **43**: p. 191–206.

[17] Rao, S.S., *Vibration of continuous systems*. 2019: John Wiley & Sons.

[18] Stasa, F.L., *Applied finite element analysis for engineers*. 1985: CBS Publishing, New York, NY.

[19] Ghadiri, M., A. Rajabpour, and A. Akbarshahi, Non-linear vibration and resonance analysis of graphene sheet subjected to moving load on a visco-Pasternak foundation under thermo-magnetic-mechanical loads: An analytical and simulation study. *Measurement*, 2018. **124**: p. 103–119.

[20] Leissa, A.W., *Vibration of plates*, Vol. 160. 1969: Scientific and Technical Information Division, National Aeronautics and Space Administration.

[21] Ghadiri, M. and S.H.S. Hosseini, Nonlinear dual frequency excited vibration of viscoelastic graphene sheets exposed to thermo-magnetic field. *Communications in Nonlinear Science and Numerical Simulation*, 2020. **83**: p. 105111.

[22] Ghadiri, M. and S.H.S. Hosseini, Nonlinear forced vibration of graphene/piezoelectric sandwich nanoplates subjected to a mechanical shock. *Journal of Sandwich Structures & Materials*, 2021. **23**(3): p. 956–987.

5 Forced Vibrations of Nonlinear Systems

5.1 INTRODUCTION

Earlier sections explored the nonlinear free vibrations of a system with one degree of freedom. The study of the free vibration of the Duffing differential equation, which is a third-order nonlinear equation, focused on determining the natural frequency of the nonlinear system. Several techniques rooted in the idea of perturbation were investigated in order to address the differential equation. An analysis was conducted on some phenomena, such as the limit cycle, that are characteristic of the nonlinear system. Subsequently, we will analyze the nonlinear forced oscillation of a system with one degree of freedom. This examination will reveal the occurrence of certain phenomena, such as jumping, under forced conditions, which will be thoroughly explored.

5.2 FORCED HARMONIC VIBRATIONS IN NONLINEAR SYSTEMS

The initial focus will be on analyzing the stable response of a one-degree freedom system under nonlinear forced vibration. This analysis will be conducted in a straightforward manner. Additionally, the phenomena of jumping will be explored in both damped and undamped settings.

The equation governing a single-degree-of-freedom system with a nonlinear spring and a damper with a constant c, subjected to a harmonic excitation with amplitude F_0, can be expressed as follows [1]:

$$m\ddot{x} + c\dot{x} + k\left(x + \alpha x^3\right) = F_0 \, cos \, \omega t \tag{5.1}$$

The expression $\left(x + \alpha x^3\right)$ represents the force exerted by a nonlinear spring. If α is positive, the spring will exhibit hardening behavior, while if α is negative, the spring will exhibit softening behavior. In the context of Equation (5.1), it is noteworthy that when the external force is represented as $\left(\varepsilon F_0 \, cos \, \omega t\right)$, the term εF_0 is referred to as soft excitation. Conversely, if simply F_0 is utilized, it is referred to as hard excitation.

By employing the frequency response approach and considering the primary equation, the stable response may be expressed as follows:

$$x(t) = A \, cos\left(\omega t - \theta\right) \tag{5.2}$$

DOI: 10.1201/9781003470694-5

To simplify the analysis, it is advisable to incorporate the phase angle θ into the input force statement. Equation (5.1) is reformulated in this instance:

$$m\ddot{x}+c\dot{x}+k\left(x+\alpha x^{3}\right)=F_{0}\cos\left(\omega t+\theta\right) \tag{5.3}$$

Evidently, in this instance, the equation undergoes alteration solely on the right side. By performing the operation of division on both sides of the Equation (5.3) with respect to the variable m, we will obtain the following:

$$\ddot{x}+2\xi\omega_{0}\dot{x}+\omega_{0}^{2}x+hx^{3}=F_{c}\cos\omega t+F_{s}\sin\omega t \tag{5.4}$$

Regarding the Equation (5.4), the forces F_c and F_s exerted on the unit of mass can be described as follows:

$$F_{c}=\frac{F_{0}}{m}\cos\theta,\ F_{s}=-\frac{F_{0}}{m}\sin\theta \tag{5.5}$$

Also, $h=\dfrac{k\alpha}{m}$, $\omega_0^2=\dfrac{k}{m}$, and $2\xi\omega_0=\dfrac{c}{m}$. The expression hx^3 in Equation (5.4) is the nonlinear expression.

Next, the issue can be analyzed in two distinct components:

1. An analysis of the resonant oscillations of a nonlinear system in the absence of damping.
2. An analysis of the resonant oscillations of a nonlinear system, considering the influence of the damping effect.

5.2.1 An Analysis of the Resonant Oscillations of a Nonlinear System in the Absence of Damping

Presently, we aim to examine the Equation (5.4) in a system without any dampening component. Thus, the differential equation for the system without a damper will be the following [1]:

$$\ddot{x}+\omega_{0}^{2}x+hx^{3}=F_{c}\cos\omega t \tag{5.6}$$

By hypothesizing the solution as $x(t)=A\cos(\omega t)$, we obtain the following:

$$\left(-A\omega^{2}+\omega_{0}^{2}A\right)\cos\omega t+hA^{3}\cos^{3}\omega t=F_{c}\cos\omega t \tag{5.7}$$

It is important to observe that in the absence of a damper, the stable state reactions, whether in phase $(\theta=0)$ or in opposition $\theta=180^{\circ}$ to the input force, exhibit a phase difference. Furthermore, during the phase difference stage, the amplitude A

undergoes a sign change. Based on the given explanation, only the term $F_c \cos \omega t$ is chosen for study in Equation (5.7) from the Equation (5.6).

Given the trigonometric equation $\left(\cos^3 \omega t = \frac{3}{4} \cos \omega t - \frac{1}{4} \cos 3\omega t\right)$, it is important to note that the term $\cos 3\omega t$ represents the third harmonic. However, since we are specifically focusing on the first harmonic, this term is disregarded. The equation is so reformulated as follows:

$$-A\omega^2 + \omega_0^2 A + \frac{3}{4} hA^3 = F_c \tag{5.8}$$

The third harmonic relationship and its accompanying reactions will be elucidated in the subsequent sections.

It is important to reiterate that if $h > 0$ in the context of Equation (5.8), we are dealing with a spring that exhibits hardening behavior. Conversely, if $h < 0$, the spring exhibits softening behavior.

Figure 5.1 depicts the frequency response of a system without a nonlinear damper. As the input frequency increases and approaches resonance, the amplitude of the system increases, leading to an addition of the spring constant. The continuous inclusion of the spring results in an increase in the natural frequency of the system and causes the natural frequency to shift towards higher values. Put simply, in the linear system, the line $F = 0$ that represents the normal frequency maintains a consistent value (Figure 5.1 a). Based on the information provided, in a nonlinear system with a stiffened spring, the spring constant and natural frequency both rise as the range expands. In the context of the spring season, if this represents the photograph, then as the distance expands, the typical frequency diminishes.

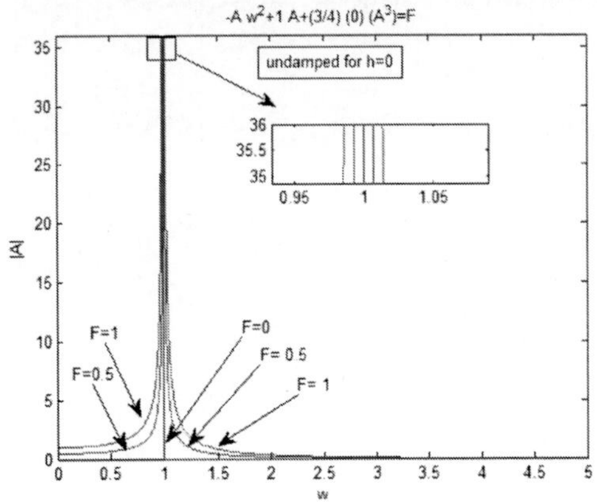

a)

FIGURE 5.1 Nonlinear frequency response curves in states: *a)* linear $(h = 0)$, *b)* hardened spring $(h > 0)$, and *c)* softened spring $(h < 0)$.

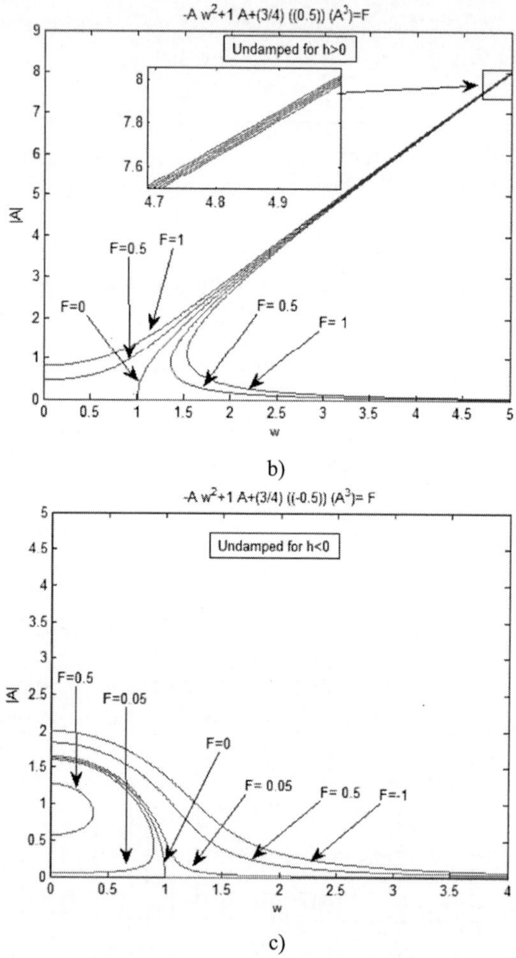

FIGURE 5.1 (Contiuned)

5.2.2 Analysis of the Resonant Oscillations of a Nonlinear System, Including the Influence of the Damping Effect

Considering the Equation (5.4) in conjunction with a damping mechanism, one may formulate the equation for the system [1]:

$$\ddot{x} + 2\beta\dot{x} + \omega_0^2 x + hx^3 = F_c \cos \omega t + F_s \sin \omega t \tag{5.9}$$

The equation $\beta = \xi\omega_0$ is true in Equation (5.9). By substituting the solution $x(t) = A\cos\omega t$ into the differential equation, we obtain the following:

$$-A\omega^2 \cos \omega t - 2\beta A\omega \sin \omega t + \omega_0^2 A \cos \omega t + hA^3 \cos^3 \omega t = F_c \cos \omega t + F_s \sin \omega t \tag{5.10}$$

By utilizing trigonometric relations, we may express the equation as $\left(\cos^3 \omega t = \frac{3}{4}\cos\omega t - \frac{1}{4}\cos 3\omega t\right)$. Removing the term $\cos 3\omega t$ eliminates the third harmonic. Additionally, by introducing coefficients such as $\sin\omega t$ and $\cos\omega t$, we can modify the equation as desired.

$$\begin{cases} cos\omega t: -A\omega^2 + \omega_0^2 A + \frac{3}{4}hA^3 = F_c \\ sin\omega t: \qquad -2\beta A\omega = F_s \end{cases} \rightarrow F_s^2 + F_c^2 = F^2 \tag{5.11}$$

However, by considering the Equation (5.9), it is noticed that $F_s^2 + F_c^2 = F^2$. Therefore:

$$\left[\left(\omega_0^2 - \omega^2\right)A + \frac{3}{4}hA^3\right]^2 + \left(2\beta A\omega\right)^2 = F^2 \tag{5.12}$$

To construct the amplitude graph by frequency, we will use the given Equation (5.12) with values $F = (0.2, 0.3, 0.5)$, $h = 1/2, \omega_0 = 1$.

Figure 5.2 represents the frequency response curve, specifically for the hardening spring mode (as shown by the rightward departure of the graph). The softening spring mode also exhibits the same behavior, but its divergence will be towards the left. Nonlinear systems exhibit distinct behavior patterns. At specific points, the amplitude of oscillation experiences abrupt rises or drops. Based on the diagram in Figure 5.2, the

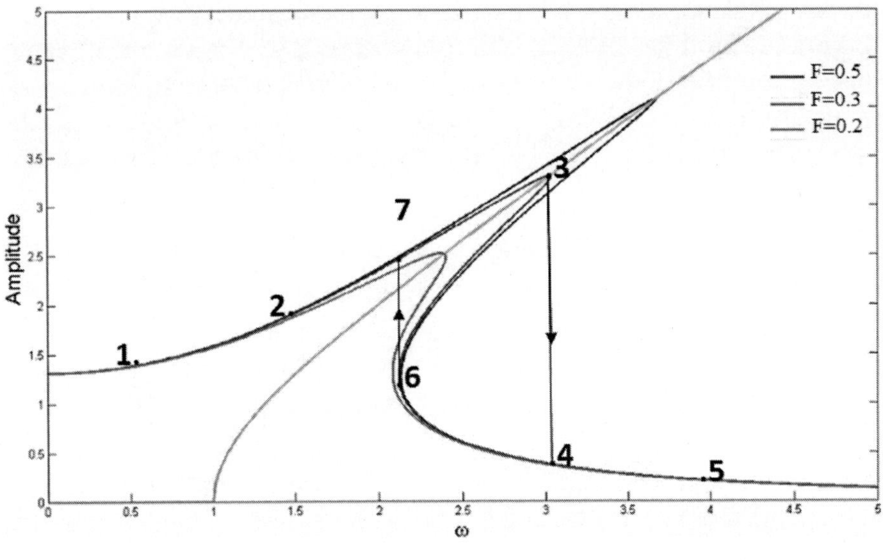

FIGURE 5.2 Frequency response curve.

jump phenomenon can be described as follows: when a constant force (F) is applied, the amplitude of vibration increases gradually as the frequency of the excitation increases. However, at point 3 on the curve, instead of moving to point 6, the oscillation amplitude jumps to point 4. Subsequently, it continues to increase along the specified path and reaches point 5. Similarly, as the frequency of excitation gradually decreases, the oscillation amplitude follows a specific curve passing through points 5, 4, 6, 7, 2, and 1. It is noteworthy that when the frequency decreases, the amplitude at point 6 deviates from its expected path towards point 3 and instead jumps to point 7. This tendency is referred to as the phenomenon of jumping. As the fall persists, it transitions from point 7 to 2 and thereafter to 1. The jump track with dimensions of 3-to-4 is commonly referred to as the jump down, while the track with dimensions of 6-to-7 is known as the jump up. Point 3 and point 6 (Figure 5.2) are saddle bifurcation. The points mentioned represent the critical boundary between stability and instability, as extensively explored in the part dedicated to the bifurcation theory.

The crucial aspect is that the trajectory from point 3 to 6, which is inherently unstable, is commonly depicted by a dash line. As depicted in Figure 5.3, there is no movement in this particular direction. Furthermore, the diagram depicted in Figure 5.3 illustrates the frequency response of a spring that exhibits softening behavior.

Figure 5.4 indicates the presence of two stable oscillation domains for a given excitation frequency. The region where oscillations are not sustained is referred to as the unstable conditions. Additionally, the path from 3 to 6 is considered unstable due to the absence of movement. In Figure 5.4, at a specific frequency, the diagram displays three distinct amplitude values denoted as a, b, and c. The domain of point b is unstable, and it is not visible. In the two remaining domains that exhibit stability and correction, they are referred to as bistable stability; depending on the beginning conditions, one of them emerges. The line that connects the vertices of the charts in the term, as shown in Figure 5.5, is referred to as the backbone curve.

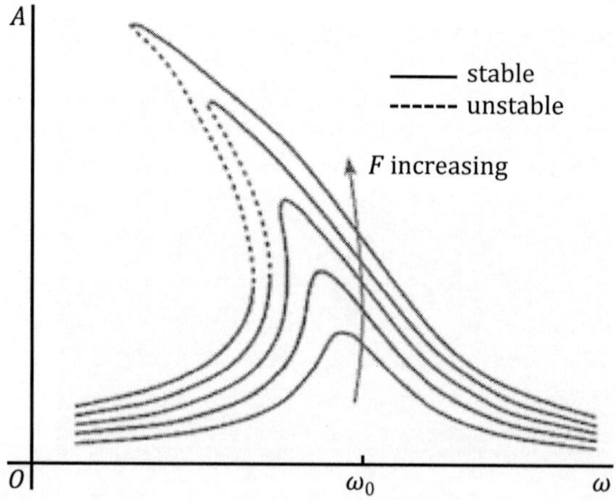

FIGURE 5.3 Frequency response curve for a softening spring.

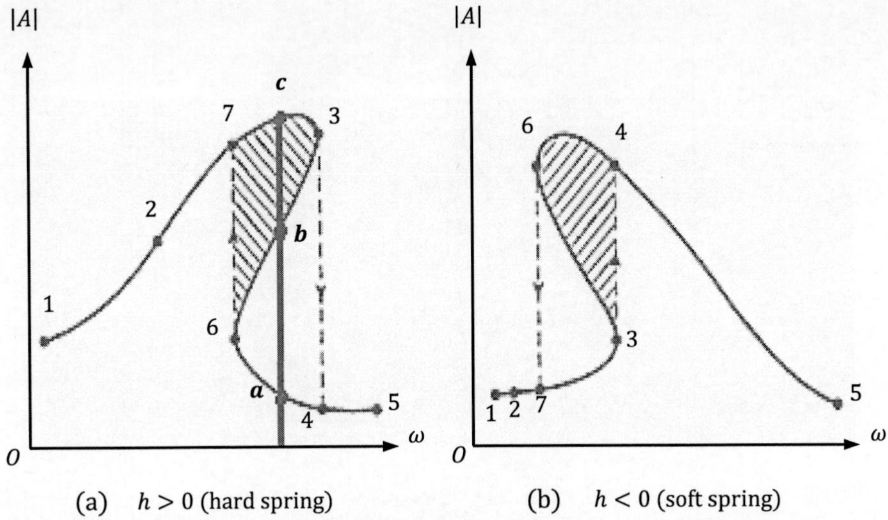

FIGURE 5.4 The phenomenon of jumping: a) hard spring and b) soft spring.

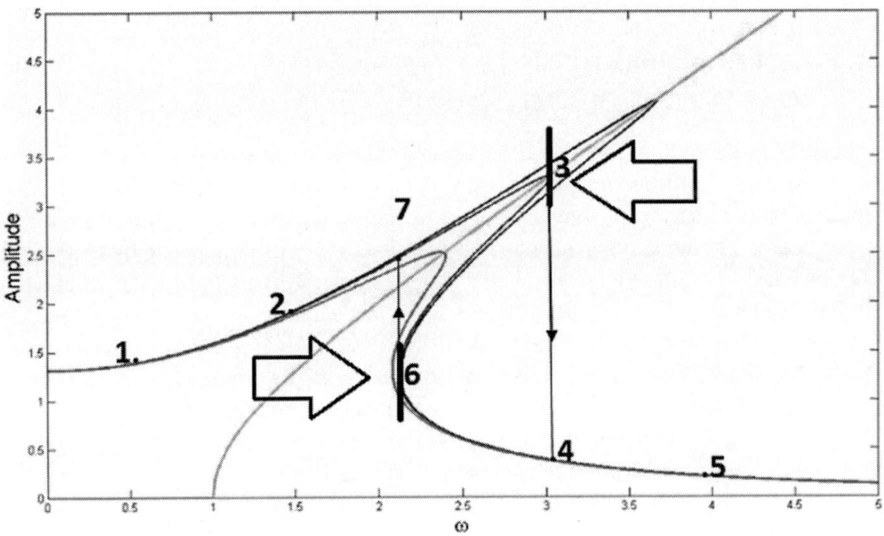

FIGURE 5.5 Backbone curve in the frequency response.

It is important to observe that the occurrence of jumping is characterized by the condition $\dfrac{dA}{d\omega} = \infty$ or $\dfrac{d\omega^2}{dA} = 0$ (as shown in Figure 5.5). To calculate $\dfrac{d\omega^2}{dA}$ by deriving the Equation (5.12) and setting it equal to zero, we obtain the following:

$$\left[\left(\omega_0^2 - \omega^2\right) + \frac{3}{4}hA^2\right]\left[\left(\omega_0^2 - \omega^2\right) + \frac{3}{4}hA^2\right] + \left(2\beta\omega\right)^2 = 0 \tag{5.13}$$

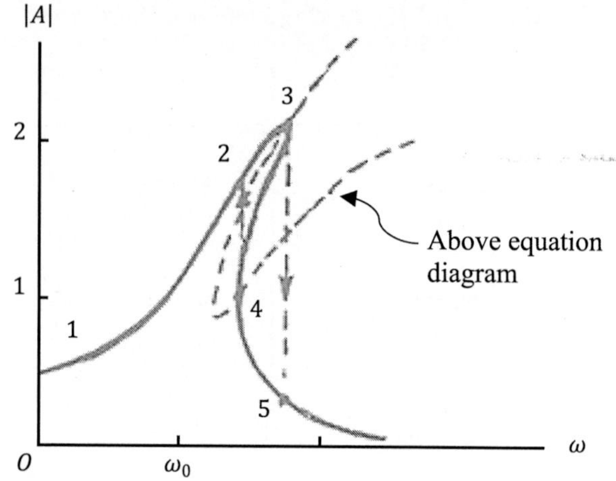

FIGURE 5.6 Jump points in the frequency response curve.

By drawing the diagrams of Equations (5.12) and (5.13), one may identify jump points, which correspond to the intersection of the two jump point diagrams (Figure 5.6).

5.3 FORCED VIBRATIONS OF SYSTEMS WITH ONE DEGREE OF FREEDOM

The preceding section analyzed the steady-state response of forced vibrations in a single-degree-of-freedom system using a straightforward approach. This section focuses on the analysis of forced vibration response using the principles of perturbation theory. The underlying phenomena of this study will be thoroughly examined. In this part, we analyze a system with a single degree of freedom that is continuously influenced by an excitation. Typically, the types of excitations include the following [1]:

1. An external excitation is a word used to describe a stimulus that appears as a non-homogenous expression in the equations regulating motion.
2. A variable coefficient that appears in the governing equations and is dependent on time is referred to as a parametric excitation.

In the following analysis, we investigate the nonlinear oscillations of a system with one degree of freedom, which are caused by an external input. To fulfill this objective, please examine the subsequent equation:

$$\ddot{u}+\omega_0^2 u=\epsilon f(u,\dot{u})+E \tag{5.14}$$

Where ε is a small parameter. f is a nonlinear function that depends on the variables u, $\dot{u}$, and E, which represent an external force applied as an excitation. At first, the excitation originates from an energy source that is presumed to be infinite or very big, allowing the stimulated system to have a minimal impact on it. Here, $E=E(t)$.

This implies that E is not dependent on u, $\dot{u}$, or $\ddot{u}$. These sources are referred to as perfect sources of energy. In the second scenario, the excitation originates from a finite energy source, resulting in a sensible impact on the stimulated system. In this scenario, $E = E(t, u, \dot{u}, \ddot{u})$. In other words, E is dependent on the state of the system. A non-ideal energy source is the term used to refer to such a resource. Systems are categorized into ideal and non-ideal energy sources based on their respective sources of energy.

In this context, we are examining an ideal system where the system's excitation is derived by the aggregate of N harmonic sentences.

$$E(t) = \sum_{n=1}^{N} K_n \cos(\Omega_n t + \theta_n) \tag{5.15}$$

If the values of K_n (domains), Ω_n (frequencies), and θ_n remain constant, the excitation is considered stationary. Otherwise, it is considered non-stationary. When domains and frequencies exhibit slow temporal changes, perturbation approaches resort to analyzing non-stationary systems.

5.3.1 Systems Exhibiting Third-Order Nonlinearity

Here, we analyze the forced vibrations of a mass that is coupled to a spring with nonlinear characteristics and is subject to viscous damping. The equation of motion for the system can be expressed as follows [2]:

$$\ddot{u} + \omega_0^2 u = -2\epsilon \mu \dot{u} - \epsilon \alpha u^3 + E(t) \tag{5.16}$$

μ is a positive value, while α might be either a positive constant (representing a hard spring) or a negative constant (representing a soft spring). As stated in the introduction, we make the following assumption:

$$E(t) = K \cos \Omega t \tag{5.17}$$

5.3.1.1 Primary Resonance, $\Omega \approx \omega_0$

Instead of utilizing the excitation frequency Ω as a parameter, we will introduce another parameter called the detuning parameter. This parameter quantitatively quantifies the proximity of Ω to ω_0. By utilizing this parameter, we can effectively distinguish between secular sentences encountered previously and other semi-secular sentences inside the words of the governing equation for u_1. Consequently, we put our thoughts into writing:

$$\Omega = \omega_0 + \epsilon \sigma \tag{5.18}$$

The value of $\sigma = O(1)$ within the Equation (5.18). When the value of σ is equal to zero ($\sigma = 0$), irrespective of the magnitude of the excitation, the nonlinear theory will forecast vibrations of infinite magnitude. The presence of damping and nonlinear effects in a real system restricts the magnitude of significant fluctuations. Hence, to provide a consistent and accurate approximation of this problem, it is necessary to reorganize the irritants such that the presence of damping and nonlinearity is also accounted for. In order to accomplish this, we define a set as $K = \epsilon k$. This way of organizing expressions aligns with the previously discussed considerations regarding the primary resonance. Under these circumstances, we anticipate that in a system exhibiting mild damping, when subjected to a low-amplitude excitation, a reaction of considerable magnitude will occur. A soft excitation is referred to as $K = \epsilon k$ when considering the domain of the excitation.

The problem can be solved using several ways of perturbation, specifically the method of multiple time scales, to obtain an approximate answer. Consequently, we express the response in sentences that pertain to various temporal frames in the following manner:

$$u(t,\epsilon) = u_0(T_0,T_1) + \epsilon u_1(T_0,T_1) + \dots \quad (5.19)$$

The values of T_0 and T_1 are defined as $T_0 = t$ and $T_1 = \epsilon t$, respectively. Furthermore, the expression of soft excitation in sentences including T_0 and T_1 is articulated in the following manner [3]:

$$E(t) = \epsilon k \cos(\omega_0 T_0 + \sigma T_1) \quad (5.20)$$

By substituting the Equations (5.19) and (5.20) into Equation (5.16) and setting the coefficients ϵ^0 and ϵ^1 equal on both sides, we derive the following:

$$\epsilon^0: D_0^2 u_0 + \omega_0^2 u_0 = 0 \quad (5.21)$$

$$\epsilon^1: D_0^2 u_1 + \omega_0^2 u_1 = -2D_0 D_1 u_0 - 2\mu D_0 u_0 - \alpha u_0^3 + k \cos(\omega_0 T_0 + \sigma T_1) \quad (5.22)$$

It is important to mention that, thus, the process involves both organizing and considering soft excitation, excitation terms, dampening, and nonlinear sentences in Equation (5.22). The Equation (5.21) has a general solution that may be expressed as follows:

$$u_0 = A(T_1) exp(i\omega_0 T_0) + \bar{A}(T_1) exp(-i\omega_0 T_0) \quad (5.23)$$

The function $A(T_1)$ is an unknown function and will be derived by eliminating secular terms from u_1. By establishing a connection between u_0 and the Equation (5.22) and expressing $\cos(\omega_0 T_0 + \sigma T_1)$ in a complex format, we obtain the following:

$$\begin{aligned} D_0^2 u_1 + \omega_0^2 u_1 = &-\left[2i\omega_0(A' + \mu A) + 3\alpha A^2 \bar{A}\right] exp(i\omega_0 T_0) - \alpha A^3 exp(3i\omega_0 T_0) \\ &+ \frac{1}{2} k\, exp\left[i(\omega_0 T_0 + \sigma T_1)\right] + cc \end{aligned} \quad (5.24)$$

The term "cc" refers to the complex-conjugate of the preceding sentences. If we select response A for the given relationship, any secular terms will be eliminated from the private response.

$$2i\omega_0\left(A'+\mu A\right)+3\alpha A^2\bar{A}-\frac{1}{2}k\,exp\left(i\sigma T_1\right)=0 \tag{5.25}$$

In order to solve the Equation (5.25), we represent A in a diagonal manner:

$$A=\frac{1}{2}a\,exp\left(i\beta\right) \tag{5.26}$$

That, a, and β are all real numbers. By decomposing the complex number into its real and imaginary components, we obtain the following:

$$\begin{aligned} a' &= -\mu a+\frac{1}{2}\frac{k}{\omega_0}sin\left(\sigma T_1-\beta\right) \\ a\beta' &= \frac{3}{8}\frac{\alpha}{\omega_0}a^3-\frac{1}{2}\frac{k}{\omega_0}cos\left(\sigma T_1-\beta\right) \end{aligned} \tag{5.27}$$

The Equation (5.26) can be substituted into Equation (5.23), and the resulting outcome can be substituted into Equation (5.19) to obtain the solution, which can be expressed as follows:

$$u=a\,cos\left(\omega_0 t+\beta\right)+O\left(\epsilon\right) \tag{5.28}$$

That, a, and β are obtained using Equation (5.27). To turn Equation (5.27) into an autonomous system, one can insert the following:

$$\gamma=\sigma T_1-\beta \tag{5.29}$$

Which results:

$$\begin{aligned} a' &= -\mu a+\frac{1}{2}\frac{k}{\omega_0}sin\,\gamma \\ a\gamma' &= \sigma a-\frac{3}{8}\frac{\alpha}{\omega_0}a^3+\frac{1}{2}\frac{k}{\omega_0}cos\,\gamma \end{aligned} \tag{5.30}$$

In order to obtain a response, it is necessary to initially determine the locations of the fixed points, followed by an examination of the motion within their surrounding areas. The system response is considered to be in a steady-state motion since the amplitude and phase remain constant at solitary points. The behavior of the trajectory line in the vicinity of the singular points reveals whether a slight perturbation in the condition of the steady-state motion leads to a decrease or an increase in movement. Indeed, they demonstrate the steadfastness of the steady-state mode's motion.

5.3.1.2 Steady-State Mode Movements

Steady-state motion occurs when $a' = \gamma' = 0$, which corresponds to the singular points of Equation (5.30). In other words, they are equivalent to the solution:

$$\mu a = \frac{1}{2}\frac{k}{\omega_0}\sin\gamma$$
$$\sigma a - \frac{3}{8}\frac{\alpha}{\omega_0}a^3 = -\frac{1}{2}\frac{k}{\omega_0}\cos\gamma \tag{5.31}$$

By combining and including these equations, we obtain the following:

$$\left[\mu^2 + \left(\sigma - \frac{3}{8}\frac{\alpha}{\omega_0}a^2\right)^2\right]a^2 = \frac{k^2}{4\omega_0^2} \tag{5.32}$$

Equation (5.32) represents an implicit equation that relates the amplitude of response, denoted as a, to the detuning parameter σ (which represents the frequency of excitation) and the amplitude of excitation, denoted as k. The equation is referred to as the frequency response in Equation (5.32).

The initial approximation for the equation answer to the steady-state answer is achieved by inserting the Equations (5.29) and (5.18) into Equation (5.28).

$$u = a\cos(\omega_0 t + \epsilon\sigma t - \gamma) + O(\epsilon) = a\cos(\Omega t - \gamma) + O(\epsilon) \tag{5.33}$$

Constants such as a and γ are fixed values. Thus, the steady-state response aligns precisely with the frequency of excitation. The response phase is determined by the magnitude of the shift in the excitation phase $-\gamma$. Graph a represents the frequency response curve, which varies with σ, for the specified values of μ and k, as mentioned before. Each point on this curve corresponds to a fixed point on a different state plane. Indeed, there exists a mode screen that accommodates any possible combination of settings.

In order to plot the frequency response curve using the Equation (5.32), we can either solve the third-order equation for the a^2 in terms of σ or solve σ based on a. The second mode is less complex and is expressed in the following manner:

$$\sigma = \frac{3}{8}\frac{\alpha}{\omega_0}a^2 \pm \left(\frac{k^2}{4\omega_0^2 a^2} - \mu^2\right)^{1/2} \tag{5.34}$$

Figure 5.7 depicts a contrast between linear $(\alpha = 0)$ and nonlinear $(\alpha > 0)$ response curves. Equation (5.34) demonstrates that the greatest range, denoted as $a_p = k/(2\omega_0\mu)$, remains unaffected by the value of α. At this level of responsiveness, the linear outcomes exhibit symmetry. The response is concentrated within a narrow range centered around the resonance frequency (denoted as $\Omega = \omega_0 + \epsilon\sigma$), resulting in an expanded frequency scale A in this region. The presence of nonlinearity results in the bending of the amplitude curve and the distortion of the phase curve.

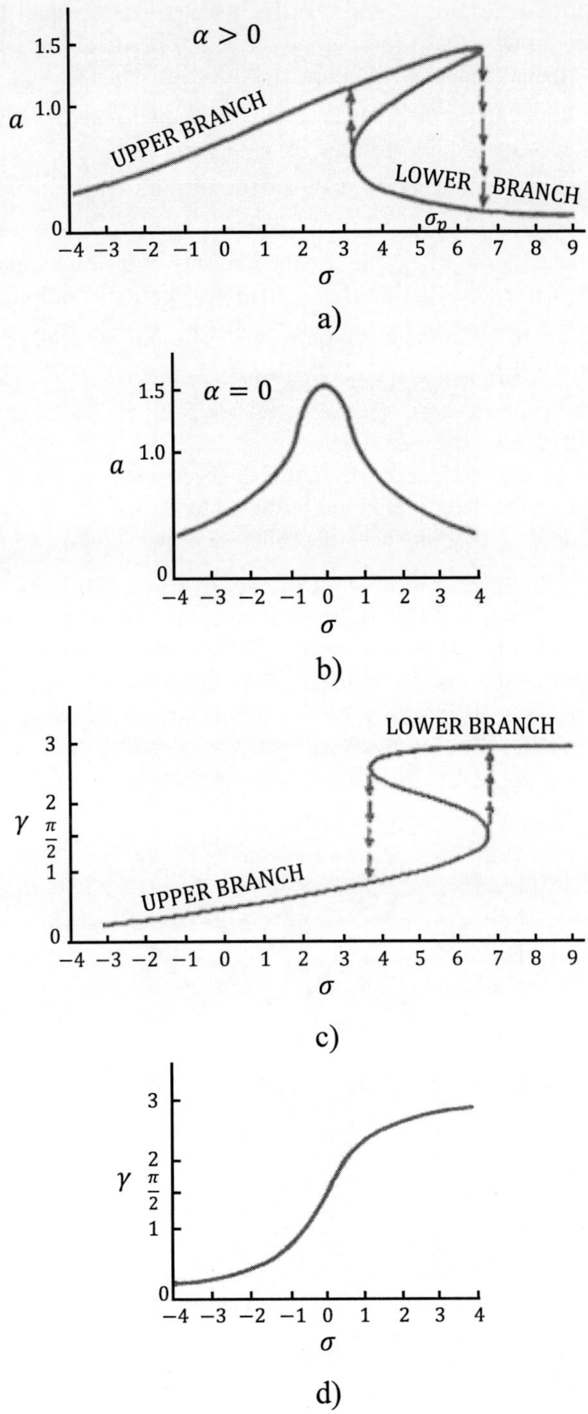

FIGURE 5.7 Comparison of linear and nonlinear response curves: (a) linear domain, (b) linear phases, (c) nonlinear domain, and (d) nonlinear phases.

Multi-quantitative zones are produced in both scenarios. We shall further explore how multi-quantitative zones contribute to the phenomenon of jumping. The arrows in Figures 5.7 (a) and (c) depict the jump.

Figure 5.8 (a) demonstrates that the presence of nonlinearity causes a deviation of the frequency response curve from the linear curve $(\alpha = 0)$. This deviation is towards the right for hard springs $(\alpha > 0)$ and towards the left for soft springs $(\alpha < 0)$. Figure 5.8 (b) illustrates the variation in frequency response curves as the excitation amplitude increases for a soft spring. As the intensity of the excitation rises, the frequency response curves diverge from the $\sigma = 0$ axis. The maximum domains are located geometrically according to the equation $\sigma = \frac{3}{8}(\alpha / \omega_0) a^2$, as seen in Figure 5.8 (b) with the fold line, commonly referred to as the backbone curve, as previously mentioned. It is observed that some frequency curves can have multiple values or only one value, depending on the value of k.

Figure 5.9 depicts the impact of the damping coefficient μ on the response curves. Without damping, the peak amplitude is unbounded, and the frequency response curve is composed of two branches on either side of its curve, with the equation $\sigma = \frac{3}{8}(\alpha / \omega_0) a^2$. In the reciprocal phase of 180°, where $\mu = 0$ and $\gamma = n\pi$, where n is an integer, the Equation (5.31) is given. Hence, the Equation (5.33) signifies that the response is either in phase or excitation. However, the existence of damping will restrict the maximum range of the peak. Furthermore, the initial Equation (5.31) demonstrates that $\gamma = \sin^{-1}(2\mu a \omega_0 / k)$. Consequently, the presence of damping in the response phase leads to displacement.

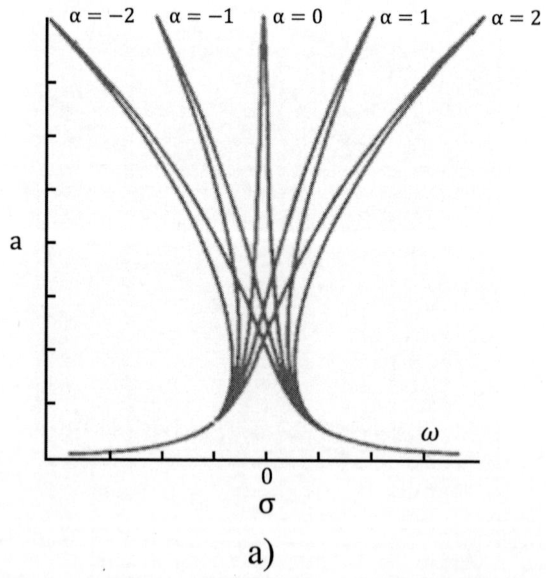

FIGURE 5.8 Frequency response curves for initial resonances of the Duffing equation: (a) nonlinear effect and (b) excitation domain effect.

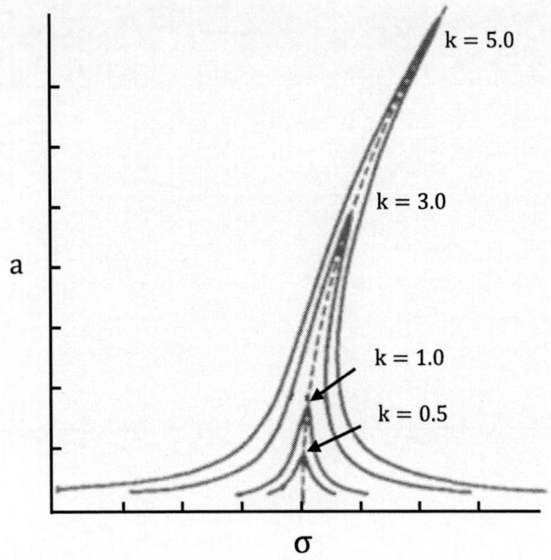

b)

FIGURE 5.8 (Continued)

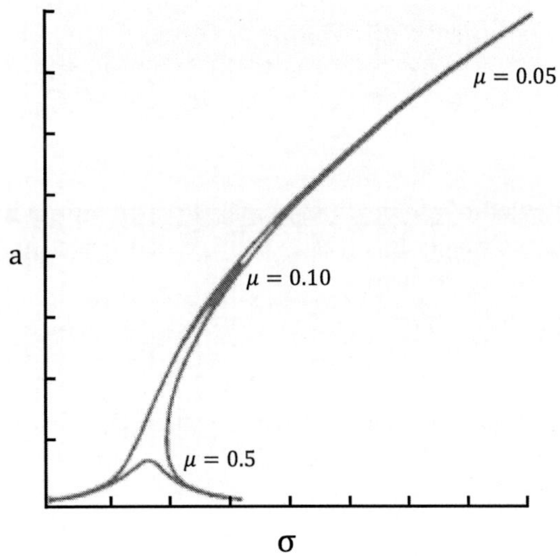

FIGURE 5.9 Damping effect in response of the Duffing equation to a primary resonance excitation.

Figure 5.10 depicts the variation in response amplitude as a function of excitation amplitude for various values of σ. The values α and μ are consistent across all curves. The curves are directly derived from the Equation (5.32). It should be noted that, based on the value of σ, certain curves can be classified as polynomials, while others can be classified as monomials.

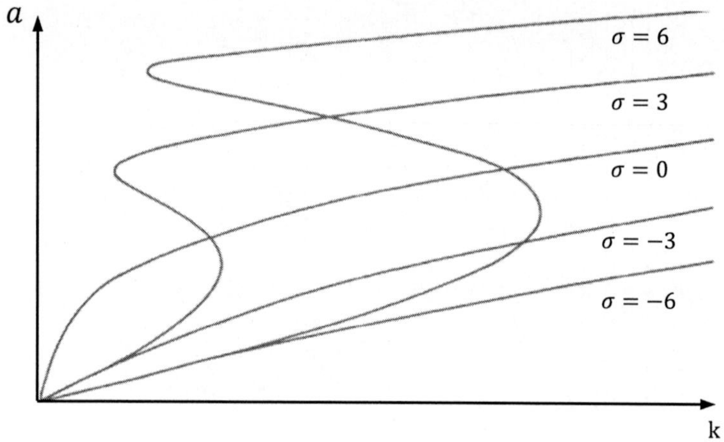

FIGURE 5.10 Response amplitude as a function of the excitation amplitude for multi values of detunig parameter.

5.4 HARD EXCITATION WITHOUT RESONANCE

When the frequency of Ω is significantly different from $\omega_\circ$, the resulting excitation effect will be minimal, unless the amplitude of the Ω is high. In other words, the complexity of K is constant, denoted as $O(1)$. Thus, we delineate the excitation in this manner [1]:

$$E(t) = K\cos \Omega T \tag{5.35}$$

We employ the method of multiple time scales to approximate the response based on the first resonance mode. In order to achieve that objective, we formulate the response in the following manner:

$$u(t,\epsilon) = u_0(T_0,T_1) + \epsilon u_1(T_0,T_1) + \dots \tag{5.36}$$

By substituting the Equation (5.36) into the Equation (5.16), utilizing the Equation (5.35), and equating the coefficients of $\in^\circ$ and $\in^1$ on both sides of the equation, we obtain the following:

$$D_\circ^2 u_\circ + \omega_\circ^2 u_\circ = K\cos \Omega T_\circ \tag{5.37}$$

$$D_\circ^2 u_1 + \omega_\circ^2 u_1 = -2D_\circ D_1 u_\circ - 2\mu D_\circ u_\circ - \alpha u_\circ^3 \tag{5.38}$$

The Equation (5.37) has a general solution that can be expressed as follows:

$$u_\circ = A(T_1)\, exp\,(i\omega_\circ T_\circ) + \Lambda\, exp\,(i\Omega T_\circ) + cc \tag{5.39}$$

Where $\Lambda = \frac{1}{2}K(\omega_\circ^2 - \Omega^2)^{-1}$ is concluded by placing u_0 in Equation (5.38).

$$\begin{aligned}D_\circ^2 u_1 + \omega_\circ^2 u_1 = &-\left[2i\omega_\circ\left(A' + \mu A\right) + 6\alpha A\Lambda^2 + 3\alpha A^2\bar{A}\right] exp\,(i\omega_\circ T_\circ)\\ &-\alpha\{A^3\, exp\,(3i\omega_\circ T_\circ) + \Lambda^3\, exp\,(3i\Omega T_\circ)\\ &+3A^2\Lambda\, exp\left[i\left(2\omega_\circ + \Omega\right)T_\circ\right] + 3\bar{A}^2\Lambda\, exp\left[i\left(\Omega - 2\omega_\circ\right)T_\circ\right]\\ &+3A\Lambda^2\, exp[i\left(\omega_\circ + 2\Omega\right)T_\circ] + 3A\Lambda^2\, exp[i\left(\omega_\circ - 2\Omega\right)T_\circ\\ &-\Lambda[2i\mu\Omega + 3\alpha\Lambda^2 + 6\alpha A\bar{A}]\, exp\,(i\Omega T_\circ) + cc\end{aligned} \tag{5.40}$$

Secular or semi-secular expressions (small derivation from secular terms), which deviate just slightly from secular expressions $exp\,(\pm i\omega_\circ T_\circ)$, may arise when $\Omega = O(\varepsilon)$. In other words, for every given time, $\omega_\circ \approx \left(m\omega_\circ + n\Omega\right)$, where m and n are integers that satisfy the equation $|m| + |n| = 3$. This phenomenon is known as superharmonic resonance when it occurs at a frequency of $\Omega \approx \frac{1}{3}\omega_\circ$ and as subharmonic resonance when it occurs at a frequency of $\Omega \approx 3\omega_\circ$. Hence, while eliminating phrases that produce secular expressions, it is necessary to differentiate between four modes:

a) The distant o is excited at frequencies of $\frac{1}{3}\omega_\circ$ and $3\omega_\circ\Omega$.
b) The frequency of the excitation is about $\Omega \approx \circ$.
c) The frequency of excitation is represented by the symbol Ω, which is equivalent to $\Omega \approx \frac{1}{3}\omega_\circ$.
d) The excitation frequency is approximately $\Omega \approx 3\omega_\circ$.

The following section will address the following:

Mode A) will discuss later.
Mode B) is offered, followed by an examination.
Mode C) superharmonic resonance.
Mode D) subharmonic resonance.

Mode A)

In the state of non resonance, secular terms are excluded.

$$2i\omega_\circ\left(A' + \mu A\right) + 6\alpha\Lambda^2 A + 3\alpha A^2\bar{A} = \circ \tag{5.41}$$

Assuming the Equation (5.41), let $A = 1/2a\, exp\left(i\beta\right)$, where a and β are real numbers. By decomposing the complex number into its real and imaginary components, we obtain the following:

$$\begin{aligned}a' &= -\mu a\\ \omega_\circ a\beta' &= 3\alpha\left(\Lambda^2 + \frac{1}{8}a^2\right)a\end{aligned} \tag{5.42}$$

Thus, for the initial estimation:

$$u = a\cos\left(\omega_0 t + \beta\right) + K\left(\omega_0^2 - \Omega^2\right)^{-1}\cos\Omega t + O\left(\epsilon\right) \tag{5.43}$$

The values of a and β are derived from the Equation (5.42). The general response for a is given by the equation $a = a_\circ\, exp\,(-\mu T_1)$, where $a_\circ$ is a constant value. Consequently, the free (homogeneous) vibration response decreases over time so that the steady-state response includes only forced (particular) response akin to the linear state. While the equation for free vibration is diminishing, its frequency is determined by the particular response domain.

The B mode will be assessed using the exercise structure provided later.

It is preferable to analyze the solution of the differential Equation (5.16) while considering the excitation force as Equation (5.35) with the assumption that the excitation frequency is $\Omega = 0$.

Please be aware that Equations (5.39) and (5.40) are also applicable. Consult the Ref. [1] for additional guidance and research.

5.4.1 Superharmonic Resonance $\left(\Omega \approx \frac{1}{3}\omega_\circ\right)$

In the case of mode C, we indicate the closeness to $\frac{1}{3}\omega_\circ$ by introducing the detuning parameter σ in the following manner:

$$3\Omega = \omega_0 + \epsilon\sigma \tag{5.44}$$

Furthermore, apart from the terms proportional to $exp\,(\pm i\omega_\circ T_\circ)$ mentioned in Equation (5.40), there exists an additional term that gives rise to a secular trend in the u_1 response. The expression is $-\alpha\Lambda^3\, exp\,(\pm 3i\Omega T_\circ)$. In order to exclude secular terms, we represent the term $3\Omega T_\circ$ in terms of $\omega_\circ T_\circ$ as follows:

$$3\Omega T_0 = (\omega_0 + \epsilon\sigma)T_0 = \omega_0 T_0 + \epsilon\sigma T_0 = \omega_0 T_0 + \sigma T_1 \tag{5.45}$$

By applying the Equation (5.45), we determine that secular sentences are excluded in u_1. If:

$$2i\omega_0(A' + \mu A) + 6\alpha\Lambda^2\bar{A} + 3A^2\bar{A} + \alpha\Lambda^3 exp(i\sigma T_1) = 0 \tag{5.46}$$

In the given Equation (5.46), let $A = \frac{1}{2}a\, exp\,(i\beta)$, where a and β are real numbers. By decomposing the complex number into its real and imaginary components, we obtain the following:

$$\begin{aligned} a' &= -\mu a - \frac{\alpha\Lambda^3}{\omega_\circ} sin\,(\sigma T_1 - \beta) \\ a\beta' &= \frac{3\alpha}{\omega_\circ}\left(\Lambda^2 + \frac{1}{8}a^2\right)a + \frac{\alpha\Lambda^3}{\omega_\circ} cos\,(\sigma T_1 - \beta) \end{aligned} \tag{5.47}$$

Equation (5.47) can be transformed into an autonomous system, which is not reliant on time (t), by incorporating:

$$\gamma = \sigma T_1 - \beta \tag{5.48}$$

Consequently, we obtain the following:

$$\begin{aligned} a' &= -\mu a - \frac{\alpha\Lambda^3}{\omega_\circ} \sin\gamma \\ a\gamma' &= \left(\sigma - \frac{3\alpha\Lambda^2}{\omega_\circ}\right) a - \frac{3\alpha}{8\omega_\circ} a^3 - \frac{\alpha\Lambda^3}{\omega_\circ} \cos\gamma \end{aligned} \tag{5.49}$$

Thus, for the initial estimation, we may express it as follows:

$$u = a\cos(3\Omega t - \gamma) + K\left(\omega_0^2 - \Omega^2\right)^{-1} \cos\Omega t + O(\epsilon) \tag{5.50}$$

The values of a and γ are derived from the Equation (5.49).

The movements of the steady state align with $a' = \gamma' = 0$. Indeed, they correspond to the subsequent answers:

$$\begin{aligned} -\mu a &= \frac{\alpha\Lambda^3}{\omega_0} \sin\gamma \\ \left(\sigma - 3\frac{\alpha\Lambda^2}{\omega_0}\right) a - \frac{3\alpha}{8\omega_0} a^3 &= \frac{\alpha\Lambda^3}{\omega_0} \cos\gamma \end{aligned} \tag{5.51}$$

The frequency response equation is generated by squaring and adding earlier equations as follows:

$$\left[\mu^2 + \left(\sigma - 3\frac{\alpha\Lambda^2}{\omega_\circ} - \frac{3\alpha}{8\omega_\circ} a^2\right)^2\right] a^2 = \frac{\alpha^2\Lambda^6}{\omega_\circ^2} \tag{5.52}$$

By solving the Equation (5.52) for σ in terms of a, the conclusion is as follows:

$$\sigma = 3\frac{\alpha\Lambda^2}{\omega_0} + \frac{3\alpha}{8\omega_0} a^2 \pm \left(\frac{\alpha^2\Lambda^6}{\omega_0^2 a^2} - \mu^2\right)^{1/2} \tag{5.53}$$

Therefore, when $\Omega \approx \frac{1}{3}\omega_\circ$, the expression representing the free vibration in response to the equation does not go towards zero, even in the presence of damping and in the comparison to the linear condition. Furthermore, the nonlinear system precisely controls the frequency of free vibration, which is exactly three times the frequency of excitation, resulting in an alternating response. The phenomenon of superharmonic occurs when the frequency of free vibration is three times higher than the frequency of excitation. The three curves in Figure 5.11 depict the formation of the response through the combination of the particular solution and free vibration, as described in Equation (5.50).

The various frequency response curves depicted in Figure 5.12. These graphs illustrate the impact of altering α, Λ, and μ. The jumping phenomenon is caused

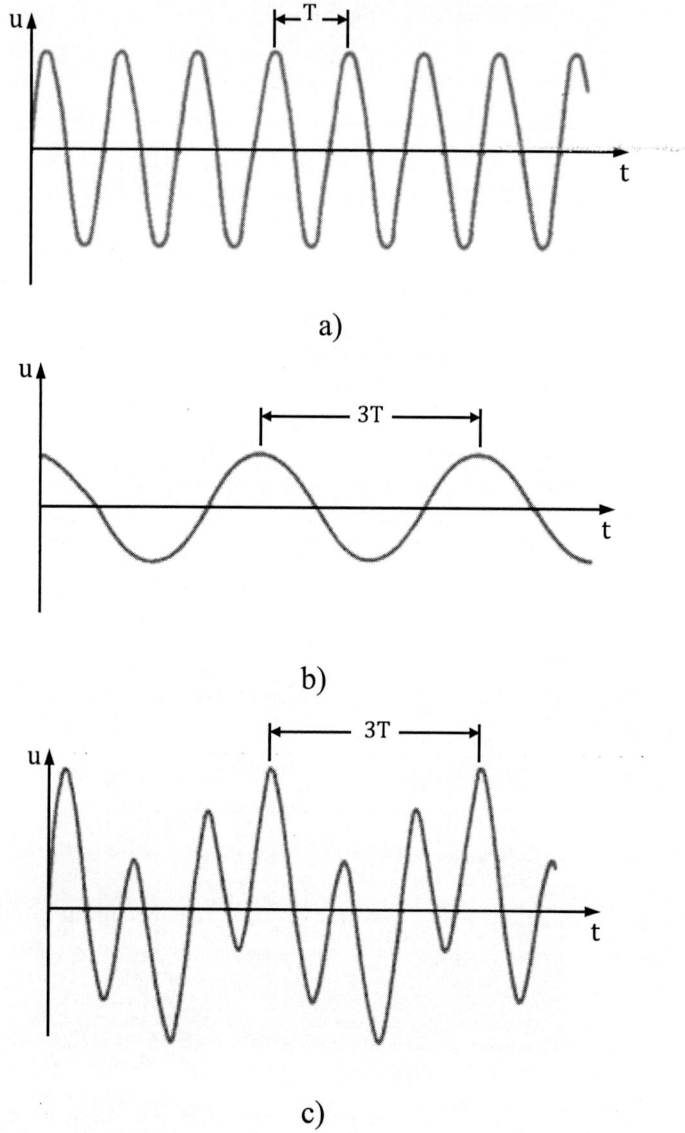

FIGURE 5.11 Combination of the Duffing equation response to superharmonic excitation: (a) free vibration response, (b) particular response, and (c) real response.

by the bending of the frequency response curves, similar to the primary resonance condition. Furthermore, altering the sign α (as shown in Figure 5.12 a) results in the creation of a symmetry centered around the line $\sigma = 0$. This symmetry is denoted by the Equation (5.53).

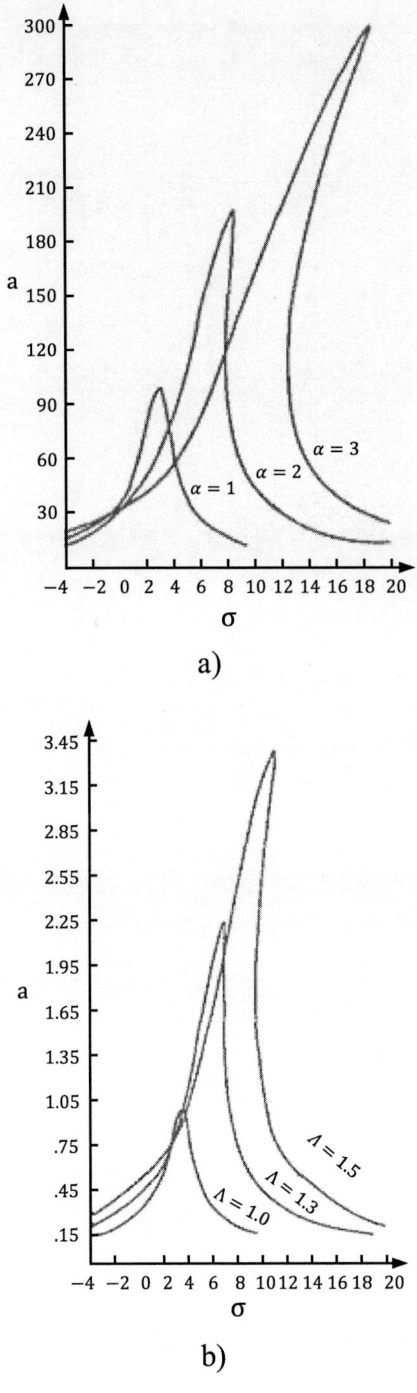

FIGURE 5.12 Superharmonic frequency response curves for the Duffing equation: (a) nonlinear effect, (b) excitation amplitude effect, and (c) damping effect.

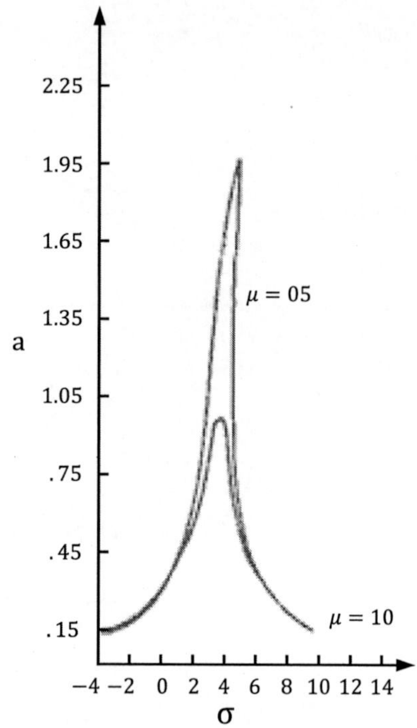

c)

FIGURE 5.12 (Continued)

5.4.2 Subharmonic Resonance $(\Omega \approx 3\omega_\circ)$

In order to examine subharmonic excitation for Equations (5.16) and (5.17) in State D, we establish the detuning parameter σ according to the following definition:

$$\Omega = 3\omega_0 + \epsilon\sigma \tag{5.54}$$

Furthermore, the expression that corresponds to $exp(\pm i\omega_\circ T_\circ)$ and the terms corresponding to $exp[\pm i(\Omega - 2\omega_\circ)T_\circ]$ are considered secular terms in u_1. The term $(\Omega - 2\omega_\circ)T_\circ$ is written in accordance with the Equation (5.54) as follows:

$$(\Omega - 2\omega_0)T_0 = \omega_0 T_0 + \epsilon\sigma T_0 = \omega_0 T_0 + \sigma T_1 \tag{5.55}$$

To eliminate these terms in the Equation (5.40) that result in secular terms in u_1, we set the following:

$$2i\omega_\circ(A' + \mu A) + 6\alpha \Lambda^2 A + 3\alpha A^2 \bar{A} + 3\alpha \Lambda \bar{A}^2\, exp(i\sigma T_1) = \circ \tag{5.56}$$

Let us establish a relationship between the complex number in Equation (5.56) and the expression $A = \frac{1}{2} a\, exp\,(i\beta)$, where a and β are integers. By performing a process of isolating the real and imaginary components, we obtain the following:

$$
\begin{aligned}
a' &= -\mu a - \frac{3\alpha\Lambda}{4\omega_0} a^2 sin\left(\sigma T_1 - 3\beta\right) \\
a\beta' &= \frac{3\alpha}{\omega_0}\left(\Lambda^2 + \frac{1}{8}a^2\right)a + \frac{3\alpha\Lambda}{4\omega_0} a^2 cos\left(\sigma T_1 - 3\beta\right)
\end{aligned}
\tag{5.57}
$$

In order to convert Equation (5.57) into an autonomous system, we establish the following definitions:

$$\gamma = \sigma T_1 - 3\beta \tag{5.58}$$

We shall possess the following:

$$
\begin{aligned}
a' &= -\mu a - \frac{3\alpha\Lambda}{4\omega_0} a^2 sin\gamma \\
a\gamma' &= \left(\sigma - \frac{9\alpha\Lambda^2}{\omega_0}\right)a - \frac{9\alpha}{8\omega_0}a^3 - \frac{9\alpha\Lambda}{4\omega_0}a^2\, cos\,\gamma
\end{aligned}
\tag{5.59}
$$

Therefore, the system's response will be determined by the initial approximation in the following manner:

$$u = a\,cos\left(\frac{1}{3}(\Omega t - \gamma)\right) + K\left(\omega_0^2 - \Omega^2\right)^{-1} cos\,\Omega t + O(\epsilon) \tag{5.60}$$

The values of a and γ are derived from the Equation (5.59).

Steady-state mode movements are associated with the following responses:

$$
\begin{aligned}
-\mu a &= \frac{3\alpha\Lambda}{4\omega_0} a^2 sin\,\gamma \\
\left(\sigma - \frac{9\alpha\Lambda^2}{\omega_0}\right)a - \frac{9\alpha}{8\omega_0}a^3 &= \frac{9\alpha\Lambda}{4\omega_0}a^2\, cos\,\gamma
\end{aligned}
\tag{5.61}
$$

The frequency response equation can be found by removing γ from earlier equations.

$$\left[9\mu^2 + \left(\sigma - \frac{9\alpha\Lambda^2}{\omega_\circ} - \frac{9\alpha}{8\omega_\circ}a^2\right)^2\right]a^2 = \frac{81\alpha^2\Lambda^2}{16\omega_\circ^2}a^4 \tag{5.62}$$

The Equation (5.62) suggests that either $a = \circ$ and/or

$$9\mu^2 + \left(\sigma - \frac{9\alpha\Lambda^2}{\omega_0} - \frac{9\alpha}{8\omega_0}a^2\right)^2 = \frac{81\alpha^2\Lambda^2}{16\omega_0^2}a^2 \tag{5.63}$$

Which is the second order of a^2. The solution is equivalent to the following:

$$a^2 = p \pm (p^2 - q)^{1/2} \tag{5.64}$$

Where:

$$p = \frac{8\omega_\circ \sigma}{9\alpha} - 6\Lambda^2, q = \frac{64\omega_\circ^2}{81\alpha^2}\left[9\mu^2 + \left(\sigma - \frac{9\alpha\Lambda^2}{\omega_\circ}\right)^2\right] \tag{5.65}$$

It is important to observe that q is consistently positive, and as a result, the non-trivial amplitude of the system's free vibration only happens when $p > \circ$ and $p^2 \geq q$. These conditions will be present in the following circumstances:

$$\Lambda^2 < \frac{4\omega_0\sigma}{27\alpha}, \quad \frac{\alpha\Lambda^2}{\omega_0}\left(\sigma - \frac{63\alpha\Lambda^2}{8\omega_0}\right) - 2\mu^2 \geq 0 \tag{5.66}$$

This necessitates that α and σ possess comparable marks. Furthermore, it can be deduced from the Equation (5.66) that, for a specific value Λ, non-trivial solutions can only arise under the following conditions:

$$\alpha\sigma \geq \frac{2\mu^2\omega_\circ}{\Lambda^2} + \frac{63\alpha^2\Lambda^2}{8\omega_\circ} \tag{5.67}$$

Non-trivial responses can only occur when certain conditions are met, given a specific value of σ.

$$\frac{\sigma}{\mu} - \left(\frac{\sigma^2}{\mu^2} - 63\right)^{1/2} \leq \frac{63\alpha\Lambda^2}{4\omega_\circ\mu} \leq \frac{\sigma}{\mu} + \left(\frac{\sigma^2}{\mu^2} - 63\right)^{1/2} \tag{5.68}$$

The boundary in the plane $(\Lambda - \sigma)$ where non-trivial responses can occur is determined by the following relation:

$$\frac{63\alpha\Lambda^2}{4\omega_0\mu} = \frac{\sigma}{\mu} \pm \left(\frac{\sigma^2}{\mu^2} - 63\right)^{1/2} \tag{5.69}$$

As depicted in Figure 5.13 for values of $\alpha > \circ$.

Under these circumstances, it is feasible for a system in this state to exhibit a response where the free vibration, despite the presence of damping, does not diminish to zero in comparison to the linear response. Furthermore, at a condition of equilibrium, the nonlinear component is directly related to the frequency of autonomous oscillation, which is one-third of the frequency of external excitation, resulting in a periodic response. Subharmonic resonance refers to the occurrences where the frequency of free vibration is one-third of the frequency of excitation.

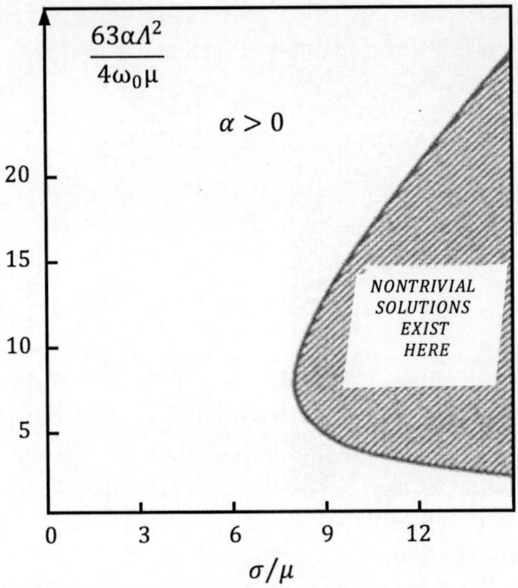

FIGURE 5.13 Areas where the subharmonic responses exist.

5.5 PARAMETRIC EXCITATION

5.5.1 Introduction

This section focuses on analyzing motions that arise as a consequence of excitation that vary with time in the system. In prior episodes, we observed that excitation manifest as variations in the governing differential equations, leading to their interpretation as external excitation. However, in this section, we familiarize ourselves with excitation that are expressed as coefficients in the governing differential equations. Mathematically, this form of excitation results in differential equations with coefficients that vary with time. In certain sections of mechanics, this phenomenon necessitates the resolution of partial differential equations with constant coefficients. These excitations are referred to as parametric excitation since they are represented as parameters in the governing equations, regardless of time. In contrast to external excitations, where a small excitation can only produce a large response if the frequency of excitation is close to one of the natural frequencies of the system, a small parametric excitation can produce a large response when the frequency of excitation is close to twice that of one of the natural frequencies of the system. This is known as principle parametric resonance.

Faraday [4] was the initial observer of the phenomena known as parametric resonance. He observed that when a fluid-filled cylinder is vertically excited, surface waves exhibit double excitation periodicities. Although the subject of parametric excitation has not been extensively studied in the past, there is a considerable amount of literature and publications dedicated to the analysis and use of this sort of excitation. One notable source is McLachlan [5], which focuses on the theory and practical use of Mathieu functions. Bondarenko [6], Magnus, and Winkler [7] examined the Hill's equation and its relevance to engineering vibration problems in a separate study. Furthermore, there are several works available that discuss parametric excitation, namely, Nayfeh [8], Evan-Iwanowski [9], and Kononenko [10].

5.5.2 Parametric Excitation in a Linear System

The Mathieu equation is the most basic form of differential equations with alternating coefficients, defined as follows:

$$\ddot{u}+\left(\delta+2\varepsilon\ cos\ 2t\right)u=0 \tag{5.70}$$

Where δ and μ are constants, the values of which are within the Equation (5.70). Equation (5.70) serves as the principal equation for numerous physical systems subjected to cosine excitation. Take note of the pendulum depicted in Figure 5.14, which is attached to the plate at Point O. The plate is vertically oscillating in a horizontal direction. The function $\cos 2t$ generates a parametric excitation in the system that varies with the variable t, representing time.

Let's consider the Equation (5.70) with a minor modification as follows:

$$\ddot{u}+\omega_0^2 u+\varepsilon\delta\ cos\ \Omega t\, u=0 \tag{5.71}$$

The solution to Equation (5.71) can be obtained using direct approaches, averaging techniques, or multiple time scales. However, in order to detect the system intensifications, we will employ a straightforward approach. Hence, the solution to Equation (5.71) can be expressed as follows:

$$u\left(t,\ \varepsilon\right)=u_0\left(t\right)+\varepsilon u_1\left(t\right)+\varepsilon^2 u_2\left(t\right)+\ldots \tag{5.72}$$

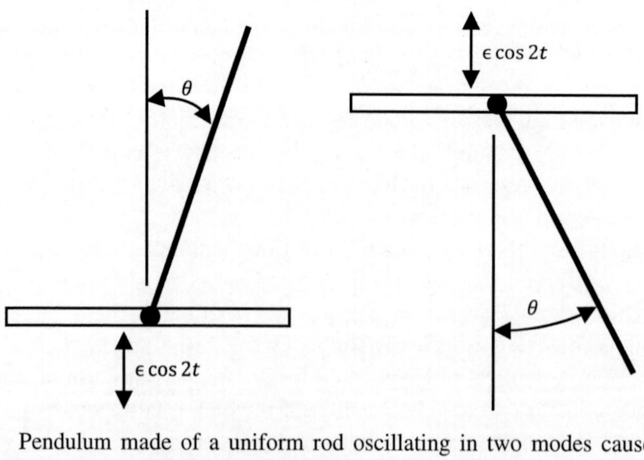

FIGURE 5.14 Pendulum made of a uniform rod oscillating in two modes caused by the vertical harmonic movement of the horizontal plane.

By substituting the solution (5.72) into the Equation (5.71) and thereafter simplifying and segregating the phrases based on the coefficients ε^0, ε^1, and ε^2, we will obtain three equations as stated:

$$\begin{aligned}
\varepsilon^0 &: \quad \ddot{u}_0 + \omega_0^2 u_0 = 0 \quad \rightarrow u_0 = a\cos(\omega_0 t + \beta) \\
\varepsilon^1 &: \quad \ddot{u}_1 + \omega_0^2 u_1 = -\delta u_0 \cos \Omega t \\
\varepsilon^2 &: \quad \ddot{u}_2 + \omega_0^2 u_2 = -\delta u_1 \cos \Omega t
\end{aligned} \tag{5.73}$$

By substituting the solution of the first Equation (5.73) into the second equation, we obtain the following:

$$\begin{aligned}
\ddot{u}_1 + \omega_0^2 u_1 &= -\delta a \cos \Omega t \cos(\omega_0 t + \beta) \\
&= -\frac{1}{2}\delta a \left\{ \cos\left[(\Omega - \omega_0)t + \beta\right] + \cos\left[(\Omega + \omega_0)t - \beta\right] \right\}
\end{aligned} \tag{5.74}$$

The particular response of u_1 can be derived from the Equation (5.74) in the following manner:

$$u_1 = -\frac{1}{2}\delta a \left\{ \frac{\cos\left[(\Omega + \omega_0)t + \beta\right]}{\Omega(\Omega + 2\omega_0)} + \frac{\cos\left[(\Omega - \omega_0)t - \beta\right]}{\Omega(\Omega - 2\omega_0)} \right\} \tag{5.75}$$

Now, by substituting the value of u_1 from the Equation (5.75) into the third Equation (5.73), we obtain the following:

$$\begin{aligned}
\ddot{u}_2 + \omega_0^2 u_2 &= -\delta u_1 \cos \Omega t \\
&= -\frac{\delta^2 a}{4\Omega(\Omega + 2\omega_0)} \left\{ \cos\left[(2\Omega + \omega_0)t + \beta\right] + \cos\left[(\omega_0 t + \beta)\right] \right\} \\
&\quad - \frac{\delta^2 a}{4\Omega(\Omega + 2\omega_0)} \left\{ \cos\left[(2\Omega - \omega_0)t - \beta\right] + \cos\left[(\omega_0 t + \beta)\right] \right\}
\end{aligned} \tag{5.76}$$

When $\Omega \simeq 2\omega_0$ in sentence $(\Omega - 2\omega_0)$ in Equation (5.75), a phenomenon known as the primary parametric resonance takes place. Furthermore, if we consider the stated Equation (5.76), when the equation $2\Omega - \omega_0 = \omega_0$ holds true, a secondary resonance will occur. This implies that when $\Omega = \omega_0$, a secondary resonance will be observed. Secondary resonance can be achieved by expressing the particular solution to Equation (5.76).

5.5.3 Primary Resonance Caused by Harmonic Excitation in the Linear System

In the last section, we determined that during the primary resonance, we will have $\Omega \simeq 2\omega_0$. Next, we will examine the primary resonance by employing the averaging technique. Once again, we will rephrase the Equation (5.71) [11, 12]:

$$\ddot{u}+\omega_0^2 u+\varepsilon\delta\cos\Omega t\,u=0 \tag{5.77}$$

Applying the aforementioned relations (Equations (4.94) to (4.98)) of the averaging approach, the following results will be obtained:

$$\dot{a}=\frac{\varepsilon}{2\pi\omega_0}\int_0^{2\pi} f\left[a\cos\phi,\,-a\omega_0\sin\phi\right]\sin\phi\,d\phi \tag{5.78}$$

$$a\dot{\beta}=\frac{\varepsilon}{2\pi\omega_0}\int_0^{2\pi} f\left[a\cos\phi,\,-a\omega_0\sin\phi\right]\cos\phi\,d\phi \tag{5.79}$$

Considering the supplied expressions $f(u,\dot{u})=\delta\cos\Omega tu$ and $\phi=(2\omega_0 t+\beta)$, we can substitute them into Equations (5.78) and (5.79) and then perform integration to obtain the following result:

$$\begin{cases}\dot{a}=\dfrac{\delta\varepsilon a}{4\omega_0}\sin\left[(2\omega_0-\Omega)t+2\beta\right]\\[2ex] a\dot{\beta}=\dfrac{\delta\varepsilon a}{4\omega_0}\cos\left[(2\omega_0-\Omega)t+2\beta\right]\end{cases} \tag{5.80}$$

Equation (5.80) has our frequency equations. Next, we aim to eliminate the temporal dependence from Equation (5.80). By defining $\gamma=(2\omega_0-\Omega)t+2\beta$ and $\dot{\gamma}=2\omega_0-\Omega+2\dot{\beta}$, Equation (5.80) can be expressed as follows:

$$\begin{cases}\dot{a}=\dfrac{\delta\varepsilon a}{4\omega_0}\sin\gamma\\[2ex] \dot{\gamma}=(2\omega_0-\Omega)+\dfrac{\delta\varepsilon}{2\omega_0}\cos\gamma\end{cases} \tag{5.81}$$

The solution to Equation (5.77) is as follows:

$$u=a(t)\cos\left[\omega_0 t+\beta(t)\right]=a(t)\cos\left[\frac{1}{2}\Omega t+\frac{1}{2}\gamma\right] \tag{5.82}$$

The values of a and γ are derived from Equation (5.81).

In order to assess the stability of the system in linear mode, we deviate from the prior procedure by initially analyzing the non-trivial solutions $(a\neq 0)$ of the system, followed by an examination of the trivial solutions $(a=0)$. Assuming that $(a\neq 0)$, let's divide both frequency Equations (5.81):

$$\frac{da}{a}=\frac{\delta\varepsilon\sin\gamma d\gamma}{4\omega_0\left[(2\omega_0-\Omega)+\dfrac{\delta\varepsilon}{2\omega_0}\cos\gamma\right]} \tag{5.83}$$

We will compute the integral of the aforementioned equation.

$$\ln a=-\frac{1}{2}\ln\left[2\omega_0(2\omega_0-\Omega)+\delta\varepsilon\cos\gamma\right]+\ln c \tag{5.84}$$

The domain "a" is formed from the Equation (5.84) as follows:

$$a = \frac{c}{\sqrt{2\omega_0 (2\omega_0 - \Omega) + \delta\varepsilon \cos\gamma}} \tag{5.85}$$

If:

$$\cos\gamma = \frac{2\omega_0 (\Omega - 2\omega_0)}{\delta\varepsilon} \tag{5.86}$$

Subsequently, the occurrence of $a \to \infty$ triggers a strong need and thus leads to an unstable state of the system. If $-1 \le \cos\gamma \le 1$, then:

$$2\omega_0 (\Omega - 2\omega_0) = \pm\delta\varepsilon \to \ \Omega = 2\omega_0 \pm \frac{\delta\varepsilon}{2\omega_0} \tag{5.87}$$

To determine the stability or instability of the system, it is essential to create a graph that illustrates the relationship between the parametric excitation domain (δ) and the excitation frequency (Ω).

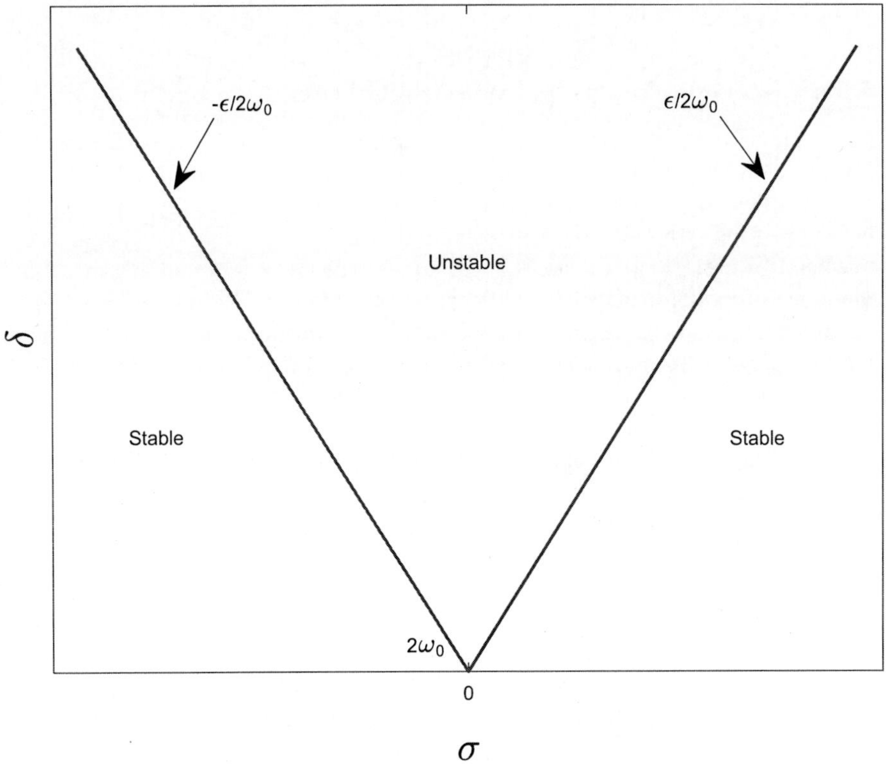

FIGURE 5.15 Stability behavior of the system in exchange for different values of the parametric excitation domain (δ).

If we make the assumption that $(a=0)$, then:

$$\begin{cases} \dot{a} = \dfrac{\delta\varepsilon a}{4\omega_0} sin\, \gamma \\ \dot{\gamma} = (2\omega_0 - \Omega) + \dfrac{\delta\varepsilon}{2\omega_0} cos\, \gamma \end{cases} \tag{5.88}$$

In order to assess the stability of the trivial solutions, assuming $(a=0)$, we examine the response as follows:

$$\begin{cases} a = 0 + a_1(t) \\ \gamma = \gamma_0 + \gamma_1(t) \end{cases} \tag{5.89}$$

Where $a_1(t)$ represents a value that is not equal to zero. Indeed, we generate a state of perturbation surrounding the origin. The value of γ_0 represents the stable state response of γ, which may or may not be zero. Additionally, we generate a perturbation $(\gamma_1(t))$ in the vicinity of γ_0. By incorporating the Equation (5.89) into the Equation (5.88), a new relationship is derived based on $a_1(t)$ and $\gamma_1(t)$, which is equivalent to the procedure used to solve the preceding ection. Further information can be found in the Nayfeh and Mook [1].

5.5.4 Nonlinear Effects on Parametric Excitation

In the previous section, we established that parametrically stimulated damper-free linear systems exhibit infinite growth in their responses over time. In practical systems, various degrees of damping can significantly influence the system's stability behavior. If the system is linear, its magnitude will increase until the system ceases to exist. However, in most systems, there are varying degrees of nonlinearity. When the range of motion reaches a significant extent, this nonlinearity comes into play and alters the system's response. Under certain circumstances, when the amplitude increases, the nonlinear impact constrains the expansion by forming a limit cycle.

In order to analyze the nonlinear effects in systems that are subjected to parametric excitation, we investigate a system that exhibits nonlinearity. Now, we will examine Equation (5.71) in conjunction with a nonlinear and damping statement as stated [13]:

$$\ddot{u} + 2\varepsilon\mu\dot{u} + \omega_0^2 u + \varepsilon\alpha u^3 + \varepsilon\delta\, cos\, \Omega t\, u = 0 \tag{5.90}$$

Within the context of Equation (5.90), μ represents the damping coefficient and α represents the nonlinear coefficient. Given that the primary resonance is approximately equivalent to $\Omega \simeq 2\omega_0$, we will address the issue by employing the method of multiple time scales. To address this, we can express the solution in the following manner:

$$u(T_0, T_1) = u_0(T_0, T_1) + \varepsilon u_1(T_0, T_1) + \ldots \tag{5.91}$$

Similarly, by substituting the value of Equation (5.91) into the Equation (5.90), we may separate the phrases based on the coefficients ε^0 and ε^1:

$$\begin{aligned} &\varepsilon^0 : D_0^2 u_0 + \omega_0^2 u_0 = 0 \rightarrow u_0 = A(T_1) e^{i\omega_0 T_0} + CC \\ &\varepsilon^1 : D_0^2 u_1 + \omega_0^2 u_1 = -2D_0 D_1 u_0 - 2\mu D_0 u_0 - \alpha u_0^3 - \delta u_0 \cos \Omega T_0 \end{aligned} \tag{5.92}$$

By substituting the solution of the first Equation (5.92) into the second equation, we obtain the following:

$$\begin{aligned} D_0^2 u_1 + \omega_0^2 u_1 = &-2i\omega_0 A' e^{i\omega_0 T_0} - 2i\mu\omega_0 A e^{i\omega_0 T_0} - \alpha A^3 e^{3i\omega_0 T_0} - 3\alpha A^2 \bar{A} e^{i\omega_0 T_0} + CC \\ &- \frac{1}{2}\delta\left(e^{i\Omega T_0} + e^{-i\Omega T_0}\right)\left(A e^{i\omega_0 T_0} + \bar{A} e^{-i\omega_0 T_0}\right) \end{aligned} \tag{5.93}$$

The expression $\cos \Omega T_0 = \frac{1}{2}\left(e^{i\Omega T_0} + e^{-i\Omega T_0}\right)$ is true in Equation (5.93). Since $\Omega \simeq 2\omega_0$, we can express the resonance conditions in a more thorough manner:

$$\Omega \simeq 2\omega_0 \rightarrow \Omega = 2\omega_0 + \sigma\varepsilon \tag{5.94}$$

The σ parameter is referred to as the detuning parameter in Equation (5.94). By including the Equation (5.94) into the Equation (5.93) and attempting to exclude secular terms in Equation (5.93), we can express it as follows:

$$\left[-2i\omega_0 A' - 2i\mu\omega_0 A - 3\alpha A^2 \bar{A}\right] e^{i\omega_0 T_0} - \frac{1}{2}\delta \bar{A} e^{i(\Omega - \omega_0) T_0} = 0 \tag{5.95}$$

Based on the earlier equations $\Omega = 2\omega_0 + \sigma\varepsilon$ and $T_1 = \varepsilon T_0$, and considering the substituting in the Equation (5.95), we may express this as follows:

$$2i\omega_0 A' + 2i\mu\omega_0 A + 3\alpha A^2 \bar{A} + \frac{1}{2}\delta \bar{A} e^{i\sigma T_1} = 0 \tag{5.96}$$

Equation (5.96) is equivalent to the equation for the frequency response. Now we must determine a solution for A expressed in polar form. To achieve this objective, we establish the definition $A = \frac{1}{2} a e^{i\beta}$, where a and β are functions dependent on T_1. The reason for utilizing the polar form of answer A is that this response format is appropriate for obtaining non-trivial system solutions $(a \neq 0)$. By considering the value A in the context of the Equation (5.96), and decomposing it into its real and imaginary components, we obtain the following:

$$\begin{cases} a' = -\mu a + \dfrac{\delta}{4\omega_0} a \sin\left(\sigma T_1 - 2\beta\right) \\ a\beta' = \dfrac{3\alpha}{8\omega_0} a^3 - \dfrac{\delta}{4\omega_0} a \cos\left(\sigma T_1 - 2\beta\right) \end{cases} \tag{5.97}$$

By introducing the variables $\gamma=(\sigma T_1-2\beta)$ and $\gamma'=(\sigma-2\beta')$, Equation (5.97) can be expressed in a simplified form as follows:

$$\begin{cases} a'=-\mu a+\dfrac{\delta}{4\omega_0}a\sin\gamma \\ a\gamma'=\sigma a-\dfrac{3\alpha}{4\omega_0}a^3+\dfrac{\delta}{2\omega_0}a\cos\gamma \end{cases} \tag{5.98}$$

In order to assess the stability of the trivial steady-state responses described by Equation (5.90) under parametric excitation, we will now employ the Cartesian form response $\left(A=\frac{1}{2}(p\pm iq)e^{i\lambda T_1}\right)$ instead of the polar form response. The Cartesian-type response is employed solely for the sake of simplifying the problem-solving process. As demonstrated later, this form of response aids in the examination of the stability condition of the system's trivial response. It is worth mentioning that both $(p+iq)$ and $(p-iq)$ can be utilized interchangeably without impacting the solving procedure. By substituting the solution of the Cartesian form into the frequency Equation (5.96) and resolving it, we obtain the following:

$$\begin{aligned} &i\,\omega_0(p'-iq')-\omega_0\lambda(p-iq)+i\mu\omega_0(p-iq)+\frac{3\alpha}{8}(p^2-q^2-2ipq)(p+iq) \\ &+\frac{\delta}{4}(p+iq)e^{i(\sigma-2\lambda)T_1}=0 \end{aligned} \tag{5.99}$$

For the given values (5.99), p and q are dependent on the variable T_1. To eliminate the time dependence of the Equation (5.99), it is imperative to set $\lambda=\frac{\sigma}{2}$. By substituting the value of λ with $\frac{\sigma}{2}$ in the Equation (5.99), we obtain the following:

$$\begin{aligned} &i\,\omega_0(p'-iq')-\omega_0\frac{\sigma}{2}(p-iq)+i\mu\omega_0(p-iq)+\frac{3\alpha}{8}(p^2-q^2-2ip)(p+iq) \\ &+\frac{\delta}{4}(p+iq)=0 \end{aligned} \tag{5.100}$$

The frequency response equations are derived by segregating the real and imaginary components in Equation (5.100).

$$\begin{cases} p'=-\dfrac{\sigma}{2}q-\mu p+\dfrac{3\alpha}{8\omega_0}q(p^2+q^2)-\dfrac{\delta}{4\omega_0}q \\ q'=\dfrac{\sigma}{2}p-\mu q-\dfrac{3\alpha}{8\omega_0}p(p^2+q^2)-\dfrac{\delta}{4\omega_0}p \end{cases} \tag{5.101}$$

5.5.4.1 Evaluating the Stability of the Trivial Solution to the Equation $(p=q=0)$

An analysis is conducted to assess the stability and instability of the system by studying its steady-state responses, regardless of whether they are trivial or non-trivial.

In order to achieve this objective, we shall get the Jacobian matrix by linearizing Equation (5.101) around the point $p = q = 0$. The resulting matrix is as follows [14]:

$$A = \begin{bmatrix} -\mu & -\frac{\sigma}{2} - \frac{\delta}{4\omega_0} \\ \frac{\sigma}{2} - \frac{\delta}{4\omega_0} & -\mu \end{bmatrix} \tag{5.102}$$

Upon performing the computation of the determinant and the trace matrix of Equation (5.102), the following result will be obtained:

$$\begin{cases} \Delta = \mu^2 + \frac{\sigma^2}{4} - \frac{\delta^2}{16\omega_0^2} \\ \tau = -2\mu \end{cases} \tag{5.103}$$

Given that $\mu > 0$ and defining $k = \frac{\delta}{2\omega_0}$, when:

$$\begin{cases} \Delta < 0 \\ \mu^2 + \frac{\sigma^2}{4} < \frac{\delta^2}{16\omega_0^2} \end{cases} \rightarrow \begin{cases} \Delta < 0 \\ 4\mu^2 + \sigma^2 < k^2 \end{cases} \tag{5.104}$$

The system is expected to exhibit instability. Furthermore:

$$\begin{cases} \Delta > 0 \\ \mu^2 + \frac{\sigma^2}{4} > \frac{\delta^2}{16\omega_0^2} \end{cases} \rightarrow \begin{cases} \Delta > 0 \\ 4\mu^2 + \sigma^2 > k^2 \end{cases} \tag{5.105}$$

The system will maintain a state of stability. By plotting the equation $\left(4\mu^2 + \sigma^2 = k^2\right)$ on a graph using k and σ as variables, we obtain the following graphical representation.

Now, one can examine the system's stability mode for trivial solutions. Based on the graph shown in Figure 5.16, it is evident that answers above the curve can be characterized as unstable, namely, when $4\mu^2 + \sigma^2 = k^2$. Conversely, responses below the curve can be considered stable. At $\sigma = 0$, the value of the parametric excitation domain (k) is precisely 2μ.

5.5.4.2 The Stability of Non-trivial Steady-State Responses

In order to retrieve the non-trivial solutions of the system, we utilize the polar form solution that we already derived (Equation (5.98)). Once we write the frequency equations of the steady state explicitly, the solving process is completed [13, 14].

$$\begin{cases} a' = 0 \rightarrow \mu a = \frac{\delta}{4\omega_0} a \sin\gamma \\ \gamma' = 0 \rightarrow \frac{1}{2}\sigma a - \frac{3\alpha}{8\omega_0} a^3 + \frac{\delta}{4\omega_0} a \cos\gamma \end{cases} \tag{5.106}$$

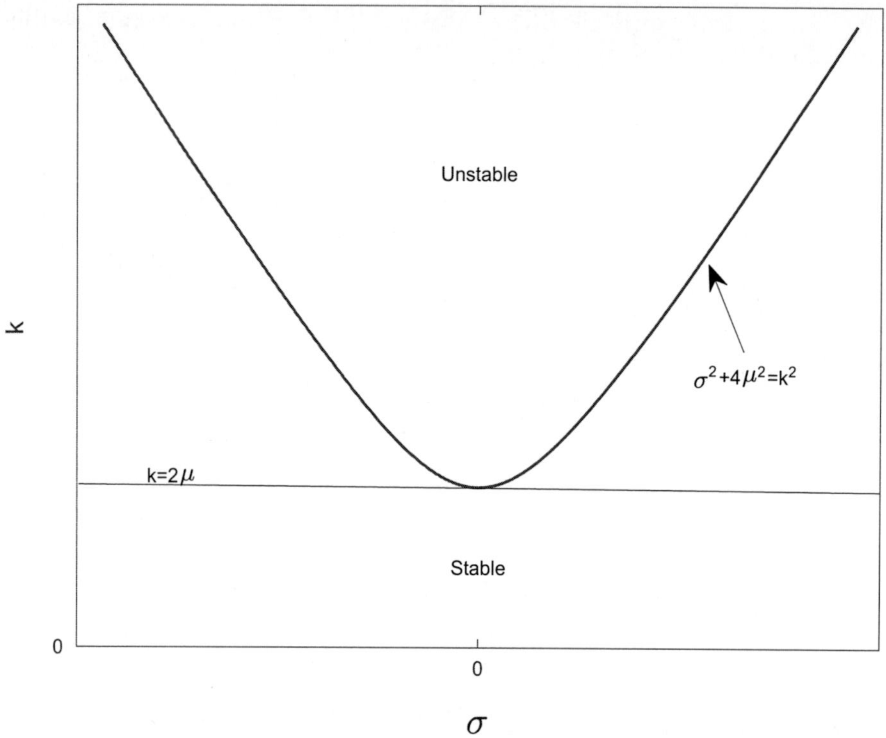

FIGURE 5.16 Behavior of the detuning parameter σ by parametric excitation domain k.

Given that $a \neq 0$ (yielding non-trivial solutions), we proceed by dividing two Equation (5.106) by a and then by utilizing the trigonometric relationship $\left(\sin^2 \gamma + \cos^2 \gamma = 1\right)$, we arrive at the following relation:

$$\mu^2 + \left(\frac{1}{2}\sigma - \frac{3\alpha}{8\omega_0}a^2\right)^2 = \frac{\delta^2}{16\omega_0^2} \tag{5.107}$$

The domain (a) is obtained as the response of the real part of the complex form from the Equation (5.107). Therefore, we will have the following:

$$a = \left[\frac{4\omega_0}{3\alpha}\left(\sigma \pm \sqrt{\frac{\delta^2}{4\omega_0^2} - 4\mu^2}\right)\right]^{\frac{1}{2}} = \left[\frac{4\omega_0}{3\alpha}\left(\sigma \pm \sqrt{k^2 - 4\mu^2}\right)\right]^{\frac{1}{2}} \tag{5.108}$$

The notable aspect of this topic is that, considering the circumstances and limitations outlined in Equation (5.108) (including the presence of negative values within the square root), it can be asserted that non-trivial solutions are not always present.

However, the trivial solution is always present. Thus, in order to verify the presence or absence of non-trivial solutions, we proceed as follows:

Given that $\alpha > 0$ and the domain of a is a real value, it follows that the term under radical in question is always positive, as this condition necessitates. The value of $k \geq 2\alpha$ is greater than or equal to 2 ($k \geq 2\mu$).

1. If $\sigma < 0$, then:

$$|\sigma| < \sqrt{k^2 - 4\mu^2} \tag{a}$$

In this scenario, we are situated on the negative side of the curve depicted in Figure 5.17 $(\sigma < 0)$, where we can expect an unstable outcome. The unstable response is located at the apex of the curve $(4\mu^2 + \sigma^2 = k^2)$ and on the negative side of the graph $(\sigma < 0)$, shown by the label II in Figure 5.23.

2. If $\sigma > 0$, then:

$$\begin{cases} (1) \rightarrow \sigma > \sqrt{k^2 - 4\mu^2} \\ (2) \rightarrow \sigma < \sqrt{k^2 - 4\mu^2} \end{cases}$$

There are two solutions in this scenario. Specifically, in this instance, we are positioned below the curve $(4\mu^2 + \sigma^2 = k^2)$ and above the line $k = 2\alpha$ on the right side of the picture shown in Figure 5.23. This region is designated with the III. Furthermore, it should be noted that in region I, namely, below the line where $k = 2\alpha$ and to the left and below the curve $(4\mu^2 + \sigma^2 = k^2)$, there is no solution available as depicted in Figure 5.23.

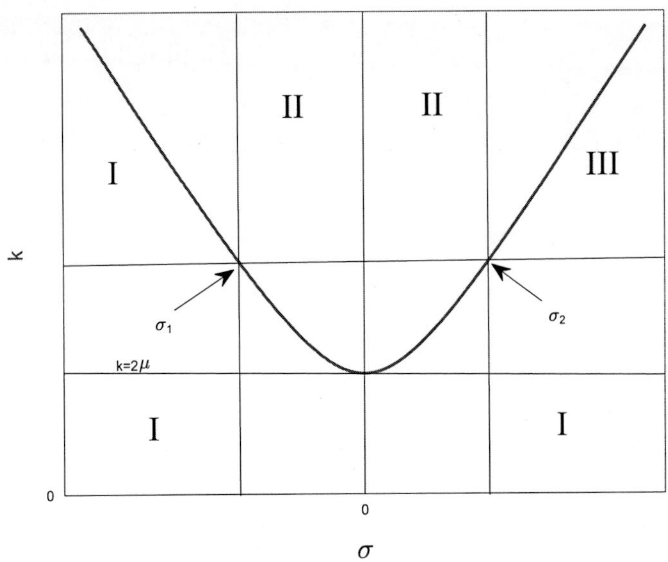

FIGURE 5.17 Different areas of stability and instability of the system.

To assess the stability of non-trivial steady responses, it is recommended to employ Equation (5.98). The following equations have been restated later for the purpose of refreshing one's memory:

$$\begin{cases} a' = -\mu a + \dfrac{\delta}{4\omega_0} a \sin\gamma \\ a\gamma' = \sigma a - \dfrac{3\alpha}{4\omega_0} a^3 + \dfrac{\delta}{2\omega_0} a \cos\gamma \end{cases} \tag{5.109}$$

Given that non-trivial responses have a non-zero amplitude, it would be beneficial to employ the following variable transformation in the process of issue solving:

$$a = a_0 + a_1 \tag{5.110}$$

$$\gamma = \gamma_0 + \gamma_1 \tag{5.111}$$

The roots of Equation (5.98) for the condition $a' = 0$ and $\gamma' = 0$ are denoted as a_0 and γ_0, respectively. Equations (5.110) and (5.111) are now inserted into Equation (5.98).

$$a_1' = -\mu a_0 - \mu a_1 + \frac{\delta}{4\omega_0}(a_0 + a_1)\sin(\gamma_0 + \gamma_1) \tag{5.112}$$

$$\gamma_1' = -\frac{3\alpha}{4\omega_0}\left(a_0 + a_1)^2 + \frac{\delta}{2\omega_0}\cos(\gamma_0 + \gamma_1\right) \tag{5.113}$$

Now, by extending the capabilities of two variables' functions, the process of linearizing Equations (5.112) and (5.113) can be carried out in the following manner:

$$\begin{aligned} a_1' &= [-\mu + \frac{\delta}{4\omega_0}\sin(\gamma_0 + \gamma_1)]_{(a_1=0,\gamma_1=0)} a_1 + [\frac{\delta}{4\omega_0}(a_0 + a_1) \\ &\cos(\gamma_0 + \gamma_1)]_{(a_1=0,\gamma_1=0)} \gamma_1 = [-\mu + \frac{\delta}{4\omega_0}\sin(\gamma_0]a_1 + [\frac{\delta}{4\omega_0}a_0\cos(\gamma_0)]\gamma_1 \end{aligned} \tag{5.114}$$

$$\begin{aligned} \gamma_1' &= [-\frac{3\alpha}{2\omega_0}(a_0 + a_1)]_{(a_1=0,\gamma_1=0)} a_1 + [-\frac{\delta}{2\omega_0}\sin(\gamma_0 + \gamma_1)]_{(a_1=0,\gamma_1=0)} \gamma_1 \\ &= [-\frac{3\alpha}{2\omega_0}a_0]a_1 + [-\frac{\delta}{2\omega_0}\sin(\gamma_0)]\gamma_1 \end{aligned} \tag{5.115}$$

In order to proceed with the solution, it would be more convenient to transform the trigonometric values of the Equations (5.114) and (5.115) into algebraic values. In order to achieve this, it is necessary for the derivatives in Equation (5.109) to be equal to zero. Once this condition is met, the resulting equations will have roots a_0 and γ_0. By substituting these roots into Equations (5.114) and (5.115), trigonometric values can be determined as follows:

$$\begin{aligned} &a' = -\mu a + \frac{\delta}{4\omega_0}a\sin(\gamma) \xrightarrow{a'=0} \mu = \frac{\delta}{4\omega_0}\sin(\gamma_0) \rightarrow \sin(\gamma_0) = \frac{4\mu\omega_0}{\delta} \\ &a\gamma' = \sigma a - \frac{3\alpha}{4\omega_0}a^3 + \frac{\delta}{2\omega_0}a\cos(\gamma) \xrightarrow{\gamma'=0} \frac{\delta}{2\omega_0}\cos(\gamma_0) = -\sigma + \frac{3\alpha}{4\omega_0}a_0^2 \end{aligned} \tag{5.116}$$

$$cos(\gamma_0) = \frac{-2\omega_0\sigma}{\delta} + \frac{3\alpha}{2\delta}a_0^{\ 2} \tag{5.117}$$

The final linearization is obtained by substituting Equations (5.116) and (5.117) into Equations (5.114) and (5.115), respectively.

$$a_1' = \left[-\frac{1}{2}\sigma a_0 + \frac{3\alpha a_0^{\ 3}}{8\omega_0}\right]\gamma_1 \tag{5.118}$$

$$\gamma_1' = \left[-\frac{3\alpha}{2\omega_0}a_0\right]a_1 + [-2\mu]\gamma_1 \tag{5.119}$$

Based on Equations (5.118) and (5.119), the Jacobian matrix is obtained in the following manner:

$$A = \begin{bmatrix} 0 & -\frac{1}{2}\sigma a_0 + \frac{3\alpha a_0^{\ 3}}{8\omega_0} \\ -\frac{3\alpha}{2\omega_0}a_0 & -2\mu \end{bmatrix} \tag{5.120}$$

The determinant and trace can be determined using the matrix of Equation (5.120) as follows:

$$\tau = -2\mu \tag{5.121}$$

$$\Delta = -\frac{3\alpha\sigma a_0^{\ 2}}{4\omega_0} + \frac{9\alpha^2 a_0^{\ 4}}{16\omega_0^{\ 2}} \tag{5.122}$$

We are commencing the examination of stability.

Initially, the values with a $\sigma < 0$ are taken into account. Indeed, inside this band, only values of $+\sqrt{k^2 - 4\mu^2}$ are deemed acceptable. Now we substitute Equation (5.108) into Equation (5.122). It is important to mention that the trace $\tau < 0$, so:

$$\begin{aligned}\Delta &= -\frac{3\alpha\sigma}{4\omega_0}(\frac{4\omega_0}{3\alpha}(\sigma + \sqrt{k^2 - 4\mu^2})) + \frac{9\alpha^2}{16\omega_0^{\ 2}}(\frac{16\omega_0^{\ 2}}{9\alpha^2}(\sigma^2 + 2\sigma\sqrt{k^2 - 4\mu^2} + k^2 - 4\mu^2)) \\ &= \sigma\sqrt{k^2 - 4\mu^2} + k^2 - 4\mu^2\end{aligned} \tag{5.123}$$

The solution is considered unstable when $\Delta < 0$. Therefore:

$$\Delta < 0 \rightarrow \sigma\sqrt{k^2 - 4\mu^2} + k^2 - 4\mu^2 < 0 \rightarrow \sigma < \frac{4\mu^2 - k^2}{\sqrt{k^2 - 4\mu^2}} \rightarrow \sigma^2 < k^2 - 4\mu^2 \tag{5.124}$$

Consequently, if $\sigma < 0$, the upper region of the curve depicted in Figure 5.23 will exhibit instability.

The answer is stable when $\Delta > 0$.

$$\Delta > 0 \rightarrow \sigma\sqrt{k^2 - 4\mu^2} + k^2 - 4\mu^2 > 0 \rightarrow \sigma^2 > k^2 - 4\mu^2 \rightarrow Unacceptable \tag{5.125}$$

Equation (5.125) is unacceptable. The equation contradicts the requirement $\sqrt{k^2-4\mu^2}>|\sigma|$ that was derived for $\sigma<0$.

Now, we will consider the $\sigma>0$, for which there will exist a steady state on the aforementioned subjects. If $\sigma<\sqrt{k^2-4\mu^2}$, then there is only one solution. Specifically, only the values of $+\sqrt{k^2-4\mu^2}$ are considered acceptable in regard to Equation (5.108). Next, we insert the Equation (5.108) into Equation (5.121) while acknowledging that $\tau<0$:

$$\Delta=\sigma\sqrt{k^2-4\mu^2}+k^2-4\mu^2 \tag{5.126}$$

The solution is considered unstable when $\Delta<0$.

$$\Delta<0\rightarrow\sigma\sqrt{k^2-4\mu^2}+k^2-4\mu^2<0\rightarrow\sigma^2<k^2-4\mu^2 \tag{5.127}$$

Thus, when $\sigma>0$, the peak of the curve depicted in Figure 5.23 will exhibit instability.

The solution is stable when $\Delta>0$:

$$\Delta>0\rightarrow\sigma\sqrt{k^2-4\mu^2}+k^2-4\mu^2>0\rightarrow\sigma^2>k^2-4\mu^2\rightarrow \textit{Unacceptable} \tag{5.128}$$

The equation denoted as Equation (5.128) is deemed unacceptable. The equation is in direct conflict with the relation $\sigma<\sqrt{k^2-4\mu^2}$.

Next, we will examine the second scenario in which $\sigma>0$ and also $\sigma>\sqrt{k^2-4\mu^2}$. There are two acceptable solutions in this scenario, and the allowable values for Equation (5.108) are $\pm\sqrt{k^2-4\mu^2}$. Hence, we will derive two equations for the determinants. The expression $+\sqrt{k^2-4\mu^2}$ is initially examined in the context of the Equation (5.108). By substituting the value of Equation (5.108) into the Equation (5.122), we obtain the following:

$$\Delta=\sigma\sqrt{k^2-4\mu^2}+k^2-4\mu^2 \tag{5.129}$$

The solution is deemed unstable when $\Delta<0$.

$$\Delta<0\rightarrow\sigma\sqrt{k^2-4\mu^2}+k^2-4\mu^2<0\rightarrow\sigma^2<k^2-4\mu^2\rightarrow \textit{Unacceptable} \tag{5.130}$$

The obtained answer is unacceptable. The requirement $\sigma>\sqrt{k^2-4\mu^2}$ is contradicted by this response. Next, we will examine the case where $\Delta>0$:

$$\Delta>0\rightarrow\sigma\sqrt{k^2-4\mu^2}+k^2-4\mu^2>0\rightarrow\sigma^2>k^2-4\mu^2 \tag{5.131}$$

One of the solutions for $\sigma>0$, located beneath the curve depicted in Figure 5.23.

For values of $\sigma>0$ and $\sigma>\sqrt{k^2-4\mu^2}$, there exist two states. The initial state was analyzed using Equations (5.129) to (5.131). Next, we will examine the second condition in which the Equation (5.108) incorporates the term $-\sqrt{k^2-4\mu^2}$. We substituted the Equation (5.108) into Equation (5.121):

$$\Delta = -\frac{3\alpha\sigma}{4\omega_0}\left(\frac{4\omega_0}{3\alpha}\left(\sigma-\sqrt{k^2-4\mu^2}\right)\right)+\frac{9\alpha^2}{16\omega_0{}^2}\left(\frac{16\omega_0{}^2}{9\alpha^2}\left(\sigma^2-2\sigma\sqrt{k^2-4\mu^2}+k^2-4\mu^2\right)\right)$$
$$= -\sigma\sqrt{k^2-4\mu^2}+k^2-4\mu^2 \tag{5.132}$$

The solution is considered unstable when " <0.

$$\Delta<0 \rightarrow -\sigma\sqrt{k^2-4\mu^2}+k^2-4\mu^2<0 \rightarrow -\sigma\sqrt{k^2-4\mu^2}<4\mu^2-k^2$$
$$\rightarrow \sigma^2\sqrt{k^2-4\mu^2}>k^2-4\mu^2 \rightarrow \sigma^2>k^2-4\mu^2 \tag{5.133}$$

Equation (5.133) reveals that when $\sigma>0$, an unstable response occurs below the curve depicted in Figure 5.23.

The solution is stable when the value of $\Delta>0$.

$$\Delta>0 \rightarrow -\sigma\sqrt{k^2-4\mu^2}+k^2-4\mu^2>0 \rightarrow \sigma^2<k^2-4\mu^2 \rightarrow \textit{Unacceptable} \tag{5.134}$$

The equation denoted as Equation (5.134) is deemed unacceptable. The equation is in direct conflict with the constraint $\sigma>\sqrt{k^2-4\mu^2}$.

In the scenario where $\sigma>0$ and $\sigma>\sqrt{k^2-4\mu^2}$, there exist two solutions, one of which is stable and the other is unstable. This section is indicated in the lower part of the curve depicted in Figure 5.24, specifically in region III. It is important to mention that the aforementioned subjects are depicted with greater elaboration in Figure 5.18 compared to Figure 4.21.

To summarize, the quantity of responses in each region (I, II, III) of graph Figure 5.18, along with their stability and instability status, can be expressed as follows:

1. Within region I, there exists a trivial and stable response.
2. Within area II, there exist two solutions: one is readily trivial but lacks stability, while the other is non-trivial and possesses stability.
3. Within district III, there exist three distinct answers. The first option is both trivial and stable, while the second option is also non-trivial and stable. However, the third option is non-trivial and unstable.

By altering the amplitude of parametric excitation (k) and the detuning parameter (σ) in the diagram shown in Figure 5.19, we can observe the different binding states inside the system.

Equation (5.108) allows us to depict the non-trivial response amplitude (a) by utilizing the frequency response (σ) as defined in Figure 5.19. The relationship between various values of the frequency response (σ) and the amplitude of the excitation (k) is illustrated in Figure 5.17. By examining the frequency response variations for a

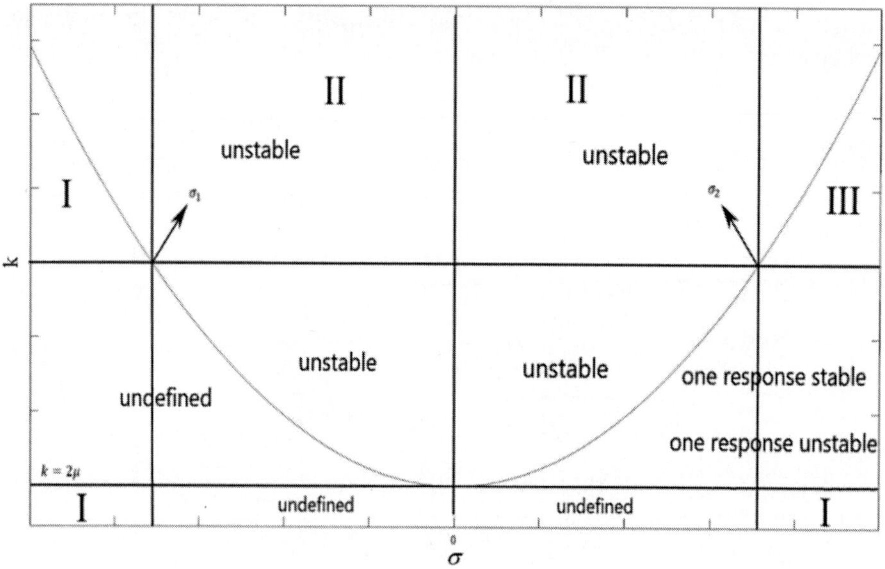

FIGURE 5.18 Examination of the stability of the parametric excitation graph in terms of frequency for the non-trivial response of the steady state.

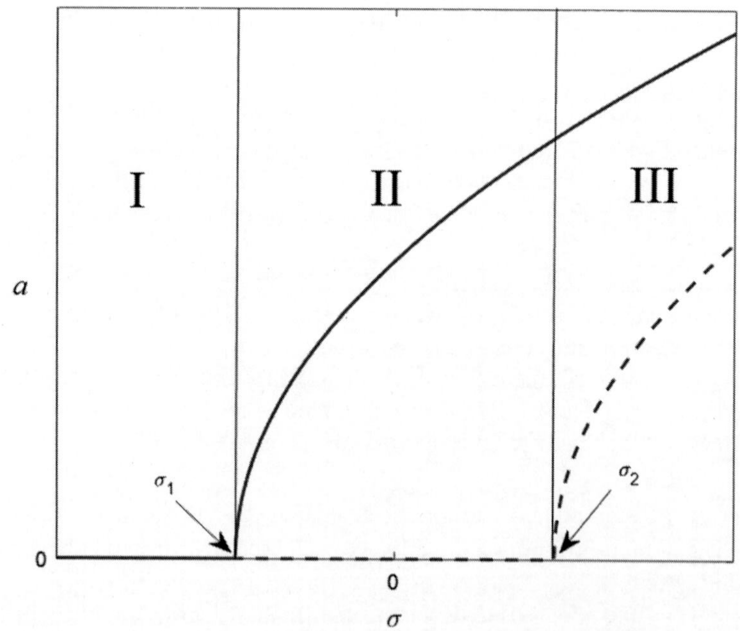

FIGURE 5.19 Behavior of the detuning parameter σ in terms of amplitude a.

constant k as depicted in Figure 5.17, we observe a progression from left to right, transitioning from area I to area II and finally entering area III. The graphic in Figure 5.19 displays the same variations for two specific values, σ_1 and σ_2, when the constant k is held constant. This is also seen in Figure 5.17. The diagram in Figure 5.19 shows two bifurcations, supercritical fork bifurcation at the σ_1 point and a subcritical bifurcation at the σ_2 position. It is important to observe that the dash line represents an unstable condition, whereas the extended line represents a stable condition inside the system.

Figure 5.20 demonstrates the relationship between fluctuations in the excitation amplitude (k) and the amplitude (a) in exchange for a constant frequency response (σ). The diagram in Figure 5.20 represents a specific frequency response value (σ_1) which is negative. By altering the value of k for σ_1, we transition from area I to area II. Consequently, we will observe a bifurcation known as a supercritical fork bifurcation.

Figure 5.20 and Figure 5.21 show modifications of the excitation amplitude (k) versus the amplitude (a) while maintaining a consistent frequency response (σ). In Figure 5.21, the graph illustrates the relationship between the excitation amplitude (k) and the amplitude (a) for the frequency response value (σ_2). The value of (σ_2) is positive. When the value of k is increased from a small value to a large value at point (σ_2), the transition occurs from area I to area III and finally to area II. From area I, characterized by a stable trivial solution, to area III, where the solution remains stable

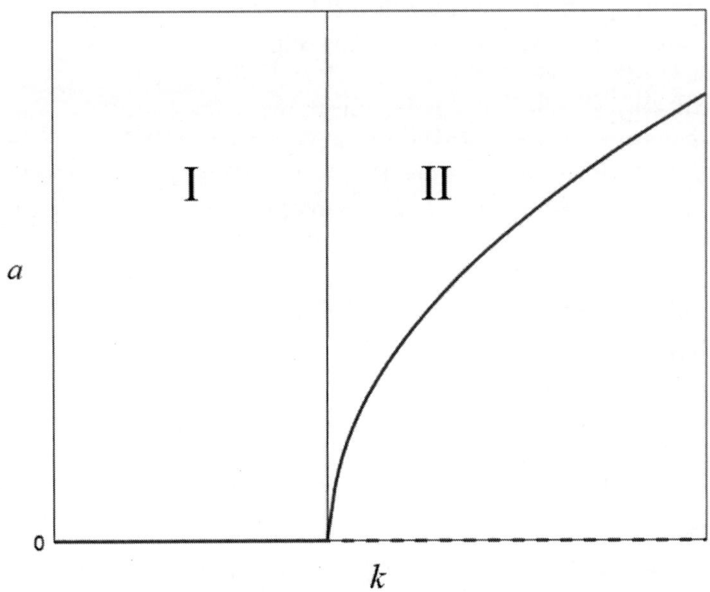

FIGURE 5.20 Parametric excitation amplitude k according to amplitude a for point σ_1.

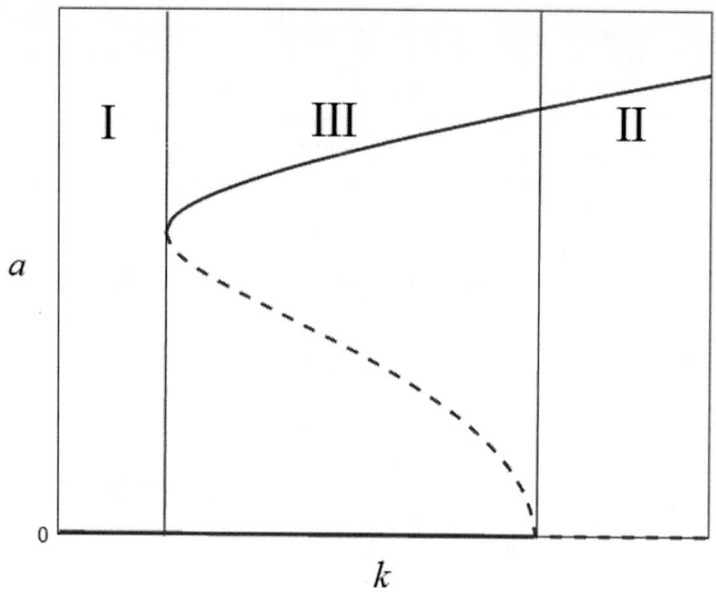

FIGURE 5.21 Parametric excitation amplitude k according to amplitude a for point σ_2.

and trivial, an unstable and non-trivial solution is introduced, along with a non-trivial and stable solution. At the point where the stable and unstable states overlap, namely, where the line and the expanded line $(k = 2\mu)$ collide, we will observe the formation of a saddle node bifurcation. This point also marks the end of the I-zone. Furthermore, when transitioning from area III to area II, the previously trivial and stable response transforms into a trivial although unstable response. However, there exists a stable and non-trivial solution (see to Figure 5.21). At the transition between area III and area II, there is a subcritical fork bifurcation with a value of $(a = 0)$. If you want to study an instance of parametric excitation of Euler-Bernoulli nanobeams, you can see the references [13].

REFERENCES

[1] Nayfeh, A.H. and D.T. Mook, *Nonlinear oscillations*. 2008: John Wiley & Sons.

[2] Saeidi, H., A. Zajkani, and M. Ghadiri, Nonlinear micromechanically analysis of forced vibration of the rectangular-shaped atomic force microscopes incorporating contact model and thermal influences. *Mechanics Based Design of Structures and Machines*, 2022. **50**(2): p. 609–629.

[3] Ghadiri, M., A. Rajabpour, and A. Akbarshahi, Non-linear forced vibration analysis of nanobeams subjected to moving concentrated load resting on a viscoelastic foundation considering thermal and surface effects. *Applied Mathematical Modelling*, 2017. **50**: p. 676–694.

[4] Faraday, M., On a peculiar class of acoustical figures; and on certain forms assumed by groups of particles upon vibrating elastic surfaces. In *Abstracts of the Papers Printed in the Philosophical Transactions of the Royal Society of London*. 1837: The Royal Society London.

[5] McLachlan, N.W., *Theory and application of Mathieu functions.* 1951: Geoffrey Cumberlege, Oxford University Press.

[6] Bondarenko, G., *The hill differential equation and its uses in engineering vibration problems.* 1936: Akademiia Nauk SSSR, Moscow.

[7] Magnus, W. and S. Winkler, *Hill's equation*. 2004: Courier Corporation.

[8] Nayfeh, A.H., *Perturbation methods*. 2008: John Wiley & Sons.

[9] Evan-Iwanowski, R., *Resonance oscillations in mechanical systems.* 1976: *Elsevier Scientific Publishing Company*.

[10] Kononenko, V.O., *Vibrating systems with a limited power supply.* 1969: Iliffe Books *Ltd, London*.

[11] Ghadiri, M. and S.H.S. Hosseini, Nonlinear forced vibration of graphene/piezoelectric sandwich nanoplates subjected to a mechanical shock. *Journal of Sandwich Structures & Materials*, 2021. **23**(3): p. 956–987.

[12] Ghadiri, M. and S.H.S. Hosseini, Nonlinear dual frequency excited vibration of viscoelastic graphene sheets exposed to thermo-magnetic field. *Communications in Nonlinear Science and Numerical Simulation*, 2020. **83**: p. 105111.

[13] Ghadiri, M. and S.H.S. Hosseini, Parametric excitation of Euler–Bernoulli nanobeams under thermo-magneto-mechanical loads: Nonlinear vibration and dynamic instability. *Composites Part B: Engineering*, 2019. **173**: p. 106928.

[14] Shayestenia, F. and M. Ghadiri, Investigation of flexoelectric effect on nonlinear vibration and dynamic instability of piezoelectric sandwich micro/nanobeam using the nonlocal strain gradient theory. *International Journal of Structural Stability and Dynamics*, 2023. **23**(04): p. 2350045.

6 Nonlocal Systems and Kinematics of the Continuous Structures

6.1 INTRODUCTION

Later, we are examining the application of the principles and information from earlier chapters to study the nonlinear characteristics of the continuous system in a precise and comprehensive manner. In this chapter, initially, an overview of the fundamental principles of the continuum mechanics is presented, followed by the utilization of the principles of continuum mechanics, we extract nonlinear strains based on the governing assumptions, and nonlinear strains are introduced based on the von Karman's theory. Next, we will introduce the nonlocal elastic theory, the modified coupled stress theory, and nonlocal strain gradient elasticity theory, which take into account the size effect and allow us to use them to study the dynamic behavior of micro and nanostructures. Finally, Hamilton's principle is introduced, which is used as a powerful tool for deriving the governing equations of nonlinear behavior of continuous structures in subsequent chapters.

6.2 EXPLANATION THE DYNAMICS OF A CONTINUOUS ENVIRONMENT

The trajectory of a particle in particle kinematics is determined by the time vector function t as follows [1]:

$$\boldsymbol{r} = r(t) \tag{6.1}$$

The position vector $\boldsymbol{r}(t)$ is defined as $\boldsymbol{r}(t) = x_1(t)\boldsymbol{e}_1 + x_2(t)\boldsymbol{e}_2 + x_3(t)\boldsymbol{e}_3$, where $x_1(t)$, $x_2(t)$, and $x_3(t)$ are the components of the vector as follows:

$$x_1 = x_1(t),\ x_2 = x_2(t),\ x_3 = x_3(t) \tag{6.2}$$

If there are N particles, there are likewise N trajectories, each of which can be represented by an equation as follows:

$$\boldsymbol{r}_n = r_n(t), \quad n = 1, 2, 3, \ldots, N \tag{6.3}$$

Specifically, the path for particle number 1 is represented by the notation $r_1(t)$, while the path for particle number 2 is designated as $r_2(t)$ and so forth. In the context of

DOI: 10.1201/9781003470694-6

a continuum environment, there exists a substantial quantity of particles. Hence, it is unfeasible to distinguish particles by assigning a numerical label to each one and following a comparable trajectory as the kinematic particles. It is possible to identify them based on their position at time t_0.

For instance, if a particle exists within a continuum environment and is located at location (1, 2, 3) at time $t = 0$, then the coordinate system $(1,2,3)$ can be employed to uniquely identify this particle. Therefore, if a particle of a continuum environment is located at position (X_1, X_2, X_3) at a specific reference time t_0, the coordinate system (X_1, X_2, X_3) can be employed to uniquely identify this particle. Therefore, the trajectories of motion for each particle in a continuum environment can be expressed using a vector equation as follows:

$$x = x(X,t), \quad X = x(X, t_0) \tag{6.4}$$

The position vector at time t for particle p (Figure 6.1) is given by $x = x_1\boldsymbol{e}_1 + x_2\boldsymbol{e}_2 + x_3\boldsymbol{e}_3$. At time t_0, particle p was at location $X = X_1\boldsymbol{e}_1 + X_2\boldsymbol{e}_2 + X_3\boldsymbol{e}_3$.

6.3 DISPLACEMENT FIELD

The displacement vector, denoted as $u(X,t)$, is the vector from the reference point $p(t_0)$ to the moving location $p(t)$ in the continuum environment with material abbreviation X according to Figure 6.2 [1, 2].

$$u(X,t) = x(X,t) - X \tag{6.5}$$

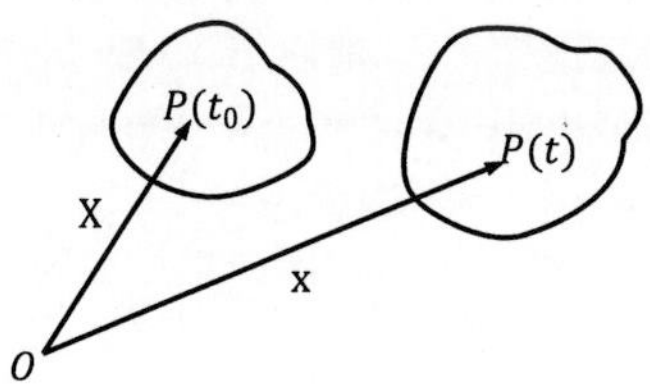

FIGURE 6.1 Position of a particle relative to the origin at time t.

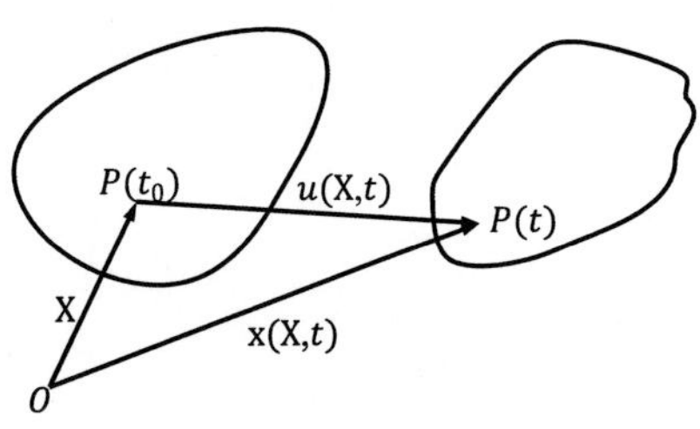

FIGURE 6.2 Vector of displacement of a particle in a continuum environment.

Equation (6.5) demonstrates that when the path lines of a continuum environment are unobstructed, the displacement field will also be unobstructed. Therefore, the movement of a continuum environment can be characterized either by the trajectories defined in Equation (6.4) or by the displacement vector field stated in Equation (6.5).

6.4 SMALL DEFORMATION

There are numerous engineering challenges that include mechanical organs or components that undergo minuscule deformations, which are analyzed in terms of mathematical foundations known as infinitesimal. In this episode, we get a tensor that characterizes the distortion of these objects [1, 2].

Let's examine an object that has a distinct shape at time t_0 and a different shape at time t (as seen in Figure 6.3). The material point p undergoes a displacement u, resulting in its relocation to a new position as follows:

$$x = X + u(X,t) \tag{6.6}$$

The Q point in the neighborhood, located at $X + dX$, reaches the $x + dx$ point, which is connected as $x + dx$.

$$x + dx = X + dX + u(X + dX, t) \tag{6.7}$$

By subtracting the value of Equation (6.6) from the value of Equation (6.7), we obtain the following:

$$dx = dX + u(X + dX, t) - u(X,t) \tag{6.8}$$

Applying the gradient definition of a vector function, the Equation (6.8) can be expressed as follows:

$$dx = dX + (\nabla u)dX \tag{6.9}$$

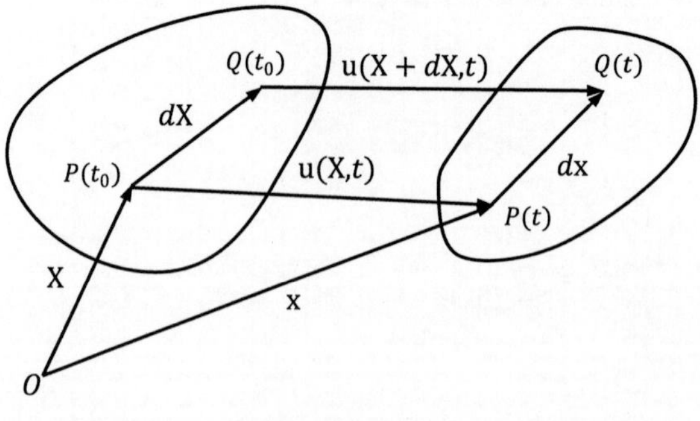

FIGURE 6.3 Very small deformations at the material point p under displacement u at time t.

The symbol ∇u represents a second-order tensor which is referred to as a displacement gradient. The ∇u matrix will be expressed with respect to the Cartesian coordinates $\left(u = u_i e_i, X = X_i e_i\right)$.

$$[\nabla u] = \begin{bmatrix} \frac{\partial u_1}{\partial X_1} & \frac{\partial u_1}{\partial X_2} & \frac{\partial u_1}{\partial X_3} \\ \frac{\partial u_2}{\partial X_1} & \frac{\partial u_2}{\partial X_2} & \frac{\partial u_2}{\partial X_3} \\ \frac{\partial u_3}{\partial X_1} & \frac{\partial u_3}{\partial X_2} & \frac{\partial u_3}{\partial X_3} \end{bmatrix} \tag{6.10}$$

The Equation (6.9) can be expressed as follows:

$$dx = FdX \tag{6.11}$$

Where:

$$F = I + \nabla u \tag{6.12}$$

The deformation gradient, denoted as F, is the gradient of the function $\hat{x}(X,t)$ that describes the motion, where $x = \hat{x}(X,t)$. In order to get the correlation between ds (a length of dx) and dS (a length of dX), we perform a dot product of the Equation (6.11) within itself:

$$dx.dx = FdX.FdX = dX.\left(F^T F\right)dX \tag{6.13}$$

It means:

$$ds^2 = dX.CdX \tag{6.14}$$

Where:

$$C = F^T F \tag{6.15}$$

The tensor C is referred to as the Caushy-Green deformation tensor. It should be noted that if $C = I$, then $ds^2 = dS^2$. Hence, the equation $C = I$ represents the motion of a physical entity, which might involve both translation and rotation. We have a form Equation (6.12):

$$C = F^T F = (I + \nabla u)^T (I + \nabla u) = I + \nabla u + (\nabla u)^T + (\nabla u)^T (\nabla u) \tag{6.16}$$

Let's:

$$E^* = \frac{1}{2}\left[\nabla u + (\nabla u)^T + (\nabla u)^T (\nabla u)\right] \tag{6.17}$$

Next, the Equation (6.16) will be expressed as follows:

$$C = I + 2E^{*} \tag{6.18}$$

Equation (6.18) states that the Tensor E^{*} represents alterations in length within a continuum environment due to the movement of material points, as $C = I$ pertains to the motion of a solid object. A Lagrange strain tensor, denoted as Tensor E^{*}, is classified as a tensor that represents finite deformations.

Typically, when examining elastic problems, we encounter three categories of strain:

1. Finite strain
2. Small strain
3. Infinitesimal strain

When discussing the divisions mentioned earlier, it is important to note that if all the sentences in the Equation (6.17) are thoroughly examined, it results in finite strain, which is essentially the same as large strains. If we simply consider the linear phrases of the given relation, specifically $\frac{1}{2}\left[\nabla u + (\nabla u)^{T}\right]$, we are dealing with infinitesimally strains. Given the current circumstances surrounding the problem, if we exclude some nonlinear sentences (sentences connected to finite strain), we are dealing with small strains. These small strains are known as von Karman strains, which will be discussed later.

6.5 RECTANGULAR PLATES SUBJECTED TO SMALL DEFORMATIONS: VON KARMAN'S THEORY

This section focuses on von Karman's theory, which assumes minimal deformation strains. To fulfill this objective, examine the rectangular plate as depicted in Figure 6.4.

The point's displacements in the central plane of the plate are denoted as u, v, and w, corresponding to the x, y, and z directions, respectively. The displacements of a point on the plate, located at a distance z from the central plane, are denoted as u_1, u_2, and u_3 (as shown in Figure 6.4a) [3].

When the displacement of the w plate is almost equal to the thickness of the h plate, the conclusions obtained from linear theories will be highly erroneous. In this context, a theory is proposed to explain significant deformations in which the magnitude of w is not negligible compared to h, resulting in noticeable differences between the original and deformed images. The theory is formulated using Cartesian coordinates, which are appropriate for rectangular plates with dimensions b and a (as seen in Figure 6.4 (b)). This theory takes into account the following assumptions [1, 3]:

$(H1)$: The plate is thin, $h \ll a, h \ll b$.

$(H2)$: The magnitude of the deformation is about equal to the thickness h of the plate. Therefore, it is relatively minor compared to the dimensions b and a of the plate for $(H1)$: $|w| = O(h)$.

$(H3)$: The slope at any point is significantly small: $|\partial w / \partial x| \ll 1, |\partial w / \partial y| \ll 1$.

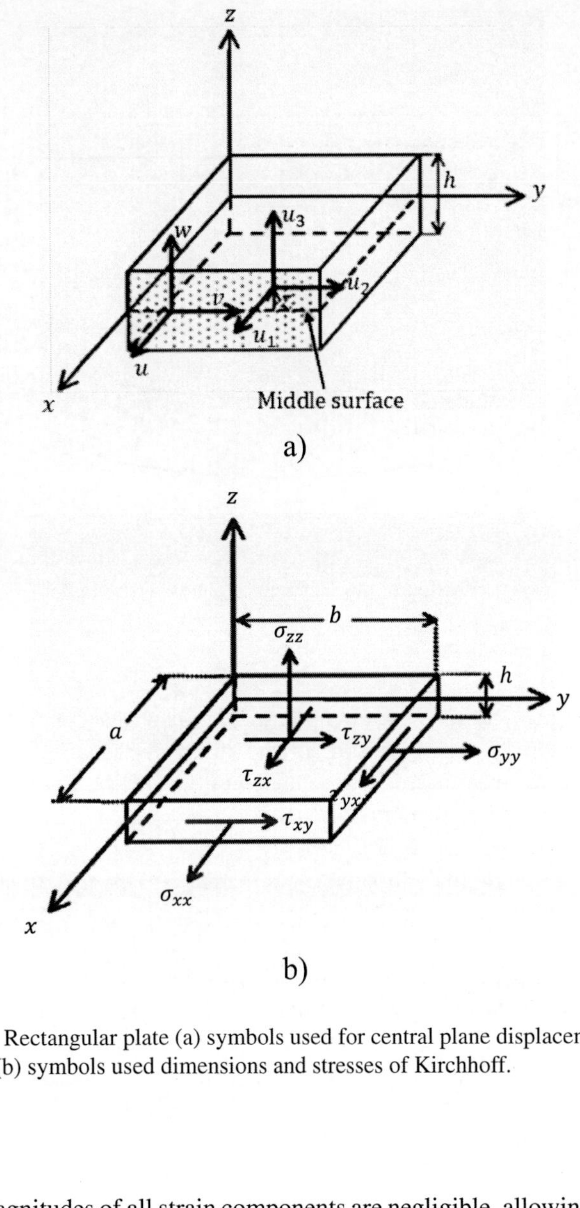

FIGURE 6.4 Rectangular plate (a) symbols used for central plane displacements of the general point and (b) symbols used dimensions and stresses of Kirchhoff.

$(H4)$: The magnitudes of all strain components are negligible, allowing for the use of linear elasticity.

$(H5)$: Kirchhoff's assumptions are confirmed, meaning that stresses are undeniable in the direction perpendicular to the central surface of the plate, and stresses vary linearly in the direction of the plate thickness. These assumptions are accurate estimations for slender plates. Despite the planned application of external loads that are perpendicular to the sheet plate, stresses are generated in the vertical direction, albeit with a magnitude smaller than the other stresses.

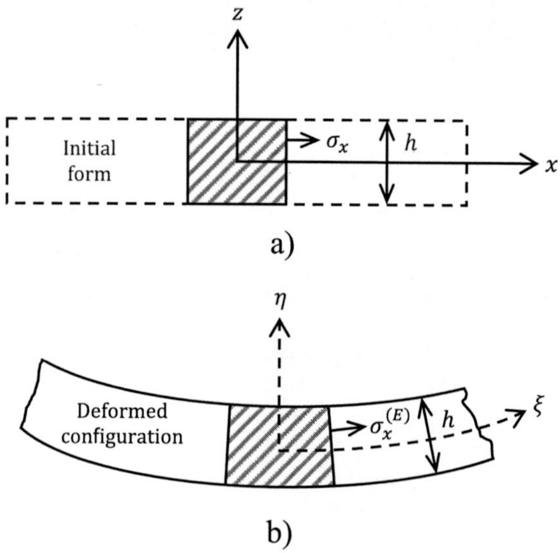

FIGURE 6.5 Transverse section of the rectangular plate: (a) the initial form, σ_x, is the normal Kirchhoff tension, and (b) the deformed form, $\sigma_x^{(E)}$, is the normal Eulerian tension.

$(H6)$: According to von Karman's hypothesis, the displacements u and v within the plane are extremely minor. In the strain-displacement relationships, only the nonlinear phrases that depend on w need to be retained. Disregard any other sentences that are not linear.

The hypothesis $(H6)$ on the omission of nonlinear plate theories to attain improved accuracy might be disregarded.

Figure 6.5 illustrates that the distorted image of the plate deviates from its initial shape. The Lagrangian approach is employed to describe the plate, ensuring that the upper and lower surfaces of the plate are consistently positioned at $z = \pm h/2$. A right-hand Cartesian coordinate system $(O; x, y, z)$ is employed, where the x and y planes align with the central surface of the primary plate (unmodulated picture), and the z axis is perpendicular to them. The Green's strain tensor is defined in the Lagrangian description.

$$\varepsilon_{ij} = \frac{1}{2}\left[\left(\sum_{k=1}^{3}\frac{\partial u_k}{\partial x_i} + \delta_{ki}\right)\left(\sum_{k=1}^{3}\frac{\partial u_k}{\partial x_j} + \delta_{kj}\right) - \delta_{ij}\right] = \frac{1}{2}\left(\frac{\partial u_i}{\partial x_j} + \frac{\partial u_j}{\partial x_i} + \sum_{k=1}^{3}\frac{\partial u_k}{\partial x_i}\frac{\partial u_k}{\partial x_j}\right) \tag{6.19}$$

It is important to highlight that when considering the first two expressions in the definition of strain, we will encounter infinitesimally small strains. If we include the expressions denoted as $\sum$ in the strain analysis, along with the first two expressions, we are dealing with a small strain. Ultimately, if all of the aforementioned relationship expressions are taken into account during stress analysis, the resulting stress will be constrained. For instance, if we replace x_1, x_2, and x_3 with x, y, and z correspondingly, we obtain the following:

$$\varepsilon_{xx} = \frac{\partial u}{\partial x} + \frac{1}{2}\left[\left(\frac{\partial u_1}{\partial x}\right)^2 + \left(\frac{\partial u_2}{\partial x}\right)^2 + \left(\frac{\partial u_3}{\partial x}\right)^2\right] \tag{6.20}$$

$$\gamma_{xy} = \frac{1}{2}\left[\frac{\partial u_1}{\partial y} + \frac{\partial u_2}{\partial x} + \left(\frac{\partial u_1}{\partial x}\frac{\partial u_1}{\partial y} + \frac{\partial u_2}{\partial x}\frac{\partial u_2}{\partial y} + \frac{\partial u_3}{\partial x}\frac{\partial u_3}{\partial y}\right)\right] \tag{6.21}$$

Applying hypothesis $(H5)$, we obtain the following:

$$u_1 = u(x, y) - z\frac{\partial w}{\partial x} \tag{6.22}$$

$$u_2 = v(x, y) - z\frac{\partial w}{\partial y} \tag{6.23}$$

$$u_3 = w(x, y) \tag{6.24}$$

The Equations (6.22), (6.23), and (6.24) are characterized by linearity. These equations are derived based on hypothesis $(H5)$, which assumes that the strains in the plate change linearly with thickness. Additionally, the equations assume that the Kirchhoff stresses are absent in the direction perpendicular to the middle plane of the plate $(\sigma_{zz} = \tau_{zx} = \tau_{zy} = 0)$, where σ_{ij} represents the vertical stress acting perpendicular to level i and in line with j. Furthermore, the symbol τ_{ij} represents the tangent stress depicted in Figure 6.4 (b). Thus, by employing linear elasticity:

$$\sigma_{xx} = a_1(x, y) + b_1(x, y)z \tag{6.25}$$

$$\sigma_{yy} = a_2(x, y) + b_2(x, y)z \tag{6.26}$$

$$\varepsilon_{zz} \simeq \frac{\partial u_3}{\partial z} = -\frac{\nu}{E}(\sigma_{xx} + \sigma_{yy}) \tag{6.27}$$

The letter E represents the Young modulus, whereas the letter ν represents the Poisson coefficient. The Equation (6.27), which approximates the linearized expression ε_{zz}, is derived from the linear Equations (6.22) to (6.24). By integrating the given Equation (6.63), we obtain the following:

$$u_3 = w(x, y) - \frac{\nu}{E}\left[a_1(x, y) + a_2(x, y)\right]z - \frac{\nu}{E}\left[b_1(x, y) + b_2(x, y)\right]\frac{z^2}{2} \tag{6.28}$$

The expression "w" represents the integral constant. Due to the insignificance of the value of $\frac{\nu}{E}$, it is possible to disregard the last two phrases on the right side of the

Equation (6.28) for thin plates. The Equation (6.24) has been altered, resulting in $\varepsilon_{zz} \simeq 0$. Furthermore, the expressions produced for hypothesis $(H5)$ are as follows:

$$\gamma_{zx} \simeq \frac{1}{2}\left(\frac{\partial u_1}{\partial z} + \frac{\partial u_3}{\partial x}\right) = \frac{\tau_{zx}}{E} - \frac{\nu}{E}\left(\tau_{zy} + \sigma_{zz}\right) = 0 \tag{6.29}$$

$$\gamma_{zy} \simeq \frac{1}{2}\left(\frac{\partial u_2}{\partial z} + \frac{\partial u_3}{\partial y}\right) = 0 \tag{6.30}$$

By substituting the expression $u_3 = w(x, y)$ derived from Equation (6.24) into Equations (6.29) and (6.30), and then integrating over those equations, we obtain the Equations (6.22) and (6.23). Now, by incorporating the correlations between Equations (6.22) and (6.24) into the Green's strain tensor, we obtain the following:

$$\varepsilon_{ij} = \frac{1}{2}\left(\frac{\partial u_i}{\partial x_j} + \frac{\partial u_j}{\partial x_i} + \sum_{k=1}^{3} \frac{\partial u_k}{\partial x_i}\frac{\partial u_k}{\partial x_j}\right) \tag{6.31}$$

The purpose of this study is to derive the strain-displacement correlations of a plate in Cartesian coordinates. The strain components ε_{xx}, ε_{yy}, and γ_{xy} at any given position on the plate are connected to the middle surface strains $\varepsilon_{x,0}$, $\varepsilon_{y,0}$, and $\gamma_{xy,0}$, as well as to the changes in curvature k_x, k_y, and k_{xy} of the middle surface, through the following relationships:

$$\varepsilon_{xx} = \varepsilon_{x,0} + zk_x \tag{6.32}$$

$$\varepsilon_{yy} = \varepsilon_{y,0} + zk_y \tag{6.33}$$

$$\gamma_{xy} = \gamma_{xy,0} + zk_{xy} \tag{6.34}$$

Where Z represents the distance between the desired location on the plate and its central surface. By employing von Karman's hypothesis $(H6)$, one can get the subsequent equations for the intermediate surface stresses, as well as the variations in curvature and torsion of said surface:

$$\varepsilon_{x,0} = \frac{\partial u}{\partial x} + \frac{1}{2}\left(\frac{\partial w}{\partial x}\right)^2 \tag{6.35}$$

$$\varepsilon_{y,0} = \frac{\partial v}{\partial y} + \frac{1}{2}\left(\frac{\partial w}{\partial y}\right)^2 \tag{6.36}$$

$$\gamma_{xy,0} = \frac{\partial u}{\partial y} + \frac{\partial v}{\partial x} + \frac{\partial w}{\partial x}\frac{\partial w}{\partial y} \tag{6.37}$$

$$k_x = -\frac{\partial^2 w}{\partial x^2} \tag{6.38}$$

$$k_y = -\frac{\partial^2 w}{\partial y^2} \tag{6.39}$$

$$k_{xy} = -2\frac{\partial^2 w}{\partial x \partial y} \tag{6.40}$$

Generally, these correlations are sufficiently precise for plate with significant variations. If a higher level of specificity is necessary for relationships, the hypothesis $(H6)$ cannot be used, and Equations (6.35) to (6.40) can be computed by incorporating all nonlinear statements. It is important to mention that the terms $\left(\frac{\partial w}{\partial x}\right)^2$, $\left(\frac{\partial w}{\partial y}\right)^2$, and $\frac{\partial w}{\partial x}\frac{\partial w}{\partial y}$, which appear in the equations earlier, are referred to as strain nonlinearity in the von Karman model.

6.6 FUNDAMENTAL PRINCIPLES UNDERLYING NON-CLASSICAL CONTINUUM MECHANICS THEORIES

6.6.1 Fundamentals of Nonlocal Theory

Eringen initially introduced the nonlocal elastic theory to account for the influence of the tiny size parameter in the continuum model. In nonlocal theory, as opposed to classical elasticity theory, the stress at a specific place in a continuous physical model is influenced by the strain of all points in that model. Put simply, the strain at a specific place is determined by the stress and its partial derivatives at that same position. The nonlocal theory examines the interaction between atoms within a molecule on a macroscopic level and links the outcomes to the dimensions of the physical model. The fundamental equation for a linear homogeneous body in the nonlocal elastic domain, disregarding volume forces, can be expressed as follows according to this theory [4–6]:

$$\begin{gathered} \sigma_{ij,j} = 0 \\ \sigma_{ij}(\mathrm{x}) = \iiint \phi\left(\left|x'-x\right|,\alpha\right) t_{ij}(x')\,dV(x'), \quad \forall x \in V \\ t_{ij} = C_{ijkl}\varepsilon_{kl}, \qquad \varepsilon_{kl} = \left(u_{k,l} + u_{l,k}\right)/2 \end{gathered} \tag{6.41}$$

In the earlier connection, the nonlocal stress tensor is represented by the symbol σ_{ij}, while the local stress tensor is represented by the symbol t_{ij}. Additionally, C_{ijkl} refers to the fourth-order elastic tensor, and ε_{kl} represents the classical strain tensor. The integral is computed across the volume inhabited by the object, and the stress equation resulting from classical elasticity is applied. Nonlocal elasticity causes stress to bind.

The kernel function $\phi\left(\left|x'-x\right|,\alpha\right)$ is a nonlinear modulus. This module functions as a mechanism for reducing the strength of a structural equation that connects the nonlocal strain impact at the source point x' to the nonlocal impact at the reference point x.

The variable $|x'-x|$ denotes the distance in the Euclidean form of the material, while α is a constant that relies on the internal properties of the material, such as the regular network parameter, grain size, and bond distance between carbon-carbon bonds. Additionally, α also depends on external longitudinal characteristics, such as crack size or wave size. The α constant is assigned the value of $e_0\ a/\ell$, while the material determines the proportionality of e_0. This allows for the adjustment and calibration of the model to align it with reliable experimental data from other models. This parameter should be capable of accurately aligning the current connections in the nonlocal field with a good approximation to the outcomes derived from the atomic diffusion curves when longitudinal waves are present in the atomic lattice dynamics laboratory. The values of the ℓ and a parameters are influenced by both the internal and external characteristics of the nanostructure. These characteristics include the regular network parameter of grain size, carbon ribbon bond distance, crack size, and wave size, which were previously discussed.

The equation is expressed as a partial integral, and its analytical solution is typically challenging, so a differential form of the elasticity equation is commonly employed. Eringen states that the equation for nonlocal elasticity can be expressed in the following manner.

$$\phi(|x|,\alpha)=\left(2\pi\ell^2\alpha^2\right)^{-1}K_0\left(\sqrt{x.x}\,/\,\ell\alpha\right) \tag{6.42}$$

In the given relation, the K_0 function has been altered to Bessel. The rationale for motion in the context of nonlocal theoretical statements can be expressed as follows.

$$\sigma_{ij,j}+f_i=\rho\ddot{u}_i \tag{6.43}$$

The variables "ρ" and "f_i" represent the mass density and volume force or applied force, respectively. The variable "u_i" represents the displacement vector. Similarly, various relations for different coordinate lines can be expressed using the same approach.

Eringen formulated the structural equation as a differential equation in nonlocal theory, assuming that the kernel function ϕ is a Green's function.

$$\sigma_{ij,j}+\mathcal{L}(f_i-\rho\ddot{u}_i)=0 \tag{6.44}$$

The linear differential operator $\mathcal{L}$ is defined by the given relation.

$$\mathcal{L}=1-(e_0a)^2\nabla^2=1-\mu\nabla^2 \tag{6.45}$$

In the given relationship, μ represents the nonlinear parameter and is measured in square nanometers (nm^2). By considering Equations (6.44) and (6.45), the differential equation associated with the opinion in Eringen's nonlocal theory can be stated as follows [7–9]:

$$\left[1-(e_0a)^2\nabla^2\right]\sigma_{ij}=\sigma_{ij}(nl) \tag{6.46}$$

6.6.2 The Hypothesis of Modified Stress Coupling

In 1909, Cozart presented the inaugural mathematical model for analyzing materials that exhibit a stress pair. Subsequently, additional researchers have further improved the stress pair theory for materials exhibiting elastic properties throughout the years. These theories introduce the gradient of the rotation vector as a curvature tensor, leading to their classification as infinite stress couple theories [10–12].

Experiments conducted on metals and polymers have observed the influence of micron-scale factors on the movement behavior of materials. The explanation for this behavior is outside the scope of classical mechanics theories. The stress pair theory has been employed to demonstrate the correlation between displacement behavior and size. Later, we will analyze the distinctions between stress couple theory and other mechanical theories.

The stress couple theory for linear elastic materials states that it encompasses not just a force that induces particle movement but also a couple that induces particle rotation.

The equilibrium relations posited in this theory are derived from the principle of momentum conservation. The modified stress couple theory introduces a set of additional equilibrium relationships for a system of material particles. This theory extends the original stress couple theory by incorporating the determination of particle rotation. According to this theory, the stress couple vector is a vector that exhibits symmetry. The primary objective of the modified stress couple theory is to demonstrate the length scale effect using only one parameter, whereas the stress couple theory requires two parameters for the same purpose [13].

6.7 THE EQUATIONS THAT GOVERN THE STRESS COUPLE THEORY FOR A CHANGEABLE BODY

We are examining a random volume ν' of a variable object and a constituent of its surface dv. t_n and μ_n are defined as the force and torque per unit area, respectively. The subscript "n" denotes the orientation of the normal vector in the out-of-plane direction. The symbols f and l represent the quantities of force and torque per unit volume, respectively. The following are the suggested relations for a continuous volume [14, 15]:

$$\int_{v'} f dv + \int_{\partial v'} t_n ds = 0$$

$$\int_{v'} (x \times f + l) dv + \int_{\partial v'} (x \times t_n + \mu_n) ds = 0 \tag{6.47}$$

Let x denote the location vector of a particle of matter.

However, as:

$$t_n = t.n \,;\, \mu_n = \mu.n \tag{6.48}$$

The stress tensor is denoted by t, while the stress couple tensor is represented by μ. The expression for the surface integral using the divergence theorem can be formulated.

$$\int_{v'} (t.\nabla + f) dv = 0$$

$$\int_{v'}\left(x\times\left(t.\nabla+f\right)-\in:t+\mu.\nabla+l\right)dv=0 \tag{6.49}$$

Given that volume v' is discretionary, the corresponding dependent volume can be excluded.

$$\begin{gathered} t.\nabla+f=0 \\ t.\nabla+l-\in:t=0 \end{gathered} \tag{6.50}$$

$\in$ is a tensor that alternates, while ∇ is Hamilton's differential operator. They collaborate to sustain a consistent equilibrium. In order to omit the temporary torque, we have the following:

$$\int_{v'}\left(x\times\left(l-\in:t\right)\right)dv+\int_{\partial v'}x\times\mu_n ds=0 \tag{6.51}$$

$l-\in:t$ represents the coupling between the object and the rest of the system. The relationships earlier can be expressed using the divergence theorem.

$$\int_{v'}\left(x\times\left(l-\in:t+\mu.\nabla\right)-\in:\mu\right)dv=0 \tag{6.52}$$

Which results as follows:

$$\in:\mu=0 \tag{6.53}$$

The stress tensor can be divided into two parts: the symmetric part and the antisymmetric part.

$$\sigma=1/2\left(t+t^T\right),\quad \tau=1/2\left(t-t^T\right) \tag{6.54}$$

Where t^T is the transposition of t.

The stress coupling tensor μ is separated into a spherical component μ_g and a deviatoric component m.

$$\begin{gathered} \mu=\mu g+m\,, \\ \mu g=\frac{1}{3}tr\left(\mu\right), \\ tr\left(m\right)=0 \end{gathered} \tag{6.55}$$

By utilizing relationships $.\nabla+l-\in:\tau=0$ and employing $t.\nabla+f=0$, and substitution of relationships, we achieve equilibrium within it.

$$\left(\sigma+\tau\right).\nabla+f=0$$

$$\mu.\nabla+\mathrm{m}.\nabla+l-\in:\tau=0 \tag{6.56}$$

By eliminating the asymmetric stress tensor from the equation mentioned earlier, the resulting equilibrium equation is as follows:

$$\sigma.\nabla+\frac{1}{2}\in:(m.\nabla\otimes\nabla+1\otimes\nabla)+f=0 \tag{6.57}$$

6.7.1 A Novel and Comprehensive Iteration of the Theory Rooted in the Concept of Strain Generalization

In this part, displacements and ε, χ, σ, and m are determined by applying the virtual work principle, and we present the obtained results here, where χ is the symmetric curvature tensor. The initial values of the displacement slope and rotation slope are provided for the symmetrical section.

$$\begin{aligned}\varepsilon&=\frac{1}{2}(u\otimes\nabla+\nabla\otimes u)\\ \chi&=\frac{1}{2}(\theta\otimes\nabla+\nabla\otimes\theta)\end{aligned} \tag{6.58}$$

Ultimately, we are employing the concept of virtual work.

$$\sigma=\frac{\partial w}{\partial\varepsilon},\quad m=\frac{\partial w}{\partial\chi},\quad \frac{\partial w}{\partial\theta}=0 \tag{6.59}$$

Here, w represents the angular density of deformation. The equations earlier demonstrate that the deformation energy density is independent of rotation θ.

The w for the second-degree function of generalized strains in linear isotropic materials follows the linear elastic law.

$$w=\frac{1}{2}\lambda(tr\,\varepsilon)^2+\mu\left(\varepsilon:\varepsilon+l^2\chi:\chi\right) \tag{6.60}$$

The substitution of the Equation (6.60) in the Equation (6.59) results [13]:

$$\begin{aligned}&\sigma=\lambda g\,tr(\varepsilon)+2\mu\varepsilon\\ &l^{-1}m=2l\mu\chi\\ &\theta=\frac{1}{2}curl\,u\\ &U=\frac{1}{2}\int(\sigma:\varepsilon+m:\chi)dv\\ &w=\frac{1}{2}\lambda(tr\,\varepsilon)^2+\mu\left(\varepsilon:\varepsilon+l^2\chi:\chi\right)\end{aligned} \tag{6.61}$$

6.8 MODIFIED COUPLE STRESS THEORY FOR NON-ISOTROPIC MATERIALS

In the field of classical elasticity, shear strains are quantified by taking the derivatives of displacements. The coefficients associated with these strains are determined by examining the relationship between stress and shear actions. Furthermore, these coefficients can be readily extended to encompass non-isotropic elasticity. The coupled stress theory's spherical inhomogeneous elasticity is determined by rotation components rather than rotational derivatives. Additionally, the modified coupled stress theory restricts the curvature to spherically isomorphic materials. Hence, the adapted couple stress theory is inapplicable to this particular material, necessitating the revision of both stress and moment components.

$$\sigma_{ij} = C_{ijkl}\varepsilon_{kl} \tag{6.62}$$

$$m_{ij} = \left(l_i^2 \mu_i \chi_{ij} + l_j^2 \mu_j \chi_{ji}\right)$$

As previously mentioned, χ represents the curvature tensor of asymmetric couple stress, m represents the moment tensor of symmetric couple stress, and l_i represents the length scale parameter. The direction i represents the direction of impurity and defects in the material. Consequently, the length scale parameter in each direction can indicate a scale of measurement. The dimensions and scale of impurities and flaws inside the underlying structure of the material. Consequently, this theory will incorporate five constants corresponding to the five directions of the material crystal. Hence, the strain energy for the modified couple theory can also be expressed in the subsequent manner [16–18].

$$U = \frac{1}{2}\int_{\pi}^{'} \left(\sigma_{ij}\varepsilon_{ij} + m_{ij}^{s}\chi_{ij}^{s}\right) dv \tag{6.63}$$

6.9 FUNDAMENTALS OF MODIFIED STRAIN GRADIENT THEORY

According to the strain gradient theory, stress is not only dependent on strain, as stated in classical mechanics, but also on the strain gradient. The notion of lowest total potential energy has been employed to derive new relationships in the strain gradient theory. The strain gradient theory incorporates a statistical component known as the length effect parameter, which reveals that the material's behavior at a small scale is influenced by its dimensions. The subject in the theory of classical elasticity could not be articulated due to the omission of this parameter in the equations. The strain gradient theory introduces a new stress element known as the total stress tensor. This tensor differs from the Cauchy stress tensor and can be utilized as the total stress tensor in the momentum equation.

The theory of strain gradient elasticity incorporates a new parameter, known as the characteristic length of the material, into the structural equations. This allows for the consideration of the material's structure when analyzing its behavior. This theory considers both the potential energy of the material and the strain. Additionally, the strain gradient influences this function, resulting in the inclusion of various novel variables in the structural equations. Put simply, the strain energy density in the strain gradient theory is determined by the normal strains ε_{ij}, which are the symmetric component of the first-order deformation gradient of the displacements, as well as the second-order deformation gradient η_{ijk}. The answer varies depending on the circumstances. The theory was initially proposed in 1963 by Medlin [14, 15, 19].

$$U = U\left(\varepsilon_{ij}.\eta_{ijk}\right)$$
$$\varepsilon_{ij} = \frac{1}{2}\left(u_{i.j} + u_{j.i}\right) \tag{6.64}$$
$$\eta_{ijk} = u_{k.ij}$$

The strain tensor ε_{ij} in the given relationships consists of six separate components, while the second-order deformation gradient tensor consists of 18 independent components. It is important to note that both of these tensors are symmetric tensors. The stresses associated with this strain can be described using the second-order gradient tensor present in this theory.

$$\sigma_{ij} = \frac{\partial U}{\partial \varepsilon_{ij}}$$
$$\tau_{ijk} = \frac{\partial U}{\partial \eta_{ijk}} \tag{6.65}$$

Hence, the overall strain energy can be represented in the following manner.

$$\Pi = W = \sigma_{ij}\varepsilon_{ij} + \tau_{ijk}\eta_{ijk}$$
$$u = \delta \int W dV = \int \left(\sigma_{ij}\varepsilon_{ij} + \tau_{ijk}\eta_{ijk}\right) dV \tag{6.66}$$

Mindlin formulated the strain energy density using the strain gradient theory for classical linear elastic isotropic materials in the following manner.

$$U = \frac{1}{2}\lambda\varepsilon_{ii}\varepsilon_{jj} + \mu\varepsilon_{ij}\varepsilon_{ij} + a_1\eta_{ijk}\eta_{ikk} + a_2\eta_{ijk}\eta_{kjj} + a_3\eta_{iik}\eta_{jjk} + a_4\eta_{jik}\eta_{ijk} + a_5\eta_{ijk}\eta_{kji} \tag{6.67}$$

The normal Lamé constants are directly related to the normal strain invariant $a_n\left(1,\ldots,5\right)$, whereas the additional second-order elastic constant is directly related to the second-order deformation gradient invariants.

Fleck and Hutchinson categorize the transformation of the second-order gradient tensor into symmetric η_{ijk}^{s} and asymmetric η_{ijk}^{a} components as follows [20, 21]:

$$\eta_{ijk}^{s} = \frac{1}{3}\left(\eta_{ijk} + \eta_{jki} + \eta_{kij}\right)$$
$$\eta_{ijk}^{a} = \frac{2}{3}\left(e_{ikl}\chi_{lj} + e_{jkl}\chi_{li}\right) \tag{6.68}$$

The terms "permutation tensor e_{ijk}" and "curvature tensor χ_{ij}" are referred to in the aforementioned equations.

By partitioning the symmetrical component of the second-order strain η^{s}_{ijk} into two components $\eta^{(0)}_{ijk}$ and $\eta^{(1)}_{ijk}$, we derive novel, distinct second-order strain metrics.

$$\begin{aligned}
\eta^{s}_{ijk} &= \eta^{(0)}_{ijk} + \eta^{(1)}_{ijk} \\
\eta^{(0)}_{ijk} &= \frac{1}{5}\left(\delta_{ij}\eta^{s}_{mmk} + \delta_{jk}\eta^{s}_{mmi} + \delta_{ki}\eta^{s}_{mmj}\right) \\
\eta^{(1)}_{ijk} &= \eta^{s}_{ijk} - \eta^{(0)}_{ijk} \\
\eta^{s}_{mmk} &= \frac{1}{3}\left(\eta_{mmk} + 2\eta_{kmm}\right)
\end{aligned} \tag{6.69}$$

The χ_{ij} -curvature tensor was decomposed into two components: a symmetric part and an asymmetric one.

$$\begin{aligned}
\chi_{ij} &= \chi^{s}_{ij} + \chi^{a}_{ij} \\
\chi^{s}_{ij} &= \frac{1}{2}\left(\chi_{ij} + \chi_{ji}\right) \\
\chi^{a}_{ij} &= \frac{1}{2}\left(\chi_{ij} + \chi_{ji}\right)
\end{aligned} \tag{6.70}$$

The initial component of the second-order symmetric deformation gradient is determined by the expansion gradient, whereas the nonsymmetric component is represented by the curvature tensor. Therefore, the expression can be formulated as follows:

$$\eta^{s}_{ipp} = \varepsilon_i + \frac{2}{3}e_{imm}\chi^{a}_{mn} = \varepsilon_i + \frac{2}{3}e_{imn}\chi_{mn} \tag{6.71}$$

In the given relationships, $\varepsilon_i = \varepsilon$ is the dilation strain, ε_i corresponds to the rate of change of dilation strain, $\eta^{(1)}_{ijk}$ is the rate of change of deviatoric tension, and χ_{ij} is the rate of change of rotation.

If $\tau^{(0)}_{ijk}$ and $\tau^{(1)}_{ijk}$ are the trace and traceless sections of the components of the symmetric part of the couple stress tensor $\eta^{(0)}_{ijk}$, $\eta^{(0)}_{ijk}$, and η^{a}_{ijk} and the related strains as $\tau^{(0)}_{ijk}$, $\tau^{(1)}_{ijk}$, and τ^{a}_{ijk}, the virtual work resulting from the second-order gradient tensors can be calculated by considering the following:

$$\delta\hat{w} = \tau^{(0)}_{ijk}\delta\eta^{(0)}_{ijk} + \tau^{(1)}_{ijk}\delta\eta^{(1)}_{ijk} + \tau^{a}_{ijk}\delta\eta^{a}_{ijk} \tag{6.72}$$

The variables $\tau^{(0)}_{ijk}$ and $\tau^{(1)}_{ijk}$ are defined according to the following specifications:

$$\begin{aligned}
\tau^{(0)}_{ijk} &= \frac{1}{5}\left(\delta_{ij}\tau^{s}_{mmk} + \delta_{jk}\tau^{s}_{mmi} + \delta_{ki}\tau^{s}_{mmj}\right) \\
\tau^{(1)}_{ijk} &= \tau^{s}_{ijk} - \tau^{(0)}_{ijk}
\end{aligned} \tag{6.73}$$

By employing ε_i, $\eta_{ijk}^{(1)}$, and χ_{ij} as the second-order strain requirements, the following can be inferred:

$$\delta\hat{w} = P_i\delta\varepsilon_i + \tau_{ijk}^{(1)}\delta\eta_{ijk}^{(1)} + m_{ij}'\delta\chi_{ij}$$
$$P_i = \frac{3}{5}\tau_{mmi}^{s} \tag{6.74}$$
$$m_{ij}' = \frac{4}{3}\tau_{ipq}^{a}e_{jpq} - \frac{2}{5}e_{ijk}\tau_{mmk}^{s}$$

Thus, based on the aforementioned relationships in the strain gradient theory, it can be concluded that the strain energy density is dependent on both the first-order gradient e_{ij} and the second-order gradient e_{ijk} of the displacements. Therefore, it may be expressed as follows:

$$W = W\left(\varepsilon_{ij}.\varepsilon_{ij,k}\right) \tag{6.75}$$

Consequently, the amount of stored energy per unit volume in Mindlin's theory of elasticity, which takes into account the variation of strain, may be mathematically represented as follows:

$$W = \frac{1}{2}\lambda\varepsilon_{ii}\varepsilon_{jj} + \mu\varepsilon_{ij}\varepsilon_{ij} + a_1\eta_{ij,j}\eta_{ik,k} + a_2\eta_{ii,k}\eta_{kj,j} + a_3\eta_{ii,k}\eta_{jj,k} + a_4\eta_{ij,k}\eta_{ij,k} + a_5\eta_{ij,k}\eta_{kj,i} \tag{6.76}$$

The strain energy density relation for strain gradient theory can be expressed as follows:

$$W = \frac{1}{2}\lambda\varepsilon_{ii}\varepsilon_{jj} + \mu\varepsilon_{ij}\varepsilon_{ij} + a_1\eta_{ijk}\eta_{ikk} + a_2\eta_{ijk}\eta_{kjj} + a_3\eta_{iik}\eta_{jjk} + a_4\eta_{ijk}\eta_{ijk} + a_5\eta_{ijk}\eta_{kji} \tag{6.77}$$

The couple stress theory can be considered as a specific instance of a higher-order stress theory, where the influence of the dilatation gradient γ_i or ε_i or the deviatoric tension gradient $\eta_{ijk}^{(1)}$ can be disregarded. Hence, the expression for the whole internal virtual work density may be formulated in this particular scenario.

$$\delta w = \sigma_{ij}\delta\varepsilon_{ij} + m_{ij}\delta\chi_{ij} \tag{6.78}$$

Without the size scale component, the equations derived from the strain gradient theory are identical to the equations found in classical mechanics. The stored strain energy U in the surroundings of linear elastic materials is expressed in Hamilton's relation, based on the modified strain gradient theory [22].

$$U = \frac{1}{2}\int\left(\sigma_{ij}\varepsilon_{ij} + p_i\gamma_i + \tau_{ijk}\eta_{ijk} + m_{ij}\chi_{ij}\right)dV \tag{6.79}$$

Regarding the aforementioned relationship:

$$\varepsilon_{ij} = \frac{1}{2}\left(u_{i,j} + u_{j,i}\right)$$

$$\eta_{ijk}^{(1)} = \frac{1}{3}\left(\varepsilon_{jk,i} + \varepsilon_{ki,j} + \varepsilon_{ij,k}\right) - \frac{1}{15}\begin{bmatrix} \delta_{ij}\left(\varepsilon_{mm,k} + 2\varepsilon_{mk,m}\right) \\ +\delta_{jk}\left(\varepsilon_{mm,i} + 2\varepsilon_{mi,m}\right) \\ +\delta_{ki}\left(\varepsilon_{mm,j} + 2\varepsilon_{mj,m}\right) \end{bmatrix} \tag{6.80}$$

$$\chi_{ij} = \frac{1}{4}\left(e_{ipq}\varepsilon_{qj,p} + e_{jpq}\varepsilon_{qi,p}\right)$$

$$\gamma_i = \varepsilon_{mm,i}$$

In the given equation, the variable u_i denotes the displacement vector along the x, y, and z axes. γ_i symbolizes the dilatation gradient tensor, η_{ijk} the deviatoric stretch gradient tensor, χ_{ij} the symmetric rotation gradient tensor, δ_{ij} the Kronecker delta, and ε_{ijk} the permutation tensor symbol.

The Cauchy stress tensor and high-order stress tensors, denoted as τ_{ijk}, p_i, and m_{ij} are defined as follows. It is important to mention that the relations that follow involve three independent length scale parameters l_0, $l_1, and\, l_2$. These parameters are the gradient vector of the deviatoric tension gradient tensor and the symmetric rotation gradient tensor, respectively, and are interdependent

$$\sigma_{ij} = k\varepsilon_{mm}\delta_{ij} + 2\mu\left(\varepsilon_{ij} - \frac{1}{3}\varepsilon_{mm}\delta_{ij}\right)$$

$$p_i = 2\mu l_0^{\,2}\gamma_i \tag{6.81}$$

$$\tau_{ijk} = 2\mu l_1^{\,2}\eta_{ijk}$$

$$m_{ij} = 2\mu l_2^{\,2}\chi_{ij}$$

6.10 NONLOCAL STRAIN GRADIENT ELASTICITY THEORY

Two basic presumptions of classical mechanics prohibit the use of classical ideas at micro- and nanoscales. At the nanoscale, the continuity of the material's mechanical field is one of these presumptions that is fully questioned because, at lower sizes, the existence of voids between atoms has been demonstrated and is unavoidable. The assumption that the stress at a location depends only on the strain at that same position in classical mechanics presents another barrier to applying the theory on smaller scales. Eringen's research indicates that, at the nanoscale, the strain of the material as a whole as well as the stress at a particular place affect the stress at that location. Consequently, non-classical theories are needed for analysis at the nanoscale. Numerous studies have been conducted recently with the goal of taking into account the longitudinal scale of materials in the micro and nano size. Some of these studies have used the strain gradient theory, which is based on the theory of Eringen, while others have used the nonlocal theory. The strain gradient theory's modified pair stress is regarded as its most crucial component. After much debate among academics over this disparate length scales, Lim and her associates finally articulated the theory of nonlocal strain gradient elasticity based on the nonlocal effects of strain field and strain gradient field. This theory takes into account both the nonlocality of strain gradients and higher-order stress gradients to offer an elastic system. This theory is entirely distinct from the other two theories because, while the widely used strain gradient theory only looks at the local effects of higher-order stresses, the nonlocal theory ignores the nonlocal effects of higher-order stresses. This theory states that the nonlocal stress at reference point x depends on the strain at that place as well as the strain at every other point in volume V. This theory takes the following into consideration for the internal potential energy density U_0 of an isotropic elastic body [23, 24].

$$\begin{aligned} U_0\left(\varepsilon_{ij},\varepsilon'_{ij},\alpha_0;\varepsilon_{ij,m},\varepsilon'_{ij,m},\alpha_1\right) &= \frac{1}{2}\varepsilon_{ij}C_{ijkl}\int_V \alpha_0\left(\left|\boldsymbol{x}-\boldsymbol{x}'\right|,\mu_0\right)\varepsilon'_{kl}dV' \\ &\quad+\frac{l^2}{2}\varepsilon_{ij,m}C_{ijkl}\int_V \alpha_1\left(\left|\boldsymbol{x}-\boldsymbol{x}'\right|,\mu_1\right)\varepsilon'_{kl,m}dV' \end{aligned} \tag{6.82}$$

The volume integral encompasses the entire object. ε_{ij} and ε'_{IJ} denote the strain tensor at the reference point x and the strain tensor at the adjacent points x', respectively. α_0 and α_1 are kernel functions associated with nonlocal effects in relation to strain fields and strain gradients. Furthermore, μ_0 and μ_1 are nonlocal parameters, while l is the length scale parameter of the strain gradient.

Utilizing Equation (6.83), the classical stress tensor σ, the higher-order stress tensor σ^1, and the total stress tensor associated with the nonlocal strain gradient theory t can be expressed as follows:

$$\begin{aligned} &\sigma = \int_V \alpha_0\left(\boldsymbol{x}',\boldsymbol{x}',\mu_0\right)\boldsymbol{C}:\varepsilon' dV' \\ &\sigma^1 = l^2\int_V \alpha_1\left(\boldsymbol{x}',\boldsymbol{x},\mu_1\right)\boldsymbol{C}:\nabla\varepsilon' dV' \\ &\boldsymbol{t} = \sigma - \nabla^{(1)} \end{aligned} \tag{6.83}$$

In this context, the notation ":" denotes the multiplication of two tensors. Given that resolving an integral equation is significantly more challenging than addressing a differential problem, Lim et al. [24] introduced the differential formulation of structural equations grounded in the high-order nonlocal strain gradient theory as follows:

$$\left[1-\mu_1^2\nabla^2\right]\left[1-\mu_0^2\nabla^2\right]t_{ij} = C_{ijkl}\left[1-\mu_1^2\nabla^2\right]\varepsilon_{kl} - C_{ijkl}l^2\left[1-\mu_0^2\nabla^2\right]\nabla^2\varepsilon_{kl} \tag{6.84}$$

$$\mu_1 = e_1 a, \quad \mu_0 = e_0 a, \quad \nabla^2 = \frac{\partial^2}{\partial x^2}+\frac{\partial^2}{\partial y^2} \tag{6.85}$$

Where ∇^2 denotes the Laplacian operator, e_0 and e_1 are nonlocal material constants, and a represents the internal characteristic length. By setting $\mu_0 = \mu_1 = \mu$ in equation

(6.84) and neglecting terms of order $O\left(\nabla^2\right)$, a more simplified representation of the structural equations for nonlocal strain gradient will be derived [25].

$$\left[1-\mu^2\nabla^2\right]t_{ij} = C_{ijkl}\left[1-l^2\nabla^2\right]\varepsilon_{kl} \tag{6.86}$$

The nonlocal strain gradient theory simplifies to classical continuous medium theory when $e_i a = l = 0$ and to strain gradient theory when $e_i a = 0$. Under the assumption of isotropic Euler-Bernoulli beam theory, Equation (6.84) is expressed as follows [26]:

$$\left[1-\mu_1^2\nabla^2\right]\left[1-\mu_0^2\nabla^2\right]\sigma_{xx} = E\left[1-\mu_1^2\nabla^2\right]\varepsilon_{xx} - El^2\left[1-\mu_0^2\nabla^2\right]\nabla^2\varepsilon_{xx} \tag{6.87}$$

Where σ_{xx} represents normal stress, ε_{xx} denotes normal strain, and $\nabla = \dfrac{\partial}{\partial x}$. Under the premise of retaining just terms of order $O\left(\nabla^2\right)$ and setting $e = e_0 = e_1$, the structural Equation (6.87) is thus simplified as follows:

$$\left[1-ea^2\nabla^2\right]\sigma_{xx} = E\left[1-l^2\nabla^2\right]\varepsilon_{xx} \tag{6.88}$$

6.11 VARIATIONAL METHOD

This section describes the extraction of differential equations that regulate the static and dynamic behavior of an elastic object, utilizing energy-compliant methods based on the principles of variation [27, 28].

Various techniques can be employed to derive differential equations that describe the behavior of the elastic item. The principles of minimum potential energy, minimum complementary energy, and principle of Reissner energy can be utilized to address static problems. The Hamilton principle refers to the fundamental concept of variation, specifically in relation to the dynamic behavior shown by systems consisting of particles, solid objects, or deformable objects. This section briefly discusses the concepts of variation, with a particular emphasis on the Hamilton principle.

Specifically, this section focuses on the equilibrium equations and equations of motion for continuous systems. Since we have knowledge of the principles of continuum environment mechanics discussed earlier, we aim to apply these principles to determine the principles governing the changes in a continuous environment. Specifically, we want to find the following:

a) Principles that are equivalent to the equilibrium relations or movement of a continuous environment. (These equations are referred to as exact equations.)
b) The extraction of engineering or technical equations and boundary conditions, which are often known as approximation equations, is relevant. Alternative approaches to solving 3D elasticity problems include employing equations of beams, plates, and shells, which are approximate theories.

c) To obtain approximate solutions, one can utilize both exact equations and approximation equations. There are three methods for deriving the equations that describe the mechanical behavior of an object, which are as follows:

1. The equilibrium approach: Newton's second law of motion
2. The integral equation method
3. Variation method

This section will outline the Hamilton principle as a highly effective technique for deriving differential equations that govern the static and dynamic behavior of a continuous system.

6.12 HAMILTON PRINCIPLE

The Hamilton principle is the fundamental concept that can be employed to address dynamic problems involving variation. Based on this approach, functional change is measured in relation to time. The Hamilton principle employs a function known as Lagrangian, which is defined as follows [27, 28]:

$$L = T - U = \text{Kinetic Energy} - \text{Potential Energy} \tag{6.89}$$

By formulating the equation of motion, we establish a continuous environment in the following manner:

$$\sigma_{ij,j} + \rho b_i = \rho \ddot{u}_i, \quad i = 1, 2, 3 \tag{6.90}$$

The equation in question is known as the Navier-Cauchy equation. By doing the multiplication of δu_i in Equation (6.90) and thereafter integrating the volume, we obtain the following:

$$\iiint_v \left[\sigma_{ij,j} + \rho b_i\right] \delta u_i \, dV = \iiint_v \rho \ddot{u}_i \, \delta u_i \, dV \tag{6.91}$$

By conducting same processes to those described in the preceding section, which resulted in the establishment of the principle of minimum potential energy, Equation (6.91) can be expressed as follows:

$$\iint_{S\sigma} \overset{\sigma}{\underset{\sim}{}}i \, \delta u_i \, dS - \iiint_v \sigma_{ij} \, \delta\varepsilon_{ij} \, dV + \iiint_v \rho b_i \, \delta u_i \, dV - \iiint_v \rho \ddot{u}_i \, \delta u_i \, dV = 0 \tag{6.92}$$

Considering the existence of time function variables and their derivatives in Equation (6.92), we can derive the aforementioned equation with respect to time integration (t) in relation to the presence of time function variables and their derivatives as follows:

$$\int_{t_1}^{t_2} \left\{ \iint_{S_\sigma} \overset{\sigma}{\underset{\sim}{}}i \, \delta u_i \, dS - \iiint_v \sigma_{ij} \, \delta\varepsilon_{ij} \, dV + \iiint_v \rho b_i \, \delta u_i \, dV - \iiint_v \rho \ddot{u}_i \, \delta u_i \, dV \right\} dt = 0 \tag{6.93}$$

We possess a connection with the periodic integral of the last Equation (6.93).

$$\int_{t_1}^{t_2}\iiint_{v} \rho\ddot{u}_i\ \delta u_i\ dVdt = \left[\rho\dot{u}_i\ \delta u_i\ dV\right]_{t_1}^{t_2} - \int_{t_1}^{t_2}\iiint_{v} \rho\dot{u}_i\ \delta\dot{u}_i\ dV\ dt \tag{6.94}$$

Given that the u_i function is specified at the time boundary, the first sentence on the right side of the Equation (6.94) is equal to zero. Therefore:

$$-\frac{1}{2}\int_{t_1}^{t_2}\iiint_{v} \rho\ \delta\left(\dot{u}_i\dot{u}_i\right) dV\ dt = -\int_{t_1}^{t_2}\delta\left[\frac{1}{2}\iiint_{v} \rho\ \dot{u}_i\dot{u}_i\ dV\right]dt = -\int_{t_1}^{t_2}\delta T\ dt \tag{6.95}$$

Means:

$$T = \frac{1}{2}\iiint_{v} \rho\ \dot{u}_i\dot{u}_i\ dV \tag{6.96}$$

The Equation (6.96) denotes the total kinetic energy of a continuous environment. Based on Equations (6.94), (6.95), and (6.96), the Equation (6.93) can be restated in the following manner:

$$\int_{t_1}^{t_2}\delta T\ dt - \int_{t_1}^{t_2}\iiint_{v} \sigma_{ij}\delta\varepsilon_{ij}\ dV\ dt + \int_{t_1}^{t_2}\int_{\sigma} \sigma_i\delta u_i\ dS\ dt + \int_{t_1}^{t_2}\iiint_{v} \rho b_i\delta u_i\ dV\ dt = 0 \tag{6.97}$$

Equation (6.97) is the general shape which refers to the mathematical representation of the equations of motion for any object and is occasionally referred to as the general form of the Hamilton principle, irrespective of material.

However, in the case of an elastic object, the following is well-established [1]:

$$\sigma_{ij} = \frac{\partial u}{\partial \varepsilon_{ij}} \tag{6.98}$$

It is important to acknowledge that the Equation (6.98) is contingent upon both elastic and plastic behavior. In the context of Equation (6.98), "u" represents the density of strain energy, which can be expressed in terms of linear elastic behavior.

$$u = \frac{1}{2}C_{ijkl}\varepsilon_{ij}\varepsilon_{kl} \tag{6.99}$$

The C_{ijkl} represents the fourth-order tensor of elasticity. Given that $u = u\left(\varepsilon_{ij}\right)$, the expression can be simplified as follows:

$$\sigma_{ij}\delta\varepsilon_{ij} = \frac{\partial u}{\partial \varepsilon_{ij}}\delta\varepsilon_{ij} = \delta u \tag{6.100}$$

And

$$\Pi = \iiint_{v} u\ dV \rightarrow \delta\Pi = \iiint_{v} \delta u\ dV = \iiint_{v} \sigma_{ij}\delta\varepsilon_{ij}\ dV \tag{6.101}$$

The symbol Π represents the potential energy associated with the strain. Given the Equation (6.101) with respect to Equation (6.97), we have the following:

$$\int_{t_1}^{t_2}\left[\delta T-\delta\Pi+\underbrace{\iint_S \sigma_i\delta u_i\ dS+\iiint_v \rho b_i\delta u_i\ dV}_{-\delta v}\right]dt=0 \qquad (6.102)$$

Let us now establish the definition of potential energy for the specified external forces, encompassing both the forces applied to the object's boundary and the physical force. This can be expressed as follows:

$$V=-W_{ext}=W_{int}=-\iint_{S_\sigma}\sigma_i\delta u_i\ dS-\iiint_v \rho b_i\delta u_i\ dV \qquad (6.103)$$

The surface integral S_σ reflects the external forces exerted on the object's surface, whereas the volume integral V represents the physical forces. Therefore, when the Equation (6.96) is supplied, the Equation (6.102) can be derived as follows [27, 28]:

$$\delta\int_{t_1}^{t_2}L\ dt\equiv\delta\int_{t_1}^{t_2}\left(T-\Pi+W_{ext}\right)\ dt=0 \qquad (6.104)$$

The Lagrangian function is denoted as L in the context of Equation (6.104). The Equation (6.104) is referred to as the Hamilton principle. The Hamilton principle can be defined as follows: the integral of the Lagrangian function over the time interval from t_1 to t_2 is minimized or maximized for all possible real displacements, while keeping the virtual displacements zero. This condition holds at all points of the object at times t_1 and t_2, as well as on the surface S_u where the displacements are specified.

The Hamilton principle, which considers the displacements $\left(u_i\left(x_1,x_2,x_3,t\right)\right.$ $\left.i=1,2,3\right)$ that generate a dynamic trajectory in space, might be subject to alternative interpretations. The Hamilton principle states that among all possible dynamic paths that meet the geometric boundary conditions on S_u at all times and the conditions specified at two arbitrary moments t_1 and t_2 at any point in the object, the actual dynamic path (response) minimizes the Lagrangian function.

6.12.1 Utilizations of the Hamilton Principle

The Hamiltonian principle can be applied to derive equations that govern intricate structures and engineering problems, encompassing the following [28]:

1. Transverse vibration of the springs
2. Longitudinal vibration of bars
3. Torsion vibration of shafts
4. Transverse vibration of the beams
5. Vibration of the membranes
6. Transverse vibration of plates
7. Vibration of the shells

In the upcoming chapters, we will explore how the Hamilton principle can be employed to derive static and dynamic equations that control geometric problems. An advantage of employing the Hamilton principle is that, along with the governing equations, it yields all feasible boundary conditions for the problem. These boundary conditions are selected based on the physics of the problem and the relevant conditions for the boundaries.

REFERENCES

[1] Lai, W.M., D. Rubin, and E. Krempl, *Introduction to continuum mechanics*. 2009: Butterworth-Heinemann.

[2] Fung, Y.C., Foundations of solid mechanics. *The Journal of the Royal Aeronautical Society*, 1966. **70**(663): p. 453–454.

[3] Amabili, M., *Nonlinear vibrations and stability of shells and plates*. 2008: Cambridge University Press.

[4] Eringen, A.C. and D. Edelen, On nonlocal elasticity. *International Journal of Engineering Science*, 1972. **10**(3): p. 233–248.

[5] Eringen, A.C., On differential equations of nonlocal elasticity and solutions of screw dislocation and surface waves. *Journal of Applied Physics*, 1983. **54**(9): p. 4703–4710.

[6] Eringen, A.C., Nonlocal polar elastic continua. *International Journal of Engineering Science*, 1972. **10**(1): p. 1–16.

[7] Ghadiri, M. and N. Shafiei, Vibration analysis of a nano-turbine blade based on Eringen nonlocal elasticity applying the differential quadrature method. *Journal of Vibration and Control*, 2017. **23**(19): p. 3247–3265.

[8] Shafiei, N., M. Kazemi, and M. Ghadiri, Nonlinear vibration behavior of a rotating nanobeam under thermal stress using Eringen's nonlocal elasticity and DQM. *Applied Physics A*, 2016. **122**: p. 1–18.

[9] Ghadiri, M. and N. Shafiei, Nonlinear bending vibration of a rotating nanobeam based on nonlocal Eringen's theory using differential quadrature method. *Microsystem Technologies*, 2016. **22**: p. 2853–2867.

[10] Kolter, W., Couple stresses in the theory of elasticity. *Proceedings of the Koninklijke Nederlandse Akademie van Wetenschappen*, 1964. **67**: p. 20.

[11] Toupin, R., Elastic materials with couple-stresses. *Archive for Rational Mechanics and Analysis*, 1962. **11**(1): p. 385–414.

[12] Doyle, J., *A* general solution for strain-gradient elasticity theory. *Journal of Mathematical Analysis and Applications*, 1969. **27**(1): p. 171–180.

[13] Yang, F., et al., Couple stress based strain gradient theory for elasticity. *International Journal of Solids and Structures*, 2002. **39**(10): p. 2731–2743.

[14] Mindlin, R.D. and H. Tiersten, Effects of couple-stresses in linear elasticity. *Archive for Rational Mechanics and Analysis*, 1962. **11**: p. 415–448.

[15] Mindlin, R., Influence of couple-stresses on stress concentrations. *Experimental Mechanics*, 1963. **3**(1): p. 1–7.

[16] Shafiei, N., A. Mousavi, and M. Ghadiri, Vibration behavior of a rotating non-uniform FG microbeam based on the modified couple stress theory and GDQEM. *Composite Structures*, 2016. **149**: p. 157–169.

[17] Ghadiri, M. and N. Shafiei, Vibration analysis of rotating functionally graded Timoshenko microbeam based on modified couple stress theory under different temperature distributions. *Acta Astronautica*, 2016. **121**: p. 221–240.

[18] Ghadiri, M., N. Shafiei, and S. Alireza Mousavi, Vibration analysis of a rotating functionally graded tapered microbeam based on the modified couple stress theory by DQEM. *Applied Physics A*, 2016. **122**: p. 1–14.

[19] Mindlin, R.D. and N. Eshel, On first strain-gradient theories in linear elasticity. *International Journal of Solids and Structures*, 1968. **4**(1): p. 109–124.
[20] Fleck, N. and J.W. Hutchinson, Strain gradient plasticity. *Advances in Applied Mechanics*, 1997. **33**: p. 295–361.
[21] Fleck, N. and J. Hutchinson, A reformulation of strain gradient plasticity. *Journal of the Mechanics and Physics of Solids*, 2001. **49**(10): p. 2245–2271.
[22] Ghanbari, B., M. Ghadiri, and H. SafarPour, A modified strain gradient shell model for vibration analysis of DWCNT conveying viscous fluid including surface effects. *Mechanics Based Design of Structures and Machines*, 2022. **50**(5): p. 1506–1536.
[23] Reddy, J.N., *Theory and analysis of elastic plates and shells*. 1999: CRC Press.
[24] Lim, C., G. Zhang, and J. Reddy, A higher-order nonlocal elasticity and strain gradient theory and its applications in wave propagation. *Journal of the Mechanics and Physics of Solids*, 2015. **78**: p. 298–313.
[25] Rashidpour, P., M. Ghadiri, and A. Zajkani, Low-velocity impact analysis of viscoelastic composite laminated nanoplate based on nonlocal strain gradient theory for different boundary conditions. *Journal of Sandwich Structures & Materials*, 2021. **23**(7): p. 3194–3233.
[26] Shayestenia, F. and M. Ghadiri, Investigation of flexoelectric effect on nonlinear vibration and dynamic instability of piezoelectric sandwich micro/nanobeam using the nonlocal strain gradient theory. *International Journal of Structural Stability and Dynamics*, 2023. **23**(04): p. 2350045.
[27] Reddy, J.N., *Energy principles and variational methods in applied mechanics*. 2017: John Wiley & Sons.
[28] Rao, S.S., *Vibration of continuous systems*. 2019: John Wiley & Sons.

7 An Introduction to Smart Materials

7.1 INTRODUCTION

In modern times, different equipment requires materials with diverse qualities that are suitable for their specific operating conditions. Hence, selecting the appropriate material for the fabrication of these devices based on their operational circumstances is a crucial engineering concern. Currently, over 50,000 different materials have been identified for the purpose of designing and producing goods in various industries. These materials span a wide spectrum, ranging from commonly used substances like copper, zinc, and brass, which have been in use for a long time, to more advanced materials such as superalloys. Engineering ceramics and composite materials are classified into separate categories. In order to ensure the appropriate design of components used in various equipment, it is important to identify the available materials and select those that possess the ideal qualities for manufacturing distinct parts. Over the past few decades, there has been a considerable rise in the use of industrial materials, including composite materials and functionally graded materials (FGM). This increase can be attributed to their exceptional performance and qualities, as well as their extensive use across many industries. Hence, this chapter focuses on comprehensively examining the fundamental principles and core ideas pertaining to composite materials, functionally graded materials (FGM), viscoelastic materials, magnetostrictive materials, dielectric materials, and metamaterials.

7.2 WHAT ARE SMART MATERIALS?

Smart materials are a class of materials that have the ability to respond to external stimuli by changing their properties. These materials are designed to exhibit unique and often complex behaviors, making them highly versatile and adaptable for various applications. One particular type of smart material that has gained significant attention is functionally graded materials (FGMs).

Functionally graded materials are a special category of smart materials that possess a gradual variation in composition, structure, or properties across their volume. Unlike traditional materials, which have uniform properties throughout, FGMs exhibit a continuous transition from one material phase to another. This gradient in properties allows FGMs to possess tailored characteristics that can be optimized for specific applications.

The concept of functionally graded materials originated from nature itself. Many biological structures, such as bones and teeth, exhibit a gradual change in composition and properties, enabling them to withstand different mechanical loads and

DOI: 10.1201/9781003470694-7

perform specific functions. Inspired by these natural examples, researchers have developed synthetic FGMs that mimic these gradient structures.

The unique feature of FGMs lies in their ability to combine the desirable properties of different materials into a single structure. By carefully controlling the composition and microstructure, FGMs can exhibit a wide range of properties, including mechanical, thermal, electrical, magnetic, and optical characteristics. This versatility makes FGMs suitable for a diverse range of applications, from aerospace and automotive industries to biomedical and energy sectors.

The composition gradient in FGMs can be achieved by various techniques, including powder metallurgy, additive manufacturing, casting, and solidification. These manufacturing methods allow for precise control over the distribution of different materials within the FGM, resulting in a tailored gradient profile. The choice of manufacturing technique depends on the specific requirements of the FGM and the desired properties.

The properties and performance of FGMs are influenced by several factors, including the composition gradient, microstructure, and processing conditions. The gradual change in properties across the FGM can lead to enhanced mechanical strength, improved thermal stability, and increased resistance to wear and corrosion. Additionally, FGMs can exhibit unique functionalities, such as shape memory effect, self-healing, and piezoelectricity, which further expand their potential applications.

The applications of FGMs are vast and diverse. In the aerospace industry, FGMs can be used to design lightweight and high-strength components, such as turbine blades and structural panels. The gradual change in properties allows for better load distribution and improved performance under extreme conditions. In the automotive sector, FGMs can be utilized to develop fuel-efficient engines, lightweight body structures, and advanced braking systems.

The energy and power sector can benefit from FGMs by utilizing their thermal and electrical properties. FGMs can be employed in the design of efficient heat exchangers, thermoelectric devices, and energy storage systems. In the biomedical field, FGMs have the potential to revolutionize tissue engineering and regenerative medicine. By mimicking the natural gradient structures found in biological tissues, FGMs can enhance the integration and functionality of implants and prosthetics.

Despite the numerous advantages offered by FGMs, there are also challenges associated with their development and implementation. The design and analysis of FGMs require advanced modeling and simulation techniques to predict their behavior under different conditions. Additionally, the characterization of FGMs is a complex task due to their gradient nature, requiring specialized techniques for microstructural analysis, mechanical testing, and thermal analysis.

In conclusion, smart materials, particularly functionally graded materials, are a fascinating class of materials that possess unique properties and behaviors. The gradual variation in composition and properties across the volume of FGMs allows for tailored characteristics that can be optimized for specific applications. With their wide range of properties and potential applications, FGMs have the potential to revolutionize various industries and pave the way for innovative technologies.

7.3 APPLICATIONS OF SMART MATERIALS

7.3.1 Biomedical Applications

Smart materials have revolutionized the field of biomedical engineering, offering new possibilities for medical devices, implants, and tissue engineering. One of the key applications is in the development of smart drug delivery systems. These systems use stimuli-responsive materials to release drugs at specific locations or in response to specific conditions in the body. This allows for targeted and controlled drug delivery, minimizing side effects and improving treatment outcomes.

Shape memory polymers (SMPs) have also found applications in biomedical engineering. They are used in the design of smart implants that can change their shape and properties in response to body temperature or other stimuli. This enables minimally invasive surgeries and improves the performance and longevity of implants.

7.3.2 Energy and Power Applications

Smart materials have the potential to revolutionize the energy and power sector by improving efficiency, reliability, and sustainability. One of the key applications is in the development of smart energy storage systems. Smart materials, such as shape memory alloys and phase change materials, are used to create energy storage devices that can store and release energy efficiently. These devices have the potential to enhance the performance of renewable energy systems and enable the widespread adoption of clean energy sources.

Another important application is in the field of energy harvesting. Smart materials, such as piezoelectric materials and thermoelectric materials, can convert mechanical or thermal energy into electrical energy. This opens up new possibilities for powering electronic devices and sensors in remote or inaccessible locations.

7.3.3 Other Applications

Smart materials have found applications in various other fields as well. In the field of civil engineering, smart materials are used in the design of smart structures that can adapt to changing environmental conditions, such as temperature, humidity, and wind loads. These structures can self-monitor, self-diagnose, and self-repair, leading to improved safety and durability.

In the field of consumer electronics, smart materials are used in the development of flexible displays, touch screen, and wearable devices. These materials offer unique properties, such as flexibility, transparency, and stretchability, enabling the creation of innovative and user-friendly electronic products.

In conclusion, smart materials have a wide range of applications across different industries. From aerospace to automotive, biomedical to energy, these materials have transformed the way we design and develop products. With their unique properties and capabilities, smart materials continue to drive innovation and open up new possibilities for the future.

7.4 ADVANTAGES AND CHALLENGES OF SMART MATERIALS

Smart materials have gained significant attention in various fields due to their unique properties and capabilities. These materials have the ability to respond to external stimuli, such as temperature, light, pressure, or electric fields, by changing their physical or chemical properties. This responsiveness makes them highly versatile and opens up a wide range of applications in engineering, medicine, and other industries. However, along with their advantages, smart materials also present certain challenges that need to be addressed for their successful implementation. In this section, we will explore the advantages and challenges associated with smart materials, with a particular focus on functionally graded materials.

7.4.1 Advantages of Smart Materials

7.4.1.1 Enhanced Functionality

One of the key advantages of smart materials is their ability to enhance the functionality of a system or device. By incorporating smart materials, engineers can design systems that can adapt, respond, or self-regulate based on changing conditions. For example, shape memory alloys (SMAs) can recover their original shape after deformation, making them ideal for applications such as actuators and sensors. This enhanced functionality allows for the development of more efficient and intelligent systems.

7.4.1.2 Improved Performance

Smart materials offer improved performance compared to traditional materials. For instance, piezoelectric materials can convert mechanical energy into electrical energy and vice versa, enabling the development of sensors, transducers, and energy-harvesting devices. The unique properties of smart materials, such as high sensitivity, fast response time, and low power consumption, contribute to the improved performance of systems and devices.

7.4.1.3 Energy Efficiency

Smart materials can contribute to energy efficiency in various ways. For example, shape memory alloys can be used in smart building systems to regulate temperature and reduce energy consumption. Similarly, piezoelectric materials can be employed in energy harvesting devices to convert mechanical vibrations into electrical energy. By utilizing the energy conversion capabilities of smart materials, energy-efficient systems can be developed, leading to reduced energy consumption and environmental impact.

7.4.1.4 Self-healing and Self-repair

Certain smart materials possess self-healing and self-repair capabilities, which can significantly extend the lifespan of structures and devices. For instance, self-healing polymers can autonomously repair minor damages, preventing the need for costly repairs or replacements. This property is particularly beneficial in applications where maintenance is challenging or expensive, such as aerospace or offshore structures.

7.4.1.5 Miniaturization and Integration

Smart materials enable the miniaturization and integration of components, leading to compact and lightweight systems. This advantage is crucial in industries such as electronics and aerospace, where size and weight reduction are critical factors. By utilizing smart materials, engineers can design smaller and more efficient devices, leading to advancements in portable electronics, wearable technology, and miniaturized sensors.

7.4.2 Challenges of Smart Materials

7.4.2.1 Cost

One of the primary challenges associated with smart materials is their cost. Many smart materials, such as shape memory alloys and piezoelectric materials, are relatively expensive compared to traditional materials. The high cost of production and limited availability of certain smart materials can hinder their widespread adoption in various industries. However, as research and development continue, the cost of smart materials is expected to decrease, making them more accessible for commercial applications.

7.4.2.2 Reliability and Durability

Reliability and durability are crucial factors in the successful implementation of smart materials. Some smart materials may exhibit degradation or fatigue over time, affecting their performance and lifespan. For example, shape memory alloys can experience fatigue failure after a certain number of shape memory cycles. It is essential to understand the long-term behavior and reliability of smart materials to ensure their safe and efficient operation.

7.4.2.3 Integration and Compatibility

Integrating smart materials into existing systems or structures can be challenging due to compatibility issues. Smart materials may have different mechanical, thermal, or electrical properties compared to traditional materials, requiring careful consideration during the design and manufacturing processes. Ensuring proper integration and compatibility between smart materials and other components is crucial to achieve the desired functionality and performance.

7.4.2.4 Manufacturing Complexity

Manufacturing smart materials can be complex and require specialized techniques. Some smart materials, such as functionally graded materials, involve the combination of different materials with varying properties. Achieving a seamless transition between different material compositions and maintaining the desired gradient can be challenging. Developing efficient and cost-effective manufacturing techniques for smart materials is essential to enable their widespread adoption.

7.4.2.5 Environmental Impact

The environmental impact of smart materials is another important consideration. Some smart materials may contain hazardous substances or require energy-intensive manufacturing processes. It is crucial to assess the environmental implications of smart materials throughout their lifecycle, from raw material extraction to disposal. Developing sustainable and environmentally friendly approaches for the production and use of smart materials is necessary to minimize their impact on the environment.

In conclusion, smart materials offer numerous advantages, including enhanced functionality, improved performance, energy efficiency, self-healing capabilities, and miniaturization. However, challenges such as cost, reliability, integration, manufacturing complexity, and environmental impact need to be addressed for their successful implementation. Overcoming these challenges will pave the way for the widespread adoption of smart materials in various industries, leading to advancements in technology, sustainability, and quality of life.

7.5 COMPOSITE MATERIALS

Composite materials have been used for a significant period of time. Flower straw is considered one of the earliest examples of man-made composites. However, the utilization of sophisticated composites dates back to the 1940s. During that period, both the American and former Soviet military achieved the production of epoxy-boron polymer composites for utilization in the aerospace industry, engaging in a fierce competition. After a period of 20 to 30 years, composites became extensively utilized in various industries including building, transportation, electronics, oil, and gas, among others. Furthermore, composites have been extensively utilized in maritime sectors, encompassing the fabrication of boats, ships, and offshore installations. Composite materials are formed by combining two or more materials on a large scale to enhance their technical features in comparison to the individual components. On a macroscopic scale, the constituents of a composite material retain their molecular structure and do not create chemical connections with each other. Contrary to composite materials, certain materials like metal alloys are formed at a microscopic level. This results in the creation of a uniform material, which means that metal alloys are not classified as composite materials when seen on a larger scale.

During the period from 1847 to 1909 AD, advancements in chemistry led to the development of resins that were well-suited for the production of composite materials. The fundamental principles of composite materials were established in 1930, followed by the development of glass fiber–reinforced polyester in 1942 and epoxy resin–based composite materials in 1946. Over time, several varieties of fields and reinforcements were employed to manufacture composite materials. In modern times, composite materials are manufactured with the incorporation of robust and elongated fibers, resulting in exceptional strength and resistance to deformation despite their low bulk. Composite materials are very ideal for constructing aircraft bodies and space equipment due to their ability to reduce the overall mass of the equipment. This reduction in mass leads to a large increase in the efficiency of these equipment. Thus far, numerous endeavors have been undertaken to substitute metals like steel and aluminum with composite materials, aiming to enhance the performance of metal components while also reducing weight. Composite materials are specifically engineered for various uses by combining two or more distinct materials and establishing a bond between them. This is done in order to ensure that the new material possesses the desired qualities. Composite materials have been employed in constructing structures that require both high strength and resistance to external loads while also needing to be lightweight. In the design of these structures, it is crucial to ensure dimensional stability and corrosion resistance, taking into account the specific operating circumstances. The objective of developing composite materials is often to attain specific physical qualities that are absent in pure materials [1].

7.5.1 Characteristics of Composite Materials

Composite materials have garnered significant attention in various industries due to their ability to possess desirable properties that are absent in metals, polymers, and ceramics. Consequently, there is a continuous effort to enhance the performance of various equipment by optimizing the utilization of composite materials. Composite materials can enhance various properties through effective design. Some of these properties include the following:

1. Strong resistance to deformation
2. Superior strength-to-mass ratio
3. Exceptional durability against fatigue and corrosion
4. Versatility in material composition
5. Proficiency in developing composite materials based on specific production and assembly requirements
6. Capacity to assimilate energy and dampen vibration
7. Affordable production tools
8. Developing intelligent materials

Some drawbacks of composite materials include the exorbitant cost of raw materials and the expensive nature of some production and assembly procedures. Composite materials exhibit reduced strength in the direction orthogonal to the fibers, and under compressive loading, the composite sheets may experience buckling in the direction perpendicular to the plane. Polymer composite materials exhibit sensitivity to temperature, ambient humidity, and flaws resulting from impact loading and layer separation. Composite material parts present greater challenges in terms of repair compared to metal parts.

7.5.2 Application of Composite Materials

Composite materials are extensively utilized in numerous industries due to their diversified qualities and the capacity to achieve desired engineering characteristics. Considerable study has been conducted in the domain of composite materials, with the aim of enhancing our understanding of these materials and their potential as viable alternatives to other materials in many applications. The utilization of composite materials may be observed in several sectors, such as military and aerospace industries, automobile industries, sports industries, marine industries, civil and building structures, industrial parts, and medicine.

The aerospace industry has a longer history of utilizing composite materials compared to other industries. This is mostly due to the significant advantage of reducing the weight of aircraft and enabling them to achieve greater heights and speeds. Composite materials incorporating glass, carbon, and Kevlar fibers are frequently employed in the design and production of diverse aircraft components (Figure 7.1 (a)).

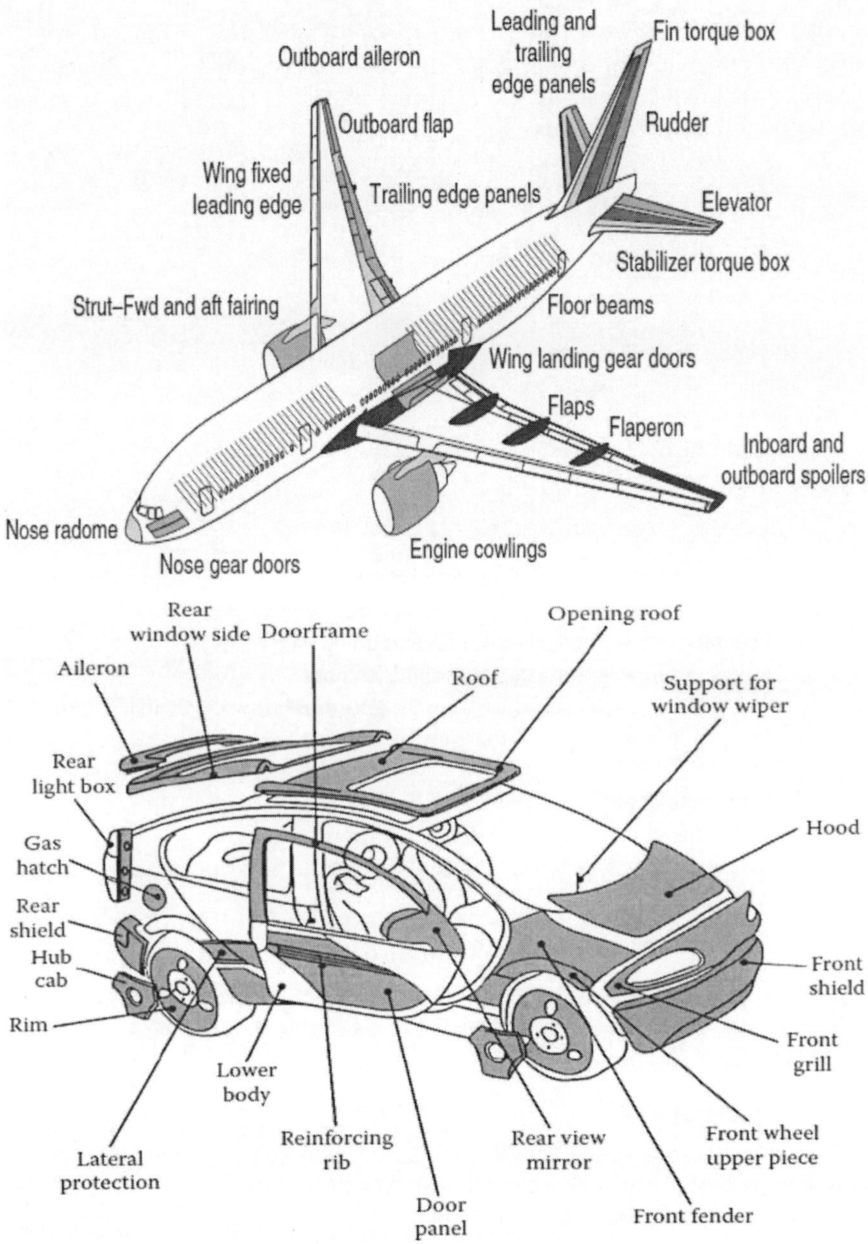

FIGURE 7.1 Composite materials: a) utilized in the Boeing 777 airplane and b) cars.

The utilization of composite materials in mass production of car parts has been driven by several key factors. These include the ability to produce parts with desired surface finishes, achieve the necessary strength while reducing weight, and enhance corrosion resistance. Additionally, the use of composite materials in marine industries has been motivated by their lightweight nature, improved efficiency, and fuel reduction. The utilization of composite materials in the construction industry has garnered significant interest due to its ability to withstand earthquakes and environmental variables. Previously, bridges have utilized plastics that are strengthened with glass or carbon fibers for constructing some components. These materials are extensively utilized in the production of industrial parts due to their ability to be tailored to specific working circumstances.

7.5.3 The Classification of Composite Materials

Composite materials are often classified into four main groups, as outlined by Jones [2]:

1. Fiber composite materials are a type of material
2. Materials composed of many layers
3. Materials composed of particles
4. Composite materials featuring a unified structure

7.5.4 Fiber Composite Materials

Fiber composite materials are composed of a substrate material that is strengthened by strands of fibers. Fiber composite materials can be composed of polymer, metal, or ceramic components, referred to as polymer composite materials, metal composite materials, and ceramic composite materials, respectively. Typically, these materials exhibit reduced density, tensile strength, and durability in comparison to fiber strands. The primary roles of the fibers are as follows:

- Resilience to external influences
- Inducing the ability to withstand changes in shape
- Generating robustness and establishing structural integrity

7.5.5 Enhancements of Fiber Composite Materials

Typically, polymer composite materials incorporate glass, carbon, and polymer fibers to enhance their technical characteristics. The subsequent text provides a description of the characteristics of the aforementioned materials.

7.5.5.1 Glass Fibers

These fibers are formed by the flow of molten glass material, driven by gravity, through the perforations of the mold. The fibers are then rapidly cooled to solidify. The diameter of the fibers is calculated based on the dimensions of the mold hole. Once the fibers are produced, a coating is applied to their surfaces to establish favorable circumstances for bonding the fibers to the substrate material, so creating composite materials. Glass fibers are extensively utilized in various composite structures due to their exceptional properties such as high flexibility, tensile strength, corrosion resistance, and fatigue resistance.

7.5.5.2 Carbon Fibers

These fibers are commonly employed in a variety of structures, particularly in aerospace applications, because of their high tensile modulus and strength. This results in suitable dimensional stability and their lightweight nature, which is advantageous compared to glass fibers. Carbon fibers are manufactured by subjecting carbon oxide polymer strands to high temperatures, followed by carbonization. Carbon fibers are utilized when glass fibers are unsuitable for reinforcing polymer materials due to their higher cost. Carbon fibers offer superior strength and lower volumetric mass, making them the preferred choice in such cases.

7.5.5.3 Polymer Fibers

In order to generate these fibers, the initial substance is liquefied and then forced into a mold to form the intended diameter of the fiber. Subsequently, once it exits the mold, the material is cooled down. Polymers consist of molecular chains, and their orientation occurs when they travel through a mold. This process greatly enhances the qualities of the polymer fibers in a specific direction. Kevlar is the most significant polymer fiber due to its superior tensile strength to volume ratio in comparison to other fibers. Other notable qualities of this product include exceptional resistance to impact and a negative coefficient of thermal expansion. The user's text is a single period. Some drawbacks of Kevlar polymer fibers include a relatively low working temperature (about 100 degrees Celsius), difficulty in cutting and machining, and weaker compressive mechanical qualities compared to its tensile strength. These fibers are less expensive than carbon fibers and more costly than glass fibers. Polymer materials can be classified into two distinct categories: thermoset materials and thermoplastic materials, each possessing unique features.

7.5.6 Composite Materials Consisting of Many Layers

This composite material is composed of a minimum of two layers of distinct materials that are joined together in a way that imparts desirable qualities to the final material, in comparison to the individual layers (Figure 7.2). The objective of developing layered composite materials is to enhance several material properties, including strength, deformation resistance, weight reduction, corrosion resistance, wear resistance, thermal properties, and surface polish. Layered composite materials can incorporate fiber materials, which introduces more design variables and allows for the creation of many properties.

7.5.7 Particulate Composite Materials

Occasionally, in order to enhance the characteristics of the material, one may introduce one or more varieties of particles from different materials into the domain.

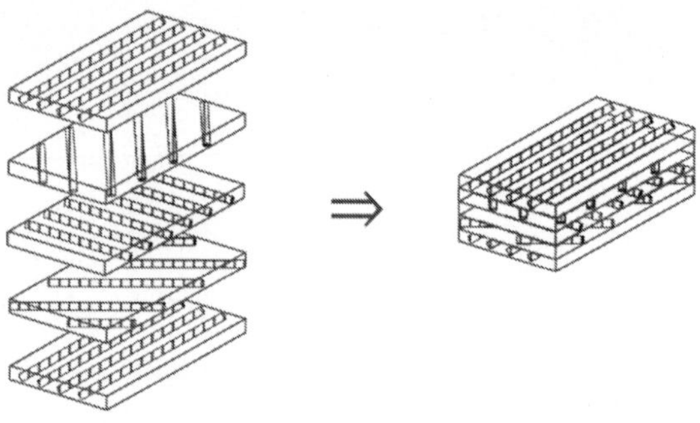

FIGURE 7.2 Schematic of the composite material reinforced with continuous fibers in a layered structure.

The particles and backdrop can have either metallic or non-metallic properties. The addition of sand particles to cement to create concrete for building construction is an instance of reinforcing a non-metallic substance with non-metallic particles, hence, enhancing the compressive strength of the composite material.

7.5.8 Composite Materials by Combining Different Elements

Occasionally, a composite material is created by combining two or three previously discussed forms of composite materials. In such cases, there are numerous factors to consider when designing the composite material for a particular purpose. When constructing multilayer composite materials, it is possible to use fibrous composite materials for each layer. In this case, the type of constituent materials, the orientation of the fibers, and the volume percentage of the fibers can be altered in relation to the adjacent layer. This allows for the creation of different desired properties in different directions.

7.5.9 The Composition of the Material

Composite materials, comprising many materials, have superior engineering qualities compared to conventional materials, such as metals. Composite materials have the potential to enhance several features such as stiffness, strength, weight reduction, corrosion resistance, thermal properties, and fatigue life. Composite materials typically comprise two components: fibers, which serve as the reinforcing material, and a matrix material, which forms the base (see Figure 7.3). The matrix material serves the purpose of binding the fibers together, as well as facilitating the transfer of loads and safeguarding the fibers against elongation caused by the environment [3].

Composite materials often exist in three distinct forms:

1. Fiber composites consisting of fibers made from one material embedded in a matrix material made from another substance.
2. Particulate composites consist of a mixture of particles of large size embedded in a matrix material.

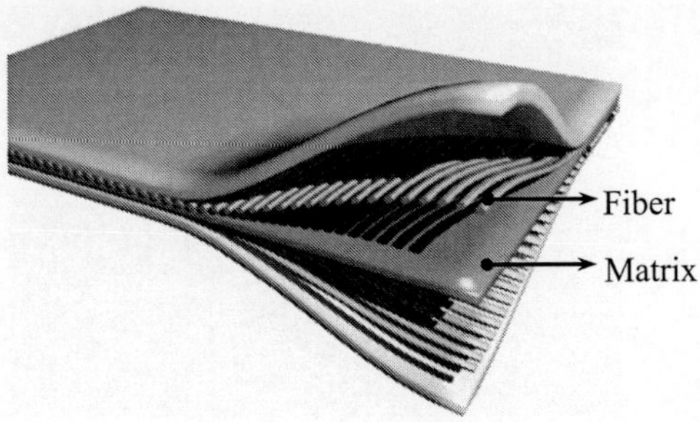

FIGURE 7.3 Schematic of the constituent elements of composite materials.

3. Multilayer composites consist of multiple layers of diverse materials, encompassing the first two types of composites. There will be four potential combinations: metal within non-metal, non-metal within metal, non-metal within non-metal, and metal within metal.

7.5.10 Multiple Layers

A fiber-reinforced layer comprises many fibers that are embedded in a matrix, which may contain a metal like aluminum or a non-metal like a thermoplastic polymer. Fibers can exhibit characteristics such as continuity, discontinuity, waviness, parallelism, non-parallelism, or random distribution. The identification of each layer in the multilayer structure can be determined based on its position, material composition, and angle of orientation with respect to the reference axis (in this case, the x-axis), as illustrated in Figure 7.4. The orientation of each layer is indicated by its angle and is distinguished from other layers by a forward slash (/). A multilayer refers to a collection of layers that are specifically designed to possess the necessary level of stiffness and strength. For instance, the layer that is strengthened with fibers of the same orientation can be arranged in a manner where the fibers in each layer are either aligned in the same direction or in varying directions (as shown in Figure 7.5). The layers are typically fused together within a homogeneous matrix material. A multilayer with fibers oriented at 30 or 45 degrees in the lamination direction can generate shear loads [2].

7.5.11 Study of the Mechanical Properties and Behavior of Composite Materials

Composite materials exhibit distinct mechanical properties in comparison to their individual constituent materials, necessitating the use of analytical and laboratory techniques for their evaluation. The majority of engineering materials are characterized by their homogeneity and isotropy, which can be defined as follows:

1. Homogeneous material has consistent properties across its entirety so that the properties are defined at every place and are not influenced by its position.

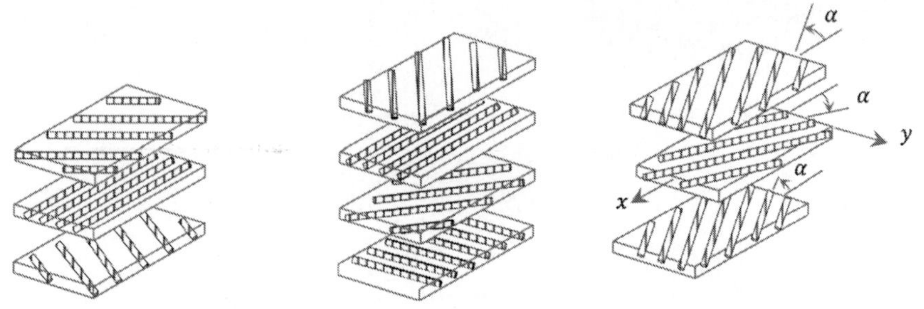

FIGURE 7.4 The orientation of fibers in the different layers of a composite material.

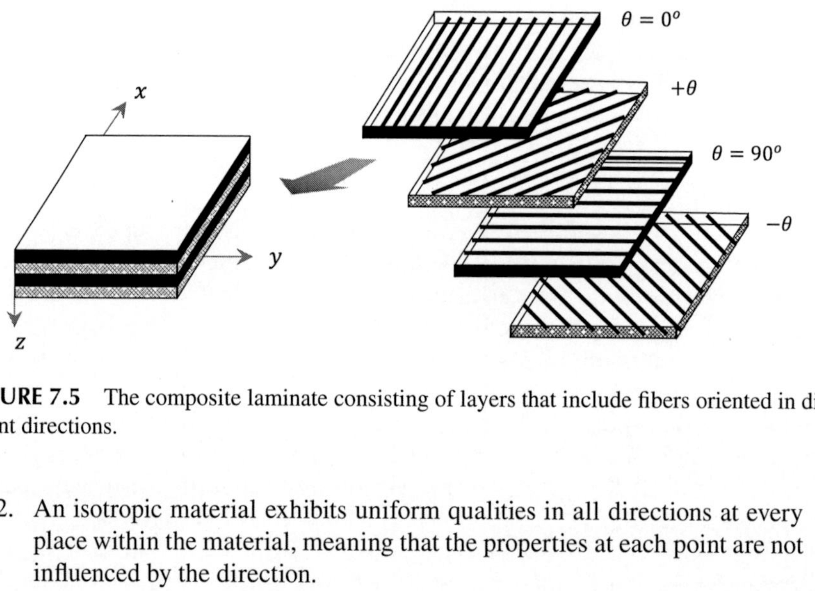

FIGURE 7.5 The composite laminate consisting of layers that include fibers oriented in different directions.

2. An isotropic material exhibits uniform qualities in all directions at every place within the material, meaning that the properties at each point are not influenced by the direction.

The majority of composite materials exhibit heterogeneous properties, resulting in non-uniform qualities that vary based on the material's position. Conversely, composite materials exhibit anisotropic behavior, meaning that their properties vary depending on the direction at each point within the material. Materials can exhibit four different types of anisotropic behavior.

7.5.11.1 Material Exhibiting Perfect Anisotropy

Matter exhibits varying qualities at different points and in different directions, lacking any plane of symmetry for these properties.

7.5.11.2 Material with a Monoclinic

Non-anisotropic materials exhibit symmetry with respect to a plane.

7.5.11.3 Orthotropic Materials

An orthotropic substance refers to a material that has different mechanical properties in different directions.

7.5.11.4 Transverse Isotropic Material

At a specific location within the material, its properties vary along three perpendicular directions. However, the material's qualities exhibit symmetry when observed from three perpendicular planes.

A transversely isotropic material is a type of material that exhibits the same mechanical properties in all directions perpendicular to a certain axis.

A material is considered transversely isotropic if its properties remain consistent in all directions within a plane, when defined at the anisotropic point.

Composite materials can be investigated from two perspectives based on their heterogeneous and anisotropic qualities.

1. The composite material's behavior is analyzed from a microscopic perspective, taking into account the qualities of its components and their interaction.
2. The macroscopic perspective is employed to analyze composite materials by seeing them as a continuous environment. This allows us to describe the behavior of the composite material as if it were a homogeneous material, exhibiting identical behavior to the composite material [2].

7.5.12 Attributes of a Layer that is Aligned in a Single Direction

An orthotropic material is formed by a reinforced layer with fibers that are positioned in a way that the symmetry planes of the material are parallel and perpendicular to the direction of the fibers. The material x_1 is aligned with the fibers, x_2 is perpendicular to the fibers, and x_3 is perpendicular to the composite layer. The characteristics of the orthotropic material in a composite layer are determined through appropriate laboratory experiments or theoretical approaches. The micromechanical approach, also known as the theoretical approach, is employed to calculate the engineering constants of fiber-reinforced composite materials. This method relies on the following assumptions:

1. The matrix and fibers exhibit perfect adherence.
2. The fibers exhibit parallel alignment and are evenly dispersed.
3. The matrix is devoid of micro-cracks and does not possess any initial tension.
4. Both fibers and matrix exhibit isotropy and behave in accordance with Hooke's law.
5. The applied loads are either parallel or perpendicular to the direction of the fibers.

Based on these postulates, the modulus and Poisson's ratio of a material reinforced with fibers can be expressed in relation to the modulus of elasticity, Poisson's ratio, and volume fraction of the structural materials in the following manner:

$$\rho = \rho_f V_f + \rho_m V_m \tag{7.1}$$

$$E_{11} = E_f V_f + E_m V_m \tag{7.2}$$

$$\nu_{12} = E_f V_f + E_m V_m \tag{7.3}$$

$$E_{22} = \frac{E_f E_m}{E_f V_m + E_m V_f} \tag{7.4}$$

$$G_{12} = \frac{G_f G_m}{G_f V_m + G_m V_f} \tag{7.5}$$

E_{11} represents the longitudinal modulus, E_{22} represents the transverse modulus, ν_{12} and ν_{21} represent Poisson's ratio, and G_{12}, G_{23}, and G_{13} represent the shear modulus for the x_1 x_2, x_2 x_3, and x_1 x_3 planes, respectively. In Equations (7.1) to (7.5), the subscripts f and m denote the distinct characteristics of the orthotropic layer, specifically the fiber (reinforcing phase) and matrix (ground phase) features, respectively. Furthermore, the symbol ρ denotes the mass density of the layer. Additionally, ν_f and ν_m are used to determine the volume fraction of each material. These volume fractions are related by the equation $\nu_f + \nu_m = 1$ and can be expressed as follows:

$$G_f = \frac{E_f}{2(1+\nu_f)}, \quad G_m = \frac{E_m}{2(1+\nu_m)} \tag{7.6}$$

The engineering parameters E_{11}, E_{22}, E_{33}, G_{12}, G_{13}, ν_{12}, ν_{13}, and ν_{23} for an orthotropic material can be found by experimental methods.

7.5.13 Study of Structural Equations in Linear Elasticity

When formulating the fundamental connections of a composite layer, the assumption is made that plastic deformations are not taken into account.

1. The composite layer is uninterrupted, meaning there are no gaps.
2. The composite layer exhibits the characteristics of a linear elastic material.

Composite materials are intrinsically heterogeneous, and the properties of the composite material are derived from the mean weight of the structural constituents (fibers and matrix). Structural equations establish the relationship between stresses and strains in the theory of elasticity. Linear elasticity is the most basic type of structural equations, encompassing Hooke's law as a broader concept. If we assume that the stresses are directly proportional to the strains, the linear elastic structural equations can be expressed in the following generic form:

$$\begin{Bmatrix} \sigma_{11} \\ \sigma_{22} \\ \sigma_{33} \\ \sigma_{12} \\ \sigma_{13} \\ \sigma_{23} \end{Bmatrix} = \begin{bmatrix} C_{1111} & C_{1122} & C_{1133} & C_{1112} & C_{1113} & C_{1123} \\ C_{2211} & C_{2222} & C_{2233} & C_{2212} & C_{2213} & C_{2223} \\ C_{3311} & C_{3322} & C_{3333} & C_{3312} & C_{3313} & C_{3323} \\ C_{1211} & C_{1222} & C_{1233} & C_{1212} & C_{1213} & C_{1223} \\ C_{1311} & C_{1322} & C_{1333} & C_{1312} & C_{1313} & C_{1323} \\ C_{2311} & C_{2322} & C_{2333} & C_{2312} & C_{2313} & C_{2323} \end{bmatrix} \begin{Bmatrix} \varepsilon_{11} \\ \varepsilon_{22} \\ \varepsilon_{33} \\ \varepsilon_{12} \\ \varepsilon_{13} \\ \varepsilon_{23} \end{Bmatrix} \tag{7.7}$$

Alternatively, it might be stated in a suggestive manner as follows:

$$\sigma_{ij} = C_{ijkl}\varepsilon_{kl} \tag{7.8}$$

The variables v_{12} represent the stress components, while ε_{kl} represent the strain components. C_{ijkl} represents the material coefficients that are determined in orthogonal Cartesian coordinates, as stated by Reddy [3]. The set of 36 C_{ijkl} coefficients are referred to as elastic constants and are required to be determined for every material. Various theories have been proposed to decrease the quantity of elastic constants in structural equations. Typically, in linear elastic analysis, it is assumed that the constituent matrix C is symmetric. By making this assumption, the total number of constants is decreased to 21. The number can be further decreased if the material qualities exhibit symmetry in specific planes. For instance, if the properties of the materials in plane 1–2 exhibit symmetry, it can be demonstrated that the matrix C takes on a certain form consisting of 13 independent coefficients.

$$\begin{Bmatrix} \sigma_{11} \\ \sigma_{22} \\ \sigma_{33} \\ \sigma_{12} \\ \sigma_{13} \\ \sigma_{23} \end{Bmatrix} = \begin{bmatrix} C_{1111} & C_{1122} & C_{1133} & C_{1112} & 0 & 0 \\ C_{2211} & C_{2222} & C_{2233} & C_{2212} & 0 & 0 \\ C_{3311} & C_{3322} & C_{3333} & C_{3312} & 0 & 0 \\ C_{1211} & C_{1222} & C_{1233} & C_{1212} & 0 & 0 \\ 0 & 0 & 0 & 0 & C_{1313} & C_{1323} \\ 0 & 0 & 0 & 0 & C_{2313} & C_{2323} \end{bmatrix} \begin{Bmatrix} \varepsilon_{11} \\ \varepsilon_{22} \\ \varepsilon_{33} \\ \varepsilon_{12} \\ \varepsilon_{13} \\ \varepsilon_{23} \end{Bmatrix} \tag{7.9}$$

An orthotropic material is characterized by symmetrical material properties in all three coordinate planes, resulting in a total of nine constants. Wood typically exhibits its orthotropic behavior. The fundamental equations for an orthotropic material are often formulated in the following manner:

$$\begin{Bmatrix} \sigma_{11} \\ \sigma_{22} \\ \sigma_{33} \\ \sigma_{12} \\ \sigma_{13} \\ \sigma_{23} \end{Bmatrix} = \begin{bmatrix} C_{1111} & C_{1122} & C_{1133} & 0 & 0 & 0 \\ C_{2211} & C_{2222} & C_{2233} & 0 & 0 & 0 \\ C_{3311} & C_{3322} & C_{3333} & 0 & 0 & 0 \\ 0 & 0 & 0 & C_{1212} & 0 & 0 \\ 0 & 0 & 0 & 0 & C_{1313} & 0 \\ 0 & 0 & 0 & 0 & 0 & C_{2323} \end{bmatrix} \begin{Bmatrix} \varepsilon_{11} \\ \varepsilon_{22} \\ \varepsilon_{33} \\ \varepsilon_{12} \\ \varepsilon_{13} \\ \varepsilon_{23} \end{Bmatrix} \tag{7.10}$$

In conclusion, if the characteristics of the material are the same in all directions, it is referred to as isotropic. In this case, just two separate constants are needed to describe the structural equations. The constants are referred to as E, which represents the modulus of elasticity, and ν, which represents Poisson's ratio. Structural equations establish a correlation between stress and strain at a certain location. If the structural equations remain constant throughout the material, it is referred to as homogeneous. Otherwise, if the structural equations vary between different points in the material, it is considered heterogeneous. Typically, in the context of linear elasticity problems, it is commonly believed that materials exhibit homogeneity and isotropy. Consequently, the characteristics of materials are typically explained based on the constant values of E and ν, which apply uniformly throughout the solid.

7.6 FUNCTIONALLY GRADED MATERIALS (FGM)

FGMs, or functionally graded materials, are a type of advanced composites. The concept of functionally graded materials (FGM) was initially proposed in 1984 by a cohort of Japanese scientists with the aim of developing heat protective materials. Subsequently, FGM has garnered significant attention and interest as a material for heat shielding purposes. The materials are acquired by the process of powder metallurgy, which involves the combination of two or more materials [4].

Local stress concentration can occur due to a sudden alteration in the composition and structure of materials and mechanical systems. This concentration of stress can be generated by either internal or external loading. By eliminating abrupt alterations in the structure and modifying the intensity of these variations, it is evident that the level of stress concentration is significantly diminished. Targeted materials have a remarkable capacity to enhance the thermomechanical properties of materials. Here are some techniques for enhancing the thermomechanical characteristics of materials using functionally graded materials (FGMs):

1. It minimizes thermal stresses and regulates their occurrence.
2. The plasticization point and failure point can be postponed for a given thermomechanical force.
3. Reducing the intensity of stress, particularly at the outside margins of the body, and completely removing it at the frontiers.
4. Enhancing the strength of connections between different materials, such as metal and ceramic, by establishing gradients.
5. To achieve the highest force necessary for crack propagation at the joint border, it is possible to enhance this by establishing a gradient in the mechanical characteristics of the material.

7.6.1 FGM Definitions

1. An FGM, or functionally graded material, is a material that exhibits a varying composition, structure, and engineering properties in a specific direction. This gradient is intentionally designed to enhance the material's engineering properties, making it superior to a homogenous material [5].

2. A substance that exhibits continuous or progressive variations in its qualities or performance in one or more directions.
3. FGMs, or functionally graded materials, are highly sophisticated composites engineered at a microscopic level to exhibit spatial and progressive variations in their material properties.

FGMs, or functionally graded materials, consist of a combination of ceramic and metal components. The specific benefits of using each of these elements, as well as the rationale for incorporating them into the structure of a desired material, are detailed in Table 7.1 [6].

Ceramics possess a high heat transfer coefficient and excellent heat resistance, allowing them to endure extremely high temperatures. Additionally, the presence of a metal framework in these materials contributes to their required flexibility. The utilization of both ceramic and metal enables seamless transitions between different levels of the construction. These materials were initially developed to create thermal shields in different structures and industries. Their key characteristic is their ability to endure extremely high temperatures, reaching up to approximately 2,500 degrees Celsius. Additionally, they exhibit resistance to significant temperature variations, corrosion, and abrasion. Furthermore, their metal composition grants them exceptional durability. There is no text provided. These materials have a significant impact on the construction of important structures, defense, and sophisticated industries.

FGM materials have microscopic heterogeneity, and their alterations occur gradually through the manipulation of the volume ratio between two constituent materials. Consequently, the mechanical properties of the structure undergo a continuous and gradual change from one side to the other. This eliminates the issues of incompatibility between metal and ceramic at their shared surface, such as warping in layered composites. Warping occurs when there is a sudden change in tension at the interface between the materials. The manufacture of FGM materials can utilize a variety of materials, including stainless steel, zirconia, nickel, silicon, nitride, titanium, tungsten, and copper.

TABLE 7.1
Materials that Are Specifically Composed of Metal and Ceramic

		Property
High-temperature surface	Pure ceramic	High thermal resistance
		Good anti-oxidation properties
		Low heat capacity
Low-temperature surface	Pure metal	High toughness and mechanical strength
		High thermal conductivity
		High fracture toughness

7.6.2 Definition and Characteristics of Functionally Graded Materials

Functionally graded materials (FGMs) are a class of smart materials that exhibit unique properties and characteristics, making them highly versatile and suitable for a wide range of applications. Unlike traditional materials, which have uniform properties throughout, FGMs are designed to have a gradual variation in composition, structure, or properties across their volume. This gradual variation allows FGMs to possess tailored properties that can be optimized for specific applications.

7.6.2.1 Composition and Structure

The composition and structure of functionally graded materials play a crucial role in determining their properties and performance. FGMs are typically composed of two or more different materials, such as metals, ceramics, polymers, or composites, which are combined in a controlled manner. The composition of the materials varies gradually from one end of the material to the other, resulting in a continuous transition of properties.

The structure of FGMs can be classified into two main types: continuous and graded. In continuous FGMs, the composition changes smoothly and continuously from one end to the other, without any distinct interfaces between the different materials. Graded FGMs, on the other hand, have distinct layers or regions with different compositions, resulting in a stepwise change in properties.

7.6.2.2 Tailored Properties

One of the key advantages of functionally graded materials is their ability to exhibit tailored properties. By carefully designing the composition and structure, FGMs can possess a wide range of properties, including mechanical, thermal, electrical, magnetic, and optical properties. This tailoring of properties allows FGMs to meet specific requirements and perform optimally in various applications.

For example, in structural applications, FGMs can be designed to have a gradient in mechanical properties, such as stiffness or strength, to optimize load-bearing capabilities. In thermal management applications, FGMs can be engineered to have a gradient in thermal conductivity, enabling efficient heat transfer across different regions. Similarly, in electrical or magnetic applications, FGMs can be tailored to exhibit varying electrical or magnetic properties, respectively.

7.6.2.3 Gradient Control

The control of the gradient in functionally graded materials is a critical aspect of their design and manufacturing. The gradient can be controlled by adjusting the composition, structure, or processing parameters during fabrication. Various techniques, such as powder metallurgy, additive manufacturing, casting, and solidification, can be employed to achieve the desired gradient.

The control of the gradient allows for precise tuning of the properties along the material's length or across its volume. This control can be achieved by adjusting the composition ratio, the thickness of the layers, or the processing conditions. The ability to control the gradient enables the customization of FGMs for specific applications, ensuring optimal performance and functionality.

7.6.3 Advantages of Functionally Graded Materials

Functionally graded materials offer several advantages over conventional materials, making them highly desirable for a wide range of applications. Some of the key advantages include the following:

1. Tailored properties: FGMs can be designed to possess specific properties, allowing for optimal performance in various applications.
2. Improved functionality: The tailored properties of FGMs enable them to perform multiple functions simultaneously, reducing the need for multiple materials or components.
3. Enhanced performance: The gradual variation in properties across FGMs can improve their overall performance, such as increased strength, improved thermal stability, or enhanced wear resistance.
4. Reduced stress concentration: The gradual transition in properties helps to minimize stress concentration at interfaces, reducing the risk of failure or damage.
5. Design flexibility: FGMs offer greater design flexibility compared to conventional materials, as their properties can be customized to meet specific requirements.
6. Cost-effectiveness: FGMs can potentially reduce material and manufacturing costs by eliminating the need for multiple materials or complex assembly processes.

7.6.3.1 Challenges of Functionally Graded Materials

While functionally graded materials offer numerous advantages, they also present certain challenges that need to be addressed during their design, manufacturing, and application. Some of the key challenges include the following:

1. Material compatibility: The selection and compatibility of different materials used in FGMs can be challenging, as they need to have similar thermal expansion coefficients and chemical compatibility to avoid delamination or degradation.
2. Manufacturing complexity: The fabrication of FGMs can be complex and require specialized manufacturing techniques, such as additive manufacturing or powder metallurgy, which may increase production costs.
3. Quality control: Ensuring consistent and uniform properties throughout the FGMs can be challenging, as any variations in composition or structure can affect their performance.
4. Limited understanding: Despite extensive research, there are still gaps in our understanding of the behavior and performance of FGMs, making their design and analysis a challenging task.
5. Scale-up and commercialization: Scaling up the production of FGMs and integrating them into commercial applications can be a significant challenge, requiring further research and development.

Despite these challenges, the unique properties and advantages offered by functionally graded materials make them a promising class of smart materials with immense potential for various engineering applications. Continued research and development in this field will further enhance our understanding and utilization of FGMs, opening up new possibilities for advanced materials design and innovation.

7.6.4 Manufacturing Techniques for Functionally Graded Materials

Functionally graded materials (FGMs) are a class of smart materials that exhibit varying properties and composition across their structure. These materials are designed to have a gradual transition in their composition, allowing for a seamless integration of different materials with distinct properties. The manufacturing techniques used to create functionally graded materials are crucial in achieving the desired properties and performance.

7.6.4.1 Powder Metallurgy

Powder metallurgy is a widely used manufacturing technique for functionally graded materials. It involves the mixing of powders with different compositions and properties, followed by compaction and sintering processes. The powders are carefully selected to achieve the desired composition gradient, and the compaction process ensures uniform distribution of the powders. Sintering then facilitates the bonding of the particles, resulting in a solid structure with a gradual change in composition.

Powder metallurgy offers several advantages for manufacturing functionally graded materials. It allows for precise control over the composition gradient, enabling the design of materials with tailored properties. Additionally, it enables the incorporation of different materials, such as metals, ceramics, and polymers, into a single structure. The flexibility of powder metallurgy makes it suitable for a wide range of applications, including aerospace, automotive, and biomedical fields.

7.6.4.2 Additive Manufacturing

Additive manufacturing, also known as 3D printing, has emerged as a promising technique for manufacturing functionally graded materials. This technique involves the layer-by-layer deposition of materials to create complex structures with varying composition and properties. Additive manufacturing offers the advantage of high precision and the ability to create intricate designs that are difficult to achieve using traditional manufacturing methods.

In the context of functionally graded materials, additive manufacturing allows for the precise control of the composition gradient. Different materials can be deposited in specific regions, resulting in a seamless transition between them. This technique also enables the incorporation of functional features, such as embedded sensors or actuators, within the structure of the material. Additive manufacturing has found applications in various industries, including aerospace, automotive, and electronics.

7.6.4.3 Casting and Solidification

Casting and solidification techniques are commonly used for manufacturing functionally graded materials, particularly in the production of metal-ceramic composites. These techniques involve the controlled solidification of a molten mixture of different materials to create a graded structure. The composition gradient is achieved by controlling the cooling rate and the distribution of the materials within the mold.

Casting and solidification techniques offer several advantages for manufacturing functionally graded materials. They allow for the production of large and complex structures with a continuous composition gradient. The process can be easily scaled up for mass production, making it suitable for industrial applications. However, the control over the composition gradient may be limited compared to other manufacturing techniques.

7.6.4.4 Joining and Bonding Techniques

Joining and bonding techniques are essential for creating functionally graded materials by combining different materials with distinct properties. Various methods, such as welding, brazing, and adhesive bonding, can be employed to achieve a seamless integration of materials. The choice of joining technique depends on the materials involved and the desired properties of the final product.

Welding is commonly used for joining metals in functionally graded materials. It involves the fusion of the materials at high temperatures, resulting in a strong bond. Brazing, on the other hand, uses a filler material with a lower melting point to join the materials. Adhesive bonding utilizes adhesives to create a bond between the materials. These joining techniques allow for the creation of functionally graded materials with tailored properties and performance.

7.6.4.5 Hybrid Manufacturing Techniques

Hybrid manufacturing techniques combine multiple manufacturing processes to create functionally graded materials. These techniques leverage the advantages of different methods to achieve the desired composition gradient and properties. For example, a combination of additive manufacturing and powder metallurgy can be used to create complex structures with precise control over the composition gradient.

Hybrid manufacturing techniques offer enhanced flexibility and control over the manufacturing process. They allow for the integration of different materials and the creation of intricate designs. However, these techniques may require more complex equipment and processes, making them suitable for specialized applications.

In conclusion, the manufacturing techniques for functionally graded materials play a crucial role in achieving the desired properties and performance. Powder metallurgy, additive manufacturing, casting and solidification, joining and bonding techniques, and hybrid manufacturing techniques offer different advantages and capabilities. The choice of manufacturing technique depends on the specific requirements of the application and the desired composition gradient.

7.6.5 Properties and Performance of Functionally Graded Materials

Functionally graded materials (FGMs) are a class of smart materials that exhibit unique properties and performance characteristics. These materials are designed to have a gradual variation in composition, microstructure, or properties across their volume, resulting in a seamless transition from one material to another. This gradual variation allows FGMs to possess tailored properties that can be optimized for specific applications.

7.6.5.1 Composition and Microstructure

The composition and microstructure of functionally graded materials play a crucial role in determining their properties and performance. FGMs are typically composed of two or more different materials, such as metals, ceramics, polymers, or composites, which are carefully selected based on their desired properties. The composition gradient can be achieved by varying the ratio or concentration of these materials along a specific direction.

The microstructure of FGMs can also vary gradually, leading to changes in grain size, phase distribution, or porosity. These microstructural variations can significantly influence the mechanical, thermal, and electrical properties of the material. By controlling the composition and microstructure, FGMs can be tailored to exhibit specific characteristics, such as enhanced strength, improved thermal stability, or superior electrical conductivity.

7.6.5.2 Mechanical Properties

One of the key advantages of functionally graded materials is their ability to exhibit a wide range of mechanical properties. By carefully designing the composition gradient, FGMs can achieve a seamless transition from one material with high strength and stiffness to another with high toughness and ductility. This unique property gradient allows FGMs to withstand varying mechanical loads and provide improved resistance to fatigue, fracture, and wear.

The mechanical properties of FGMs can be further enhanced by optimizing the microstructure. For example, controlling the grain size distribution can improve the material's strength and hardness, while introducing specific phases or reinforcements can enhance its toughness and impact resistance. These tailored mechanical properties make FGMs suitable for a wide range of applications, including structural components, load-bearing parts, and protective coatings.

7.6.5.3 Thermal and Electrical Properties

Functionally graded materials also exhibit exceptional thermal and electrical properties due to their composition and microstructure gradients. The gradual variation in material composition allows FGMs to have a controlled thermal expansion coefficient, which can reduce thermal stresses and improve thermal stability. This property is particularly advantageous in high-temperature applications where thermal mismatch can lead to premature failure.

Moreover, FGMs can possess unique electrical conductivity characteristics. By incorporating conductive materials in the composition gradient, FGMs can exhibit varying electrical conductivity along their length. This property can be utilized in applications such as electrical contacts, sensors, or electromagnetic shielding.

7.6.5.4 Durability and Reliability

The properties and performance of functionally graded materials contribute to their overall durability and reliability. The tailored mechanical, thermal, and electrical properties of FGMs allow them to withstand harsh operating conditions, such as high temperatures, corrosive environments, or dynamic loading. The gradual variation in material composition and microstructure also reduces the likelihood of stress concentration and material degradation, enhancing the material's overall reliability.

However, it is important to note that the performance of FGMs can be influenced by factors such as manufacturing defects, material degradation over time, and environmental exposure. Therefore, proper design, characterization, and testing are essential to ensure the long-term durability and reliability of functionally graded materials.

7.6.5.5 Performance Optimization and Design Considerations

To fully exploit the properties and performance of functionally graded materials, careful consideration must be given to their design and optimization. The selection of appropriate materials, composition gradients, and microstructural variations should be based on the specific requirements of the intended application. Computational modeling and simulation techniques can aid in predicting the material's behavior and optimizing its performance.

Furthermore, the manufacturing process plays a crucial role in achieving the desired properties and performance of FGMs. The choice of manufacturing technique, such as additive manufacturing, powder metallurgy, or casting, can influence the material's microstructure and properties. Therefore, a comprehensive understanding of the manufacturing techniques and their impact on the material's performance is essential for successful FGM fabrication.

In conclusion, functionally graded materials possess unique properties and performance characteristics due to their composition and microstructure gradients. These materials offer tailored mechanical, thermal, and electrical properties, making them suitable for a wide range of applications. However, careful design, optimization, and manufacturing considerations are necessary to fully exploit the potential of functionally graded materials and ensure their durability and reliability.

When utilizing FGM materials, it is imperative to express the relationships that govern this specific type of materials in mathematical equations and incorporate them into the computations. Hence, it is imperative to initially elucidate the classifications of these materials from a mathematical perspective. The majority of researchers employ power functions, exponential functions, or hyperbolic functions, all of which will be analyzed in the next sections. To fulfill this objective, let us examine a rectangular piece as depicted in Figure 7.6. The image illustrates that the

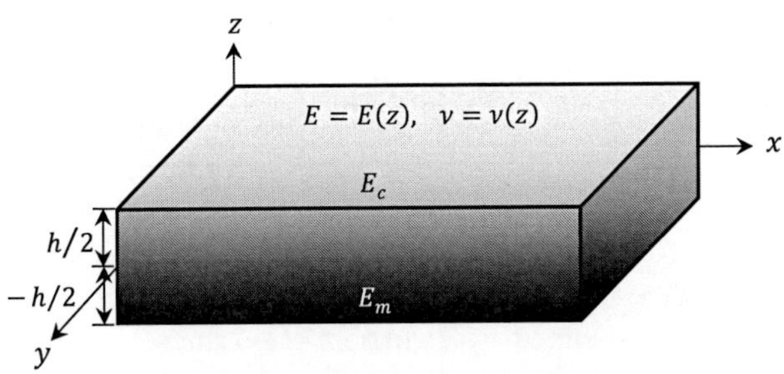

FIGURE 7.6 Displaying the physical properties of a functionally graded material (FGM) sheet.

x and y coordinates determine the plane of the sheet, while the z coordinate indicates the central plane of the sheet in the direction of its thickness. The upper and bottom sides of the sheet have distinct material properties, including differing elastic moduli and Poisson's ratios. These qualities only vary in the direction perpendicular to the surface, specifically in the z-coordinate direction (where z=h/2 and z=−h/2). A sheet with a specific focus is referred to as a targeted sheet, also known as FGM (focused growth matrix).

Delale and Erdogan [7] have demonstrated that the influence of Poisson's ratio on the deformation of the sheet is significantly smaller compared to the effect of elastic modulus. However, it is possible to consider Poisson's ratio as a constant, while other parameters such as density, Young's modulus, and the coefficient of thermal expansion are taken into account. The direction of sheet thickness was presumed to be changeable. These alterations can manifest as one of the following functions.

7.6.6 Material Attributes Modeled as a Power Function (P-FGM)

The volume percent of the target plane is defined as follows [8, 9]:

$$g(z)=\left(\frac{1}{2}+\frac{z}{h}\right)^{p} \tag{7.11}$$

Regarding this matter, p represents one of the material's properties, while h denotes the material's thickness. Thus, the material properties can be represented by a combination of volume fraction and process properties. Thus, the elastic modulus is determined by applying the law of combination of characteristics in the following manner:

$$E(z)=g(z)E_m+\left[1-g(z)\right]E_c \tag{7.12}$$

In the earlier connection, E_m represents the elastic modulus of the metal at the lowest level of the FGM sheet ($z=-h/2$), whereas E_c represents the elastic modulus of the ceramic at the highest level of the FGM sheet ($z = -h/2$). The variations in the elastic modulus across the thickness of the target plate, as influenced by the power distribution, are illustrated in Figure 7.7. It is imperative to clarify that in the aforementioned relationship, the temperature is assumed to be constant, which is a fundamental assumption underlying the obtained results. This diagram illustrates that the elastic modulus for $p > 1$ in the vicinity of the upper surface of the sheet undergoes rapid changes. In other words, the characteristics of a significant portion of the sheet's core thickness tend to align with those of the upper surface. Conversely, for $p < 1$ near the lower surface of the sheet, this tendency tends to be reversed.

7.6.7 S-Shaped Functionally Graded Material

At the scenario when a sheet is made up of multiple composite layers and the power distribution function is defined by Equation (7.12), there is a presence of stress concentration at one of the joint surfaces. This occurs specifically in areas where the material is continuous yet undergoes quick changes. Chung and Chi [10] have

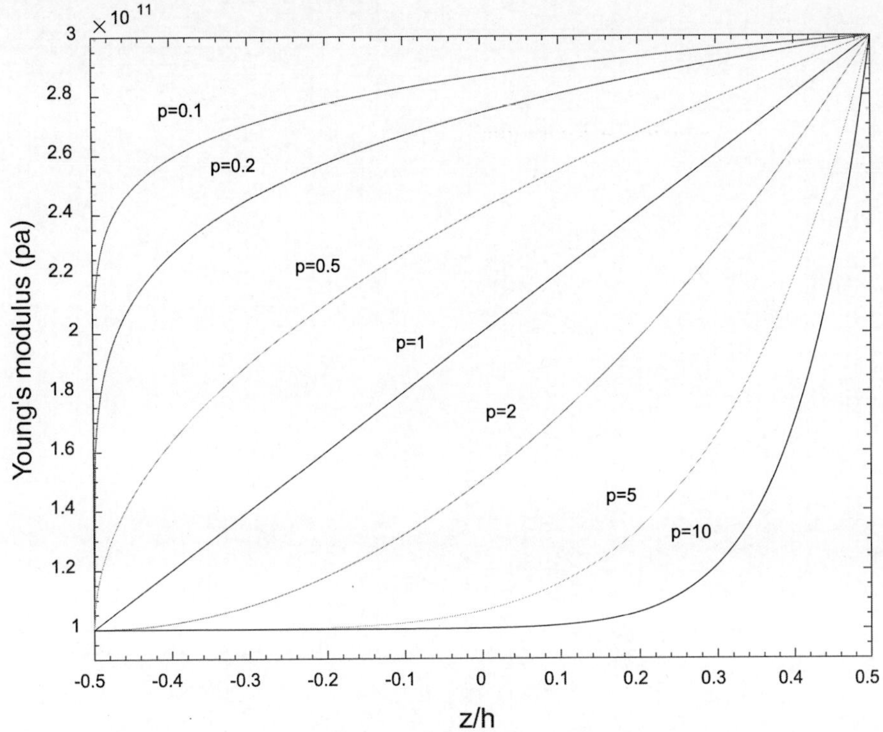

FIGURE 7.7 Diagram of the variations in the elastic modulus of a P-FGM plate.

established a volume fraction by employing two power functions to guarantee the evenness of stress distribution across all joint surfaces. The definitions of these two power functions are as follows:

$$g_1(z) = 1 - \frac{1}{2}\left(1 - \frac{2z}{h}\right)^p, \quad for \ \ 0 \le z \le h/2 \tag{7.13}$$

$$g_2(z) = \frac{1}{2}\left(1 - \frac{2z}{h}\right)^p, \quad for - h/2 \le z \le 0 \tag{7.14}$$

The elastic modulus of the sheet (FGM-S) is determined by utilizing these two combinations in the following manner:

$$E(z) = g_1(z)E_m + \left[1 - g_1(z)\right]E_c, \ \ for \ \ 0 \le z \le h/2 \tag{7.15}$$

$$E(z) = g_2(z)E_m + \left[1 - g_2(z)\right]E_c, \ \ for - h/2 \le z \le 0 \tag{7.16}$$

The S-shaped distribution function described in Equations (7.15) and (7.16) is used to illustrate the variations in the elastic modulus of the sheet, as illustrated in Figure 7.8.

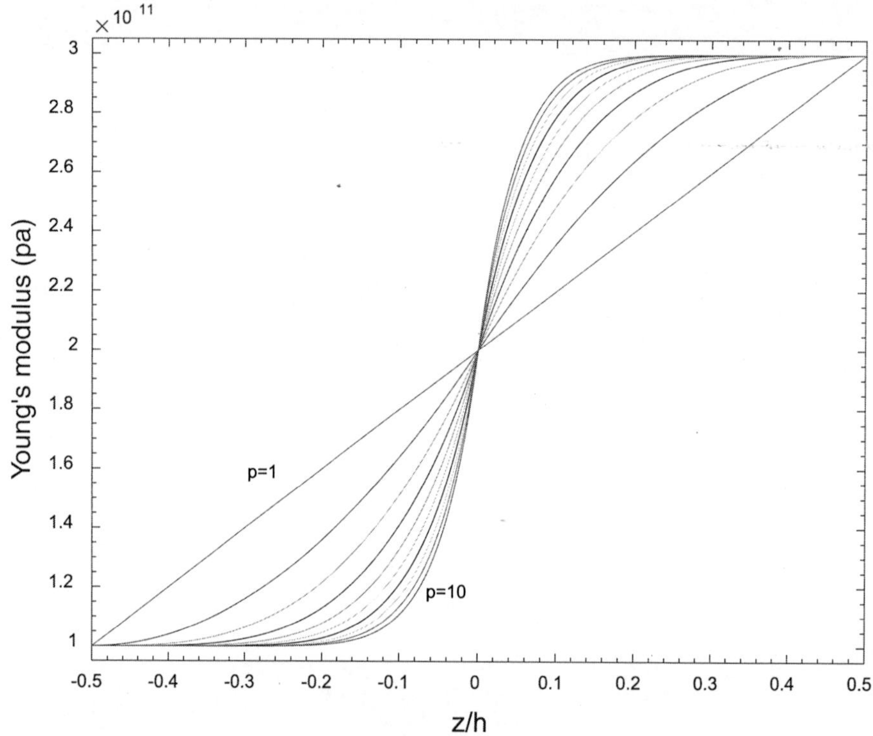

FIGURE 7.8 Diagram of the variations in elastic modulus within an S-FGM plate.

7.6.8 Material Qualities in the Context of an Exponential Function (E-FGM)

Many researches utilize the exponential function, specifically Equation (7.17), to accurately depict the characteristics of FGM materials.

$$E(z) = Ae^{B(z+h/2)} \tag{7.17}$$

Where:

$$A = E_m \;\; and \;\; B = \frac{1}{h}\ln\left(\frac{E_c}{E_m}\right) \tag{7.18}$$

Figure 7.9 displays the variations in the elastic modulus across the thickness of the specific plate.

7.6.9 Mori-Tanaka Homogenization Method

The homogenization method developed by Mori Tanaka is used to calculate the effective properties of a material that consists of a continuous background phase and

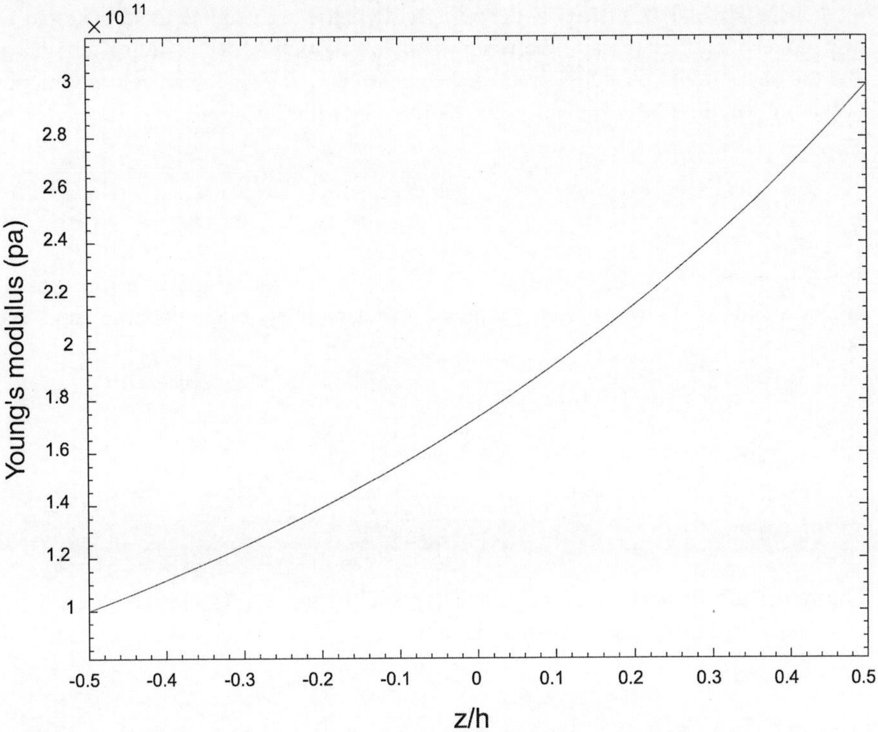

FIGURE 7.9 Diagram of the variations in Young's modulus within an E-FGM plate.

a discontinuous reinforcement phase [11]. In the case of a material consisting of two phases, Mori Tanaka's method involves labeling the background phase as "1" and the spherical particles of the reinforcing phase as "2". The concepts of volumetric modulus and local effective shear modulus are defined as follows:

$$\frac{K-K_1}{K_2-K_1}=\frac{V_2}{1+\frac{3(1-V_2)(K_2-K_1)}{3K_1}+4G_1} \tag{7.19}$$

$$\frac{G-G_1}{G_2-G_1}=\frac{V_2}{1+\frac{3(1-V_2)(G_2-G_1)}{G_1+f_1}} \tag{7.20}$$

Where:

$$f_1=G_1\frac{(9k_1+8G_2)}{6(k_1+2G_2)} \tag{7.21}$$

The volume modulus, shear modulus, and volume fraction of the background phase are denoted as K_1, G_1, and V_1, respectively. Similarly, the volume modulus, shear modulus, and volume fraction of the reinforcing phase are denoted as K_2, G_2, and V_2, respectively. The relationship between the volume fraction of the amplifying phase and the background phase is given by the equation $V_1 + V_2 = 1$. The volume fraction of the reinforcing phase is determined by the following formula:

$$V_2 = \left(\frac{2z + h}{2h} \right)^n \tag{7.22}$$

The relationship between the elastic modulus and Poisson's ratio of a functionally graded material (FGM) can be expressed in terms of its bulk modulus and shear modulus as follows:

$$E = \frac{9KG}{3K + G} \tag{7.23}$$

$$\nu = \frac{3K - G}{6K + 2G} \tag{7.24}$$

The values for the heat transfer coefficient k and the thermal expansion coefficient α are determined using the following method:
(7.25)

$$\frac{k - k_1}{k_2 - k_1} = \frac{V_2}{1 + \frac{(1 - V_2)(k_2 - k_1)}{3K_1}} \tag{7.25}$$

$$\frac{\alpha - \alpha_1}{\alpha_2 - \alpha_1} = \frac{\left(\frac{1}{K} - \frac{1}{K_1} \right)}{\left(\frac{1}{K_2} - \frac{1}{K_1} \right)} \tag{7.26}$$

The values of density (ρ), specific heat capacity (C_p), and other properties are determined using the law of mixtures method in the following manner:

$$\rho = \rho_1 V_1 + \rho_1 V_1 \tag{7.27}$$

$$C_p = C_{p1} \rho_1 V_1 + C_{p2} \rho_1 V_1 \tag{7.28}$$

7.7 VISCOELASTIC MATERIALS

When designing structures and machines, it is frequently essential to compute the impact of intricate conditions of stress, strain, and environment on the mechanical properties of various types of materials. To determine the mechanical response of a structure to various stress or strain conditions and environmental factors, it is necessary to derive the following fundamental relationships [12]:

1. Equilibrium equations represent the various stress relationships at each specific point to ensure equilibrium.
2. The kinematic equations relate the strain components to displacement, providing a description of the body's deformation.
3. Consistency equations involve multiple strain components and ensure continuity at a continuous level without any discontinuities.
4. The fundamental equations express the connections between stress, strain, and time using material constants specific to the given material.
5. Boundary conditions specify the stresses experienced by the displacements at the boundaries.

Hooke's law describes fundamental relationships when the behavior of materials under stress is linear and not influenced by time. Every book on the theory of elasticity contains a comprehensive explanation of the equations pertaining to assumptions 1, 2, 3, 4, and 5. This section provides a concise overview of the fundamental equations governing nonlinear and time-dependent materials.

7.7.1 Elastic Behavior

The majority of materials exhibit elastic behavior or a behavior that closely resembles it when subjected to small stresses. Figure 7.10 depicts the curves representing the elastic strain response of an object. Constant strains persist consistently when subjected to a continuous stress and vanish promptly upon the removal of the load. Reversibility is the primary attribute of elastic strain. The majority of elastic materials exhibit linear elasticity, meaning that a decrease in stress results in a corresponding decrease in strain.

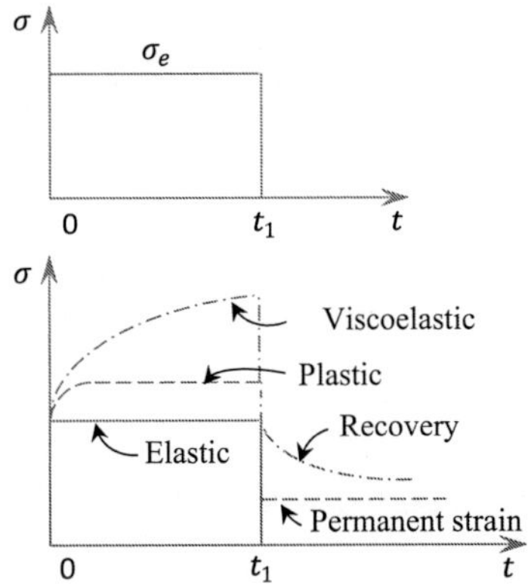

FIGURE 7.10 Comparing the various reaction to a constant load along the time.

7.7.2 Plastic Behavior

If the level of stress exceeds a certain threshold, the behavior will not exhibit elasticity. The point at which the material's behavior transitions from elastic to nonelastic, due to the stress exceeding a certain threshold, is referred to as the elastic limit. Inelastic strain refers to the strain that persists even after the removal of stress. Certain materials exhibit a temporary, continuous deformation under a brief load, while the strain remains constant when the load is fully applied and maintained. However, once the stress is alleviated, a strain will persist indefinitely. The permanent deformation is referred to as plastic strain (Figure 7.10). Plastic strain is considered to be time-independent, although it is common to observe some time-dependent strains along with plastic strain [13].

7.7.3 Viscoelastic Behavior

Certain materials initially demonstrate elastic behavior when subjected to rapid loading. However, the rate at which strain increases gradually slows down, resulting in a steady increase in strain with a decreasing growth rate. Upon the removal of stress, the strain undergoes a continuous decrease and exhibits an initial elastic rebound (Figure 7.10). Viscoelastic materials are those that are notably influenced by the rate of stress or strain. For instance, when the rate of stress is higher than the corresponding strain, it takes a longer time for these materials to reach their final stress value at a constant rate of stress. Refer to Figure 7.10. Plastics, wood, natural and synthetic fibers, concrete, and metals are examples of materials that exhibit viscoelastic behavior, particularly at elevated temperatures. Viscoelastic materials are commonly referred to as time-dependent materials due to the significant influence of time on their behavior. Viscoelasticity can be defined as the amalgamation of elasticity and viscosity, as described by Findley in 1976. Figure 7.11 displays the stress-strain curves for a viscous material and an ideal material. A material exhibiting linear viscoelastic behavior causes the stress-strain curve to shift towards contraction, as shown in the left diagram of Figure 7.11 [14].

Several phenomena that manifest in a viscoelastic material include the following:

1. When the stress remains constant, the strain exhibits a progressive increase over time.
2. When the strain is maintained at a constant level, the stress gradually diminishes over time.
3. The level of stiffness that is actually achieved depends on the rate at which the load is applied.
4. The utilization of a loading cycle results in hysteresis, causing the dissipation of mechanical energy.
5. Sound waves undergo attenuation and degradation.
6. The magnitude of the jump and the elasticity of the movement resulting from the impact is less than 100%.
7. Frictional resistance arises during the process of rolling.

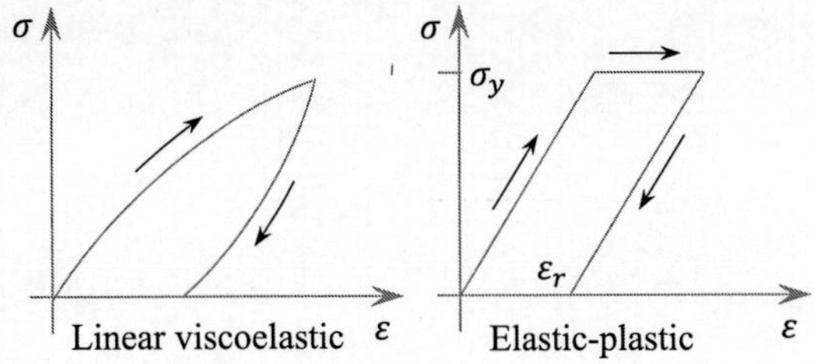

FIGURE 7.11 The stress-strain diagrams in constant strain rate: a) linear viscoelastic and b) elastic-plastic.

In general, all materials have some of these characteristics in response to a viscoelastic material. Common metals, such as steel and aluminum, exhibit behavior similar to quartz when subjected to small strain at room temperature. Their behavior remains consistent with that of linear elastic materials. Composite polymers, wood, and human tissues exhibit significant viscoelastic behavior similar to metals at elevated temperatures. A minor viscoelastic response can hold significance in certain applications. In order to achieve this objective, it is necessary to conduct an examination of the viscoelastic behavior of any material during its analysis or design process.

7.7.4 The Stress-Strain Relationship in a Viscoelastic Material

In order to examine the correlation between stress and strain in viscoelastic materials, three models can be taken into account: the Maxwell model, the Voigt model, and the standard linear solid model [14]. Maxwell's model consists of a spring and a damper arranged in series with each other (Figure 7.12). This model assumes that the deformation is quasi-static. The calculation of total deformation or strain in Maxwell's model is determined by the following equation:

$$\frac{d\varepsilon}{dt} = \frac{d\varepsilon_s}{dt} + \frac{d\varepsilon_d}{dt} = \frac{1}{E}\frac{d\sigma}{dt} + \frac{\sigma}{\eta} \tag{7.29}$$

Regarding this matter, ε represents the overall strain, ε_s denotes the strain of the spring, ε_d signifies the strain of the damper, η represents the damping coefficient of the damper, and σ refers to the stress. By substituting the equation $\tau = \frac{\eta}{E}$ into Equation (7.29), we can express it in the following form:

$$E\frac{d\varepsilon}{dt} = \frac{d\sigma}{dt} + \frac{\sigma}{\tau} \tag{7.30}$$

The Voigt model, also known as the Kelvin-Voigt model, consists of a spring and damper that are arranged in parallel (see Figure 7.12). In this model, the total stress

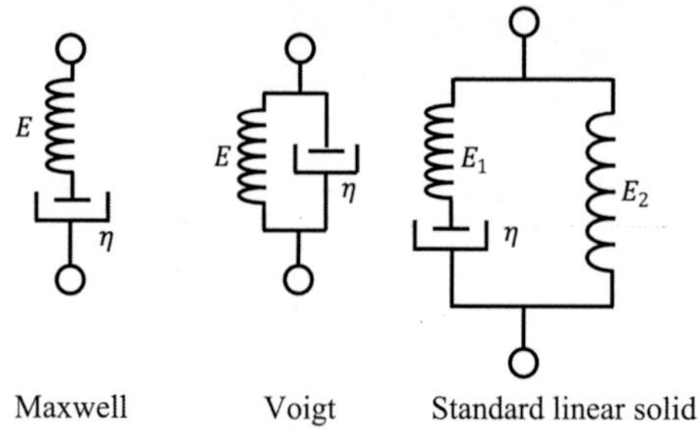

FIGURE 7.12 Schematic of different spring-damper models: a) Maxwell, b) Voigt, and c) standard linear solid.

is obtained from the sum of the stresses of each element separately. The tension relationship in this model will be as follows [15, 16]:

$$\frac{\sigma}{\tau} = \varepsilon + \tau_c \frac{d\varepsilon}{dt} \tag{7.31}$$

Or

$$\sigma = E\varepsilon + \eta \frac{d\varepsilon}{dt} \tag{7.32}$$

Where $\tau = \frac{\eta}{E}$ refers to the delay time. According to the two models explained earlier, it can be said that the most realistic model can be formed from the combination of both Maxwell and Kelvin models, which is known as the standard linear solid. This model includes three components. On the left side of Figure 7.12 is the model, which is the same as Maxwell's model:

$$E_1 \frac{d\varepsilon}{dt} = \frac{d\sigma_1}{dt} + \frac{\sigma_1}{\tau} \tag{7.33}$$

And similarly, $\tau = \frac{\eta}{E_1}$ a. For the right side of the model, we have the following:

$$E_2 \frac{d\varepsilon}{dt} = \frac{d\sigma_2}{dt} \tag{7.34}$$

In the earlier relations, E_1 and E_1 show the elastic modulus for each of the spring models and the sum of the total stress $\sigma = \sigma_1 + \sigma_2$. In general, the stress sum is as follows:

$$\left(E_1 + E_2\right) \frac{d\varepsilon}{dt} + \frac{E_2 \varepsilon}{\tau} = \frac{d\sigma}{dt} + \frac{\sigma}{\tau} \tag{7.35}$$

7.8 MAGNETOSTRICTIVE MATERIALS

It is not an exaggeration to say that the study of magnetostrictive materials is one of the areas of interest related to the exploration of smart materials. In point of fact, these materials are capable of transforming magnetic excitations into elastic reactions, which may take the form of either stretching or contracting. It is important to take note of the fact that the magneto-elastic interaction in question manifests itself when the structure in question is subjected to a magnetic field. Structures that have a magnetic effect are great candidates for use as ultrasonic generators, ultrasonic receivers, and echo detectors due to the qualities listed earlier.

Active metallic compounds known as magnetostrictive materials change shape in response to magnetic fields. Magnetoelastic coupling and the associated dependence of magnetic moment orientation on interatomic distance are responsible for these distortions. The linear or Joule magnetostriction is the most prevalent kind of magnetoelastic coupling, and it refers to the situation in which strains are recorded parallel to the direction of the magnetic field. It should be observed that irrespective of the direction of rotation of magnetic moments, the material will elongate if the magnetostriction is positive. Moreover, the transverse dimension will decrease, which will result in the volume remaining the same. When the magnetostriction occurs in the negative direction, the sample diameter grows while the sample length becomes shorter. After this, a magnetostriction curve that is symmetric is formed as a result of cycling the magnetic field. Although though Joule magnetostriction is present in the majority of magnetic materials, only a select few compounds containing rare earth elements may produce stresses that are more than $1{,}000 \times 10^{-6}$. Magnetostrictive materials alter their magnetic state in response to stresses because of the symmetric magnetoelastic coupling. The Villari effect allows for the measurement methods of force and displacement.

Magnetostriction is a property that is shared by all ferromagnetic materials; nevertheless, its value is often rather low for the majority of these materials. When combined with alloys that raise their Curie temperature above the surrounding temperature, certain transition metals and rare elements exhibit a greater degree of magnetostriction than others do. This is especially true when the Curie temperature of the alloys is raised above the surrounding temperature. The Terfenal-D alloy, which is a mixture of Terbion, iron, and Dispersium, is the magnetostrictive material that has the best circumstances in terms of the amount of strain and the Curie temperature. This material is the winner among the magnetostrictive materials. As the field strength is increased, the positive value of the magnetostrictive coefficient in iron flips to a negative value. Magnetostriction is a phenomenon that only happens in materials when temperatures drop below the Curie point; nevertheless, the temperature of the surrounding environment is almost always lower than the temperature of the furnace. This allows magnetostriction to have a practical use. Currently, magnetostrictive materials that have the best possible characteristics are continuing to increase and improve [17, 18].

Table 7.1 is a compilation of the nominal longitudinal strain for a variety of materials. It is interesting to note that some materials, like nickel, have a negative

TABLE 7.2
Nominal Saturation Strain and Curie Temperature of Magnetostrictive Materials

Material	Saturation Strain (ppm)	Curie Temperature (K)
$SmFe_2$	−1560	—
$CoFe_2O_4$	−110	793
Cobalt	−62	1388
Nickel	−40	630
Iron	−14	1040
Co_{72} Fe_3 B_6 A_{13} (Amorphous)	0	—
82%Ni-18%Fe, Permalloy	0	—
Fe-3.2%Si	+9	1015
45%Ni-55%Fe, Permalloy45	+27	—
87%Fe-13%Al	+30	673
Fe_{66} Co_{18} B_{15} Si	+35	—
Fe_3 O_4, Magnetite	+40	860
49%Co-49%Fe-2%V, permendur	+70	1253
Fe_{100-x} Ga_x, $15 \leq x \leq 28$ (Amorphous)	+250	—
Nickel-Cobalt	+186	—
$TbFe_2$	+1753	—
$Tb_{0.3}$ $D_{0.7}$ $Fe_{1.93,}$ Terfenol-D	+2000	650
$Tb_{0.5}$ Dy_x Zn	+5000	200
$Tb_{0.5}$ $Zn_{0.5}$	+5500	180

magnetostrictive coefficient, meaning that their length shortens when exposed to a magnetic field. In contrast, other materials, like Terfenol D, have a positive magnetostrictive coefficient, and their length lengthens when exposed to a magnetic field [19].

7.8.1 The Origin of Magnetism in Materials

Magnetic materials are made up of various different domains that are together referred to as magnetic domains (see Figure 7.13). Magnetic dipoles in a region of matter are said to be parallel and aligned in the same direction to form a domain. The total of all vectors from each domain is considered to be zero when the material is non-magnetic; however, when the material is magnetic, this value is considered to be something other than zero. When an external magnetic field is applied to magnetic materials, they get magnetized in a manner that is specific to their structure. Likewise, when the exposure to the magnetic field is removed, the magnetic materials demagnetize in a manner that is specific to their structure. The magnetic and non-magnetic properties of this process allow for the classification of materials into the following four groups: paramagnetic, ferromagnetic, antiferromagnetic, and ferrimagnetic, each of which will be discussed in further detail in the following paragraphs. When it comes to paramagnetic materials, the magnetic vectors are dispersed randomly throughout the item and are aligned in the direction of the external field; however, once they exit the field,

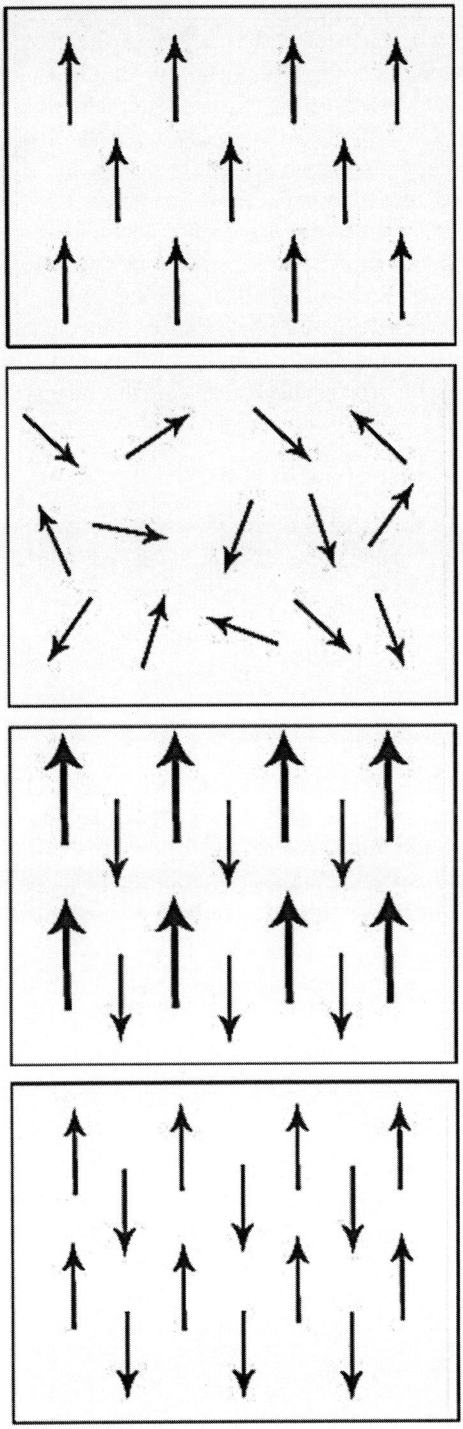

FIGURE 7.13 Vectors in materials: (a) paramagnetic, (b) ferromagnetic, (c) antiferromagnetic, and (d) ferrimagnetic.

they revert back to their initial configuration. If there is a magnetic field available, the magnetic vectors in ferromagnetic materials have a tendency to line up in a parallel fashion with one another. On the other hand, in contrast to materials that are paramagnetic, they do not change position when an external field is removed. Because the magnetic vectors in antiferromagnetic materials are parallel to each other but point in opposite directions, their magnitudes are typically the same, and the magnetic intensity is, therefore, zero throughout the entire object when an external field is not present. Antiferromagnetic materials have a magnetic intensity of zero. The third and final classification is known as ferromagnetic materials. Once the vectors in these materials are aligned parallel to one another, as indicated in the figure, the amplitude and strength of these vectors are not equal, which results in the production of a magnetic field that permeates the whole item. In most cases, the presence of more than one ion in the chemical is responsible for causing this quality [17, 19].

7.8.2 Basic Equations of the Material

The basic equations for magnetostrictive materials are outlined as follows. Considering a Cartesian coordinate system, O-x_1 x_2 x_3, the equilibrium equations are given by the following [20–22]:

$$\begin{aligned} \sigma_{ji,j} &= 0; \\ \varepsilon_{ijk} H_{k,j} &= 0; \\ B_{i,i} &= 0 \end{aligned} \tag{7.36}$$

Where σ_{ji}, H_i, and B_i are respectively the components of the stress tensor, the intensity vector of the magnetic field, and the magnetic induction vector, whereas ε_{ikj} is the Levi-Civita symbol. A comma followed by an index denotes partial differentiation with respect to the spatial coordinate x_i, and the Einstein's summation convention for repeated tensor indices is applied. The constitutive laws are given as follows:

$$\begin{aligned} \varepsilon_{ij} &= s^H_{ijkl}\sigma_{kl} + d_{kij} H_k \\ B_i &= d_{ikl}\sigma_{kl} + \mu^T_{ik} H_k \end{aligned} \tag{7.37}$$

Where ε_{ij} are the components of the strain tensor and s^H_{ijkl}, d_{ikl}, and μ^T_{ij} are respectively the magnetic field elastic compliance, the magnetoelastic constants, and the magnetic permittivity. Valid symmetry conditions are the following:

$$\begin{aligned} s^H_{ijkl} &= s^H_{jikl} = s^H_{ijlk} = s^H_{klij} \\ d_{kij} &= d_{kji} \\ \mu^T_{ij} &= \mu^T_{ji} \end{aligned} \tag{7.38}$$

The relation between the strain tensor and the displacement vector u_i is the following:

$$\varepsilon_{ij} = \frac{1}{2}\left(u_{j,i} + u_{i,j}\right) \tag{7.39}$$

The magnetic field intensity, named φ the potential, is written as follows:

$$H_i = \varphi_{,i} \tag{7.40}$$

For Terfenol-D, the constitutive relations can be written as follows:

$$\begin{Bmatrix} \varepsilon_{11} \\ \varepsilon_{22} \\ \varepsilon_{33} \\ 2\varepsilon_{23} \\ 2\varepsilon_{31} \\ 2\varepsilon_{12} \end{Bmatrix} = \begin{bmatrix} s_{11}^H & s_{12}^H & s_{13}^H & 0 & 0 & 0 \\ s_{12}^H & s_{11}^H & s_{13}^H & 0 & 0 & 0 \\ s_{13}^H & s_{13}^H & s_{33}^H & 0 & 0 & 0 \\ 0 & 0 & 0 & s_{44}^H & 0 & 0 \\ 0 & 0 & 0 & 0 & s_{44}^H & 0 \\ 0 & 0 & 0 & 0 & 0 & s_{66}^H \end{bmatrix} \begin{Bmatrix} \sigma_{11} \\ \sigma_{22} \\ \sigma_{33} \\ \sigma_{23} \\ \sigma_{31} \\ \sigma_{12} \end{Bmatrix} + \begin{bmatrix} 0 & 0 & d_{31} \\ 0 & 0 & d_{31} \\ 0 & 0 & d_{33} \\ 0 & d_{15} & 0 \\ d_{15} & 0 & 0 \\ 0 & 0 & 0 \end{bmatrix} \begin{Bmatrix} H_1 \\ H_2 \\ H_3 \end{Bmatrix} \tag{7.41}$$

$$\begin{Bmatrix} B_1 \\ B_2 \\ B_3 \end{Bmatrix} = \begin{bmatrix} 0 & 0 & 0 & 0 & d_{15} & 0 \\ 0 & 0 & 0 & d_{15} & 0 & 0 \\ d_{31} & d_{31} & d_{33} & 0 & 0 & 0 \end{bmatrix} \begin{Bmatrix} \sigma_{11} \\ \sigma_{22} \\ \sigma_{33} \\ \sigma_{23} \\ \sigma_{31} \\ \sigma_{12} \end{Bmatrix} + \begin{bmatrix} \mu_{11}^T & 0 & 0 \\ 0 & \mu_{11}^T & 0 \\ 0 & 0 & \mu_{33}^T \end{bmatrix} \begin{Bmatrix} H_1 \\ H_2 \\ H_3 \end{Bmatrix} \tag{7.42}$$

Where:

$$\left.\begin{array}{l} \sigma_{23} = \sigma_{32},\ \sigma_{31} = \sigma_{13},\ \sigma_{12} = \sigma_{21} \\ \varepsilon_{23} = \varepsilon_{32},\ \varepsilon_{31} = \varepsilon_{13},\ \varepsilon_{12} = \varepsilon_{21} \end{array}\right\}$$

$$\left.\begin{array}{l} s_{11}^H = s_{1111}^H = s_{2222}^H,\ s_{12}^H = s_{1122}^H,\ s_{13}^H = s_{1133}^H = s_{2233}^H,\ s_{33}^H = s_{3333}^H \\ s_{44}^H = 4s_{2323}^H = 4s_{3131}^H,\ s_{66}^H = 4s_{1212}^H = 2\left(s_{11}^H - s_{12}^H\right) \end{array}\right\}$$

$$d_{15} = 2d_{131} = 2d_{223},\ d_{31} = d_{311} = d_{332},\ d_{33} = d_{333}$$

7.9 FLEXOELECTRIC MATERIALS

The flexoelectric effect in dielectric materials can cause a substantial and non-uniform deformation, leading to the disruption or substantial alteration of the material's inversion symmetry and the generation of a net polarization. This phenomenon, known as electrostriction, differs from other types of electromechanical coupling, such as piezoelectricity and electrification, in that it is an inherent and widespread effect that becomes stronger as the sample size decreases. Research conducted over the past decade indicates that the emergence of flexoelectricity has not only introduced a novel method for achieving electromechanical coupling but has also had a significant impact on other domains, such as smart materials, electronics, and even physics. The diagram illustrates the direct flexoelectric effect in a dielectric beam, which results in the generation of electrical output when the beam is bent (see Figure 7.14) [23].

Flexoelectricity is a phenomenon in which the polarization and strain gradients are interconnected in an electromechanical process that is influenced by the size of the system. This feature is present in a diverse array of materials and becomes more prominent in objects at the nanoscale, where the strain gradient is greater. Simulations play a crucial role in comprehending flexoelectricity due to the challenges of conducting experiments at extremely small sizes and the limited availability of analytical solutions [24].

Furthermore, it has the ability to violate the principles of central symmetry and can be observed in a wide range of substances, such as insulators, liquid crystals, biological materials, and semiconductors. Recent research indicates that certain biological substances, including bones, hair, and biomembranes, have a notable flexoelectric reaction. The initial investigations on flexoelectricity in biological materials were conducted in 1975 by Williams and Berger. This preliminary investigation examined certain electromechanical characteristics of bones that may be attributed to "gradient polarization"; however, the precise mechanism remains incompletely comprehended.

7.9.1 Applications of Flexoelectric Materials

Some of the recent applications of flexoelectricity include flexoelectric energy harvesters, flexoelectric sensors and actuators, domain engineering, liquid crystal display, and the adjustment of the photovoltaic effect using flexoelectricity [25].

7.9.1.1 Flexoelectric Energy Harvester

Energy harvesters are machines capable of extracting energy from external sources, including wind energy, solar energy, thermal energy, and more. Specifically, the

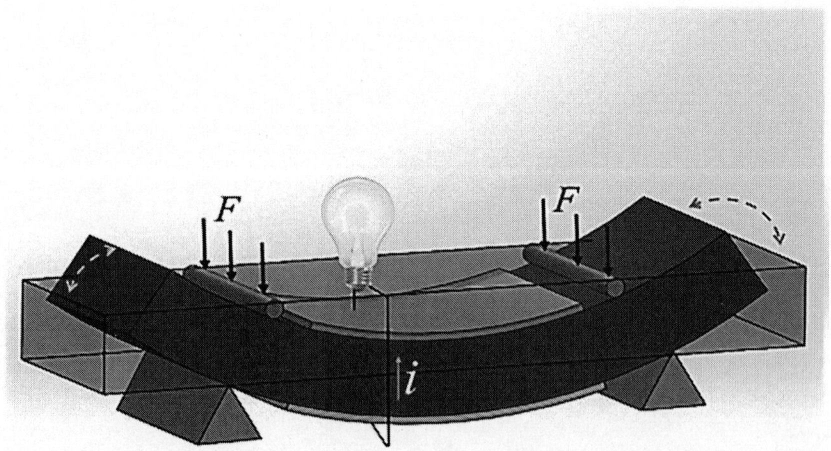

FIGURE 7.14 Schematic of the electric energy generation during bending of the flexoelectric beam.

electromechanical energy harvester is capable of gathering energy from mechanical vibrations. The flexoelectric energy harvester is well-suited for integration in small sizes due to its size-dependent unique effect, which allows for the creation of a substantial strain gradient. The flexoelectric energy harvesting device has achieved an energy conversion efficiency of 6.6% with appropriate design, as reported in Refs. [26–28].

7.9.1.2 Actuators

Piezoelectric materials have been extensively utilized for both sensing and actuating purposes. Furthermore, flexoelectricity can be utilized in sensor and actuator applications, particularly for devices of nanoscale dimensions. Flexoelectric sensors have an advantage over piezoelectric sensors and actuators because they are not restricted by the crystal symmetry of materials or operating temperature. The image later illustrates the design of a curved flexoelectric actuator made from non-polarized PVDF (Figure 7.15). This actuator has demonstrated excellent actuation characteristics, achieving a displacement resolution of up to 1.0 nm and a maximum displacement of 63.6 nm (Figure 7.16) [29].

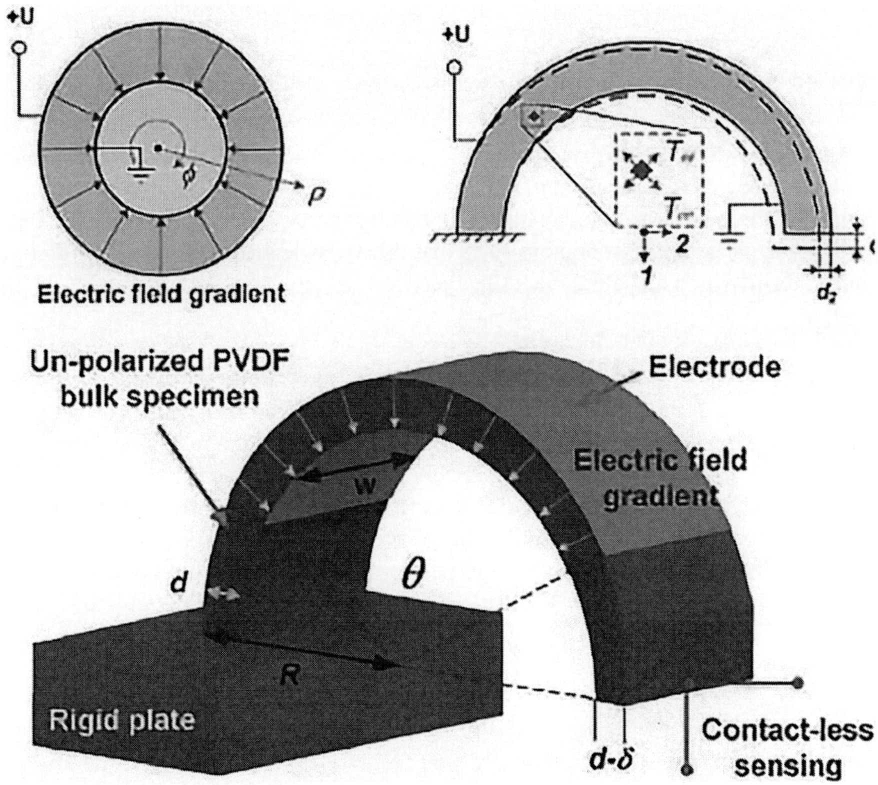

FIGURE 7.15 Curved flexoelectric actuator using PVDF material.

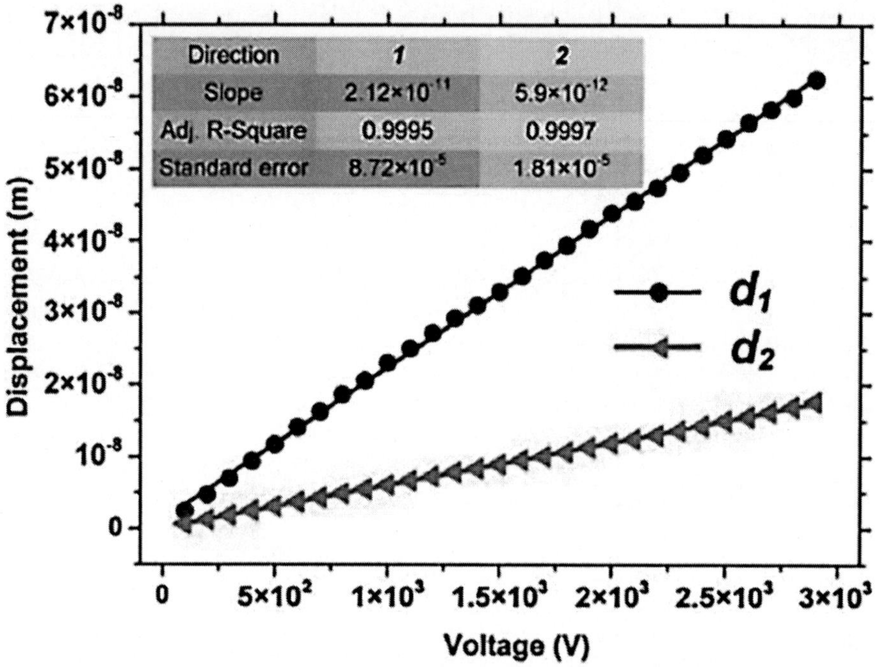

FIGURE 7.16 Proper operation characteristic with acceptable displacement.

Bhaskar has successfully included a flexoelectric actuator composed of barium titanate particles that are on a micron scale according to Figure 7.17. This actuator is totally compatible with semiconductor silicon technology. According to Figure 7.18, the performance analysis demonstrates that this flexoelectric actuator exhibits similar performance to actuators that utilize piezoelectric materials containing lead [30].

7.9.1.3 Flexoelectric Sensors

Flexoelectric sensors are gaining popularity because of their compact size, absence of discharge and deterioration issues, and use of lead-free materials. The sophisticated flexoelectric sensor has proven to be effective in various applications, including structural health monitoring, crack detection, and curvature measurement. Yan et al. [31] developed a flexoelectric bending sensor using barium strontium titanate (BST) that can directly convert bending deflections into output charge. The operational mechanism of this sensor is illustrated in Figure 7.19. In order to precisely measure the bending deflection, two BST micro-curvature sensors were affixed to the center lateral surfaces of an aluminum beam. These sensors were positioned symmetrically in relation to the beam's neutral axis.

7.9.1.4 Amplitude Adjustment and Polarity Switching

Another noteworthy application of flexoelectricity that warrants consideration is the ability to regulate amplitude and switch polarity. Ferroelectric materials are defined by their inherent polarization, which may be altered by the application

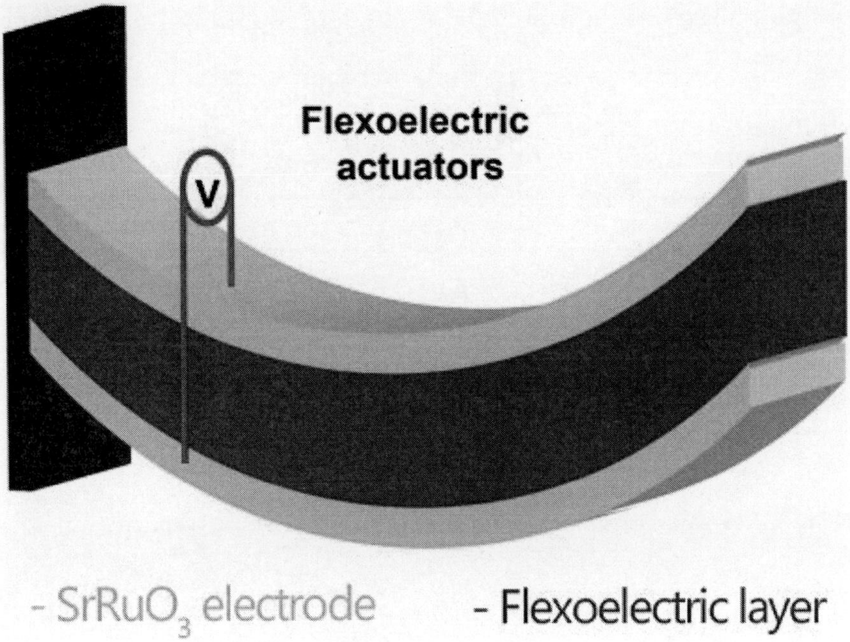

FIGURE 7.17 Schematic of the flexoelectric actuator.

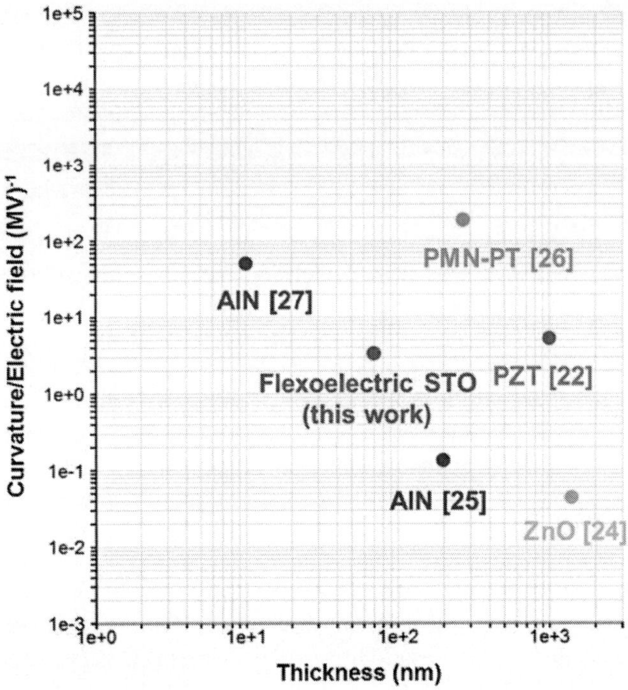

FIGURE 7.18 Proper performance of the flexoelectric actuator in comparison with piezoelectric actuators containing lead.

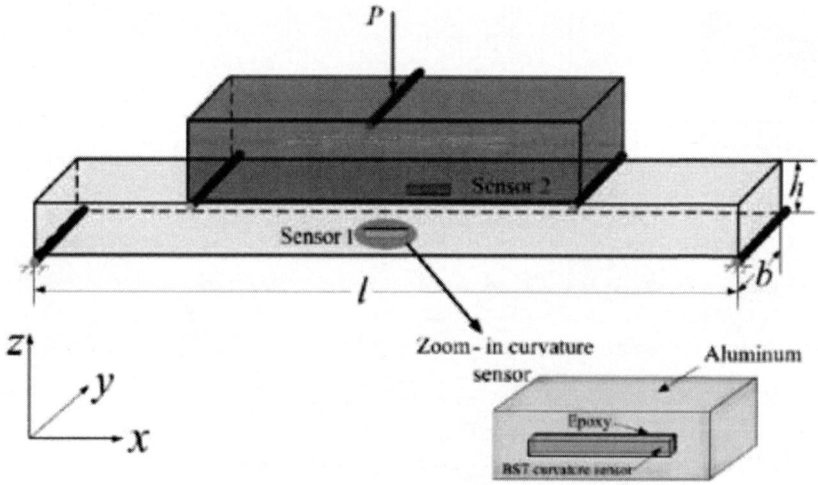

FIGURE 7.19 Working principles of BST curvature sensor attached to beam.

of an external electric field. Within a material that is extremely small, such as an ultrathin ferroelectric film, the inherent polarization can be altered by a gradient in mechanical strain. The impressive performance of this material has generated growing interest, making the inhomogeneous thin film an ideal candidate for flexoelectric applications [25].

The phenomenon of coupling between polarization and strain gradient is referred to as flexoelectricity. This phenomenon is observed in all dielectric materials exhibiting any form of symmetry. This study investigates the Timoshenko beam energy harvesting system, specifically focusing on the effects of flexoelectricity and strain gradient. The governing equations and boundary conditions have been obtained using Hamilton's principle. The flexoelectric effect is characterized by the variation in both normal and shear strain, resulting in a more comprehensive model. This article presents a model that examines and explores the impact of flexoelectricity on the dielectric beam and the energy harvesting system derived from the beam's fundamental harmonic. An extensive analysis was conducted to assess the flexoelectric coefficients, gravity gradient constants, base acceleration, and the impact of an additional concentrated mass on Timoshenko's energy harvesting system. The findings indicate that flexoelectricity significantly influences the efficiency of the energy harvesting system, particularly at the nanoscale. Typically, this impact results in a smoother beam behavior and alters the harvester's first resonance frequency [32].

Compact wireless devices are fully autonomous, using ambient energy to power themselves, and have extensive practical uses. This study presents the development of an analytical model that incorporates the strain gradient effect for nanoscale flexoelectric energy harvesters. The proposal suggests finding the closed form of the output voltage, which will allow for a rapid evaluation of the efficiency of flexoelectric energy harvesting at the nanoscale. The study reveals that the ideal load resistance is not significantly influenced by the impacts of transverse shear and strain gradient. The impact of transverse cutting on the output power is significantly greater in the nanoscale flexoelectric energy harvesting system with a smaller length-to-thickness ratio and a bigger patched mass. The modelling employing the Euler-Bernoulli beam has yielded much more output power in comparison to the Timoshenko beam. Furthermore, the strain gradient effect amplifies the frequency and diminishes the output power of the beam. This paper provides guidance for mechanical engineers and material scientists in designing a nanoscale flexoelectric energy harvesting device with excellent performance [33]. Figures 7.20, 7.21, and 7.22 depict the relationship between output power and frequency for piezoelectric and flexoelectric energy harvesting systems. The analysis considers both the deformation and non-deformation of the transverse section, with a length-to-thickness ratios (*l/h*) of 5, 10, and 20. It is evident that the output power of the flexoelectric energy harvesting system surpasses that of the flexoelectric energy harvesting system. The flexoelectric energy harvesting system has a maximum output power that is tenfold greater than the maximum output power of the piezoelectric energy harvesting system. Furthermore, it has been noted that the primary frequency estimated by the Euler-Bernoulli beam is higher than the frequency of the Timoshenko beam. The estimated output power according to the Euler-Bernoulli beam model exceeds that of the Timoshenko beam model. When the length-to-thickness ratio of a structure reaches 20, both the Timoshenko and Euler-Bernoulli beam models yield identical predictions for output power.

Flexoelectricity, which is characterized by its significant size-dependent feature, is a highly advantageous use of piezoelectricity in the field of micro/nanoscale energy harvesting. Nevertheless, there has been limited investigation into energy harvesters made from functionally graded flexoelectric materials. This work presents a theoretical

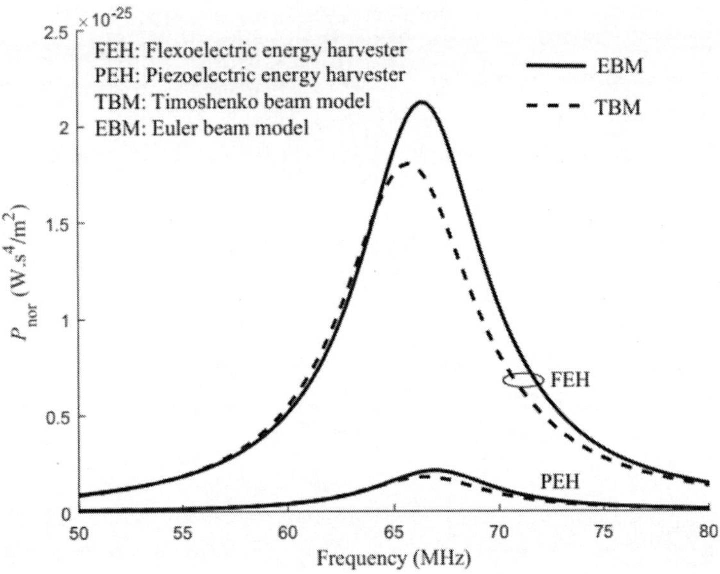

FIGURE 7.20 Output power in terms of frequency for different beam models with $l/h = 5$.

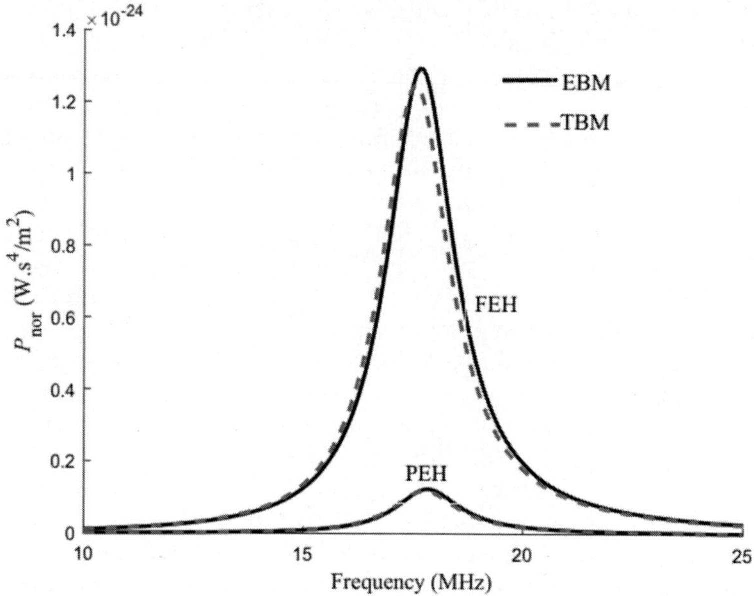

FIGURE 7.21 Output power according to frequency for different beams with *l/h = 10.*

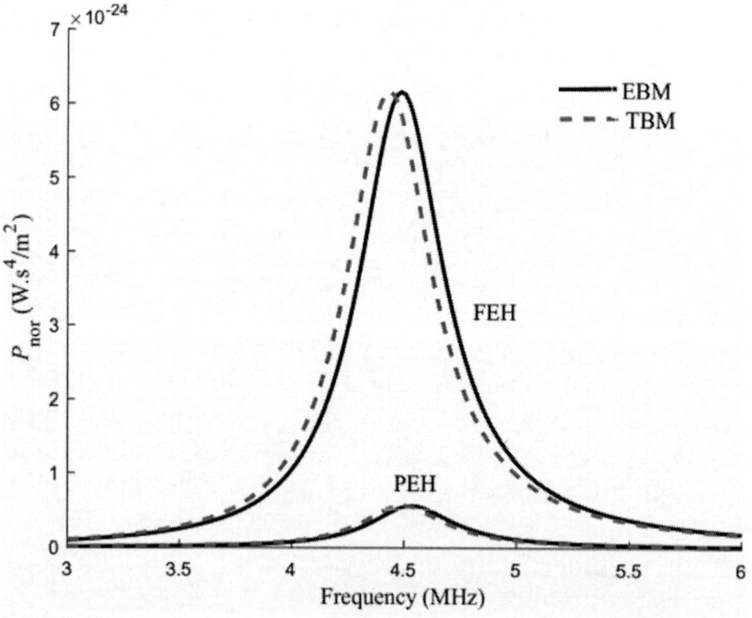

FIGURE 7.22 Output power according to frequency for different beams with *l/h = 20.*

analysis of the nonlinear vibration of a flexoelectric energy harvesting nanobeam with a concentrated mass at the free end. The analysis takes into account the electromechanical coupling generated by the strain gradient in the beam. The Galerkin approach is used to derive the equations for the coupled system and to provide approximate closed solutions for the output power. The study discovered that the voltage output and power density of the functionally graded flexoelectric energy harvester are significantly influenced by the material components, gradient index, size scale, and loading resistance [34].

7.9.2 Fundamental Equations of Flexoelectric Materials

To consider the flexoelectric effect, the concept of Gibbs disturbance energy is used [35].

$$G_b = -\frac{1}{2}k_{ij}E_iE_j - \frac{1}{2}b_{ijkl}E_{i,j}E_{k,l} + \frac{1}{2}C_{ijkl}\varepsilon_{ij}\varepsilon_{kl} - e_{ijk}E_i\varepsilon_{jk} - \mu_{ijkl}\left(E_k\varepsilon_{ij,l} - \varepsilon_{ij}E_{k,l}\right) \quad (7.43)$$

In the context of Equation (7.43), the variables E_i and E_j represent the electric field, ε_{ij} represents a component of strain, $\varepsilon_{ij,l}$ represents the strain gradient, $E_{k,l}$ represents the electric field gradient, k_{ij} represents the dielectric constant tensor, b_{ijkl} represents the nonlocal electric coupling coefficient tensor, C_{ijkl} represents the elastic stiffness tensor, e_{ijk} represents the piezoelectric parameter, and μ_{ijkl} represents the flexoelectric parameter.

Furthermore, the expressions for electrostatic potential, electric field, and electric field gradient are as follows:

$$\varphi = -\left(z^2\frac{e_{31}}{2k_{33}}\frac{\partial^2 w}{\partial x_2}\right) - \left(z\frac{\mu_{31}}{2k_{33}}\frac{\partial^2 w}{\partial x_2}\right) + z\frac{V_P}{h_P} + \left(\frac{e_{31}}{8\,k_{33}}h^2\frac{\partial^2 w}{\partial x_2}\right) + \frac{V_P}{2} \quad (7.44)$$

$$E_z = -\frac{\partial\varphi}{\partial z} = z\frac{e_{31}}{k_{33}}\frac{\partial^2 w}{\partial x^2} + \frac{\mu_{31}}{2k_{33}}\frac{\partial^2 w}{\partial x^2} - \frac{V_P}{h_P} \quad (7.45)$$

$$E_{z,z} = -\frac{\partial^2\varphi}{\partial z^2} = \frac{e_{31}}{k_{33}}\frac{\partial^2 w}{\partial x^2}, \quad (7.46)$$

In the given context, φ represents the electric potential, E_z represents the electric field, and $E_{z,z}$ represents the gradient of the electric field. Subsequently, fundamental equations are derived using the nonlocal strain gradient theory [36, 37].

$$\left(1-(ea^2)\nabla^2\right)\sigma_{ij} = \left(1-l^2\nabla^2\right)\left(C_{ijkl}\varepsilon_{kl}\right) + \gamma\left(-e_{ijk}E_i + \mu_{ijkl}E_{k,l}\right) \quad (7.47)$$

In the context of Equation (7.47), the symbol ∇, which represents the gradient, is defined as the differential operator $\partial/\partial x$. Furthermore, C_{ijkl} represents the elastic coefficient, *ea* denotes the nonlocal parameter, and l signifies the length scale parameter. If the size scale is taken into account for piezoelectric and flexoelectric characteristics, the value of γ is given by $\gamma = 1 - l^2\nabla^2$. Otherwise, if the size scale is not considered, $\gamma = 1$.

7.10 METAMATERIALS

The extraordinary capabilities shown by metamaterials are attributed to their innovative structure, which is achieved by combining different materials with a repetitive micro-unit structure. The Poisson's ratio of most materials is inherently positive; nevertheless, a specific approach was used on the structure of metamaterials to render their Poisson's ratio negative. Auxetics, derived from the Greek term "auxetikos", meaning "to grow", refer to a class of metamaterials characterized by a negative or zero Poisson's ratio. Certain mechanical materials possess negative Poisson's ratio (NPR), which results in enhanced properties, such as improved compression and shear resistance, increased energy absorption, better sound insulation (acoustic energy), synclastic behavior (elastic recoil), anisotropy, high elasticity (in auxetic materials that adhere to Hooke's law), and high damping resistance. Due to this distinctive characteristic, researchers conducted an experiment on auxetic materials.

7.10.1 Auxetic Materials

The auxetic core layer is composed of cells that are organized in a honeycomb pattern. The mechanical properties of the honeycomb auxetic core layer, such as Young's modulus, shear modulus, negative Poisson's ratio, and thermal expansion coefficients, are believed to be influenced by the geometrical parameters of the individual unit cell (Figure 7.23) [38, 39].

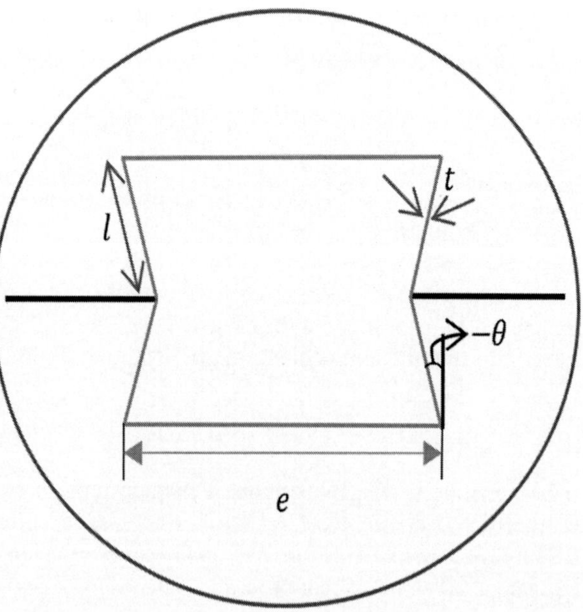

FIGURE 7.23 Schematic of the geometry of the cell of a honeycomb core layer.

7.10.1.1 Auxetic Core

The potential consequences pertaining to the effective mechanical features of the auxetic core are as follows [40, 41]:

$$E_{11}^{a} = E_{Al}\left(\frac{t}{l}\right)^{3}\frac{\cos\theta}{\left(\frac{e}{l}+\sin\theta\right)\sin^{2}\theta} \tag{7.48}$$

$$E_{22}^{a} = E_{Al}\left(\frac{t}{l}\right)^{3}\frac{\left(\frac{e}{l}+\sin\theta\right)}{\cos^{3}\theta} \tag{7.49}$$

$$G_{12}^{a} = E_{Al}\left(\frac{t}{l}\right)^{3}\frac{\left(\frac{e}{l}+\sin\theta\right)}{\left(\frac{e}{l}\right)^{2}\left(1+\frac{2e}{l}\right)\cos\theta} \tag{7.50}$$

$$\nu_{12}^{a} = \frac{\cos^{2}\theta}{\left(\frac{e}{l}+\sin\theta\right)\sin\theta} \tag{7.51}$$

$$\rho_{a} = \rho_{Al}\frac{\left(\frac{t}{l}\right)\left(\frac{e}{l}+2\right)}{2\cos\theta\left(\frac{e}{l}+\sin\theta\right)} \tag{7.52}$$

The equations presented in Equations (7.48) to (7.38) provide a means to determine the mechanical characteristics of the used auxetic core. The thickness of the auxetic cell's ribs is represented by the variable t, while the length of its horizontal ribs is signified by e, and the length of its vertical inclined ribs is represented by l. There are two efficacious auxetic core moduli, namely, the Young's modulus denoted as E_{11}^{a} and E_{22}^{a}, as well as the shear modulus referred to as G_{11}^{a} and G_{22}^{a}. Furthermore, both v_{11}^{a} and v_{22}^{a} exhibit high efficiency as auxetic variations of Poisson's ratio.

REFERENCES

[1] Ashton, J., J.C. Halpin, and P.H. Petit, *Primer on composite materials: Analysis*. 1969: Technomic Publishing Company.

[2] Jones, R.M., *Mechanics of composite materials*. 2018: CRC Press.

[3] Reddy, J.N., *Mechanics of laminated composite plates and shells: Theory and analysis*. 2003: CRC Press.

[4] Loy, C., K. Lam, and J. Reddy, Vibration of functionally graded cylindrical shells. *International Journal of Mechanical Sciences*, 1999. **41**(3): p. 309–324.

[5] Niino, M. and S. Maeda, Recent development status of functionally gradient materials. *Isij International*, 1990. **30**(9): p. 699–703.

[6] Ilschner, B., Processing-microstructure-property relationships in graded materials. *Journal of the Mechanics and Physics of Solids*, 1996. **44**(5): p. 647–656.

[7] Delale, F. and F. Erdogan, The crack problem for a nonhomogeneous plane. *Journal of Applied Mechanics* **50**(3).

[8] Bao, G. and L. Wang, Multiple cracking in functionally graded ceramic/metal coatings. *International Journal of Solids and Structures*, 1995. **32**(19): p. 2853–2871.

[9] Ghadiri, M., et al., Studying the influence of surface effects on vibration behavior of size-dependent cracked FG Timoshenko nanobeam considering nonlocal elasticity and elastic foundation. *Applied Physics A*, 2016. **122**: p. 1–21.

[10] Chung, Y.-L. and S. Chi, The residual stress of functionally graded materials. *Journal of the Chinese Institute of Civil and Hydraulic Engineering*, 2001. **13**: p. 1–9.

[11] Prakash, T., M.K. Singha, and M. Ganapathi, Thermal postbuckling analysis of FGM skew plates. *Engineering Structures*, 2008. **30**(1): p. 22–32.

[12] Hosford, W.F., *Mechanical behavior of materials*. 2010: Cambridge University Press.

[13] Meyers, M.A. and K.K. Chawla, *Mechanical behavior of materials*. 2008: Cambridge University Press.

[14] Lakes, R.S., *Viscoelastic materials*. 2009: Cambridge University Press.

[15] Ghadiri, M. and S.H.S. Hosseini, Nonlinear forced vibration of graphene/piezoelectric sandwich nanoplates subjected to a mechanical shock. *Journal of Sandwich Structures & Materials*, 2021. **23**(3): p. 956–987.

[16] Ghadiri, M. and S.H.S. Hosseini, Nonlinear dual frequency excited vibration of viscoelastic graphene sheets exposed to thermo-magnetic field. *Communications in Nonlinear Science and Numerical Simulation*, 2020. **83**: p. 105111.

[17] Engdahl, G. and I.D. Mayergoyz, *Handbook of giant magnetostrictive materials*, Vol. 107. 2000: Elsevier.

[18] Ebrahimi, F. and M.F. Ahari, *Mechanics of magnetostrictive materials and structures,* First Edition. 2023: CRC Press.

[19] Dapino, M.J., On magnetostrictive materials and their use in adaptive structures. *Structural Engineering and Mechanics*, 2004. **17**(3.4): p. 303–330.

[20] Dapino, M.J., et al., A coupled structural-magnetic strain and stress model for magnetostrictive transducers. *Journal of Intelligent Material Systems and Structures*, 2000. **11**(2): p. 135–152.

[21] Wang, L. and F. Yuan, Vibration energy harvesting by magnetostrictive material. *Smart Materials and Structures*, 2008. **17**(4): p. 045009.

[22] Zanjanchi, M., et al., Bifurcation point analysis of a magnetostrictive sandwich composite plate subjected to magnetic field and axial force. *International Journal of Structural Stability and Dynamics*, 2023. **23**(15): p. 2350172.

[23] Deng, Q., et al., The impact of flexoelectricity on materials, devices, and physics. *Journal of Applied Physics*, 2020. **128**(8).

[24] Abdollahi, A., et al., Computational evaluation of the flexoelectric effect in dielectric solids. *Journal of Applied Physics*, 2014. **116**(9).

[25] Shu, L., et al., Flexoelectric materials and their related applications: A focused review. *Journal of Advanced Ceramics*, 2019. **8**: p. 153–173.

[26] Liang, X., S. Hu, and S. Shen, Nanoscale mechanical energy harvesting using piezoelectricity and flexoelectricity. *Smart Materials and Structures*, 2017. **26**(3): p. 035050.

[27] Qi, L., et al., On the mechanics of curved flexoelectric microbeams. *International Journal of Engineering Science*, 2018. **124**: p. 1–15.

[28] Yan, Z., Modeling of a piezoelectric/piezomagnetic nano energy harvester based on two dimensional theory. *Smart Materials and Structures*, 2017. **27**(1): p. 015016.

[29] Zhang, S., et al., A curved resonant flexoelectric actuator. *Applied Physics Letters*, 2017. **111**(8).

[30] Bhaskar, U.K., et al., A flexoelectric microelectromechanical system on silicon. *Nature Nanotechnology*, 2016. **11**(3): p. 263–266.

[31] Yan, X., et al., Design of a curvature sensor using a flexoelectric material. In *Sensors and smart structures technologies for civil, mechanical, and aerospace systems 2013*. 2013: SPIE.

[32] Managheb, S., S. Ziaei-Rad, and R. Tikani, Energy harvesting from vibration of Timoshenko nanobeam under base excitation considering flexoelectric and elastic strain gradient effects. *Journal of Sound and Vibration*, 2018. **421**: p. 166–189.

[33] Wang, K., B. Wang, and J. Li, Electromechanical model of layered flexoelectric energy harvesters with strain gradient effect. *Energy*, 2020. **191**: p. 116560.

[34] Chu, L., Y. Li, and G. Dui, Nonlinear analysis of functionally graded flexoelectric nanoscale energy harvesters. *International Journal of Mechanical Sciences*, 2020. **167**: p. 105282.

[35] Zhang, D., Y. Lei, and S. Adhikari, Flexoelectric effect on vibration responses of piezoelectric nanobeams embedded in viscoelastic medium based on nonlocal elasticity theory. *Acta Mechanica*, 2018. **229**: p. 2379–2392.

[36] Lim, C., G. Zhang, and J. Reddy, A higher-order nonlocal elasticity and strain gradient theory and its applications in wave propagation. *Journal of the Mechanics and Physics of Solids*, 2015. **78**: p. 298–313.

[37] Shayestenia, F. and M. Ghadiri, Investigation of flexoelectric effect on nonlinear vibration and dynamic instability of piezoelectric sandwich micro/nanobeam using the nonlocal strain gradient theory. *International Journal of Structural Stability and Dynamics*, 2023. **23**(04): p. 2350045.

[38] Grujicic, M., et al., Multi-physics modeling of the fabrication and dynamic performance of all-metal auxetic-hexagonal sandwich-structures. *Materials & Design*, 2013. **51**: p. 113–130.

[39] Duc, N.D., et al., New approach to study nonlinear dynamic response and vibration of sandwich composite cylindrical panels with auxetic honeycomb core layer. *Aerospace Science and Technology*, 2017. **70**: p. 396–404.

[40] Ghadiri, M., M.F. Ahari, and M. Marvi, Auxetic metamaterial pre-twisted helical nanobeams: Vibrational characteristics. *Journal of the Brazilian Society of Mechanical Sciences and Engineering*, 2024. **46**(7): p. 436.

[41] Zanjanchi, M., M. Ghadiri, and S. Sabouri-Ghomi, Nonlinear parametric excitation and dynamic stability of auxetic honeycombs core with CNTRC face sheets sandwich plate. *European Journal of Mechanics-A/Solids*, 2023. **102**: p. 105109.

8 Dynamics of Nonlinear Smart Continuous Structures—Beams

8.1 INTRODUCTION

This part investigates the analysis of the beam's transverse nonlinear vibration, both in free and forced conditions. The equations that describe the nonlinear vibration of the beam are derived based on the principle of Euler-Bernoulli theories. The governing equations of the transverse vibration of the beams can be expressed as fourth-order partial differential equations. These equations are subject to four boundary conditions at both ends. The beam's boundary conditions can encompass partial derivatives of up to the third order. This chapter focuses on analyzing the free and forcede nonlinear vibration response of beams. In order to explain the principles that control the displacement field of the beam, it is important to establish a proper coordinate system. This system consists of the x coordinate, which represents the length of the beam, the z coordinate, which represents the thickness or height of the beam, and the y coordinate, which represents the width of the beam. When dealing with beam problems, the applied loads and geometry are arranged in a way that the displacement (u, v, w) along the coordinates (x, y, z) is solely dependent on the x and z coordinates. In the theory of beam, it is postulated that the displacement of v is precisely zero [1].

8.2 NONLINEAR EQUATION OF TRANSVERSE VIBRATION OF BEAMS

8.2.1 Background

Currently, there is a growing need for engineering structures, with spaceships, bridges, and autos serving as prime examples of such structures. When developing these structures, it is necessary to investigate many variables in order to enhance their performance and prolong their lifespan. One component of the design process involves analyzing the dynamic reaction of these structures to various and different stimuli. Real engineering problems, conversely, entail challenges and intricacies during the problem-solving phase. Therefore, it is imperative to tackle the process of simplifying these matters. This is achieved by constructing a mathematical representation of the engineering structure. However, in several instances, this will not be a straightforward task. However, this may be achieved mostly by streamlining the framework. For instance, the study of the behavior of basic structures like beams,

DOI: 10.1201/9781003470694-8

plates, and shells, which are considered as continuous systems, is highly significant in this process of simplification.

8.2.2 Nonlinear Governing Equation of the Euler-Bernoulli Beam (Thin Beam)

The equation regulating nonlinear motion and the boundary conditions of a thin beam (Figure 8.1) are derived in this section using the Hamilton principle. The beam is positioned on a viscoelastic substrate including a combination of a spring and a damper. In the theory of the thin or Euler-Bernoulli beam, the rotational movement of the cross-sectional surface perpendicular to the beam is neglected in comparison to its translational motion. Furthermore, unlike the bending deformation, the incision does not result in angular distortion. It is worth noting that the thin beam theory is only valid for beams that have a length-to-thickness ratio above 10 [1].

Generally speaking, the Euler-Bernoulli beam theory can be described as the most basic beam theory, where the displacement field is specified as follows:

$$u_1(x,z,t)=u(x,t)-z\frac{\partial w(x,t)}{\partial x},\qquad u_2(x,z,t)=0,\qquad u_3(x,z,t)=w(x,t) \tag{8.1}$$

The displacement components in the x, y, and z directions are given by u_1 and u_3, respectively, with respect to the Equation (8.1).

The displacement field described in Equation (8.1) states that straight lines parallel and perpendicular to the neutral axis stay unchanged before and after deformation, as depicted in Figure 8.2. Put simply, both normal transverse stresses and transverse stresses are not included.

Now, considering the Green's strain-displacement connections in the Lagrangian perspective and taking into account the assumptions that govern von Karman's theory [2]:

$$\varepsilon_{ij}=\frac{1}{2}\left(u_{i,j}+u_{j,i}+u_{k,i}u_{k,j}\right);\qquad u_{i,j}=\frac{\partial u_i}{\partial x_j} \tag{8.2}$$

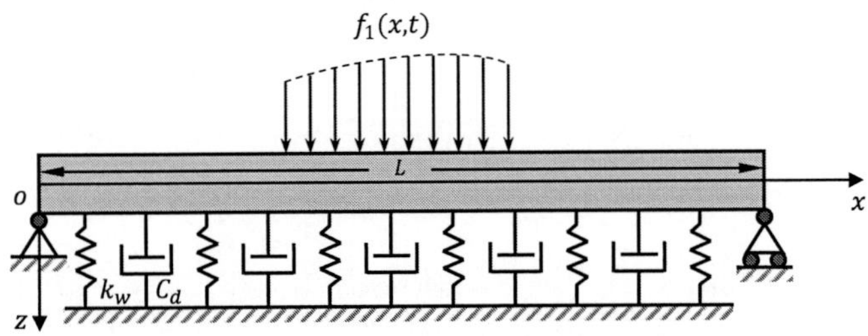

FIGURE 8.1 Schematic of the Euler-Bernoulli equation (thin beam) on a viscoelastic substrate.

Regarding the Equation (8.2), the variables i and j can represent $x, y,$ or z $(i, j = x, y, z)$. Specifically, $i, j = 1,2,3$. Hence, the Equation (8.2) can be expressed alternatively as follows:

$$\varepsilon_{xx} = \varepsilon_{11} = \frac{1}{2}\left(u_{1,1} + u_{1,1} + u_{3,1}u_{3,1}\right) = u_{1,1} + \frac{1}{2}\left(u_{3,1}\right)^2 = \frac{\partial u_1}{\partial x} + \frac{1}{2}\left(\frac{\partial u_3}{\partial x}\right)^2$$
$$\varepsilon_{xx} = \frac{\partial u}{\partial x} - z\frac{\partial^2 w}{\partial x^2} + \frac{1}{2}\left(\frac{\partial w}{\partial x}\right)^2 \tag{8.3}$$

$$\varepsilon_{yy} = \varepsilon_{22} = \frac{1}{2}\left(u_{2,2} + u_{2,2} + u_{3,2}u_{3,2}\right) = 0 \tag{8.4}$$

$$\varepsilon_{zz} = \varepsilon_{33} = \frac{1}{2}\left(u_{3,3} + u_{3,3} + u_{3,3}u_{3,3}\right) = 0 \tag{8.5}$$

$$\varepsilon_{xy} = \varepsilon_{12} = \frac{1}{2}\left(u_{1,2} + u_{2,1} + u_{3,1}u_{3,2}\right) = 0 \tag{8.6}$$

$$\varepsilon_{xz} = \varepsilon_{13} = \frac{1}{2}\left(u_{1,3} + u_{3,1} + u_{3,1}u_{3,3}\right) = 0 \tag{8.7}$$

$$\varepsilon_{yz} = \varepsilon_{23} = \frac{1}{2}\left(u_{2,3} + u_{3,2} + u_{3,2}u_{3,3}\right) = 0 \tag{8.8}$$

The equations governing movement are derived using the Hamilton principle, as stated in the preceding chapter. Thus, the Hamilton principle can be expressed in the following manner:

$$\delta\int_{t_1}^{t_2}\left(T - \pi + W_{ext}\right)dt = 0 \tag{8.9}$$

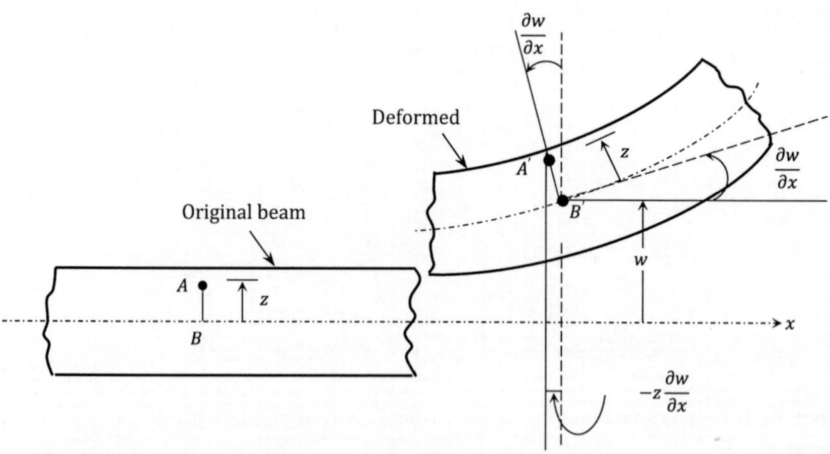

FIGURE 8.2 Transverse displacement of the beam.

In the context of Equation (8.9), T represents the kinetic energy, π represents the strain energy, and W_{ext} represents the work done by the external force.

$$\to \delta\pi = \int_V \sigma_{ij}\delta\varepsilon_{i,j} dV$$
$$= \int_V (\sigma_{xx}\delta\varepsilon_{xx} + \sigma_{yy}\delta\varepsilon_{yy} + \sigma_{zz}\delta\varepsilon_{zz} + \sigma_{xy}\delta\varepsilon_{xy} + \sigma_{xz}\delta\varepsilon_{xz}$$
$$+ \sigma_{yz}\delta\varepsilon_{yz})dV \tag{8.10}$$
$$\to \delta\pi = \int_V \sigma_{xx}\delta\varepsilon_{xx} dV = \int_v \sigma_{xx}\delta(\frac{\partial u}{\partial x} - z\frac{\partial^2 w}{\partial x^2} + \frac{1}{2}\left(\frac{\partial w}{\partial x}\right)^2)\, dV = \int_v \sigma_{xx}(\frac{\partial\ \delta u}{\partial x} -$$
$$z\frac{\partial^2\ \delta w}{\partial x^2} + \frac{\partial w}{\partial x}\frac{\partial\ \delta w}{\partial x})dv = \int_0^l \int_A \sigma_{xx}(\frac{\partial\ \delta u}{\partial x} - z\frac{\partial^2\ \delta w}{\partial x^2} + \frac{\partial w}{\partial x}\frac{\partial\ \delta w}{\partial x})dAdx$$

The definitions of N_{xx} and M_{xx} are as follows:

$$N_{xx} = \int_A \sigma_{xx} dA$$
$$M_{xx} = \int_A z\sigma_{xx} dA \tag{8.11}$$

Regarding Equation (8.11), N_{xx}, normal force, and M_{xx} represent the magnitudes exerted on the A-cross-sectional area. By incorporating Equation (8.11) into the ultimate Equation (8.10), we obtain the following:

$$\delta\pi = \int_0^l (N_{xx}\frac{\partial\ \delta u}{\partial x} - M_{xx}\frac{\partial^2\ \delta w}{\partial x^2} + N_{xx}\frac{\partial w}{\partial x}\frac{\partial\ \delta w}{\partial x})dx \tag{8.12}$$

By performing the process of integration on both sides of the equation, we establish a mathematical connection between the Equation (8.12) and the variable of time, denoted as t, inside the time span from t_1 to t_2.

$$\int_{t_1}^{t_2} \delta\pi = \int_{t_1}^{t_2}\int_0^l (N_{xx}\frac{\partial \delta u}{\partial x} - M_{xx}\frac{\partial^2 \delta w}{\partial x^2} + N_{xx}\frac{\partial w}{\partial x}\frac{\partial \delta w}{\partial x})\ dxdt \tag{8.13}$$

We shall establish a correlation with the integral of each of the statements of Equation (8.13) as follows:

$$\int_{t_1}^{t_2}\int_0^l N_{xx}\frac{\partial\ \delta u}{\partial x} dxdt = \int_{t_1}^{t_2}\left\{N_{xx}\delta u\Big]_0^l - \int_0^l \frac{\partial N_{xx}}{\partial x}\delta u dx\right\}dt \tag{8.14}$$

$$\int_{t_1}^{t_2}\int_0^l M_{xx}\frac{\partial^2\ \delta w}{\partial x^2} dxdt = \int_{t_1}^{t_2}\left\{M_{xx}\frac{\partial\ \delta w}{\partial x}\Big]_0^l - \int_0^l \frac{\partial M_{xx}}{\partial x}\frac{\partial\ \delta w}{\partial x} dx\right\}dt =$$
$$\int_{t_1}^{t_2}\left\{M_{xx}\frac{\partial\ \delta w}{\partial x}\Big]_0^l - \frac{\partial M_{xx}}{\partial x}\delta w\Big]_0^l + \int_0^l \frac{\partial^2 M_{xx}}{\partial x^2}\delta w dx\right\}dt \tag{8.15}$$

$$\int_{t_1}^{t_2}\int_0^l N_{xx}\frac{\partial w}{\partial x}\frac{\partial\,\delta w}{\partial x}dxdt = \int_{t_1}^{t_2}\left\{N_{xx}\frac{\partial w}{\partial x}\delta w\Big]_0^l - \int_0^l \frac{\partial}{\partial x}\left(N_{xx}\frac{\partial w}{\partial x}\right)\delta w dx\right\}dt \quad (8.16)$$

By utilizing the derived relationships from Equations (8.14), (8.15), and (8.16), the Equation (8.13) can be reformulated as follows:

$$\begin{aligned}\int_{t_1}^{t_2}\delta\pi = &\int_{t_1}^{t_2}\left\{N_{xx}\delta u\Big]_0^l - \int_0^l \frac{\partial N_{xx}}{\partial x}\delta u dx\right\}dt \\ &-\int_{t_1}^{t_2}\left\{M_{xx}\frac{\partial\,\delta w}{\partial x}\Bigg]_0^l - \frac{\partial M_{xx}}{\partial x}\delta w\Big]_0^l + \int_0^l \frac{\partial^2 M_{xx}}{\partial x^2}\delta w dx\right\}dt \\ &+\int_{t_1}^{t_2}\left\{N_{xx}\frac{\partial w}{\partial x}\delta w\Big]_0^l - \int_0^l \frac{\partial}{\partial x}\left(N_{xx}\frac{\partial w}{\partial x}\right)\delta w dx\right\}dt\end{aligned} \quad (8.17)$$

We will utilize a T to compute the kinetic energy:

$$\begin{aligned}\delta T = \int_v \dot{u}_i\,\delta\dot{u}_i\,dv = \int_0^l\int_A \rho\left(\dot{u}_1\delta\dot{u}_1 + \dot{u}_2\delta\dot{u}_2 + \dot{u}_3\delta\dot{u}_3\right)dAdx = \int_0^l (I\dot{u}_1\delta\dot{u}_1 \\ + I\dot{u}_3\delta\dot{u}_3)dx = \int_0^l\left\{I\left(\dot{u} - z\frac{\partial^2 w}{\partial x\partial t}\right)\delta\left(\dot{u} - z\frac{\partial^2 w}{\partial x\partial t}\right) + I\dot{w}\delta\dot{w}\right\}dx\end{aligned} \quad (8.18)$$

In the Euler-Bernoulli theory, the influence of the beam's periodic inertia is disregarded. So in the Equation (8.18), we have the following:

$$\frac{\partial^2 w}{\partial x\partial t} = 0 \quad (8.19)$$

Therefore, the Equation (8.19) can be expressed as follows:

$$\delta T = \int_0^l \left\{I\dot{u}\delta\dot{u} + I\dot{w}\;\delta\dot{w}\right\}dx \quad (8.20)$$

By performing the process of integration on both sides of the equation, we obtain a mathematical Equation (8.20) in regard to the variable of time, denoted as t, inside the time span from t_1 to t_2.

$$\begin{aligned}\int_{t_1}^{t_2}\delta T = \int_0^l\int_{t_1}^{t_2}\left\{I\dot{u}\delta\dot{u} + I\,\dot{w}\;\delta\dot{w}\right\}dtdx = \int_0^l\left\{I\dot{u}\delta\,u\Big]_{t_1}^{t_2} - \int_{t_1}^{t_2} I\,\ddot{u}\,\delta u dt\right\}dx + \\ \int_0^l\left\{I\dot{w}\delta\,w\Big]_{t_1}^{t_2} - \int_{t_1}^{t_2} I\,\ddot{w}\,\delta w dt\right\}dx = \int_{t_1}^{t_2}\int_0^l\left\{-I\,u\,\delta u - I\,\ddot{w}\,\delta w\right\}dxdt\end{aligned} \quad (8.21)$$

In order to determine the work done by the external force (W_{ext}), we express it in the following manner:

$$\delta\int_{t_1}^{t_2} W_{ext}dt = \int_{t_1}^{t_2}\int_0^l f_1\delta w dx dt \tag{8.22}$$

By incorporating the obtained Equations (8.17), (8.20), and (8.22) into the overarching Equation (8.9), we will obtain the following:

$$\begin{gathered}\delta\int_{t_1}^{t_2}(T-\pi+W_{ext})=0\\ \rightarrow\int_{t_1}^{t_2}\int_0^l\{-I\ddot{u}\delta u - I\ddot{w}\delta w+\frac{\partial N_{xx}}{\partial x}\delta u+\frac{\partial^2 M_{xx}}{\partial x^2}\delta w+\frac{\partial}{\partial x}\left(N_{xx}\frac{\partial w}{\partial x}\right)\delta w+\\ f\delta w dx dt+\int_{t_1}^{t_2}\{-N_{xx}\delta u]_0^l+M_{xx}\frac{\partial\,\delta w}{\partial x}\bigg]_0^l-\frac{\partial M_{xx}}{\partial x}\delta w]_0^l-N_{xx}\frac{\partial w}{\partial x}\delta w]_0^l\}\,dt=0\end{gathered} \tag{8.23}$$

The result of the Equation (8.23) consists of the governing equations and their corresponding boundary conditions, expressed as follows:

$$\begin{cases}\dfrac{\partial N_{xx}}{\partial x}=I\ddot{u}\\ \dfrac{\partial^2 M_{xx}}{\partial x^2}+\dfrac{\partial}{\partial x}\left(N_{xx}\dfrac{\partial w}{\partial x}\right)+f_1=I\ddot{w}\end{cases} \tag{8.24}$$

$$\begin{cases}N_{xx}\delta u]_0^l=0\\ M_{xx}\dfrac{\partial\,\delta w}{\partial x}\bigg]_0^l=0\\ (\dfrac{\partial M_{xx}}{\partial x}-N_{xx}\dfrac{\partial w}{\partial x})\delta w]_0^l=0\end{cases} \tag{8.25}$$

Considering the initial Equation (8.24), where $\ddot{u}=0$, and also taking into account that the integral of zdA is equal to 0 ($\int_A zdA=0$), the tension ratio (N_{xx}) can be derived as follows:

$$\frac{\partial N_{xx}}{\partial x}=0\ \Rightarrow N_{xx}=C_1 \tag{8.26}$$

By substituting the calculated value for N_{xx} into the initial Equation (8.11), we obtain the following:

$$\begin{gathered}N_{xx}=\int_A\sigma_{xx}dA=\int_A E\varepsilon_{x,x}dA=\int_A E\left\{\frac{\partial u}{\partial x}-z\frac{\partial^2 w}{\partial x^2}+\frac{1}{2}\left(\frac{\partial w}{\partial x}\right)^2\right\}dA=\\ EA\left\{\frac{\partial u}{\partial x}+\frac{1}{2}\left(\frac{\partial w}{\partial x}\right)^2\right\}=C_1\end{gathered} \tag{8.27}$$

In the context of Equation (8.27), E represents the modulus of elasticity of the beam, while A represents its transverse cross-sectional area. The u response is acquired in the following manner:

$$\begin{gathered}\frac{\partial u}{\partial x}+\frac{1}{2}\left(\frac{\partial w}{\partial x}\right)^2=\frac{C_1}{EA}\quad\Rightarrow\quad\frac{\partial u}{\partial x}=\frac{C_1}{EA}-\frac{1}{2}\left(\frac{\partial w}{\partial x}\right)^2\\ \Rightarrow\quad u=\frac{C_1 x}{EA}-\frac{1}{2}\int_0^x\left(\frac{\partial w}{\partial x}\right)^2 dx+C_2\end{gathered}\tag{8.28}$$

We have a boundary given as follows:

$$\begin{cases}u(0)=0\Rightarrow C_2=0\\ u(l)=0\;\Rightarrow C_1=\dfrac{EA}{2l}\displaystyle\int_0^l\left(\frac{\partial w}{\partial x}\right)^2 dx\;\rightarrow\end{cases}\quad\begin{cases}u(x,t)=\dfrac{x}{2l}\displaystyle\int_0^l\left(\frac{\partial w}{\partial x}\right)^2 dx-\frac{1}{2}\int_0^x\left(\frac{\partial w}{\partial x}\right)^2 dx\\ \dfrac{\partial u}{\partial x}=\dfrac{1}{2l}\displaystyle\int_0^l\left(\frac{\partial w}{\partial x}\right)^2 dx-\frac{1}{2}\left(\frac{\partial w}{\partial x}\right)^2\end{cases}\tag{8.29}$$

Alternatively, considering that $I_{xx}=\int_A z^2 dA$, we can get the following:

$$\begin{gathered}M_{xx}=\int_A z\sigma_{xx}dA=\int_A zE\varepsilon_{xx}dA=\int_A zE\left\{\frac{\partial u}{\partial x}-z\frac{\partial^2 w}{\partial x^2}+\frac{1}{2}\left(\frac{\partial w}{\partial x}\right)^2\right\}dA=\\ \int_A z^2 dA\left\{-E\frac{\partial^2 w}{\partial x^2}\right\}=-EI_{xx}\frac{\partial^2 w}{\partial x^2}\end{gathered}\tag{8.30}$$

By substituting the resulting Equation (8.30) into the governing Equation (8.24), we obtain the following:

$$\frac{\partial^2 M_{xx}}{\partial x^2}+\frac{\partial}{\partial x}\left(N_{xx}\frac{\partial w}{\partial x}\right)+f_1=I\ddot{w}\quad\Rightarrow\quad -EI_{xx}\frac{\partial^4 w}{\partial x^4}+\frac{\partial}{\partial x}\left(N_{xx}\frac{\partial w}{\partial x}\right)+f_1=I\ddot{w}\tag{8.31}$$

The specified requirements that define the limits or constraints of the system are as follows:

$$\begin{cases}N_{xx}\delta u]_0^l=0\\ M_{xx}\left.\dfrac{\partial\,\delta w}{\partial x}\right]_0^l=0\\ \left(\dfrac{\partial M_{xx}}{\partial x}-N_{xx}\dfrac{\partial w}{\partial x}\right)\delta w]_0^l=0\end{cases}\tag{8.32}$$

The Equation (8.33) indicates that the first boundary condition is $\delta u]_0^l = 0$ while extracting equations. Put simply, $u(0) = u(l) = 0$. The nonlinear equation describing the motion of the Euler-Bernoulli beam, based on the earlier equation (8.34), can be expressed as follows:

$$-EI_{xx}\frac{\partial^4 w}{\partial x^4} + \frac{EA}{2l}\left(\int_0^l \left(\frac{\partial w}{\partial x}\right)^2 dx\right)\frac{\partial^2 w}{\partial x^2} + f_1 = I\ddot{w} \tag{8.33}$$

For considering the influence of the viscoelastic substrate (consisting of a spring and damper) along with the external force f_1 as shown in Figure 8.1, Equation (8.34) can express as follows:

$$f = -k_w w - c_d \frac{\partial w}{\partial \tilde{t}} + f_1 \tag{8.34}$$

By substituting the aforementioned Equation into the Equation (8.33), we obtain the following:

$$\rightarrow\ I\ddot{w} + EI_{xx}\frac{\partial^4 w}{\partial x^4} - \frac{EA}{2L}\left(\int_0^L \left(\frac{\partial w}{\partial x}\right)^2 dx\right)\frac{\partial^2 w}{\partial x^2} + k_w w + c_d \frac{\partial w}{\partial \tilde{t}} = f_1 \tag{8.35}$$

8.3 CASE STUDY 1

8.3.1 Introduction

Researchers have a complex task when attempting to solve the nonlinear governing equations of a non-uniform micro- and nanobeam. This work investigates the nonlinear, size-dependent vibration of a non-uniform axially functionally graded (AFG) microbeam for the first time. The microbeam is simulated using the Euler-Bernoulli beam theory and the modified couple stress theory, including von Karman's geometric nonlinearity.

8.3.2 Formulation

8.3.2.1 Functionally Graded Materials

A microbeam is defined by its length, L, height, h, and width, b. The height is given by the equation $h = h_1(1+\beta_h x)$, where h_1 is the initial height and $\beta_h x$ is a parameter that determines how the height changes with respect to x. Similarly, the width is given by the equation $b = b_1(1+\beta_b x)$, where b_1 is the initial width and $\beta_b x$ is a parameter that determines how the width changes with respect to x. The issue is regarded as the problem (Figure 8.3). The expressions $\beta_h = 1 - \frac{h_2}{h_1}$ and $\beta_b = 1 - \frac{b_2}{b_1}$ denote the longitudinal and transverse cross sections, respectively. The moment of cross-sectional area (I) and cross-sectional area (A) are defined as follows [3]:

$$\mathrm{I}(x)=\int_{-\frac{h}{2}}^{\frac{h}{2}}\int_{-\frac{b}{2}}^{\frac{b}{2}} z^2 dydz \tag{8.36}$$

$$A(x)=\int_{-\frac{h}{2}}^{\frac{h}{2}}\int_{-\frac{b}{2}}^{\frac{b}{2}} dydz \tag{8.37}$$

The Euler-Bernoulli axially functionally graded microbeam is a structure made of metal and ceramic materials. It has different mechanical and geometrical properties along the x-axis, which can be described by a function of the microbeam's length ($F(x)$). This relationship is presented next [4]:

$$F(x)=F_1+(F_2-F_1)\left(\frac{x}{L}\right)^p \tag{8.38}$$

Here, p represents the non-negative variable parameter, which is the power-law exponent. It determines the material volume percentage along the axis of the beam. On the other hand, x represents the distance from the left end of the AFG beam. When the value of p is zero, it indicates that the material of the AFG beam is completely made of metal. In our current investigation, we used the assumption that p is equal to 1. The subscripts 1 and 2 indicate the starting point ($x = 0$) and ending position ($x = L$) of the microbeam, respectively. Equation (8.38) demonstrates that the material property at the beginning cross section ($x = 0$), the middle cross section ($x = L/2$), and the final cross section ($x = L$) may be defined as $F(0)=F_1$, $F(L/2)=F_1/2+F_2/2$ and $F(L)=F_2$, respectively. Put simply, the microbeam starts out as pure ceramic, then becomes a mixture of ceramic and metal in the intermediate segment and finally ends as pure metal. This is seen in Figure 8.3. The function F(x) may be expressed as Equation (8.39):

$$F(x)=F_1\left(1-\beta\frac{x}{L}\right) \tag{8.39}$$

FIGURE 8.3 Schematic of a non-uniform AFG microbeam.

Where:

$$\beta = \left(1 - \frac{F_2}{F_1}\right) \tag{8.40}$$

The microbeam's mechanical characteristics, including density (ρ), Young's modulus (E), shear modulus (μ), and Poisson's ratio (v), may be defined as follows:

$$\rho(x) = \rho_1\left(1 - \beta_\rho \frac{x}{L}\right) \;;\; \beta_\rho = \left(1 - \frac{\rho_2}{\rho_1}\right) \tag{8–41a}$$

$$E(x) = E_1\left(1 - \beta_E \frac{x}{L}\right) \;;\; \beta_E = \left(1 - \frac{E_2}{E_1}\right) \tag{8–41b}$$

$$\mu(x) = \mu_1\left(1 - \beta_\mu \frac{x}{L}\right) \;;\; \beta_\mu = \left(1 - \frac{\mu_2}{\mu_1}\right) \tag{8–41c}$$

$$v(x) = v_1\left(1 - \beta_v \frac{x}{L}\right) \;;\; \beta_v = \left(1 - \frac{v_2}{v_1}\right) \tag{8–41d}$$

8.3.2.2 The Modified Couple Stress Theory

The strain energy U^S of an isotropic linear elastic material may be mathematically represented as follows [5]:

$$U^S = \frac{1}{2}\iiint_{\Lambda} (\sigma : \epsilon + \boldsymbol{m} : \chi)\, dv \tag{8.42}$$

Where Λ represents the occupied zone and the other parameters are specified as follows:

Cauchy stress:

$$\boldsymbol{\sigma} = \lambda tr(\epsilon)\boldsymbol{I} + 2\mu\epsilon \tag{8.43}$$

The expression "$tr(\in)$" refers to the trace tensor of the ($\in_{kk}$) strain. Classical strain $\in$is defined as follows:

$$\epsilon = \frac{1}{2}\left(\nabla \boldsymbol{u} + (\nabla \boldsymbol{u})^T\right) \tag{8.44}$$

Symmetric curvature **m**:

$$\boldsymbol{m} = 2l^2\mu\chi \tag{8.45}$$

Deviation part of couple stress:

$$\chi = \frac{1}{2}\left(\nabla\boldsymbol{\theta} + (\nabla\boldsymbol{\theta})^T\right) \tag{8.46}$$

The material length scale parameter, denoted as "l", controls the couple stress at a certain site. The displacement field components are represented by "u_i", while the rotation vector components are denoted as "θ_i". The rotation vector is precisely specified as follows:

$$\boldsymbol{\theta} = \frac{1}{2} curl\, \boldsymbol{u} \tag{8.47}$$

Furthermore, in Equations (8.43) and (8.45), λ and μ represent Láme's constants, which are determined using the following formulas:

$$u(x) = \frac{E(x)}{2(1+v(x))} \tag{8.48}$$

$$\lambda(x) = \frac{E(x)v(x)}{(1+v(x))(1-2v(x))} \tag{8.49}$$

8.3.3 The Governing Equation and Boundary Conditions

The governing equation and boundary conditions may be obtained by using the Hamiltonian principle, which is represented as follows:

$$\delta H = 0 \tag{8.50}$$

$$H = \frac{1}{\tau}\int_0^{\tau} T - U + W_{ext}\, dt \tag{8.51}$$

The symbol τ represents the value of 2π divided by $\omega. T, U$, and W_{ext} represents the vibration period, kinetic energy, strain energy, and external work induced by an external force, respectively.

The displacement field is defined in the Euler-Bernoulli theory as the following:

$$\begin{aligned} u_x(x,z,t) &= u(x,t) - z\frac{\partial w(x,t)}{\partial x} \\ u_y &= 0 \\ u_z &= w(x,t) \end{aligned} \tag{8.52}$$

The variables u_x, u_y, and u_z represent the displacement components along the x, y, and z axes, respectively. The term $\partial w / \partial x$ refers to the rotation angle around the y-axis, while $w(x)$ represents the transverse deflection of the beam. The kinetic energy of a microbeam may be determined using the following equation:

$$\begin{aligned} T &= \frac{1}{2}\int_0^L \int_A \rho(x)\left[\left(\frac{\partial u_x}{\partial t}\right)^2 + \left(\frac{\partial u_y}{\partial t}\right)^2 + \left(\frac{\partial u_z}{\partial t}\right)^2\right] dAdx \\ &= \frac{1}{2}\int_0^L \left[\begin{array}{l} m_0\left(\dfrac{\partial u(x,t)}{\partial t}\right)^2 + m_2\left(\dfrac{\partial^2 w(x,t)}{\partial x \partial t}\right)^2 \\ -2m_1 \dfrac{\partial u(x,t)}{\partial t}\dfrac{\partial^2 w(x,t)}{\partial x \partial t} + m_0\left(\dfrac{\partial w(x,t)}{\partial t}\right)^2 \end{array}\right] dx \end{aligned} \tag{8.53}$$

Where:

$$\left(m_0, m_1, m_2\right) = \int_{-\frac{h}{2}}^{\frac{h}{2}} \int_{-\frac{b}{2}}^{\frac{b}{2}} \rho(x)\left(1, z, z^2\right) dydz \tag{8.54}$$

The von Karman nonlinear strain-displacement relation is defined as a straight Euler-Bernoulli beam, assuming massive transverse displacements, moderate rotations, and modest strains.

$$\epsilon_{xx} = \frac{\partial u(x,t)}{\partial x} + \frac{1}{2}\left(\frac{\partial w(x,t)}{\partial x}\right)^2 - z\frac{\partial^2 w(x,t)}{\partial x^2} \tag{8.55}$$

The explicit formulations of the constituents of the rotation vector are obtained by using Equation (8.47) as follows:

$$\theta_y = -\frac{\partial w}{\partial x} \tag{8.56}$$

The non-zero components of the couple stress and Cauchy stress tensors may be computed by using the following three equations.

$$\sigma_{xx} = E(x)\epsilon_{xx} \tag{8.57}$$

$$\chi_{xy} = -\frac{1}{2}\frac{\partial^2 w(x,t)}{\partial x^2} \tag{8.58}$$

$$m_{xy} = -l^2 u \frac{\partial^2 w}{\partial x^2} \ ; \ m_{xx} = m_{yy} = m_{zz} = m_{zy} = m_{xz} = 0 \tag{8.59}$$

The microbeam's strain energy, arising from the beam's strains and stresses, is precisely specified as follows:

$$U = U^s == \frac{1}{2}\int_0^L \left\{ \begin{array}{l} A_x\left[\left(\frac{\partial u(x,t)}{\partial x}\right)^2 + \left(\frac{1}{2L}\int_0^L \left(\frac{\partial w(x,t)}{\partial x}\right)^2 dx\right)^2\right] + C_x\left(\frac{\partial^2 w(x,t)}{\partial x^2}\right)^2 \\ A_x \frac{\partial u(x,t)}{\partial x}\frac{1}{2L}\int_0^L \left(\frac{\partial w(x,t)}{\partial x}\right)^2 dx + 2D_x l^2\left(\frac{1}{2}\frac{\partial^2 w(x,t)}{\partial x^2}\right)^2 \\ -B_x\left(\frac{\partial u(x,t)}{\partial x}\frac{\partial^2 w(x,t)}{\partial x^2} + \frac{1}{2L}\int_0^L \left(\frac{\partial w(x,t)}{\partial x}\right)^2 dx \frac{\partial^2 w(x,t)}{\partial x^2}\right) \end{array} \right\} dx \tag{8.60}$$

Where:

$$\begin{aligned} \left(A_x, B_x, C_x\right) &= \int_{-\frac{h}{2}}^{\frac{h}{2}} \int_{-\frac{b}{2}}^{\frac{b}{2}} E(x)\left(1, z, z^2\right) dydz \\ D_x &= \int_{-\frac{h}{2}}^{\frac{h}{2}} \int_{-\frac{b}{2}}^{\frac{b}{2}} \mu(x) dydz \end{aligned} \tag{8.61}$$

When analyzing free vibration, it is necessary to treat the external work done by applied forces as zero.

$$W_{ext} = 0 \tag{8.62}$$

By inserting equations (8.53), (8.60), and (8.62) into equation (8.50) and using the basic lemma of calculus, one may deduce the first variation of total energy. This process allows us to obtain the governing equations and boundary conditions of the microbeam, based on the Euler-Bernoulli model. The microbeam's nonlinear vibration equations have solutions that may be separated into spatial (x) and temporal (t) components. Therefore, we may infer the following:

$$\begin{aligned} w(x,t) &= aW(x)\cos\omega t, \\ u(x,t) &= U(x)\cos\omega t \end{aligned} \tag{8.63}$$

By using the Ritz-Galerkin approach and integrating with respect to time, the governing equations for Euler-Bernoulli beams may be derived using Equation (8.63).

$$\frac{1}{T}\int_0^T\int_0^L \left\{ \begin{array}{l} \dfrac{\partial}{\partial x}\left[\begin{array}{c} A_x\left(\cos\omega t\dfrac{\partial U(x)}{\partial x} + \dfrac{(a\cos\omega t)^2}{2L}\displaystyle\int_0^L\left(\dfrac{\partial W(x)}{\partial x}\right)^2 dx\right) \\ -B_x a\cos\omega t\dfrac{\partial^2 W(x)}{\partial x^2} \end{array} \right] \\ = \omega^2\left[-m_0\cos\omega t\, U(x) + m_1 a\cos\omega t\dfrac{\partial W(x)}{\partial x}\right] \end{array} \right\} \delta u\, dxdt \tag{8.64}$$

$$\frac{1}{T}\int_0^T\int_0^L \left\{ \begin{array}{l} \dfrac{\partial^2}{\partial x^2}\left[\begin{array}{c} B_x\left(\cos\omega t\dfrac{\partial U(x)}{\partial x} + \dfrac{(a\cos\omega t)^2}{2L}\displaystyle\int_0^L\left(\dfrac{\partial W(x)}{\partial x}\right)^2 dx\right) \\ -a\cos\omega t(C_x - D_x l^2)\dfrac{\partial^2 W(x)}{\partial x^2} \end{array} \right] \\ +\dfrac{\partial}{\partial x}\left[\displaystyle\int_0^L \dfrac{A_x}{l}\left(\left(\begin{array}{l} \cos\omega t\dfrac{\partial U(x)}{\partial (x)} \\ +\left(\dfrac{(a\cos\omega)}{2}\right)^3\left(\dfrac{\partial W(x)}{\partial x}\right)^2 \\ -a\cos\omega t\dfrac{B_x}{L}\dfrac{\partial^2 W(x)}{\partial x^2} \end{array} \right) dx\right)\dfrac{\partial W(x)}{\partial x}\right] \\ = \omega^2\left[\begin{array}{l} -m_0\cos\omega t\, W(x) - m_1 a\cos\omega t\dfrac{\partial U(x)}{\partial x} \\ +m_2 a\cos\omega t\dfrac{\partial^2 W(x)}{\partial x^2} \end{array} \right] \end{array} \right\} \delta u\, dxdt \tag{8.65}$$

Since the AFG materials have uniform properties throughout their thickness, the values of B_x and m_1 are assumed to be zero. Consequently, the governing equations are formulated in terms of displacements.

$$\delta u : \frac{\partial N}{\partial x} + f(x) = m_0\frac{\partial^2 u}{\partial t^2} \tag{8.66}$$

$$\delta w:\ \frac{\partial^2}{\partial x^2}(M+Y)+\frac{\partial}{\partial x}\left(N\frac{\partial w}{\partial x}\right)+q(x)=m_0\frac{\partial^2 w}{\partial t^2}-m_2\frac{\partial^4 w}{\partial t^2\partial x^2} \tag{8.67}$$

The boundary conditions are as follows:

$$U=0 \quad or \quad N=0 \tag{8.68}$$

$$W=0 \quad or \quad \frac{\partial}{\partial x}(M+Y)+N\frac{\partial W}{\partial x}=0 \tag{8.69}$$

$$\frac{\partial W}{\partial x}=0 \quad or \quad M+Y=0 \tag{8.70}$$

The values of "$q=0$" and "$f=0$" represent the distributed transverse and axial load, respectively. The stress resultants (N,M,Y) in the equations earlier are provided as follows:

$$\begin{aligned} N&=\int \sigma_{xx}dA=\epsilon_{xx}=A_x\frac{\partial u(x,t)}{\partial x}+\frac{1}{2}A_x\left(\frac{\partial w(x,t)}{\partial x}\right)^2\\ M&=\int \sigma_{xx}zdA=-C_x\frac{\partial^2 w(x,t)}{\partial x^2}\\ Y&=\int_A m_{xy}dA=-l^2D_x\frac{\partial^2 w}{\partial x^2}\end{aligned} \tag{8.71}$$

8.3.4 Solution Methodology

Differential quadrature method (DQM) is an accurate and effective numerical method presented in early 1970s. The accuracy of this method depends on the precision of weighting coefficients controlled by the number of grid points. In primary formulations of DQM, an algebraic equation system was employed to calculate weighting coefficients which determined the number of grid points. An explicit formulation for the weighting coefficients was later presented and led to generalized differential quadrature method (GDQM). Many regular domain problems are solved using this procedure. The GDQ technique defines the *rth* order derivative of function $f(x_i)$ as follows [5]:

$$\frac{\partial^r f(x)}{\partial x^r}=\Bigg|_{x=x_p}=\sum_{j=1}^{n}C_{ij}^{(r)}f(x_i) \tag{8.72}$$

The value of $C_{ij}n$ may be determined using the following equation, where n is the number of grid points along the x direction:

$$\begin{aligned} C_{ij}^{(1)}&=\frac{M(x_i)}{(x_i-x_j)M(x_j)} \quad ; \ i,j=1,2,\ldots,n \ \ and \ \ i\neq j\\ C_{ij}^{(1)}&=-\sum_{j=1,i\neq j}^{n}C_{ij}^{(1)} \quad ; \ i=j\end{aligned} \tag{8.73}$$

Where $M(x)$ is defined as follows:

$$M(x_i) = \prod_{j=1, j\neq i}^{n} (x_i - x_j) \tag{8.74}$$

The coefficient $C(r)$ for weighting, in the x direction, may be determined as follows:

$$\begin{gathered} C_{ij}^{(r)} = r\left[C_{ij}^{(r-1)}C_{ij}^{(1)} - \frac{C_{ij}^{(r-1)}}{(x_i - x_j)}\right] \;;\; i,j = 1,2,\ldots,n \;, i\neq j \quad and \quad 2\leq r\leq n-1 \\ C_{ii}^{(r)} = -\sum\nolimits_{j=1,i\neq j}^{n} C_{ij}^{(1)} \quad ; \quad i,j = 1,2,\ldots,n \quad and \quad 1\leq r\leq n-1 \end{gathered} \tag{8.75}$$

The Chebyshev-Gauss-Lobatto approach is used to achieve a more uniform distribution of mesh points.

$$x_i = \frac{L}{2}\left(1 - \cos\left(\frac{(i-1)}{(N-1)}\pi\right)\right); i = 1,2,3,\ldots,n \tag{8.76}$$

By combining the three mass matrices, we can get the linear and nonlinear stiffness of the functionally graded microbeam, as described by Equations (8.66) and (8.67) in terms of nonlinear motion [6].

$$\left\{[K]_{Linear} + [K]_{non-Linear} - \omega^2 [M]\right\}\{\lambda\} = 0 \tag{8.77}$$

In order to solve the governing motion equation using the GDQM, we begin by disregarding the nonlinear stiffness matrix. To achieve this objective, we may use the weight coefficients (Equation (8.75)) and apply them to the linear motion equations, resulting in the following:

$$\sum\nolimits_{s=1}^{n} C_{rs}^{(1)}\left(A_x \sum\nolimits_{s=1}^{n} C_{rs}^{(1)} U_s\right) = \omega^2{}_{Linear} m_0 U_s \tag{8.78}$$

$$-\sum\nolimits_{s=1}^{n} C_{rs}^{(2)}\left((C_x + D_x l^2)\sum\nolimits_{s=1}^{n} C_{rs}^{(2)} W_s\right) = \omega^2{}_{Linear}\left(m_0 W_s - m_2 \sum\nolimits_{s=1}^{n} C_{rs}^{(2)} W_s\right) \tag{8.79}$$

By applying the boundary conditions (Equations (8.66) to (8.70)) to Equations (8.78) and (8.79), and combining the corresponding matrices for the boundary conditions and governing equations, the linear fundamental frequency may be determined as shown next:

$$\begin{bmatrix} [K_{dd}] & [K_{db}] \\ [K_{bd}] & [K_{bb}] \end{bmatrix} \begin{Bmatrix} \{\lambda_d\} \\ \{\lambda_b\} \end{Bmatrix} = \omega^2{}_{Linear} \begin{bmatrix} [M_{dd}] & [M_{db}] \\ [M_{bd}] & [M_{bb}] \end{bmatrix} \begin{Bmatrix} \{\lambda_d\} \\ \{\lambda_b\} \end{Bmatrix} \tag{8.80}$$

The indices "b" and "d" represent the border and domain, respectively, whereas "λ" represents the mode shape. The linear mode forms are necessary for solving the nonlinear vibration equations of the functionally graded microbeam. The U and W mode forms may be obtained by using Equation (8.90). By inserting the obtained mode shapes into the nonlinear stiffness matrix and using Equation (8.87), as well as connecting the linear and nonlinear stiffness matrices with the mass matrix, it is possible to compute the nonlinear frequency and mode shape. Next, by using the iteration approach, the nonlinear frequency is recalculated in order to get a convergent nonlinear frequency.

$$\sum_{s=1}^{n} C_{rs}^{(1)}\left[A_x\sum_{s=1}^{n} C_{rs}^{(1)}U_s+\frac{1}{2}A_x\sum_{s=1}^{n} C_{rs}^{(1)}W_s\sum_{s=1}^{n} C_{rs}^{(1)}W_s\right]=\omega^2{}_{Non-Linear}m_0U_s$$
$$-\sum_{s=1}^{n} C_{rs}^{(2)}\left(\left(C_x+D_xl^2\right)\sum_{s=1}^{n} C_{rs}^{(2)}W_s\right) \tag{8.81}$$

$$+\sum_{s=1}^{n} C_{rs}^{(1)}\left[\left(A_x\sum_{s=1}^{n} C_{rs}^{(1)}U_s+\frac{3}{4}\frac{a^2}{2}A_x\sum_{s=1}^{n} C_{rs}^{(1)}W_s\sum_{s=1}^{n} C_{rs}^{(1)}W_s\right)\sum_{s=1}^{n} C_{rs}^{(1)}W_s\right]$$
$$=\omega^2{}_{Non-Linear}\left(m_0W_s-m_2\sum_{s=1}^{n} C_{rs}^{(2)}W_s\right) \tag{8.82}$$

8.3.5 Numerical Results

The generalized differential quadrature technique is used to solve the governing equations of motion. The findings are computed for the first two linear and nonlinear frequencies of the AFG microbeams made of pure ceramic and pure metal, with microbeams that are clamped, simply supported, and clamped-simply supported. Subsequently, the impacts of the nonlinearity and the rates of cross sections throughout the thickness and breadth of the microbeam are showcased in various figures and tables.

In order to provide a clear understanding of how various factors impact the nonlinear frequencies, the outcomes are provided in relation to the normalized frequency. The normalized frequency is a term that is used to describe the frequency of an event or phenomenon that has been adjusted or standardized in some way.

$$Normalized\ frequency=\frac{Non-linear\ frequency\ of\ micro-beam}{Linear\ frequency\ of\ ceramic\ uniform\ beam} \tag{8.83}$$

In addition, non-dimensional parameters are specified in the following manner, which facilitates the analysis of the results:

$$l_0=\frac{l}{h_1}$$
$$Amp=a\sqrt{\frac{h_1^2}{12}} \tag{8.84}$$

The terms "l_0" and "Amp" represent a non-dimensional, small-scale parameter and a nonlinear amplitude, respectively.

Table 8.1 displays the mechanical characteristics of ceramic and metal. Figures 8.4 and 8.5 show the initial and subsequent normalized frequencies of a microbeam in relation to the amplitude. The microbeam is subject to simply supported, clamped, and clamped-simply supported boundary conditions. The normalized frequencies are computed for microbeams made of pure ceramic, pure metal, and AFG. It is evident that the normalized frequencies of the microbeam rise in conjunction with the amplitude. Additionally, the normalized frequencies of the AFG microbeam are lower than those of pure ceramic and higher than those of pure metal. Figures 8.4 and 8.5 demonstrate that the normalized frequencies of the clamped-simply supported microbeam are lower than those of the simply supported microbeam and higher than those of the clamped microbeam.

Tables 8.2 to 8.4 show the linear and nonlinear fundamental and second normalized frequencies of ceramic, metal, and AFG microbeams. The tables correspond to different boundary conditions: clamped-simply supported, simply supported, and clamped, respectively. It is observed that the fundamental and second frequencies of the microbeams exhibit a rise in relation to the amplitude. Moreover, the primary frequencies of the microbeams drop as the value of β_h decreases. The frequencies shown in Tables 8.2 to 8.4, printed in italic bold, are the frequencies that rise as the β_b decreases. Table 8.2 shows that the second normalized frequency of the clamped-simply supported microbeams drops as β_h increases, when the Amp is less than or equal to 3. However, when the amplitude is set to 4, the second frequencies of the pure metal and ceramic uniform microbeams rise as the β_h decreases from zero to −0.1. The increase in AFG microbeam occurs at $\beta_b = -0.2$ and falls from −0.1 to −0.2.

Table 8.2 demonstrates that reducing the β_b value leads to a rise in the linear and nonlinear fundamental frequencies of AFG, pure ceramic, and pure metal clamped-simply supported microbeams. When the amplitude of the microbeam is modest (Amp = 0 and 1) and the cross section is almost uniform over the thickness ($\beta_h = 0, -0.1$), reducing the β_b results in an increase in the second frequencies of the microbeam. However, when the nonlinearity and the rate of cross section change rise in the z direction, the influence of β_b on the second frequencies of the microbeam changes. Specifically, the second frequencies of ceramic and metal microbeams

TABLE 8.1
The Coefficients of Young's Modulus, Mass Density, and Poisson's Ratio of Ceramic (Al2 O3) and Metal (SUS304)

Material	Properties	Value
SUS304	E (Pa)	2.0104e + 11
	ρ (kg/m^3)	8166
	N	0.3262
Al_2O_3	E (Pa)	3.4955e + 11
	ρ (kg/m^3)	3800
	N	0.24

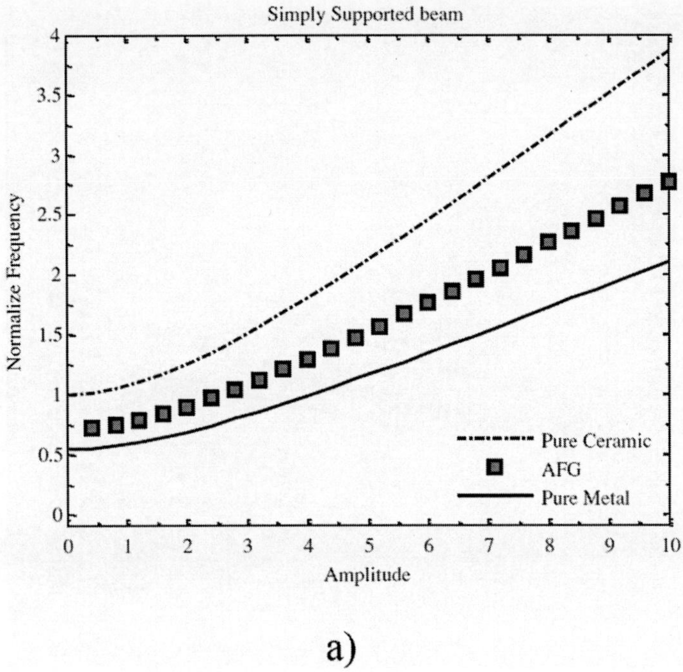

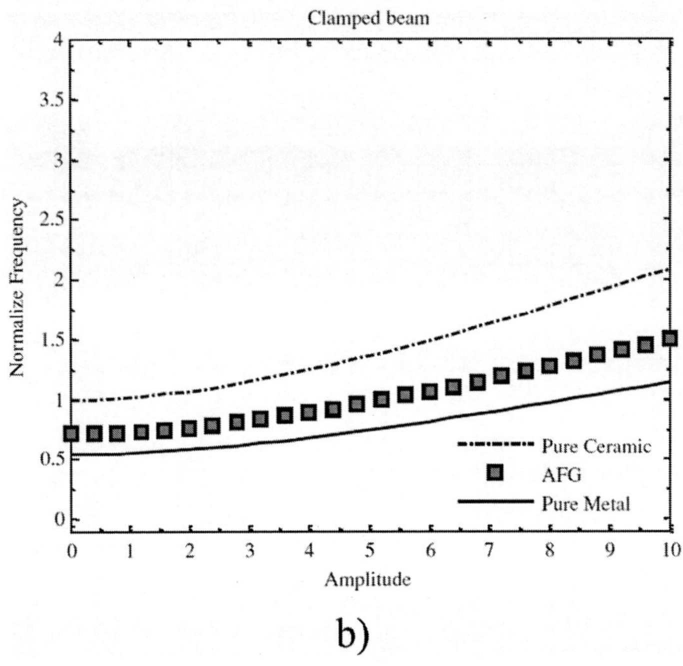

FIGURE 8.4 Relationship between the normalized fundamental frequency and the nonlinear amplitude of several types of microbeams.

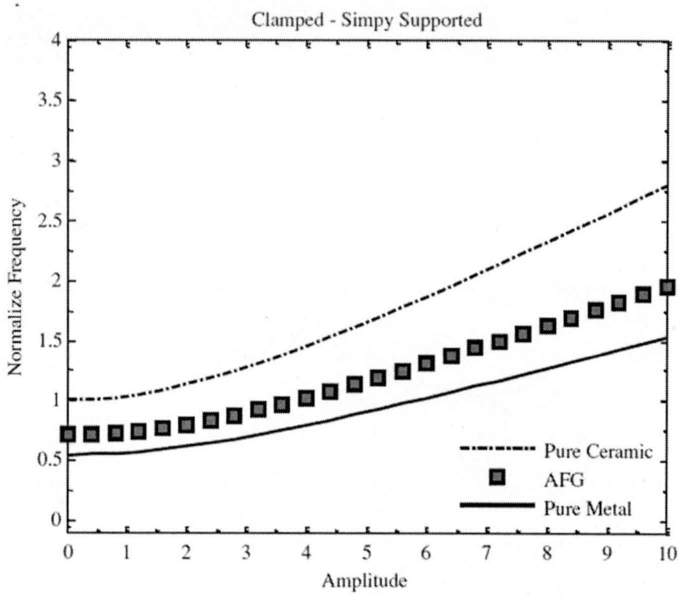

c)

FIGURE 8.4 (Continued)

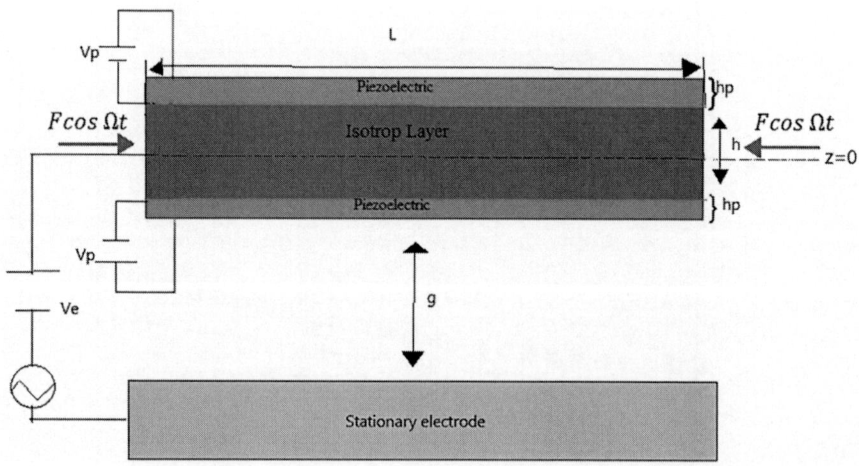

FIGURE 8.5 Schematic representation of a micro/nanobeam that is being exposed to external parametric excitation.

TABLE 8.2
Normalized Fundamental and Second Frequencies of Clamped-Simply Supported Microbeams in Different Nonlinear Amplitudes and Rates of Cross Section Change, $l_0 = 0.1$

	Fundamental Normalize Frequency					Second Normalize Frequency				
	Amp = 0	Amp = 1	Amp = 2	Amp = 3	Amp = 4	Amp = 0	Amp = 1	Amp = 2	Amp = 3	Amp = 4
$\beta b = -0$ Uniform beam Pure ceramic	*1*	*1.042961*	*1.161219*	*1.33338*	*1.540894*	*1*	*1.056035*	*1.206947*	*1.420946*	*1.673513*
AFG	*0.709033*	*0.738624*	*0.820251*	*0.939406*	*1.083356*	*0.717173*	*0.75417*	*0.854536*	*0.998067*	*1.168587*
Pure metal	*0.546804*	*0.570295*	*0.634959*	*0.729097*	*0.842567*	*0.546804*	*0.577444*	*0.659964*	*0.776979*	*0.915083*
$\beta h = -0.1$ Pure ceramic	*0.965141*	*1.010685*	*1.135145*	*1.314716*	*1.529567*	*0.955747*	*1.017631*	*1.182025*	*1.411586*	*1.679419*
AFG	*0.682953*	*0.714074*	*0.799357*	*0.922818*	*1.070945*	*0.683818*	*0.723896*	*0.831388*	*0.983101*	*1.161517*
Pure metal	*0.527743*	*0.552646*	*0.620702*	*0.718892*	*0.836373*	*0.522606*	*0.556445*	*0.646336*	*0.771861*	*0.918313*
$\beta h = -0.2$ Pure ceramic	*0.929176*	*0.977313*	*1.107851*	*1.294452*	*1.516065*	*0.910547*	*0.977371*	1.152593	1.393911	1.672685
AFG	*0.656103*	*0.689091*	*0.778773*	*0.907358*	*1.060433*	*0.649728*	*0.693798*	*0.810319*	*0.972206*	*1.160388*
Pure metal	*0.508077*	*0.534398*	*0.605777*	*0.707811*	*0.82899*	*0.497891*	*0.53443*	0.630242	0.762196	0.914631
$\beta h = -0.4$ Pure ceramic	*0.853261*	*0.90799*	*1.053515*	*1.256935*	*1.494446*	*0.816624*	*0.887576*	1.070079	1.316614	1.59769
AFG	*0.599665*	*0.637962*	*0.739837*	*0.882308*	*1.048714*	*0.578917*	*0.63486*	0.776876	0.966336	1.180592
Pure metal	*0.466566*	*0.496493*	*0.576066*	*0.687297*	*0.817169*	*0.446533*	*0.48533*	0.585123	0.71993	0.873623
$\beta b = -0.2$ $\beta h = 0$ Pure ceramic	1.014645	1.057204	1.174562	1.345794	1.552577	1.004254	1.063432	1.222072	1.445847	1.708895
AFG	0.720048	0.749389	0.830468	0.949084	1.09265	0.720434	0.759325	0.864414	1.014008	1.19109
Pure metal	0.554812	0.578083	0.642255	0.735885	0.848955	0.54913	0.581489	0.668234	0.790595	0.93443
$\beta h = -0.1$ Pure ceramic	0.979388	1.024469	1.147909	1.326437	1.54046	0.960115	1.022491	1.188136	1.41936	1.689062
AFG	0.693668	0.72461	0.80954	0.932744	1.08081	0.687142	0.729374	0.842137	1.000488	1.186007
Pure metal	0.535533	0.560184	0.627681	0.725301	0.842329	0.524995	0.559102	0.649677	0.776112	0.923586

(Continued)

TABLE 8.2 *(Continued)*
Normalized Fundamental and Second Frequencies of Clamped-Simply Supported Microbeams in Different Nonlinear Amplitudes and Rates of Cross Section Change, $l_0 = 0.1$

	Fundamental Normalize Frequency					Second Normalize Frequency				
	Amp = 0	Amp = 1	Amp = 2	Amp = 3	Amp = 4	Amp = 0	Amp = 1	Amp = 2	Amp = 3	Amp = 4
$\beta h = -0.2$ Pure ceramic	0.94304	0.990572	1.119771	1.304967	1.525392	0.915017	0.978739	1.146763	1.379525	1.64953
AFG	0.666532	0.699428	0.789	0.917664	1.071059	0.653106	0.69967	0.822148	0.991373	1.187311
Pure metal	0.515658	0.541649	0.612295	0.713561	0.834091	0.500335	0.535179	0.627054	0.75433	0.901969
$\beta h = -0.4$ Pure ceramic	0.866431	0.920883	1.06596	1.269201	1.50689	0.821266	0.889462	1.065817	1.305304	1.5793
AFG	0.609572	0.648047	0.750512	0.893992	1.061733	0.582395	0.636066	0.773128	0.957007	1.165704
Pure metal	0.473768	0.503542	0.582871	0.694004	0.823973	0.449071	0.486361	0.582793	0.713745	0.863567

TABLE 8.3
Normalized Fundamental and Second Frequencies of Simply Supported Microbeams in Different Nonlinear Amplitudes and Rates of Cross Section Change, $l_0 = 0.1$

			Fundamental Normalize Frequency					Second Normalize Frequency				
			Amp = 0	Amp = 1	Amp = 2	Amp = 3	Amp = 4	Amp = 0	Amp = 1	Amp = 2	Amp = 3	Amp = 4
$\beta b = -0$	Uniform beam	Pure ceramic	1	1.08534	1.305322	1.603108	1.943083	*1*	1.08562	1.306237	1.604769	1.94551
		AFG	***0.717417***	***0.778781***	***0.936916***	***1.150925***	***1.395211***	***0.721144***	***0.777235***	***0.923378***	***1.123329***	*1.353263*
		Pure metal	0.546804	0.593468	0.713755	0.876586	1.062486	***0.546804***	0.593621	0.714256	0.877494	1.063813
	$\beta h = -0.1$	Pure ceramic	0.951496	1.040855	1.26852	***1.573272***	***1.918655***	***0.951878***	1.038088	1.258723	1.555326	1.892406
		AFG	0.680852	***0.745581***	***0.91024***	***1.130333***	***1.379539***	***0.684561***	***0.746131***	***0.903839***	***1.116022***	*1.357291*
		Pure metal	0.520282	0.569144	0.693632	***0.860271***	***1.049128***	***0.520491***	0.567631	0.688274	0.850458	1.034775
	$\beta h = -0.2$	Pure ceramic	0.90135	0.995457	1.231955	1.544568	***1.896049***	***0.902954***	0.990198	1.211677	1.507152	1.841302
		AFG	0.643126	0.711496	***0.882899***	***1.108977***	***1.362825***	***0.647354***	0.715842	0.887655	1.114404	1.369095
		Pure metal	0.492862	0.54432	0.673638	***0.844576***	***1.036767***	***0.493739***	0.541444	0.662549	0.824117	1.006831
	$\beta h = -0.4$	Pure ceramic	0.794972	0.900852	1.15787	***1.487744***	***1.85218***	***0.80212***	0.891825	1.115162	1.407951	1.735479
		AFG	0.563393	0.640542	***0.827024***	***1.065558***	***1.328579***	***0.570634***	0.640525	0.812304	1.035036	1.282563
		Pure metal	0.434694	0.49259	0.633128	***0.813504***	***1.012779***	***0.438603***	0.487653	0.609775	0.769873	0.948967
$\beta b = -0.2$	$\beta h = 0$	Pure ceramic	0.999513	1.084874	1.304891	1.602698	1.942679	1.000118	1.080845	1.290293	1.575647	1.902841
		AFG	0.717513	0.779664	0.939588	1.155699	1.402145	0.721148	0.780907	0.935483	1.145442	1.385696
		Pure metal	0.546537	0.593213	0.713519	0.876361	1.062264	0.546868	0.59101	0.705537	0.86157	1.040481

(Continued)

TABLE 8.3 *(Continued)*
Normalized Fundamental and Second Frequencies of Simply Supported Microbeams in Different Nonlinear Amplitudes and Rates of Cross Section Change, $l_0 = 0.1$

		Fundamental Normalize Frequency					Second Normalize Frequency				
		Amp = 0	Amp = 1	Amp = 2	Amp = 3	Amp = 4	Amp = 0	Amp = 1	Amp = 2	Amp = 3	Amp = 4
$\beta h = -0.1$	Pure ceramic	0.950176	1.039819	1.268078	1.573459	1.919436	0.9522	1.033735	1.243827	1.528111	1.852589
	AFG	0.68033	0.746065	0.912939	1.135566	1.387333	0.684703	0.75047	0.917554	1.140617	1.392988
	Pure metal	0.51956	0.568577	0.69339	0.860374	1.049555	0.520666	0.56525	0.680129	0.835577	1.013003
$\beta h = -0.2$	Pure ceramic	0.899197	0.993909	1.231647	1.545569	1.898291	0.903475	0.986272	1.197862	1.481914	1.804439
	AFG	0.641991	0.711258	0.884563	1.112743	1.368671	0.647625	0.713593	0.879917	1.100422	1.348799
	Pure metal	0.491684	0.543473	0.673469	0.845123	1.037993	0.494024	0.539297	0.654996	0.810316	0.986674
$\beta h = -0.4$	Pure ceramic	0.791162	0.897929	1.156583	1.488025	1.853859	0.803024	0.888582	1.103011	1.385766	1.703198
	AFG	0.561051	0.639801	0.829429	1.071258	1.337458	0.57115	0.638022	0.803448	1.019117	1.259572
	Pure metal	0.43261	0.490991	0.632424	0.813658	1.013697	0.439096	0.48588	0.603131	0.757742	0.931315

TABLE 8.4
Normalized Fundamental and Second Frequencies of Clamped Microbeams in Different Nonlinear Amplitudes and Rates of Cross Section Change, $l_0 = 0.1$

	Fundamental Normalize Frequency					Second Normalize Frequency				
	Amp = 0	Amp = 1	Amp = 2	Amp = 3	Amp = 4	Amp = 0	Amp = 1	Amp = 2	Amp = 3	Amp = 4
$\beta b = -0$ Uniform beam Pure ceramic	1	1.021209	1.081996	1.175573	1.294351	1	1.044291	1.16593	1.342503	1.554826
AFG	0.714029	0.729358	0.773268	0.840813	0.926479	0.718212	0.746532	0.824973	0.940073	1.079741
Pure metal	0.546804	0.558401	0.59164	0.642808	0.707756	0.546804	0.571022	0.637535	0.734086	0.850185
$\beta h = -0.1$ Pure ceramic	0.951769	0.974085	1.037763	1.135139	1.257907	0.951764	0.996693	1.119472	1.296609	1.508541
AFG	0.678467	0.694646	0.740775	0.811234	0.899964	0.682174	0.712908	0.797206	0.919374	1.066081
Pure metal	0.520431	0.532633	0.567453	0.620698	0.687829	0.520428	0.544996	0.612132	0.708991	0.824876
$\beta h = -0.2$ Pure ceramic	0.902513	0.92623	0.993533	1.095626	1.223322	0.902478	0.947565	1.070206	1.246159	1.455731
AFG	0.642194	0.659135	0.707202	0.780094	0.871241	0.645356	0.679216	0.770936	0.901881	1.057238
Pure metal	0.493498	0.506466	0.543268	0.599093	0.668917	0.493478	0.518132	0.585193	0.681405	0.795999
$\beta h = -0.4$ Pure ceramic	0.800373	0.827181	0.902234	1.013936	1.151185	0.800063	0.846755	0.972039	1.148941	1.357051
AFG	0.567146	0.586426	0.640351	0.720509	0.81889	0.568856	0.606481	0.706216	0.845166	1.007007
Pure metal	0.437647	0.452306	0.493345	0.554424	0.629473	0.437477	0.463009	0.531515	0.628245	0.742041
$\beta b = -0.2$ $\beta h = 0$ Pure ceramic	0.999085	1.020313	1.08115	1.174794	1.293641	0.999552	1.042171	1.159553	1.330558	1.536802
AFG	0.714239	0.729728	0.774078	0.842253	0.92866	0.718329	0.748018	0.829975	0.949718	1.094488
Pure metal	0.546304	0.557911	0.591177	0.642382	0.707368	0.546559	0.569863	0.634048	0.727554	0.840329
$\beta h = -0.1$ Pure ceramic	0.951694	0.974101	1.038026	1.135753	1.258928	0.951738	0.994551	1.111991	1.282217	1.486657
AFG	0.679235	0.695409	0.741527	0.811977	0.900703	0.682562	0.714878	0.803167	0.930508	1.082828
Pure metal	0.52039	0.532642	0.567597	0.621034	0.688387	0.520414	0.543824	0.608041	0.701121	0.81291
$\beta h = -0.2$ Pure ceramic	0.90328	0.927177	0.994968	1.097751	1.226248	0.90287	0.945984	1.063696	1.233332	1.436104
AFG	0.64352	0.660515	0.708733	0.781848	0.873266	0.646009	0.67955	0.770489	0.900455	1.05478
Pure metal	0.493917	0.506984	0.544053	0.600255	0.670517	0.493693	0.517268	0.581633	0.674391	0.785267
$\beta h = -0.4$ Pure ceramic	0.80283	0.829698	0.90492	1.016882	1.15446	0.801271	0.846299	0.967535	1.139401	1.342196
AFG	0.569591	0.589095	0.643621	0.724624	0.823987	0.570017	0.606518	0.703589	0.839303	0.997782
Pure metal	0.438991	0.453682	0.494814	0.556035	0.631263	0.438138	0.46276	0.529052	0.623029	0.733918

drop with β_b. The second frequency of the AFG microbeam experiences a similar decrease when $\beta_h = 0.4$ and Amp is greater than or equal to 2.

Table 8.3 demonstrates that the second normalized frequencies of the simply supported pure ceramic and metal microbeams decline as β_h decreases. The second resonance frequency of the AFG (axially functionally graded) simply supported microbeam reduces as β_h decreases, given that the amplitude (Amp) is less than or equal to 3. Observations indicate that when the amplitude is 4 and the rate of cross section change of the microbeam along the width is = 0, decreasing β_h from zero to −0.2 results in an increase in the second frequency of the AFG microbeam. However, when the rate of cross section change along the thickness is high ($\beta h \geq 0.2$), the second frequency decreases as β_h decreases. Table 8.3 shows that reducing the β_b leads to a rise in the fundamental frequency of the uniform AFG microbeam, while causing a drop in the fundamental frequencies of the pure metal and ceramic microbeams. In addition, reducing β_b leads to a drop in the linear fundamental frequencies of non-uniform AFG, pure metal, and pure ceramic microbeams. Alternatively, when we enhance the nonlinearity and reduce the rate of cross section change along the thickness of the microbeam (β_h), we see a distinct impact of β_b on the frequencies under various situations. Specifically, the frequencies rise as β_b decreases. At an Amp value of 4, the fundamental frequency of the non-uniform microbeams increases as the β_b decreases. The second linear frequencies of the AFG, pure metal, and pure ceramic microbeams are inversely proportional to the decrease in β_b. When the microbeam is uniform or the rate of cross section variation along the thickness is modest ($\beta_h < -0.1$), the second nonlinear frequency of the AFG microbeam grows as the β_b drops. Observations indicate that the second frequencies of the AFG microbeam fall when β_h is less than or equal to −0.2. Similarly, the nonlinear second frequencies of the pure metal and ceramic microbeams likewise decrease with β_b.

Table 8.4 demonstrates that reducing the β_b leads to a drop in the fundamental frequency of uniform pure metal and pure ceramic microbeams, whereas it results in an increase in the fundamental frequency of uniform AFG microbeam. Table 8.4 demonstrates that the nonlinear fundamental frequencies of non-uniform AFG, pure ceramic, and metal microbeams rise as βb decreases. Furthermore, it has been shown that altering the rate at which the cross section varies throughout the thickness has an impact on the influence of β_b on the linear fundamental frequency of microbeams made of pure ceramic and metal.

Table 8.4 demonstrates that the frequency of the clamped microbeams reduces as the value of β_h decreases. The frequency of the AFG microbeam rises as the value of β_b decreases, provided that the microbeam is uniform or has a low rate of cross section change throughout the thickness ($\beta h > -0.1$). Furthermore, when the value of β_h is less than or equal to −0.2, the linear second frequency of the clamped microbeams grows as β_b decreases. Table 8.4 demonstrates that the nonlinear second frequencies of both pure ceramic and metal microbeams drop as β_b increases. The influence of β_b on the frequencies of the microbeam is contingent upon the nonlinearity and β_h. By modifying these parameters, the impact of β_b is altered.

Tables 8.5 show the normalized fundamental and second frequencies of microbeams with clamped, simply supported, and clamped-simply supported boundary

TABLE 8.5
Normalized Fundamental and Second Frequencies of Clamped Microbeams in Different Nonlinear Amplitudes and Small-Scale Parameters, $\beta_h = -0.3$, $\beta_b = -0.2$.

		Fundamental Normalize Frequency					Second Normalize Frequency				
		$l_0 = 0$	$l_0 = 0.1$	$l_0 = 0.2$	$l_0 = 0.3$	$l_0 = 0.4$	$l_0 = 0$	$l_0 = 0.1$	$l_0 = 0.2$	$l_0 = 0.3$	$l_0 = 0.4$
Amp = 0	Pure ceramic	1	1.033057	1.125864	1.264486	1.435294	1	1.033154	1.126254	1.265351	1.436764
	AFG	0.710923	0.734785	0.801723	0.901588	1.024498	0.713307	0.737369	0.804869	0.905569	1.029489
	Pure metal	0.546804	0.56488	0.615627	0.691426	0.784824	0.546804	0.564932	0.61584	0.691899	0.785628
Amp = 2	Pure ceramic	1.119012	1.148685	1.232959	1.360951	1.521217	1.199483	1.227673	1.308065	1.430903	1.585709
	AFG	0.794602	0.815994	0.876741	0.968959	1.084351	0.87013	0.889927	0.946683	1.033947	1.144427
	Pure metal	0.61188	0.628105	0.674187	0.744173	0.831807	0.655882	0.671297	0.715255	0.782424	0.867072
Amp = 4	Pure ceramic	1.412733	1.436529	1.505372	1.612828	1.751212	1.652242	1.673757	1.736141	1.833854	1.96037
	AFG	1.001446	1.01851	1.067957	1.145271	1.244928	1.219743	1.233888	1.275684	1.342601	1.430657
	Pure metal	0.772488	0.7855	0.823143	0.881901	0.957569	0.903452	0.915217	0.949329	1.002759	1.071938
Amp = 8	Pure ceramic	2.220503	2.23591	2.281716	2.355972	2.455347	2.801016	2.815541	2.858107	2.925466	3.013625
	AFG	1.571296	1.582057	1.614321	1.667139	1.738327	2.094046	2.10185	2.126377	2.168211	2.225871
	Pure metal	1.21418	1.222604	1.247651	1.288255	1.342593	1.531606	1.539549	1.562824	1.599656	1.647862

conditions. The tables provide the frequencies for various amplitudes and small-scale characteristics. The small-scale parameter has the effect of increasing the linear and nonlinear frequencies of pure ceramic, pure metal, and AFG microbeams. The stiffness of the microbeam is directly proportional to the small-scale parameter, resulting in a rise in the frequency of the microbeams.

8.3.6 Conclusions

In summary, the key findings of this research may be summarized as follows:

- The study demonstrates that the primary and secondary frequencies of clamped, simply supported, and clamped-simply supported microbeams rise in response to the nonlinear amplitude, small-scale parameter, and the rate of cross section change along the thickness (βh).
- The impact of the rates of cross section changes (β_h and β_b) on the AFG microbeam may vary in certain situations.
- The impact of these factors on microbeams made of pure metal and ceramic.

8.4 CASE STUDY 2

8.4.1 Introduction

The suggested approach involves using a micro/nanobeam that follows the principles of nonlocal strain gradient theory and von Karman hypothesis to study nonlinear vibration and instability zones. The behavior of the beam is influenced by its size. The micro/nano piezoelectric sandwich beam is affected in its axial direction by the parametric excitation. Furthermore, the electric enthalpy energy density is used to analyze the impact of flexoelectricity.

8.4.2 Mathematical Modeling

Figure 8.5 depicts the schematics of a micro/nanobeam, showcasing its axial length (L) and breadth (b). Furthermore, the distance between the micro/nanobeam and stationary electrode is shown as g. The micro/nanobeam is exposed to an external parametric excitation load at a frequency of Ω [7].

The electric Gibbs free enthalpy density is used to analyze the impact of flexoelectricity.

$$G_b = -\frac{1}{2}k_{ij}E_iE_j - \frac{1}{2}b_{ijkl}E_{i,j}E_{k,l} + \frac{1}{2}C_{ijkl}\varepsilon_{ij}\varepsilon_{kl} - e_{ijk}E_i\varepsilon_{jk} - \mu_{ijkl}(E_k\varepsilon_{ij,l} - \varepsilon_{ij}E_{k,l}) \quad (8.85)$$

E_i and E_j represent electrical field variables, ε_{ij} denotes strain components, $\varepsilon_{ij,l}$ and $E_{k,l}$ represent the strain and electrical field gradients, respectively. In addition, k_{ij}, b_{ijkl}, and C_{ijkl} define the tensor for the dielectric constant, nonlocal electrical coupling coefficient, and elastic stiffness, respectively. On the other hand, e_{ijk} and u_{ijkl} represent the parameters for piezoelectric and flexoelectric effects.

The electrostatic potential is calculated using the following equation:

$$\varphi=-\left(z^2\frac{e_{31}}{2k_{33}}\frac{\partial^2 w}{\partial x^2}\right)-\left(z\frac{\mu_{31}}{2k_{33}}\frac{\partial^2 w}{\partial x^2}\right)+z\frac{V_P}{h_P}+\left(\frac{e_{31}}{8k_{33}}h^2\frac{\partial^2 w}{\partial x^2}\right)+\frac{V_P}{2}, \tag{8.86}$$

The variables V_P and h_P represent the piezoelectric voltage and thickness of the piezoelectric layers, respectively. Furthermore, the electric field and electric field gradient may be obtained in the following manner:

$$E_z=-\frac{\partial\varphi}{\partial z}=z\frac{e_{31}}{k_{33}}\frac{\partial^2 w}{\partial x^2}+\frac{\mu_{31}}{2k_{33}}\frac{\partial^2 w}{\partial x^2}-\frac{V_P}{h_P} \tag{8.87}$$

$$E_{z,z}=-\frac{\partial^2\varphi}{\partial z^2}=\frac{e_{31}}{k_{33}}\frac{\partial^2 w}{\partial x^2}, \tag{8.88}$$

8.4.2.1 Nonlocal Strain Gradient Theory

The equation may be expressed using the nonlocal strain gradient theory (NSGT) as follows:

$$\left(1-(ea^2)\nabla^2\right)\sigma_{ij}=\left(1-l^2\nabla^2\right)\left(C_{ijkl}\varepsilon_{kl}\right)+\gamma\left(-e_{ijk}E_i+\mu_{ijkl}E_{k,l}\right) \tag{8.89}$$

The symbol $\nabla=\frac{\partial}{\partial x}$ represents the differential operator, C_{ijkl} is the elastic coefficient, ea is a nonlocal parameter, l is the length scale parameter, and ε_{kl} represents the strain. It is important to note that when considering the length scale for piezoelectric and flexoelectric effects, the value of γ is equal to $\gamma=1-l^2\nabla^2$. Otherwise, when not considering these effects, γ is equal to 1 ($\gamma=1$).

This study uses the Euler-Bernoulli beam theory to determine the displacement field in the following manner:

$$\bar{u}=u-z\frac{\partial w}{\partial x},\bar{v}=0,\bar{w}=w \tag{8.90}$$

Here, "u" represents the axial displacement and "w" represents the transverse displacement. The von Karman strain-displacement theory yields the following expressions:

$$\varepsilon_{xx}=\frac{\partial u}{\partial x}-z\frac{\partial^2 w}{\partial x^2}+\frac{1}{2}\left(\frac{\partial w}{\partial x}\right)^2 \tag{8.91}$$

$$\varepsilon_{xx,z}=\frac{\partial\varepsilon_{xx}}{\partial z}=-\frac{\partial^2 w}{\partial x^2} \tag{8.92}$$

It is important to note that in the Euler-Bernoulli beam theory, the strain gradient in the x direction is negligible compared to the strain gradient in the z direction and may be disregarded. The constitutive equations for the piezoelectric Euler-Bernoulli beam theory, including the flexoelectric effect, are derived within the context of the nonlocal strain gradient theory as follows:

$$\left(1-(ea)^2\nabla^2\right)\sigma_{xx}=\left(1-l^2\nabla^2\right)E\varepsilon_{xx}+\gamma\left(e_{31}\frac{\partial\varphi}{\partial z}-\mu_{31}\frac{\partial^2\varphi}{\partial z^2}\right)\tag{8.93}$$

$$\left(1-(ea)^2\nabla^2\right)\tau_{xxz}=\gamma\left(\mu_{31}\frac{\partial\varphi}{\partial z}\right)\tag{8.94}$$

By substituting equations (8.87), (8.88), (8.91), and (8.92) into equations (8.93) and (8.94), we get the following expressions.

$$\left(1-(ea)^2\nabla^2\right)\sigma_{xx}=\left(1-l^2\nabla^2\right)E\left(\frac{\partial u}{\partial x}-z\frac{\partial^2 w}{\partial x^2}+\frac{1}{2}\left(\frac{\partial w}{\partial x}\right)^2\right)+\gamma(e_{31}(-z\frac{e_{31}}{k_{33}}\frac{\partial^2 w}{\partial x^2}-\frac{\mu_{31}}{2k_{33}}\frac{\partial^2 w}{\partial x^2}+\frac{V_P}{h_P})+\mu_{31}\frac{e_{31}}{k_{33}}\frac{\partial^2 w}{\partial x^2})\tag{8.95}$$

$$\left(1-(ea)^2\nabla^2\right)\tau_{xxz}=\gamma\left(\mu_{31}\left(-z\frac{e_{31}}{k_{33}}\frac{\partial^2 w}{\partial x^2}-\frac{\mu_{31}}{2k_{33}}\frac{\partial^2 w}{\partial x^2}+\frac{V_P}{h_P}\right)\right)\tag{8.96}$$

Where N_{xx}, M_{xx}, and P_{xxz} represent the stress resultants and are calculated as follows:

$$N_{xx}=\int_{-\left(\frac{h}{2}+h_P\right)}^{-\frac{h}{2}}\sigma_{xx}dz+\int_{-\frac{h}{2}}^{\frac{h}{2}}\sigma_{xx}dz+\int_{\frac{h}{2}}^{\left(\frac{h}{2}+h_P\right)}\sigma_{xx}dz\tag{8.97}$$

$$M_{xx}=\int_{-\left(\frac{h}{2}+h_P\right)}^{-\frac{h}{2}}z\sigma_{xx}dz+\int_{-\frac{h}{2}}^{\frac{h}{2}}z\sigma_{xx}dz+\int_{\frac{h}{2}}^{\left(\frac{h}{2}+h_P\right)}z\sigma_{xx}dz\tag{8.98}$$

$$P_{xxz}=\int_{-\left(\frac{h}{2}+h_P\right)}^{-\frac{h}{2}}\tau_{xxz}dz+\int_{\frac{h}{2}}^{\left(\frac{h}{2}+h_P\right)}\tau_{xxz}dz\tag{8.99}$$

By inserting equations (8.95) and (8.96) into equations (8.97)–(8.99), we may get that

$$\left(1-(ea)^2\nabla^2\right)N_{xx}=\left(1-l^2\nabla^2\right)\left((EA)_{eff}\left(\frac{\partial u}{\partial x}+\frac{1}{2}\left(\frac{\partial w}{\partial x}\right)^2\right)\right)+\gamma\left(2be_{31}V_P\right)\tag{8.100}$$

$$\left(1-(ea)^2\nabla^2\right)M_{xx}=\left(1-l^2\nabla^2\right)\left(-(EI)_{eff}\frac{\partial^2 w}{\partial x^2}\right)+\gamma\left(-\acute{I}\frac{{e_{31}}^2}{k_{33}}\frac{\partial^2 w}{\partial x^2}\right)\tag{8.101}$$

$$\left(1-(ea)^2\nabla^2\right)P_{xxz}=\gamma\left(-h_p b\frac{{\mu_{31}}^2}{k_{33}}\frac{\partial^2 w}{\partial x^2}+2\mu_{31}bV_P\right)\tag{8.102}$$

It is important to clarify that these phrases are introduced in the following manner:

$$(EA)_{eff}=\left(E_{Iso}\acute{A}+E_P A''\right),\ (EI)_{eff}=\left(E_{Iso}I+E_P\acute{I}\right)\tag{8.103}$$

$$\acute{A}=bh,\ A''=bh_P\tag{8.104}$$

$$I = \frac{bh^3}{12}, \quad \acute{I} = b\left(\frac{1}{2}h^2 h_P{}^2 + hh_p{}^2 + \frac{2}{3}h_p{}^3\right) \tag{8.105}$$

It is important to note that E_{Iso} and E_P represent the Young's modulus of the elastic and piezoelectric layers, respectively. The governing equation is found using Hamilton's variational technique.

$$\delta \int_0^t \left[K - \left(U - W\right)\right] dt = 0 \tag{8.106}$$

Here, variations in external forces are defined as follows:

$$\delta \int_0^t W dt = \int_0^t \int_0^L q \delta W \; dxdt \tag{8.107}$$

The equation is given by $q = F_e + F_c + F cos \Omega t$. Furthermore, the electrostatic and Casimir forces are shown below:

$$F_e = \frac{\varepsilon_0 b \left(V_{dc}\right)^2}{2\left(g - w\right)^2}\left(1 + 0.65\frac{g - w}{b}\right) \tag{8.108}$$

$$F_c = \frac{\pi^2 \bar{h} c}{240\left(g - w\right)^4} \tag{8.109}$$

The symbol ε_0 represents the vacuum permittivity, which has a value of $8.854 \times 10^{-12} C^2 N^{-1} m^{-2}$. V_dc refers to the applied voltage on the elastic layer. Moreover, the Planck constant is $\bar{h} = 1.055 \times 10^{-3}$, and $c = 2.998 \times 10^8 \frac{m}{s}$ represents the velocity of light.

The first form of strain energy may be expressed as:

$$\delta \int_0^t U dt = \int_0^t \int_0^L \int_{-\left(\frac{h}{2}+h_P\right)}^{\left(\frac{h}{2}+h_P\right)} \left(\sigma_{xx} \delta \varepsilon_{xx} + \tau_{xxz} \delta \varepsilon_{xx,z}\right) dx\, dz\, dt \tag{8.110}$$

The variation of kinetic energy is defined as follows:

$$\int_0^t K dt = \int_0^t \int_0^L \left(\left(I_A \left(\frac{\partial^2 u}{\partial x^2}\right)\right) \delta u + \left(I_0 \left(\frac{\partial^2 u}{\partial x^2}\right) - I_1 \left(\frac{\partial^4 w}{\partial x^2 \partial t^2}\right)\right) \delta w \right) dxdt \tag{8.111}$$

Here, I_A, represents the axial inertia, I_0 represents the translatory inertia, and I_1 represents the rotary inertia. These values are calculated using the formulas $I_0 = b\left(\rho_{Iso} h + \rho_P 2h_P\right)$ and $I_1 = \frac{b}{12}\left(\rho_{Iso} h^3 + \rho_P \left(2h_p\right)^3\right)$. It is worth mentioning that ρ_{Iso} and ρ_P denote the density of the elastic and piezoelectric layers, respectively.

The governing equation for the micro/nanobeam may be obtained as follows:

$$\frac{\partial N_{xx}}{\partial x} = I_0 \left(\frac{\partial^2 u}{\partial x^2}\right) \tag{8.112}$$

$$\frac{\partial^2 M_{xx}}{\partial x^2}+\frac{\partial^2 P_{xxz}}{\partial x^2}+N_{xx}\frac{\partial^2 w}{\partial x^2}+F_e+F_c+F\cos\Omega t=I_0\left(\frac{\partial^2 w}{\partial t^2}\right)-I_1\left(\frac{\partial^4 w}{\partial x^2\partial t^2}\right) \tag{8.113}$$

The expression N_{xx} may be represented in the following manner:

$$N_{xx}=(EA)_{eff}\,(\frac{1}{2}\int_0^L\left(\frac{\partial w}{\partial x}\right)^2-l^2(\int_0^L\frac{\partial^2 w}{\partial x^2}+\frac{\partial w}{\partial x}\frac{\partial^3 w}{\partial x^3}))dx \tag{8.114}$$

Within the context of nonlocal strain gradient theory, it may be expressed as:

$$\begin{aligned}&(1-l^2\nabla^2)\left(-(EI)_{eff}\frac{\partial^4 w}{\partial x^4}\right)+(1-l^2\nabla^2)\left(-\acute{\text{I}}\frac{e31^2}{k_{33}}-bh_p\frac{\mu_{31}{}^2}{k_{33}}\right)\left(\frac{\partial^4 w}{\partial x^4}\right)+\\&\left((EA)_{eff}\left(\frac{1}{2}\int_0^L\left(\frac{\partial w}{\partial x}\right)^2dx-l^2\left(\int_0^L(\frac{\partial^2 w}{\partial x^2})^2dx+(\frac{\partial w}{\partial x})(\frac{\partial^3 w}{\partial x^3})dx\right)\right)(\frac{\partial^2 w}{\partial x^2}-\right.\\&(ea^2))\left(\frac{\partial^4 w}{\partial x^4}\right)+(1-(ea^2)\nabla^2)(2be_{31}V_P\frac{\partial^2 w}{\partial x^2}-\frac{\varepsilon_{0b}(V_{dc})^2}{2(g-w)^2}\left(1+0.65\frac{g-w}{b}\right)+\\&\frac{\pi^2\bar{h}C}{240(g-w)^4}+F\cos\Omega t-I_0\left(\frac{\partial^2 w}{\partial t^2}\right)+I_1\left(\frac{\partial^4 w}{\partial x^2\partial t^2}\right)=0\end{aligned} \tag{8.115}$$

The Taylor series expansions for electrostatic and Casimir forces are implemented as follows:

$$\frac{1}{(g-w)^2}=\frac{2w}{g^3}+\frac{1}{g^2}+\frac{3w^2}{g^4}+\frac{4w^3}{g^5}+\frac{5w^4}{g^6}+\frac{5w^4}{g^6}+\frac{6w^5}{g^7} \tag{8.116}$$

$$\frac{1}{(g-w)^4}=\frac{4w}{g^5}+\frac{1}{g^4}+\frac{10w^2}{g^6}+\frac{20w^3}{g^7}+\frac{35w^4}{g^8}+\frac{56w^5}{g^9} \tag{8.117}$$

$$\frac{1}{(g-w)^6}=\frac{6w}{g^7}+\frac{1}{g^6}+\frac{21w^2}{g^8}+\frac{56w^3}{g^9}+\frac{126w^4}{g^{10}}+\frac{252w^5}{g^{11}} \tag{8.118}$$

8.4.3 Solution Method

8.4.3.1 Galerkin Technique

The Galerkin approach may be used to get an approximate solution for the simply supported boundary condition. This solution can be written as follows:

$$w(x,t)=\phi(t)\sin\frac{k\pi}{L}x \tag{8.119}$$

One may argue that $\phi(t)$ represents the time-dependent unknown parameter, whereas $\sin\frac{k\pi}{L}x$ represents the spatial basis function. By replacing the Equation (8.119) with equation (8.115), we get the following:

$$B_1\frac{\partial^2\phi}{\partial t^2}+B_2\phi+B_3\phi^2+B_4\phi^3+B_5=0 \tag{8.120}$$

B_1, B_2, B_3, B_4, and B_5 are defined as follows:

$$B_1=(-I_0-I_1\left(\frac{n\pi}{L})^2\right)\left(1+(ea^2)\left(\frac{n\pi}{L})^2\right)\right. \tag{8.121}$$

$$\begin{aligned}B_2=&-(EI)_{eff}(\frac{n\pi}{L})^4(1+l^2\frac{n\pi}{L})^2)-\acute{I}\frac{e_{31}{}^2}{k_{33}}(\frac{n\pi}{L})^4(1+l^2(\frac{n\pi}{L})^2-\\ &bh_p\frac{\mu_{31}{}^2}{k_{33}}\left(\frac{n\pi}{L}\right)^4(1+l^2\left(\frac{n\pi}{L}\right)^2)-2be_{31}V_P\left(\frac{n\pi}{L}\right)^4\left(1+ea^2\left(\frac{n\pi}{L}\right)^2\right)+\frac{\varepsilon_0 bV_{dc}{}^2}{g^3}+\\ &\frac{0.65\ {}_0bV_{dc}{}^2}{g^2}-\frac{12ea^2\varepsilon_0 bV_{dc}{}^2}{g^5}-\frac{5.85ea^2\varepsilon_0 bV_{dc}{}^2}{g^4}+\frac{\pi^2\bar{h}c}{60\ ^5}-\frac{ea^2\pi^2\bar{h}c}{2g^7}\\ &F\cos\Omega t\left(\frac{n\pi}{L}\right)^2\left(1+ea^2\left(\frac{n\pi}{L}\right)^2\right)\end{aligned} \tag{8.122}$$

$$\begin{aligned}B_3=&\frac{6\varepsilon_0 bV_{dc}{}^2}{2g^4}\left(\frac{8}{3\pi}\right)-\frac{0.65\varepsilon_0 bV_{dc}{}^2}{g^3}\left(\frac{8}{3\pi}\right)-\frac{30ea^2\varepsilon_0 bV_{dc}{}^2}{g^6}\left(\frac{8}{3\pi}\right)-\frac{11.7ea^2\varepsilon_0 bV_{dc}{}^2}{g^5}\left(\frac{8}{3\pi}\right)+\\ &\frac{\pi^2\bar{h}c}{24g^6}\left(\frac{8}{3\pi}\right)-\frac{21ea^2\pi^2\bar{h}c}{12g^8}\left(\frac{8}{3\pi}\right)\end{aligned} \tag{8.123}$$

$$B_4=-(EA)_{eff}\left(\frac{n\pi}{L}\right)^4\frac{L}{4}\left(1+ea^2\left(\frac{n\pi}{L}\right)^2\right)-\frac{1.95\varepsilon_0 V_{dc}{}^2}{2g^4}\left(\frac{3}{4}\right)+\frac{19.5ea^2\varepsilon_0 V_{dc}{}^2}{g^6}\left(\frac{3}{4}\right) \tag{8.124}$$

$$\begin{aligned}B_5=&\frac{\varepsilon_0 bV_{dc}{}^2}{2g^2}\left(\frac{4}{L}\right)+\frac{0.65\varepsilon_0 V_{dc}{}^2}{2g}\left(\frac{4}{L}\right)-\frac{3ea^2\varepsilon_0 bV_{dc}{}^2}{g^4}\left(\frac{4}{L}\right)-\frac{1.95ea^2\varepsilon_0 V_{dc}{}^2}{g^3}\left(\frac{4}{L}\right)+\\ &\frac{\pi^2\bar{h}c}{240g^4}\left(\frac{4}{L}\right)-\frac{ea^2\pi^2\bar{h}c}{12g^6}\left(\frac{4}{L}\right)\end{aligned} \tag{8.125}$$

In addition, dimensionless parameters are added in the following manner:

$$\tau=\frac{\Omega}{2}t,\qquad \phi=r\psi,\qquad r=g \tag{8.126}$$

By using these dimensionless parameters and substituting them into Equation (8.120), the governing equation for the micro/nanobeam may be derived as follows:

$$\ddot{\psi}+S^2\psi=-\varepsilon\eta_1-\varepsilon\eta_2\psi^2+\varepsilon\eta_3\psi^3+\varepsilon\eta_4\psi\cos 2\tau \tag{8.127}$$

In addition, S, η_1, η_2, η_3, and η_4 may be defined as follows:

$$S^2 = \frac{4\omega_0^{\,2}}{\Omega^2},\quad \omega_0 = \sqrt{\frac{B_2}{B_1}},\quad \eta_1 = \frac{4B_5}{B_1 r\Omega^2},\quad \eta_2 = \frac{4B_3 r}{B_1\Omega^2},\quad \eta_3 = \frac{4B_4 r^2}{B_1\Omega^2},$$
$$\eta_4 = \frac{4F\left(\frac{n\pi}{L}\right)^2\left(1+ea^2\left(\frac{n\pi}{L}\right)^2\right)}{B_1\Omega^2} \tag{8.128}$$

To investigate the parametric resonance of the system, the dimensionless frequency is defined as follows:

$$S = 1+\varepsilon\sigma \tag{8.129}$$

Here, σ is defined as the parameter that represents the detuning.

8.4.3.2 Multiple Time Scales Method

The Nayfeh and Mook technique [8], which involves various time scales, is used to tackle the fundamental issue. Subsequently, a collection of initial approximations of the first order are taken into account as follows:

$$\psi(\tau,\varepsilon) = \psi_0(T_0,T_1) + \varepsilon\psi_1(T_0,T_1) + \ldots. \tag{8.130}$$

In which:

$$T_0 = \tau,\qquad T_1 = \varepsilon\tau \tag{8.131}$$

T_0 and T_1 were seen to differentiate between the rapid and gradual time scales. Using Equation (8.131), the following determination may be made:

$$\frac{d}{dT} = \frac{d}{dT_0} + \varepsilon\frac{d}{dT_1} + \ldots = D_0 + \varepsilon D_1 + \ldots \tag{8.132}$$

$$\frac{d^2}{dT^2} = \frac{d^2}{dT_0^{\,2}} + 2\varepsilon\frac{d}{dT_0}\frac{d}{dT_1} + \ldots = D_0^2 + 2\varepsilon D_0 D_1 + \ldots \tag{8.133}$$

Therefore, by replacing Equations (8.132) and (8.133) into Equation (8.127) and setting the coefficients of ε^0 and ε^1 equal to zero, the following may be obtained:

$$\varepsilon^0 : D_0^{\,2}\psi_0 + \omega_0^{\,2}\psi_0 \tag{8.134}$$

$$\varepsilon^1 : D_0^{\,2}\psi_1 + \psi_1 = -2D_0D_1\psi_0 - 2\sigma\psi_0 - \eta_2\psi_0^{\,2} - \eta_1 D_0\psi_0 + \eta_3\psi_0 cos2T_0. \tag{8.135}$$

The solution to Equation (8.134) may be obtained as follows:

$$\psi_0 = A(T_1)\exp(iT_0) + \bar{A}(T_1)\exp(iT_0). \tag{8.136}$$

The function A exhibits an unknown complex behavior, whereas $\bar{A}$ represents its complex conjugate version. Substituting Equation (8.136) into Equation (8.135) yields the following result:

$$D_0^{\ 2}\psi_1+\psi_1=[-2i\acute{A}-2\sigma A-3\eta_3A^2\bar{A}+\frac{\eta_3}{2}\bar{A}]\exp(iT_0)-\eta_3A^3\exp(3iT_0)+ \frac{\eta_4}{2}A\exp(3iT_0)+c.c \tag{8.137}$$

Hence, the secular factor must be removed from the equation.

$$-2i\acute{A}(T_1)-2\sigma A(T_1)-3\eta_3A^2\bar{A}+\frac{\eta_4}{2}\bar{A}(T_1)=0 \tag{8.138}$$

It is essential to take into account the polar form of function $A(T_1)$ as shown next:

$$A(T_1)=\frac{1}{2}\alpha(T_1)\exp[i\beta(T_1)] \tag{8.139}$$

Where $\alpha(T_1)$ and $\beta(T_1)$ are actual functions. By substituting Equation (8.139) into Equation (8.138) and then separating the real and imaginary parts, the following may be obtained:

$$\acute{a}=\frac{\eta_4}{4}\alpha\sin(2\beta) \tag{8.140}$$

$$\alpha\beta=\sigma\alpha+\frac{3\eta_3}{8}\alpha^3-\frac{\eta_4}{4}\alpha\cos(2\beta). \tag{8.141}$$

Furthermore, in order to assess the steady-state response, it is suggested that $\acute{\alpha}=0$ and $\alpha\beta'=0$. The modulation equations for the major parametric resonance are obtained as follows:

$$-\left(\frac{\eta_4}{4}\right)\sin(2\beta)=0 \tag{8.142}$$

$$\sigma+\frac{3\eta_3}{8}\alpha^2-\frac{\eta_4}{4}\cos(2\beta) \tag{8.143}$$

8.4.3.3 Trivial Steady-State Response

To assess the stability of the system, we use the Cartesian form represented by the equation $A=\frac{1}{2}(p-iq)e^{i\beta T_1}$), We then substitute A into Equation (8.138) as shown [7, 9]:

$$\mathrm{i}(p'-iq')-\beta(p-iq)+\frac{3\eta_3}{8}(p^2-q^2-2ipq)(p+iq)+\frac{\eta_4}{4}(p+iq)=0 \tag{8.144}$$

To characterize the autonomous form of the system, it is important to take into account the value of η_3, which is equal to $\frac{\sigma}{2}$. By substituting this value into Equation (8.144), the following derivation may be obtained:

$$p' = -\frac{\sigma}{2}q + \frac{3\eta_3}{8}q\left(p^2 + q^2\right) - \frac{\eta_4}{4}q \tag{8.145}$$

$$q' = \frac{\sigma}{2}p - \frac{3\eta_3}{8}p\left(p^2 + q^2\right) - \frac{\eta_4}{4}p \tag{8.146}$$

The last phase involves assessing the instability analysis of the system, specifically focusing on the unremarkable steady-state response of the micro/nanobeam. It is assumed that both p and q are equal to zero. Moreover, the Jacobian matrix of Equations (8.145) and (8.146) is defined as follows:

$$\mathrm{A} = \begin{bmatrix} \frac{dp'}{dp} & \frac{dp'}{dq} \\ \frac{dq'}{dp} & \frac{dq'}{dq} \end{bmatrix} = \begin{bmatrix} 0 & -\frac{\sigma}{2} - \frac{\eta_4}{4} \\ \frac{\sigma}{2} - \frac{\eta_4}{4} & 0 \end{bmatrix} \tag{8.147}$$

8.4.3.4 Non-trivial Steady-State Response

In order to ensure the stability of the piezoelectric micro/nanobeam, it is necessary to reset the matrix determination to zero. In a non-trivial, steady-state solution, the value of α is not equal to zero ($\neq 0.$). Therefore, it may be worded as follows [7]:

$$-\left(\frac{\eta_4}{4}\right)\sin\left(2\beta\right) = 0 \tag{8.148}$$

$$\sigma + \frac{3\eta_3}{8}\alpha^2 - \frac{\eta_4}{4}\cos\left(2\beta\right) \tag{8.149}$$

The trigonometric function $sin^2\beta + cos^2\beta = 1$) may be used in Equations (8.148) and Equation (8.149) and so can be obtained as follows:

$$\left[\left(\sigma + \frac{3\eta_3}{8}\alpha^2\right)^2\right] = \left(\frac{\eta_4}{4}\right)^2 \tag{8.150}$$

$$\left(\sigma + \frac{3\eta_3}{8}\alpha^2\right)^2 = \left(\frac{\eta_4}{4}\right)^2 \tag{8.151}$$

The expression for the detuning parameter at the positive bifurcation point is obtained by setting α equal to zero.

$$\sigma = +\sqrt{\left(\frac{\eta_4}{4}\right)^2} \tag{8.152}$$

8.4.4 Results and Discussion

In this part, the outcomes and numerical findings are thoroughly examined and deliberated upon.

8.4.4.1 Numerical Results

The material qualities and geometrical requirements reported in Table 8.6 are assumed for the current investigation. In addition, a simply supported border condition is used for this study. Figure 8.6 demonstrates the influence of flexoelectricity

TABLE 8.6
Material and Geometrical Properties of the Isotropic and Piezoelectric Layers [10]

	Silicon	PZT-5H
Young's Modulus	210 GPa	126 GPa
Gap	92 μm	–
Thickness	57 μm	0.57 μm
Width	5 mm	5 mm
Length	20 mm	20 mm
Piezoelectric Constant	–	$-6.35\frac{C}{m^2}$
Flexoelectric Constant	–	$-10^{-7}\frac{C}{m}$
Dielectric Constant	–	$1.3\times10^{-8}\frac{C}{Vm}$

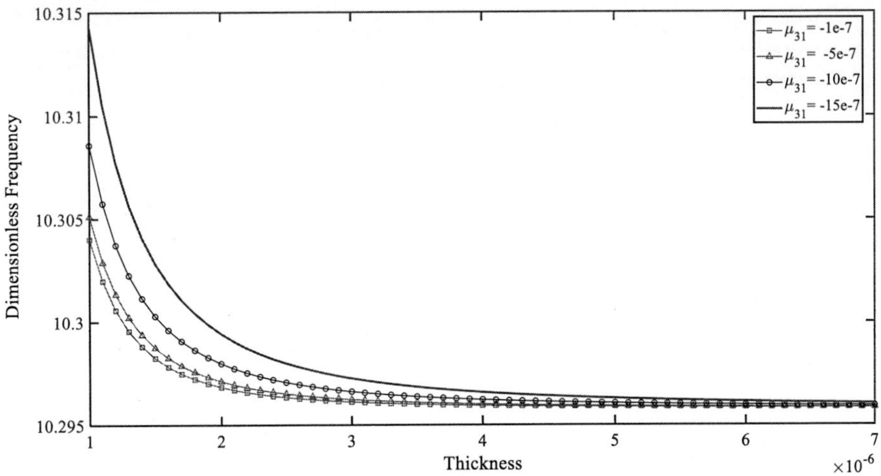

FIGURE 8.6 The dimensionless frequency against the thickness of the simply supported beam considering different flexoelectric parameter.

on the nonlinear vibrational characteristics of a micro/nanobeam that is simply supported. The phenomenon of flexoelectricity results in an increase in the stiffness of the beam. Additionally, when the flexoelectric parameter rises, the dimensionless frequency also increases. It is important to acknowledge that flexoelectricity plays a significant role in thin beams. When the thickness of the beam grows, this impact may be ignored.

$$\left(ea = 800e - 6, l = 1.8e - 3\right)$$

Figure 8.7 illustrates the relationship between the dimensionless frequency and the V_{dc}. When the direct current voltage (V_{dc}) rises, there is a drop in the dimensionless frequency. Furthermore, the dimensionless frequency decreases as the V_P increases. With the rise in V_P, the structure exhibits a higher degree of flexibility.

Figure 8.8 demonstrates that the pull-in voltage falls as the space between the sandwich beam and stationary electrode shrinks. Moreover, Figure 8.8 showed that when V_{dc} increased, there was a corresponding drop in dimensionless quantity.

Figure 8.9 demonstrates the impact of the piezoelectric parameter on the frequency that is not expressed in specific units. It can be seen that the piezoelectric parameter enhances the structural rigidity and raises the dimensionless frequency. The impact is more pronounced for the slender beams.

A graph is drawn to examine the impact of the length scale parameter on the nonlinear behavior of an electromechanical micro/nanobeam by comparing the dimensionless frequency with the nonlocal parameter. Figure 8.10 demonstrates that when

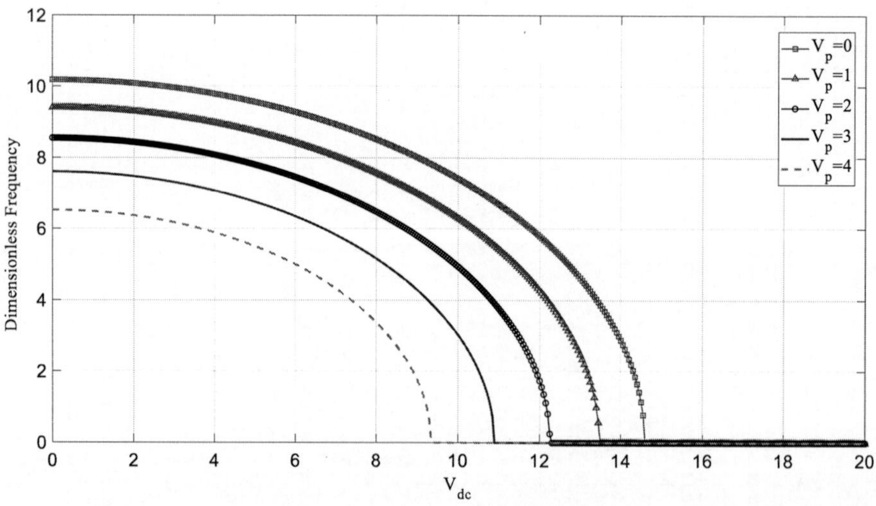

FIGURE 8.7 The dimensionless frequency versus the considering different piezoelectric voltage V_{dc}.(*thickness of the beam assumed* $(1-7\,\mu m)$) for S-S BC.

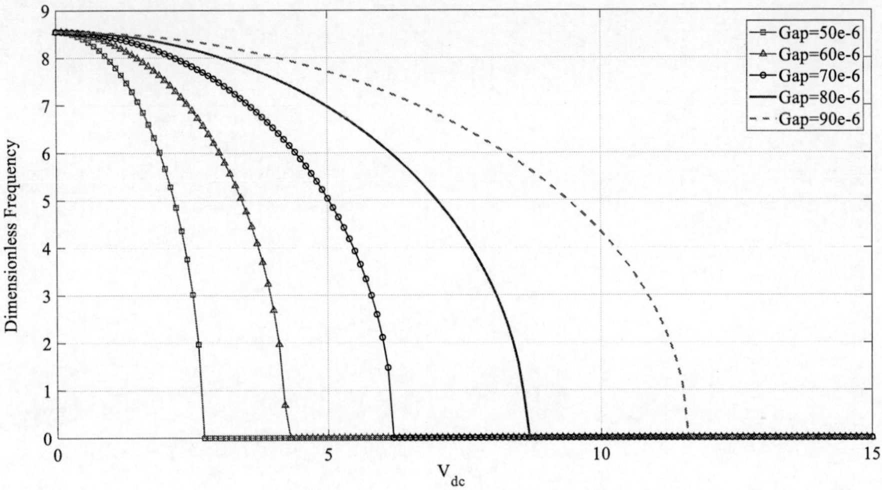

FIGURE 8.8 The relation between the dimensionless frequency and the V_{dc} considering a different gap between the sandwich beam and stationary electrode for S-S BC ($V_P = 2, ea = 800e-6, l = 1.8e-3$).

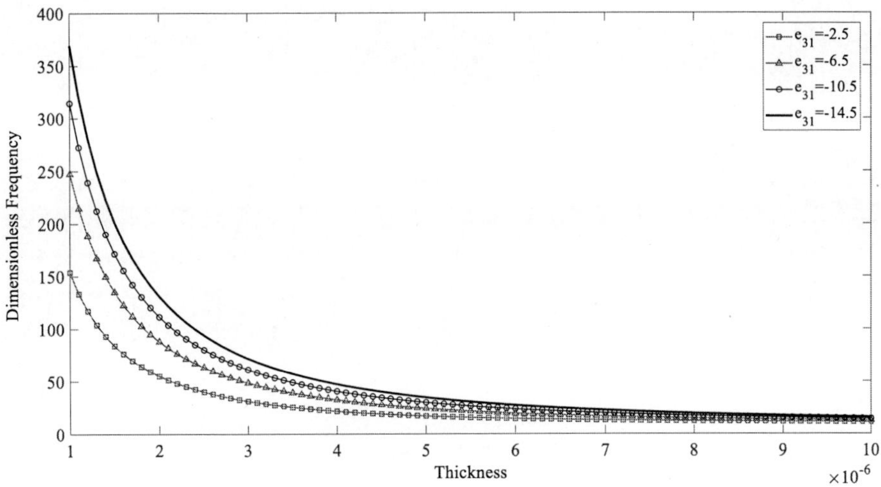

FIGURE 8.9 The relation between the dimensionless frequency and the thickness of the beam for different piezoelectric parameters $\left(ea = 800e-6, l = 1.8e-3, V_P = 0, V_{dc} = 0\right)$.

the length scale grows, the dimensionless frequency also increases. Furthermore, the dimensionless frequency drops as the nonlocal parameter increases.

Figure 8.11 illustrates the impact of adding different parametric loads. Increasing the magnitude of external parametric stresses results in the expansion of stable areas.

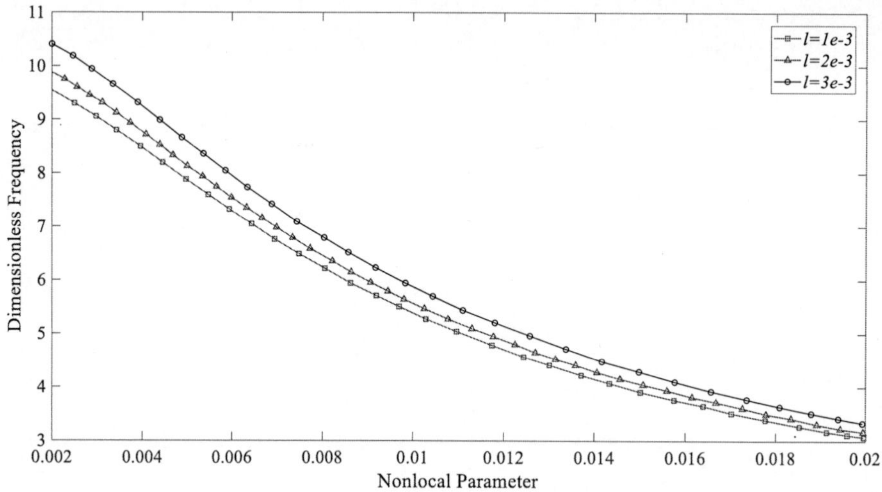

FIGURE 8.10 The relation between the dimensionless frequency and the nonlocal parameter $(V_P = V_{dc} = 0)$.

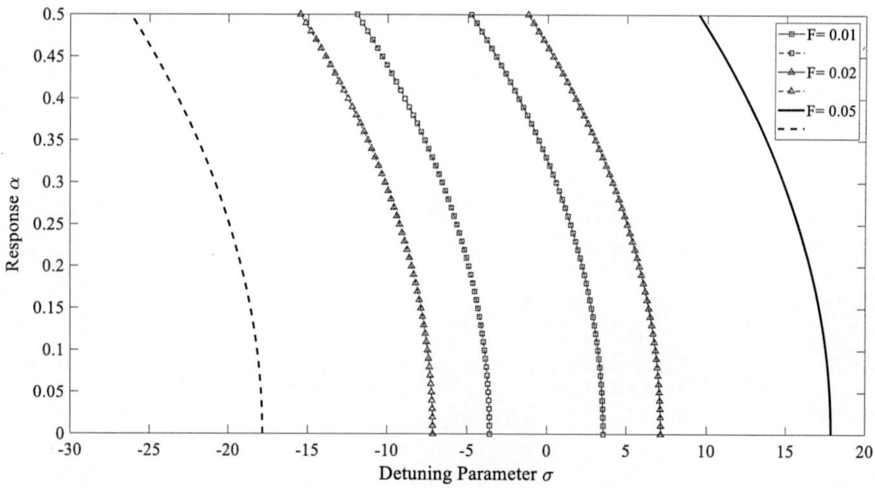

FIGURE 8.11 The amplitude response against the detuning parameter applying a different parametric excitation $\left(ea = 800e-6, l = 1.8e-3\right)$.

Figure 8.12 illustrates the influence of the piezoelectric parameter on the amplitude response and the zones of instability. The gap between the stable and unstable solution decreases as the piezoelectric parameter rises. Therefore, it can be seen that the piezoelectric parameter significantly influences the dynamic instability of electromechanical micro/nanobeams.

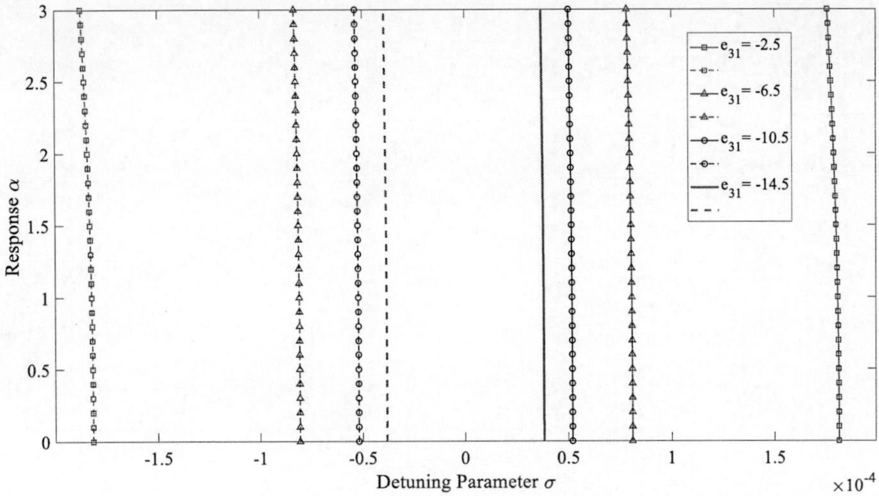

FIGURE 8.12 The effect of piezoelectric parameter on the amplitude response against the detuning parameter ($V_P = V_{dc} = 1, ea = 1.5e-3, l = 1.8e-3, F = 0.05$).

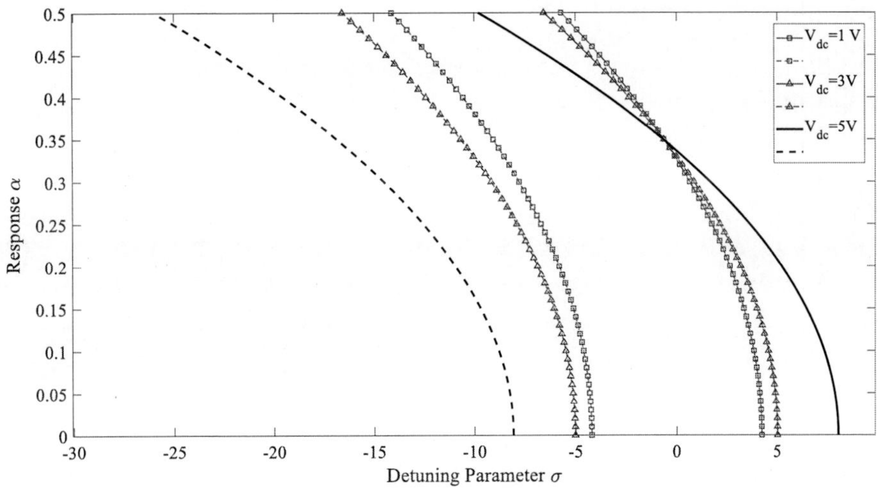

FIGURE 8.13 The effect of V_{dc} on the amplitude response against the detuning parameter ($V_P = 1V, ea = 1.5e-3, l = 1.8e-3$).

To analyze the impact of varying voltage on the elastic layer, a graph is generated to illustrate the amplitude response in relation to the detuning parameter. Figure 8.13 demonstrates that raising the applied voltage on the elastic layer expands the area where stability is maintained and results in a decrease in the stiffness of the beam.

Figure 8.14 illustrates the impact of the spacing between the sandwich beam and stationary electrode on the areas of dynamic instability. While the gap does not

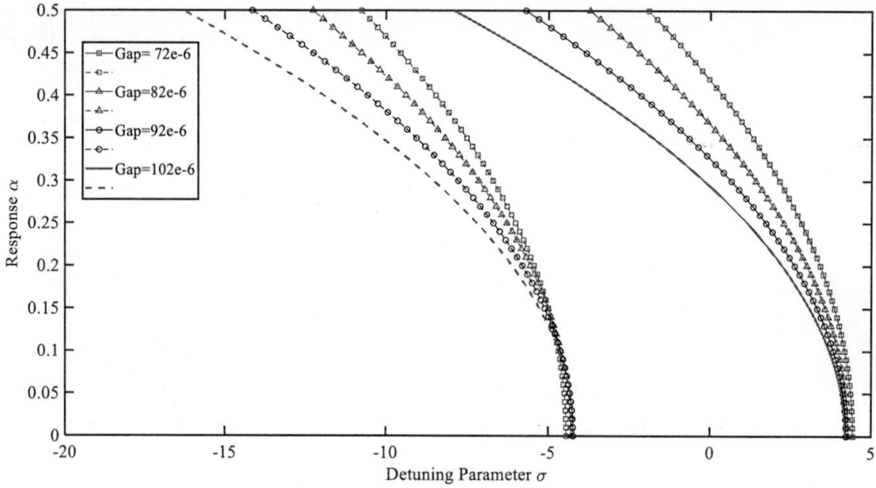

FIGURE 8.14 The effect of gap on the amplitude response against the detuning parameter $(ea = 1.5e - 3, l = 1.8e - 3, F = 0.01)$.

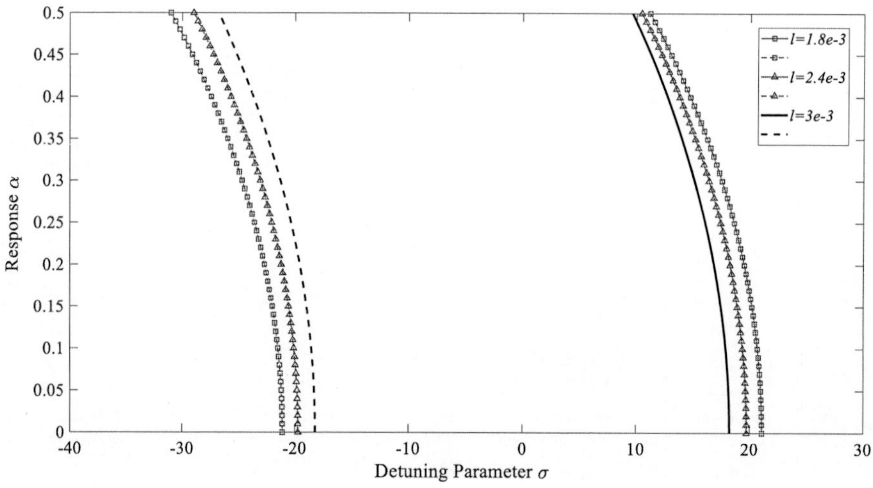

FIGURE 8.15 The effect of length scale parameter on the amplitude response against the detuning parameter $(ea = 1.5e - 3, V_P = V_{dc} = 1V)$.

significantly alter the instability of the system, widening the gap reduces the amplitude response.

Furthermore, in order to examine the impact of the length scale parameter on the vibrational characteristics of the electromechanical system, the amplitude is graphed against the detuning parameter in Figure 8.15. By raising the length scale parameter, it can be seen that the stable zone decreases in size.

Figure 8.16 illustrates that the size of the stable area expands as the piezoelectric voltage rises. The softening behavior might be identified or acknowledged.

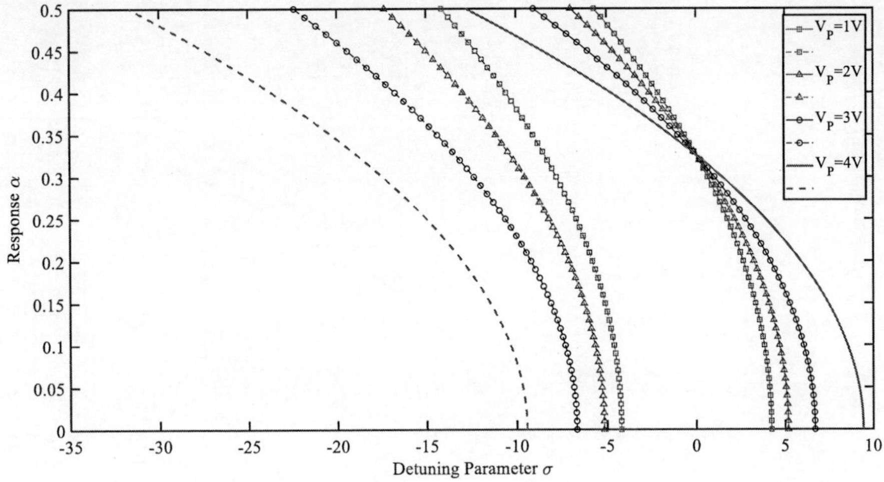

FIGURE 8.16 The effect of piezoelectric voltage on the amplitude response against the detuning parameter $\left(l = 1.8e-3, ea = 1.5e-3, F = 0.05, V_{dc} = 1V\right)$.

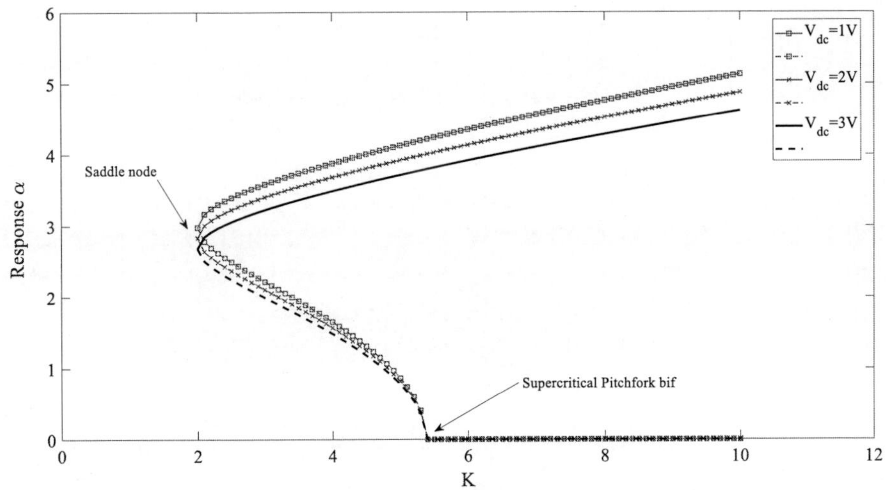

FIGURE 8.17 The effect of direct current voltage on saddle node bifurcation points for different values of force amplitude against the amplitude response $\left(\sigma = +5\right)$.

Figure 8.17 illustrates the saddle node bifurcation point for different levels of force amplitude in relation to the amplitude response. It is evident that the amplitude response diminishes as the V_{dc} rises. Put simply, changing the direct current voltage (V_{dc}) would result in the occurrence of a saddle node bifurcation point.

Figure 8.18 depicts the relationship between the force amplitude and the amplitude response for various V_{dc} levels. Furthermore, it can be inferred that the amplitude

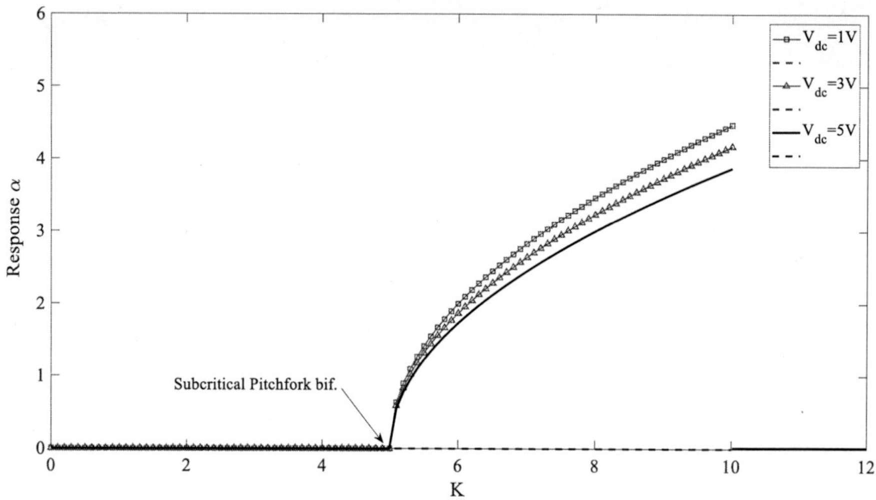

FIGURE 8.18 The force amplitude against the response amplitude for different values of V_{dc} $(\sigma = -5)$.

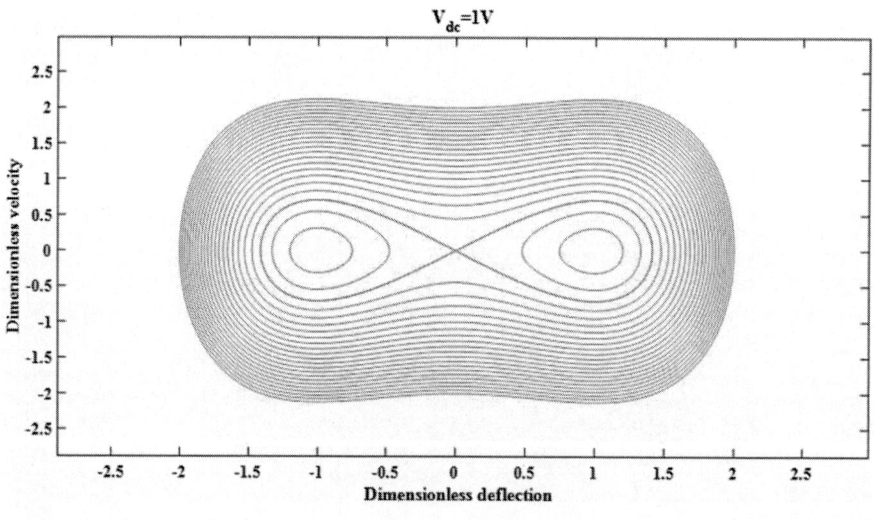

FIGURE 8.19 The dimensionless velocity against the dimensionless deflection for $V_{dc} = 1V$.

response diminishes as the V_{dc} grows, and the location of the subcritical pitchfork bifurcation point remains the same.

Figures 8.19 and 8.20 show the phase portrait of the electromechanical micro/nanobeam for various V_{dc} values. The location of the saddle node and center varies as the V_{dc} increases.

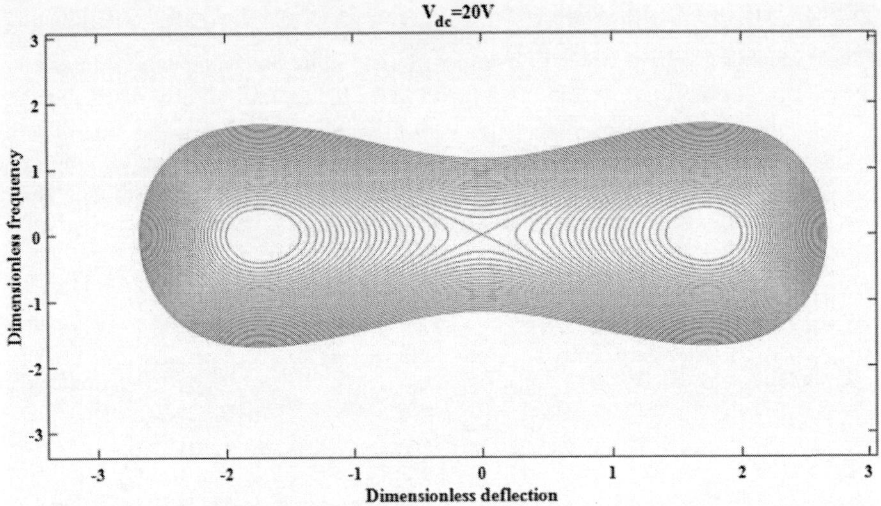

FIGURE 8.20 The dimensionless velocity against the dimensionless deflection for $V_{dc} = 20V$.

8.4.5 Conclusion

The primary outcome of this investigation may be obtained in the following manner:

- The dimensionless frequency of narrow beams rises as the flexoelectric and piezoelectric parameters increase.
- The pull-in voltage diminishes as the distance between the sandwich beam and stationary electrode decreases.
- The dimensionless frequency and instability areas are significantly affected by the length scale and nonlocal factors.
- By increasing the piezoelectric parameter, the distance between the bifurcation sites is reduced.
- The stable area expands as the applied voltage on the elastic core and piezoelectric layers increases.

8.5 CASE STUDY 3

8.5.1 Introduction

The study focuses on investigating the dynamic instability and vibration of functionally graded (FG) porous sandwich nanobeams, which are supported by a viscoelastic foundation and subjected to an axial harmonic load, based on the nonlocal theory. This article utilizes the Timoshenko and nonlocal continuum theories to include shear deformation, rotational bending, and small-scale effects.

8.5.2 Theory and Formulation

The problem's layout is shown in Figure 8.21 and Figure 8.22. The nanobeam is composed of a porous core (h_c) and two face layers (h_f) that are tightly fused together. The sandwich *FG* porous nanobeam, with a length denoted as L, is experiencing an axial harmonic load. The load has a frequency denoted as Ω and an amplitude denoted as N_d. The Timoshenko nanobeam under investigation is positioned atop a viscoelastic foundation characterized by the parameters k_w and c_d [11].

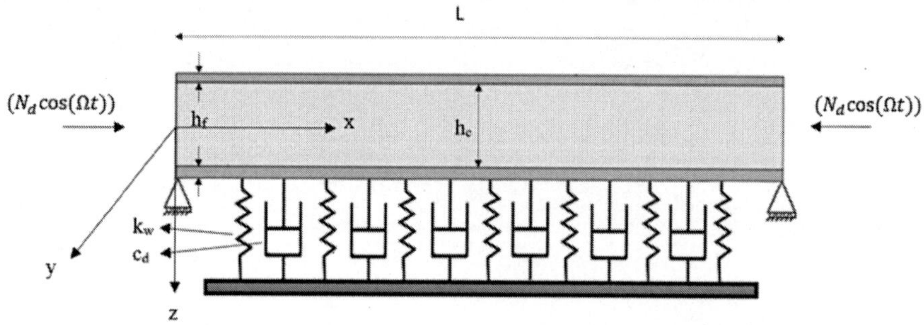

FIGURE 8.21 A simply supported FG sandwich nanobeam on a viscoelastic foundation.

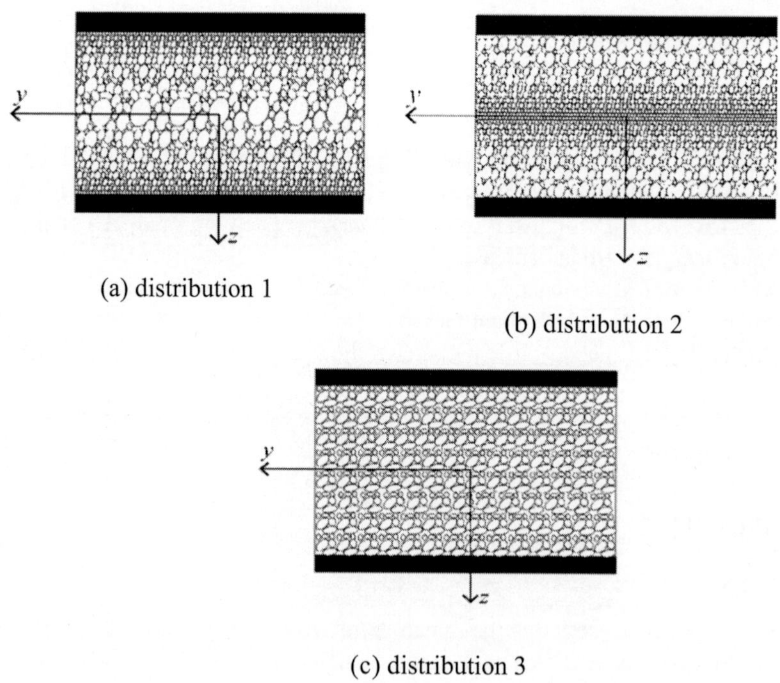

FIGURE 8.22 The cross section of FG porous nanobeam with different porosity distribution pattern.

8.5.2.1 Nonlocal Nanobeam Model

In the classical continuum theory, it is understood that the stress at a certain point x is only connected to the strain at that same location. Contrarily, according to the nonlocal elasticity theory, the stress field may be expressed as follows:

$$\sigma = \int_{v} \alpha\left(\left|x' - x\right|, \tau\right) T\left(x'\right) dx', \tag{8.153}$$

$$\tau = \frac{e_0 a}{L} \tag{8.154}$$

The symbol $T(x)$ represents the classical stress tensor at a certain position x, whereas the symbol $\alpha\left(\left|x' - x\right|\right)$ stands for the nonlocal modulus. Furthermore, e_0 represents a coefficient associated with the material, whereas a and L correspond to the internal and exterior characteristic lengths, respectively. The nonlocal constitutive relations of a beam may be mathematically represented as follows:

$$\sigma_{xx} - \left(e_0 a\right)^2 \frac{\partial^2 \sigma_{xx}}{\partial x^2} = E\varepsilon_{xx}, \tag{8.155}$$

$$\tau_{xz} - \left(e_0 a\right)^2 \frac{\partial^2 \tau_{xz}}{\partial x^2} = \frac{E(z)}{2(1+\upsilon)} \gamma_{xz}, \tag{8.156}$$

The variables σ_{xx} and τ_{xz} represent the axial and shear stress, whereas $\varepsilon_{xx}, \gamma_{xz}$ represent the axial and shear strain. Additionally, E and υ denote Young's modulus and Poisson's ratio, respectively.

8.5.2.2 Porosity Distribution

The material properties along the z-direction of the porous core exhibit fluctuation, which allows for the determination of material parameters, such as Young's modulus (E), mass density (ρ), and thermal expansion coefficient (α) [12].

$$E(z) = E_1\left[1 - e_0 \lambda(z)\right] \tag{8.157}$$

$$\rho(z) = \rho_1\left[1 - e_m \lambda(z)\right] \tag{8.158}$$

$$\alpha(z) = \alpha_1\left[1 - e_m \lambda(z)\right] \tag{8.159}$$

Where:

$$\lambda(z) = \begin{cases} \cos\left[\pi \dfrac{z}{h_c}\right], & \text{distribution 1} \\ \cos\left[\pi \dfrac{z}{2h_c} + \dfrac{\pi}{4}\right], & \text{distribution 2} \\ \lambda, & \text{distribution 3} \end{cases} \tag{8.160}$$

The values for e_0 and e_m are obtained from the open-cell metal foam model.

$$e_0 = 1 - \frac{E_2}{E_1} \tag{8.161}$$

$$e_m = 1 - \sqrt{1 - e_0} \tag{8.162}$$

$$\lambda = \frac{1}{e_0} - \frac{1}{e_0}\left(\frac{2}{\pi}\sqrt{1 - e_0} - \frac{2}{\pi} + 1\right)^2 \tag{8.163}$$

The variables E_1, ρ_1, and α_1 correspond to the highest values of Young's modulus, mass density, and thermal expansion coefficient of the porous core, respectively. On the other hand, E_2 represents the minimum value of Young's modulus.

8.5.3 Equations of Motion

This work uses the Timoshenko beam theory to analyze the impact of shear deformation on the FG nanobeam.

$$\begin{gathered} u_1(x,z,t) = u(x,t) + z\varphi(x,t) \\ u_2(x,z,t) = 0 \\ u_3(x,z,t) = w(x,t) \end{gathered} \tag{8.164}$$

The displacements along the x, y, and z axes are represented by u_1, u_2, and u_3, respectively, in Equation (8.164). The displacements of the beam are characterized by the axial displacement (u), transverse displacement (w), and rotation of the cross section (Æ). The normal and shear strains and stresses may be determined using the nonlinear von Karman strain-displacement theory.

$$\begin{gathered} \varepsilon_{xx}(x,t) = \frac{\partial u}{\partial x} + \frac{1}{2}\left(\frac{\partial w}{\partial x}\right)^2 + z\frac{\partial \varphi}{\partial x} \\ \gamma_{xz}(x,t) = \varphi + \frac{\partial w}{\partial x} \end{gathered} \tag{8.165}$$

$$\begin{gathered} \sigma_{xx} - (e_0 a)^2 \frac{\partial^2 \sigma_{xx}}{\partial x^2} = E(z)\left[\varepsilon_{xx}(x,t) - \alpha(z)\Delta T(z)\right], \\ \tau_{xz} - (e_0 a)^2 \frac{\partial^2 \tau_{xz}}{\partial x^2} = \frac{E(z)}{2(1+\upsilon)}\gamma_{xz} \end{gathered} \tag{8.166}$$

Where:

$$\begin{gathered} \Delta T(z) = T(z) - T_B \\ T(z) = \frac{T_B + T_T}{2} + \frac{T_{TB}}{h}z \;,\; T_{TB} = T_T - T_B \end{gathered} \tag{8.167}$$

The variables T_T and T_B denote the temperature of the upper and lower layers of the nanobeam, respectively. The strain energy of the Timoshenko nanobeam is expressed as follows:

$$\begin{aligned} U &= \int_0^L \int_A (\sigma_{xx}\varepsilon_{xx} + \tau_{xz}\gamma_{xz}) dA dx \\ &= \int_0^L \left\{ N_x \left(\frac{\partial u}{\partial x}\right) + M_x \left(\frac{\partial \varphi}{\partial x}\right) + N_x \frac{\partial^2 w}{\partial x^2} + Q_x \left(\varphi + \frac{\partial w}{\partial x}\right) \right\} dx \end{aligned} \tag{8.168}$$

The stress resultants (N_x, M_x, and Q_x) may be represented in the following manner:

$$N_x - (e_0 a)^2 \frac{\partial^2 N_x}{\partial x^2} = A_{11} \left[\frac{\partial u}{\partial x} + \frac{1}{2} \left(\frac{\partial w}{\partial x} \right)^2 \right] + B_{11} \frac{\partial \varphi}{\partial x} - N_x^T \tag{8.169}$$

$$M_x - (e_0 a)^2 \frac{\partial^2 M_x}{\partial x^2} = B_{11} \left[\frac{\partial u}{\partial x} + \frac{1}{2} \left(\frac{\partial w}{\partial x} \right)^2 \right] + D_{11} \frac{\partial \varphi}{\partial x} - M_x^T \tag{8.170}$$

$$Q_x - (e_0 a)^2 \frac{\partial^2 Q_x}{\partial x^2} = K_s A_{55} \left(\varphi + \frac{\partial w}{\partial x} \right) \tag{8.171}$$

$$\begin{aligned} &\left(N_x^T, M_x^T\right) = \int_A E(z)\alpha(z)\Delta T(z)(1,z)\, dA = \\ &\int_{\frac{h_c}{2}}^{\frac{h_c}{2}+h_f} E_1\alpha_1 \Delta T(z) b(1,z)\, dz + \int_{-\frac{h_c}{2}}^{\frac{h_c}{2}} E(z)\alpha(z)\Delta T(z) b(1,z)\, dz + \\ &\int_{-\frac{h_c}{2}-h_f}^{-\frac{h_c}{2}} E_1\alpha_1 \Delta T(z)(1,z)\, dz \end{aligned} \tag{8.172}$$

K_s represents the correction factor for shear. A_{ij}, B_{ij}, and D_{ij} represent the stiffnesses associated with extensional, bending, and bending-extensional coupling, respectively. The values are computed using the following formula:

$$\begin{aligned} A_{11} &= \int_{\frac{h_c}{2}}^{\frac{h_c}{2}+h_f} E_1 b\, dz + \int_{-\frac{h_c}{2}}^{\frac{h_c}{2}} E(z) b\, dz + \int_{-\frac{h_c}{2}-h_f}^{-\frac{h_c}{2}} E_1 b\, dz \\ B_{11} &= \int_{\frac{h_c}{2}}^{\frac{h_c}{2}+h_f} E_1 bz\, dz + \int_{-\frac{h_c}{2}}^{\frac{h_c}{2}} E(z) bz\, dz + \int_{-\frac{h_c}{2}-h_f}^{-\frac{h_c}{2}} E_1 bz\, dz \\ D_{11} &= \int_{\frac{h_c}{2}}^{\frac{h_c}{2}+h_f} E_1 bz^2\, dz + \int_{-\frac{h_c}{2}}^{\frac{h_c}{2}} E(z) bz^2\, dz + \int_{-\frac{h_c}{2}-h_f}^{-\frac{h_c}{2}} E_1 bz^2\, dz \\ A_{55} &= \frac{1}{2(1+\upsilon)} \left[\int_{\frac{h_c}{2}}^{\frac{h_c}{2}+h_f} E_1 b\, dz + \int_{-\frac{h_c}{2}}^{\frac{h_c}{2}} E(z) b\, dz + \int_{-\frac{h_c}{2}-h_f}^{-\frac{h_c}{2}} E_1 b\, dz \right] \end{aligned} \tag{8.173}$$

The kinetic energy of a FG nanobeam is determined by the following:

$$K = \frac{1}{2}\int_0^l \int_A \rho \left[\left(\frac{\partial u_x}{\partial t}\right)^2 + \left(\frac{\partial u_y}{\partial t}\right)^2 + \left(\frac{\partial u_z}{\partial t}\right)^2 \right] dAdx \tag{8.174}$$

The potential energy resulting from external forces and the elastic basis is mathematically represented by the following:

$$W = \int_0^L \left[\frac{1}{2}(N_d \cos(\Omega t))\left(\frac{\partial w}{\partial x}\right)^2 - c_d\left(\frac{\partial w}{\partial t}\right)w - \frac{1}{2}k_w w^2 - \frac{1}{2}k_p\left(\frac{\partial w}{\partial x}\right)^2 \right] dx \tag{8.175}$$

The equation of motion is derived using Hamilton's concept.

$$\int_{t_1}^{t_2} \delta(U - K - W)\, dt = 0 \tag{8.176}$$

By replacing the Equations (8.168), (8.174), and (8.175) with Equation (8.176), we get the equations that describe the motion:

$$\delta u: \ \frac{\partial N_x}{\partial x} = I_0 \frac{\partial^2 u}{\partial t^2} + I_1 \frac{\partial^2 \varphi}{\partial t^2} \tag{8.177}$$

$$\delta w: \frac{\partial}{\partial x}\left(N_x \frac{\partial w}{\partial x}\right) + \frac{\partial Q_x}{\partial x} = I_0 \frac{\partial^2 w}{\partial t^2} + k_w w - k_p \frac{\partial^2 w}{\partial x^2} + (N_d \cos(\Omega t))\frac{\partial^2 w}{\partial x^2} + c_d \frac{\partial w}{\partial t} \tag{8.178}$$

$$\delta\varphi: \ \frac{\partial M_x}{\partial x} - Q_x = I_1 \frac{\partial^2 u}{\partial t^2} + I_2 \frac{\partial^2 \varphi}{\partial t^2} \tag{8.179}$$

Where:

$$\begin{aligned}(I_0, I_1, I_2) = {} & \int_{\frac{h_c}{2}}^{\frac{h_c}{2}+h_t} \rho_1 (1, z, z^2)\, bdz + \int_{\frac{-h_c}{2}}^{\frac{h_c}{2}} \rho_{(z)} (1, z, z^2)\, bdz + \\ & \int_{\frac{-h_c}{2}-h_b}^{\frac{-h_c}{2}} \rho_1 (1, z, z^2)\, bdz \end{aligned} \tag{8.180}$$

By excluding the rotational and in-plane inertias, Equation (8.169) yields the stress resultants as follows:

$$N_x = (e_0 a)^2 \left[I_0 \frac{\partial^3 u}{\partial x \partial t^2} \right] + A_{11}\left[\frac{\partial u}{\partial x} + \frac{1}{2}\left(\frac{\partial w}{\partial x}\right)^2 \right] + B_{11}\frac{\partial \varphi}{\partial x} - N_x^T \tag{8.181}$$

$$M_x = B_{11}\left[\frac{\partial u}{\partial x}+\frac{1}{2}\left(\frac{\partial w}{\partial x}\right)^2\right]+D_{11}\frac{\partial \varphi}{\partial x}-M_x^T+$$

$$(e_0 a)^2\left[\begin{array}{c} I_0\frac{\partial^2 w}{\partial t^2}+k_w w-k_p\frac{\partial^2 w}{\partial x^2}+\left(N_d\cos(\Omega t)\right)\frac{\partial^2 w}{\partial x^2}+c_d\frac{\partial w}{\partial t}-\frac{\partial}{\partial x}\left(N_x\frac{\partial w}{\partial x}\right)+ \\ I_2\frac{\partial^3 \varphi}{\partial x\partial t^2}\end{array}\right] \tag{8.182}$$

By inserting the expression for Q_x from Equation (8.179) into Equation (8.178) and using the stress resultants, the dimensionless equations of motion are obtained, relying on the subsequent non-dimensional parameters:

$$x = XL\,,\quad w = Wr,\quad t=\tau t_m,\quad \gamma=\frac{e_0 a}{L},$$

$$\left(D_{11}-\frac{B_{11}{}^2}{A_{11}}\right)=\mu,\quad t_m=\sqrt{\frac{I_0L^4}{\mu}} \tag{8.183}$$

$$\frac{L^4}{\mu}k_w=K_w,\quad \frac{k_pL^2}{\mu}=K_p,\quad N_d\frac{L^2}{\mu}=N,\quad P=\Omega t_m,\quad c_d\frac{L^4}{\mu t_m}=C_d$$

Notably, the impact of the nonlinear damping factor is disregarded, resulting in the dimensionless equations:

$$\begin{gathered}\frac{\partial^2 W}{\partial\tau^2}-\frac{A_{11}r^2}{2\mu}\int_0^1\left[\left(\frac{\partial W}{\partial X}\right)^2\right]dX\frac{\partial^3 W}{\partial X^3}-\frac{B_{11}L}{\mu}\int_0^1\frac{\partial\varphi}{\partial X}dX\frac{\partial^3 W}{\partial X^3}+\frac{L^2}{\mu}\int_0^1 N_x^T dX\frac{\partial^3 W}{\partial X^3} \\ -\frac{L}{r}\frac{\partial^3\varphi}{\partial X^3}+K_wW-K_p\frac{\partial^2 W}{\partial X^2}+\left(N\cos\left(P\tau\right)\right)\frac{\partial^2 W}{\partial X^2}+C_d\frac{\partial W}{\partial\tau} \\ -\gamma^2\left[\begin{array}{c}-\frac{A_{11}r^2}{2\mu}\int_0^1\left[\left(\frac{\partial W}{\partial X}\right)^2\right]dX\frac{\partial^4 W}{\partial X^4}-\frac{B_{11}L}{\mu}\int_0^1\frac{\partial\varphi}{\partial X}dX\frac{\partial^4 W}{\partial X^4}+\frac{L^2}{\mu}\int_0^1 N_x^T dX\frac{\partial^4 W}{\partial X^4}+ \\ \frac{\partial^4 W}{\partial X^2\partial\tau^2}+K_w\frac{\partial^2 W}{\partial X^2}-K_p\frac{\partial^4 W}{\partial X^4}+\left(N\cos(P\tau)\right)\frac{\partial^4 W}{\partial X^4}+C_d\frac{\partial^3 w}{\partial x^2\partial\tau}\end{array}\right]=0\end{gathered} \tag{8.184}$$

$$k_sA_{55}r\frac{\partial W}{\partial X}+Lk_sA_{55}\varphi-\frac{\mu}{L}\frac{\partial^2\varphi}{\partial X^2}=0 \tag{8.185}$$

In order to solve the equations of motion, the Galerkin method is used.

$$W(X,\tau)=W(\tau)\phi(X) \tag{8.186}$$

$$\varphi(X,\tau)=S(\tau)\psi(X) \tag{8.187}$$

By replacing Equations (8.186) and (8.187) with Equations (8.184) and (8.185), we get the following:

$$M\ddot{W}+K_0\left(N\cos(P\tau)\right)W+C_1W^3+C_2SW+K_1W+K_4S+K_2W+K_3W+C_3\dot{W}=0 \tag{8.188}$$

$$C_4W+C_5S=0 \tag{8.189}$$

Where:

$$M=\left[\int_0^1\phi^2dX\right]-\gamma^2\left[\int_0^1\phi''\phi\,dX\right],\quad K_0=\left[\int_0^1\phi''\phi\,dX\right]-\gamma^2\left[\int_0^1\phi^{(4)}\phi\,dX\right]$$

$$K_1=\frac{L^2}{\mu}N_x^T\left(\left[\int_0^1\phi''\phi\,dX\right]-\gamma^2\left[\int_0^1\phi^{(4)}\phi\,dX\right]\right),\ K_2$$

$$K_w\left(\left[\int_0^1\phi^2dX\right]-\gamma^2\left[\int_0^1\phi''\phi\,dX\right]\right)$$

$$K_3=K_p\left(-\left[\int_0^1(\phi'')\phi\,dX\right]-\gamma^2\left[\int_0^1\phi^{(4)}\phi\,dX\right]\right),\qquad K_4=-\left[\frac{L}{r}\int_0^1(\psi'')\phi\,dX\right]$$

$$C_1=\frac{A_{11}r^2}{2\mu}\left(-\left[\int_0^1(\phi)^2dX\int_0^1(\phi'')\phi\,dX\right]\right.$$

$$\left.+\gamma^2\left[\int_0^1(\phi'')^2dX\int_0^1(\phi^{(4)})\phi\,dX\right]\right)$$

$$C_2=\frac{B_{11}L}{\mu}\left(-\left[\int_0^1(\psi')\,dX\int_0^1(\phi'')\phi\,dX\right]+\gamma^2\left[\int_0^1(\psi')\,dX\int_0^1(\phi^{(4)})\phi\,dX\right]\right)$$

$$C_3=\left[\int_0^1C_d\phi^2\,dX\right]-\gamma^2\left[C_d\int_0^1\phi''\phi\,dX\right]$$

$$C_4=\left[\int_0^1k_sA_{55}r\phi\phi'\,dX\right],\qquad C_5=\left[\int_0^1\left(Lk_sA_{55}\psi-\frac{\mu}{L}\psi''\right)\phi dX\right] \tag{8.190}$$

By substituting Equation (8.189) into Equation (8.188), we get the nonlinear equation of motion.

$$M\ddot{W}+K_0\left(N\cos\left(P\tau\right)\right)W+KW+C_3\dot{W}+C_6W^2+C_1W^3=0 \tag{8.191}$$

Where:

$$K=K_1+K_2+K_3-\frac{C_4}{C_5}K_4 \quad , \quad C_6=-C_2\frac{C_4}{C_5} \tag{8.192}$$

The eigen functions corresponding to hinged-hinged boundary conditions are formally defined as follows:

$$\phi\left(X\right)=sin\left(\pi X\right)$$

and

$$\psi\left(X\right)=cos\left(\pi X\right) \tag{8.193}$$

The dimensionless natural frequency of the Timoshenko nanobeam is determined in relation to its free vibration.

$$\hat{\omega}^2=\frac{K}{M} \tag{8.194}$$

By using the mathematical technique of perturbation theory with the use of a perturbation parameter ε, the governing equation produces the following result:

$$\ddot{W}+2\varepsilon\mu\dot{W}+\omega^2W+\varepsilon c_1\,cos\left(P\tau\right)W+\varepsilon c_2W^2+\varepsilon^2c_3W^3=0 \tag{8.195}$$

Where:

$$\hat{\mu}=\frac{C_3}{2\varepsilon M} \quad , \quad c_1=\frac{K_0N}{\varepsilon M} \quad , \quad c_2=\frac{C_6}{\varepsilon M} \quad , \quad c_3=\frac{C_1}{\varepsilon^2M} \tag{8.196}$$

8.5.4 Solution Methods

8.5.4.1 Multiple Scale Method

The technique of several scales is a very effective mathematical tool for solving various nonlinear equations. It involves adding trial variables as distinct timescales, which frequently have physical significance on their own. The multiple scale approach, developed by Mook and Nayfeh [8], allows us to get an approximate solution for Equation (8.195) in the following form:

$$W\left(\tau\right)=W_0\left(T_0,T_1,T_2\right)+\varepsilon W_1\left(T_0,T_1,T_2\right)+\varepsilon^2W_2\left(T_0,T_1,T_2\right) \tag{8.197}$$

Where:

$$T_0=\tau \quad , \quad T_1=\varepsilon\tau \quad , \quad T_2=\varepsilon^2\tau \tag{8.198}$$

By replacing the equation represented by Equation (8.197) into Equation (8.195), and subsequently for the identical power of ε, the differential equations are as follows:

$$\varepsilon^0: \left(\hat{\omega}^2 + D_0^{\ 2}\right)W_0 = 0 \tag{8.199}$$

$$\varepsilon^1: \left(\hat{\omega}^2 + D_0^{\ 2}\right)W_1 = -2D_0D_1W_0 - c_2W_0^{\ 2} - c_1\cos\left(P\tau\right)W_0 - 2\mu D_0W_0 \tag{8.200}$$

$$\begin{aligned}\varepsilon^2: \left(\hat{\omega}^2 + D_0^{\ 2}\right)W_2 = &-2D_0D_1W_1 - 2D_0D_2W_0 - D_1^{\ 2}W_0 - 2c_2W_0W_1 - c_3W_0^{\ 3} - \\ &c_1\cos\left(P\tau\right)W_1 - 2\hat{\mu}D_0W_1 - 2\hat{\mu}D_1W_0\end{aligned} \tag{8.201}$$

The complex version of Equation (8.202) is used to solve Equation (8.199).

$$W_0 = A\left(T_1,T_2\right)exp\left[i\hat{\omega}T_0\right] + \bar{A}\left(T_1,T_2\right)exp\left[-i\hat{\omega}T_0\right] \tag{8.202}$$

Where A and $\bar{A}$ are complex conjugates. Equation (8.200) is expressed in the following format:

$$\begin{aligned}\left(\hat{\omega}^2 + D_0^{\ 2}\right)W_1 = &-2i\hat{\omega}\left(D_1A + \mu A\right)exp\left[i\hat{\omega}T_0\right] - c_2\left(A^2exp\left[2i\hat{\omega}T_0\right] + A\bar{A}\right) \\ &-\frac{1}{2}c_1\left[A\,exp\left(i\left(\hat{\omega}+P\right)T_0\right) + \bar{A}\,exp\left(i\left(P-\hat{\omega}\right)T_0\right)\right] + cc\end{aligned} \tag{8.203}$$

Where cc represents the complex conjugate of the previously given variables. By establishing the detuning parameter (σ) and setting $P = 2\hat{\omega} + \varepsilon\sigma$, we eliminate the secular and small divisor components of Equation (8.203) in the following manner:

$$-2i\hat{\omega}\left(D_1A + \mu A\right) - \frac{1}{2}c_1\bar{A}\ e^{i\sigma T_1} = 0 \tag{8.204}$$

The answer to Equation (8.201) is as follows:

$$W_1 = \frac{c_1A}{2p\left(p + 2\hat{\omega}\right)}exp\left[i\left(\hat{\omega}+P\right)T_0\right] + \frac{c_2}{3\hat{\omega}^2}\left[A^2exp\left[2i\hat{\omega}T_0\right] - 3A\bar{A}\right] + cc \tag{8.205}$$

By substituting equations (8.202) and (8.205) into Equation (8.201), the secular terms of the differential equation of motion may be eliminated. This yields the following:

$$D_1^2A + 2\hat{\mu}D_1A + 2i\hat{\omega}D_2A + \frac{c_1^2A}{2\left(2p^2 + 4p\hat{\omega}\right)} - \frac{10c_2^2A^2\bar{A}}{3\hat{\omega}^2} + 3c_3A^2\bar{A} = 0 \tag{8.206}$$

Consequently, we combine Equations (8.204) and (8.206) in order to eliminate the terms D_1^2A and D_1A.

$$2i\hat{\omega}D_2A + \left[\frac{3c_1^2}{32\hat{\omega}^2} - \hat{\mu}^2\right]A + \left[3c_3 - \frac{10c_2^2}{3\hat{\omega}^2}\right]A^2\bar{A} - \left[\frac{\sigma c_1}{4\hat{\omega}}\right]\bar{A}\,exp\left[i\sigma T_1\right] = 0 \tag{8.207}$$

Using the multiple-scale technique, Equations (8.204) and (8.207) represent the first two terms.

$$2i\hat{\omega}D_2A+\left[\frac{3c_1^2}{32\hat{\omega}^2}-\hat{\mu}^2\right]A+\left[3c_3-\frac{10c_2^2}{3\hat{\omega}^2}\right]A^2\bar{A}-\left[\frac{\sigma c_1}{4\hat{\omega}}\right]\bar{A}\,exp\left[i\sigma T_1\right]=0 \quad (8.208)$$

In order to solve Equation (8.208), we must evaluate the following solutions for A and $\bar{A}$:

$$A=\frac{1}{2}a\;exp\left[i\beta\right], \qquad \bar{A}=\frac{1}{2}a\;exp\left[\text{-}i\beta\right] \quad (8.209)$$

Here, "α" and "β" represent the amplitude and phase angle of "A" and "$\bar{A}$". By substituting equation (8.209) into equation (8.208) and then separating the real and imaginary components, we get the governing equations for variables α and β.

$$a\beta' = H_1a+H_2a^3+H_3a\,cos\left[\gamma\right] \quad (8.210)$$

$$a' = H_4a-H_3a\,sin\left[\gamma\right] \quad (8.211)$$

Where:

$$H_1=\frac{3\varepsilon^2c_1^2}{64\hat{\omega}^3}-\frac{\varepsilon^2\hat{\mu}^2}{2\hat{\omega}}, \quad H_2=\frac{3\varepsilon^2c_3}{8\hat{\omega}}-\frac{5\varepsilon^2c_2^2}{12\hat{\omega}^3}, \quad H_3=\left(\frac{\varepsilon c_1}{4\hat{\omega}}\right)\left(1-\frac{\varepsilon\sigma}{2\hat{\omega}}\right), \quad (8.212)$$

$$H_4=-\varepsilon\hat{\mu}, \qquad \gamma=\varepsilon\sigma t-2\beta$$

Finally, by assuming that $a'=0$ and $\gamma'=0$ $\left(\gamma'=\varepsilon\sigma-2\beta'\right)$, the steady-state response of the nanobeam may be expressed as follows:

$$\left(H_1-\frac{\varepsilon\sigma}{2}\right)a+H_2a^3+H_3a\cos\left[\gamma\right]=0 \quad (8.213)$$

$$H_4a-H_3a\,sin\left[\gamma\right]=0 \quad (8.214)$$

The investigation of the instability and bifurcation zones of the Timoshenko nanobeam heavily relies on both the simple and complex solutions of the beam. In order to get a solution that is not trivial (where an is not equal to zero), the trigonometric function $cos^2\left[\gamma\right]+sin^2\left[\gamma\right]=1$ is used in Equations (8.213) and (8.214).

$$\left(\left(H_1-\frac{\varepsilon\sigma}{2}\right)a+H_2a^3\right)^2+\left(H_4a\right)^2=\left(H_3a\right)^2 \quad (8.215)$$

The non-trivial solution has a response amplitude (a) given by the following:

$$a=\left[\frac{1}{H_2}\left(\frac{\varepsilon\sigma}{2}-H_1\pm\sqrt{H_3^{\,2}-H_4^{\,2}}\right)\right]^{\frac{1}{2}} \quad (8.216)$$

The simple solution of the system is found by using the Cartesian form of the solution of Equation (8.208), which is given by $A=\frac{1}{2}(p-iq)exp[i\beta T_1]$.

$$\begin{gathered}-32\varepsilon^2\hat{\mu}^2\hat{\omega}^2p-64\beta\varepsilon\hat{\omega}^3p+64i\varepsilon\hat{\mu}\hat{\omega}^3p-8\varepsilon^2\sigma\hat{\omega}c_1p+16\varepsilon\hat{\omega}^2c_1p+3\varepsilon^2c_1^2p \\ +32i\varepsilon^2\hat{\mu}^2\hat{\omega}^2q+ \\ 64i\beta\varepsilon\hat{\omega}^3q+64\varepsilon\hat{\mu}\hat{\omega}^3q-8i\varepsilon^2\sigma\hat{\omega}c_1q+16i\varepsilon\hat{\omega}^2c_1q-3i\varepsilon^2c_1^2q+ \\ 64i\hat{\omega}^3p'+64\hat{\omega}^3q'=0\end{gathered} \tag{8.217}$$

In order to get a self-governing version of Equation (8.217), it is necessary to substitute β with $\sigma/2$. The imaginary and real components of Equation (8.217) are once again separated, leading to the following:

$$p'=-\frac{1}{2}\varepsilon\sigma q-\frac{\varepsilon^2\hat{\mu}^2q}{2\hat{\omega}}+\frac{\varepsilon^2\sigma c_1q}{8\hat{\omega}^2}-\frac{\varepsilon c_1q}{4\hat{\omega}}+\frac{3\varepsilon^2c_1^2q}{64\hat{\omega}^3}-\varepsilon\hat{\mu}p \tag{8.218}$$

$$q'=\frac{1}{2}\varepsilon\sigma p+\frac{\varepsilon^2\hat{\mu}^2p}{2\hat{\omega}}+\frac{\varepsilon^2\sigma c_1p}{8\hat{\omega}^2}-\frac{\varepsilon c_1p}{4\hat{\omega}}-\frac{3\varepsilon^2c_1^2p}{64\hat{\omega}^3}-\varepsilon\hat{\mu}q \tag{8.219}$$

The Jacobian matrix of equations (8.218) and (8.219) may be obtained.

$$J=\begin{bmatrix}-\varepsilon\hat{\mu} & -\frac{1}{2}\varepsilon\sigma-\frac{\varepsilon^2\hat{\mu}^2}{2\hat{\omega}}+\frac{\varepsilon^2\sigma c_1}{8\hat{\omega}^2}-\frac{\varepsilon c_1}{4\hat{\omega}}+\frac{3\varepsilon^2c_1^2}{64\hat{\omega}^3} \\ \frac{1}{2}\varepsilon\sigma+\frac{\varepsilon^2\hat{\mu}^2}{2\hat{\omega}}+\frac{\varepsilon^2\sigma c_1}{8\hat{\omega}^2}-\frac{\varepsilon c_1}{4\hat{\omega}}-\frac{3\varepsilon^2c_1^2}{64\hat{\omega}^3} & -\varepsilon\hat{\mu}\end{bmatrix} \tag{8.220}$$

By computing the determinant and trace of matrix J, we may determine the stability zones of the system by assuming positive values for $\hat{\alpha}$.

$$\Delta=4\hat{\mu}^2\hat{\omega}^2+\left(\sigma\hat{\omega}+\varepsilon\hat{\mu}^2-\frac{3\varepsilon c_1^2}{32\hat{\omega}^2}\right)^2-\frac{c_1^2}{4}\left(1-\frac{\varepsilon\sigma}{2\hat{\omega}}\right)^2 \tag{8.221}$$

$$\tau=-2\varepsilon\hat{\mu} \tag{8.222}$$

8.5.5 Results and Discussion

This section will provide numerical data illustrating the instability and nonlinear vibration of a Timoshenko FG nanobeam subjected to a harmonic axial load. The findings will be presented in the form of figures and tables.

8.5.5.1 Numerical Results

The factors studied for analyzing the stability and bifurcation zones of FG sandwich porous nanobeams include porosity, nonlocal coefficient, viscous-Winkler parameters of foundation, temperature, slenderness ratio, thickness ratio, and amplitude of parametric excitation.

$$E=200\,GPa,\quad \upsilon=0.3,\quad \rho=7850\frac{Kg}{m^3},\quad \alpha=1.2\times10^{-5}\,^{\circ}c^{-1}$$

$$b=2\text{ nm},\quad h=4\text{ nm},\quad L=10h,\quad h_c=0.7h,\quad k_w=5\times10^{8}\,Pa,\quad \gamma^2=1,$$

$$N=8\text{ nN},\quad e_0=0.5,\quad k_s=\frac{5}{6},\quad c_d=5\times10^{-5}\ Pa.s,$$

$$T_b=100\ ^{\circ}c,\quad T_t=200\ ^{\circ}c\quad \varepsilon=0.05$$

Nevertheless, these factors might be modified in order to analyze their impact. Figure 8.23 demonstrates the impact of porosity distribution on the curve of the amplitude-detuning parameter. Based on Figure 8.23, it can be deduced that modifying the structure of the porous core will not affect the behavior of the nanobeams. Given this circumstance, an additional study will be undertaken just focusing on the initial porosity distribution. This work used a continuous line to represent stable solutions, whereas dashed lines are utilized to represent unstable solutions. When observed, two stationary points are formed for the negative and positive values of the detuning parameter. Initially, when the value of σ is increased, the stable branch of the trivial solution undergoes a transformation, resulting in the presence of both stable and unstable branches. This indicates that the point is undergoing a supercritical pitchfork bifurcation. Regarding the second issue, specifically with subcritical pitchfork bifurcation, the reduction of the detuning value leads to the creation of two unstable solutions. Subcritical pitchfork bifurcation involves the first appearance of unstable equilibrium points when a parameter is modified, while supercritical pitchfork bifurcation entails the emergence of stable equilibrium points initially. These bifurcations represent significant changes in the behavior of dynamical systems when parameters are modified.

Figures 8.24 and 8.25 illustrate the impact of the porosity factor e_0 on the bifurcation and instability states of the FG sandwich porous nano Timoshenko beam. Due to the inverse link between response amplitude (α) and stiffness, it can be inferred that an increase in the porosity coefficient leads to a rise in the non-dimensional amplitude (α). Furthermore, the region of instability will become more pronounced as the value of e_0 increases. As seen in Figure 8.26, an augmentation in the porosity coefficient will cause the supercritical and subcritical pitchfork bifurcation points to migrate towards the right.

Figures 8.26 and 8.27 demonstrate the impact of the nonlocal component on the nonlinear stability of a nanobeam. This is done by examining the effects of fixed values for the detuning parameter (A) and forced amplitude (c_1), as shown in the corresponding figures. Both images demonstrate that increasing the value of the nonlocal parameter results in a decrease in the non-dimensional amplitude (α).

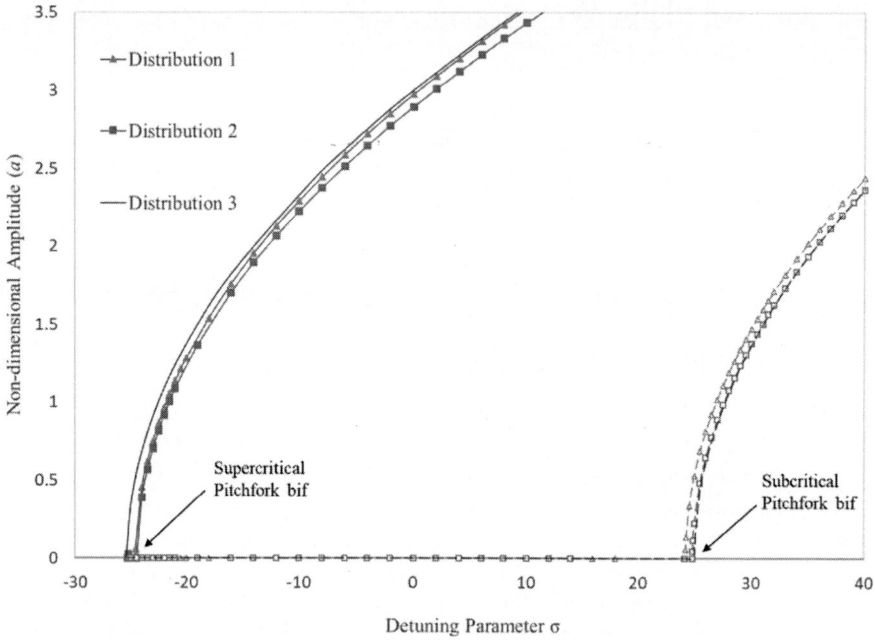

FIGURE 8.23 Non-dimensional amplitude versus detuning parameter curves different porosity distributions.

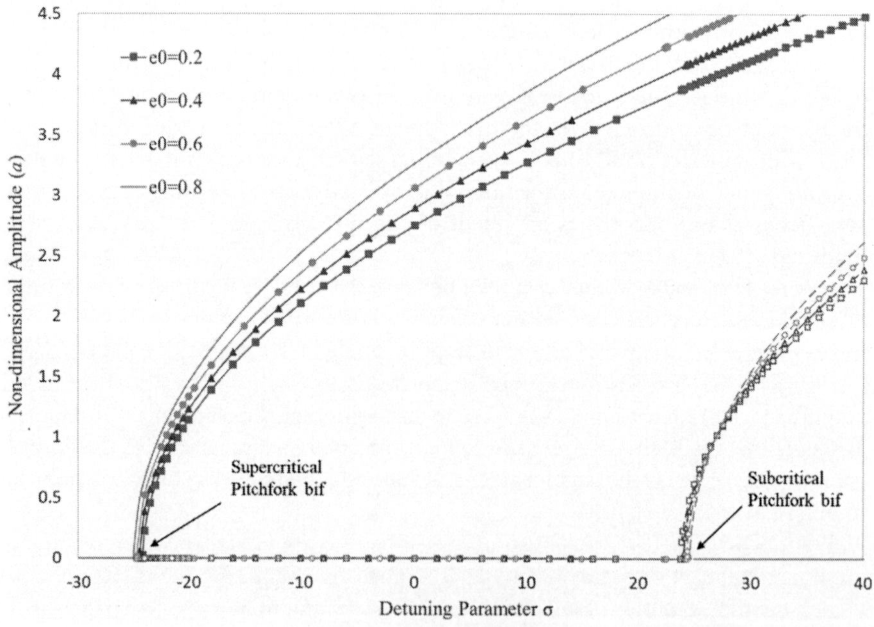

FIGURE 8.24 Parametric curves illustrating the relationship between non-dimensional amplitude and detuning parameter, specifically examining the impact of the porosity coefficient.

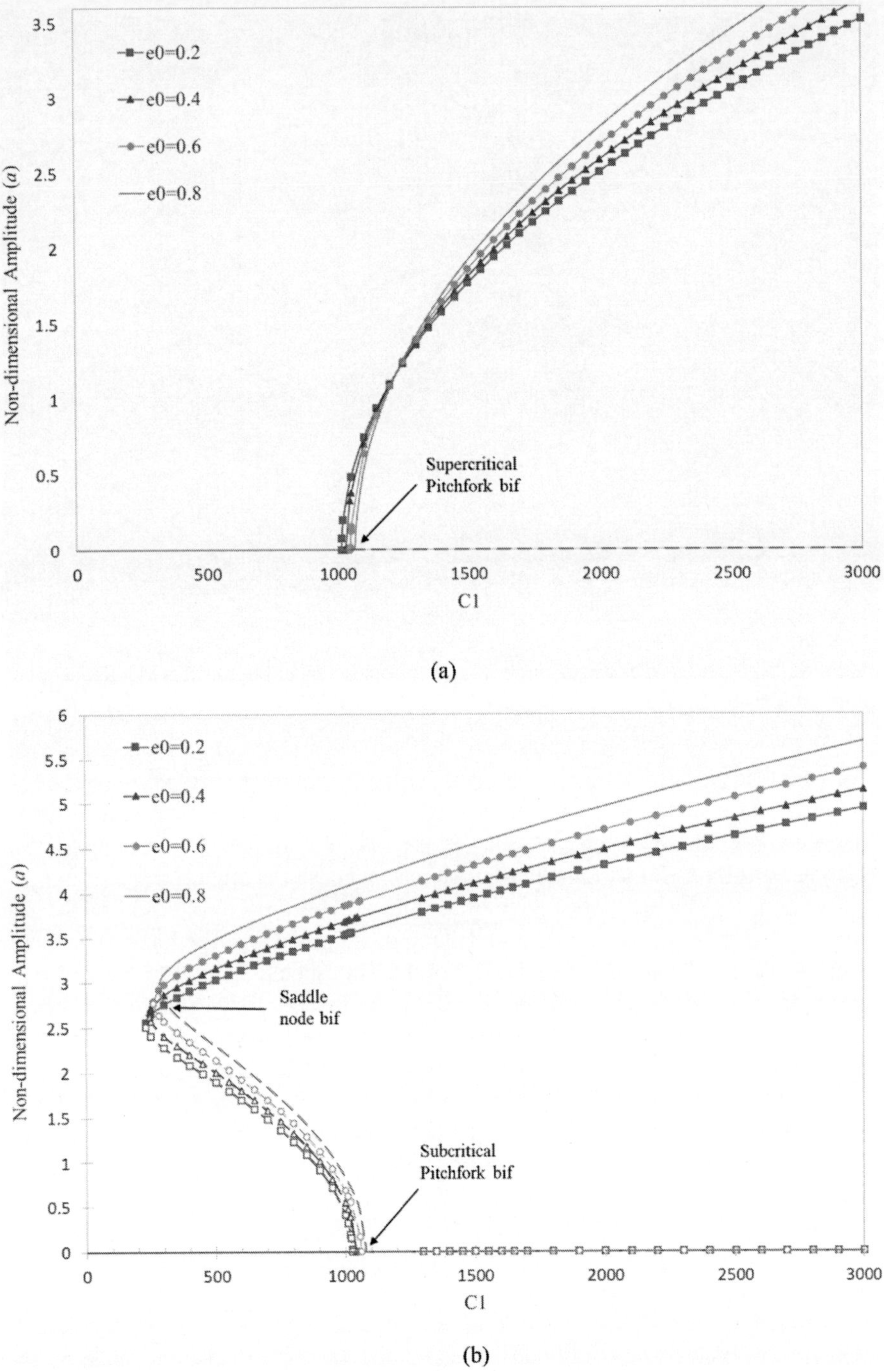

FIGURE 8.25 Non-dimensional amplitude versus force amplitude of parametric excitation curves: effects of porosity coefficient at (a) $\sigma = -20$ and (b) $\sigma = 20$.

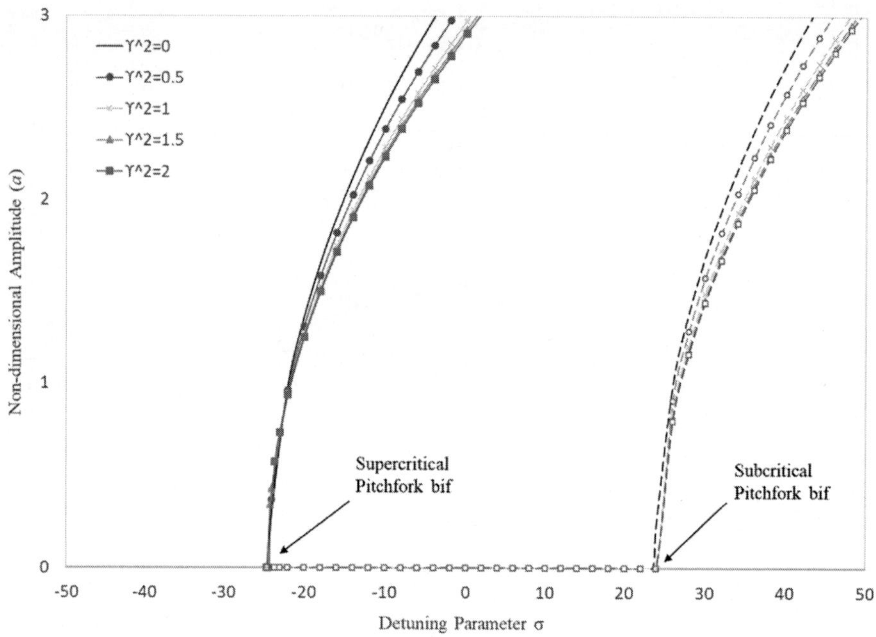

FIGURE 8.26 Non-dimensional amplitude versus detuning parametric curves: effects of nonlocal parameter.

Another evident observation is that modifying this parameter has no impact on the stability and bifurcation points of the system. Based on the information provided in Figure 8.26 and Figure 8.27, it can be seen that taking into account the nonlocal parameter will result in a reduction in the non-dimensional amplitude.

While prior characteristics have little influence on the stability and bifurcation states of a nanobeam, the Winkler and damping parameters of the foundation have a crucial role in determining the behavior of the Timoshenko nanobeam.

Figure 8.28 displays the relationship between the non-dimensional amplitude (a) and the forced amplitude (c_1) for different Winkler coefficients (k_w). As seen in Figure 8.28(a), the supercritical bifurcation point will exhibit a greater level of excitation as the values of the Winkler coefficient grow. Similarly, Figure 8.28(b) exhibits the similar behavior, and as the forced amplitude increases, the bifurcation points move towards the right. Furthermore, when the value of k_w rises, the size of the non-trivial unstable branch expands, leading to a corresponding increase in the location of another bifurcation point associated with a saddle node.

Various values of the damping coefficient (C_d) are examined to assess its impact on the instability of a sandwich nanobeam. These findings are then visualized in Figure 8.29. It is possible that a decrease in the damping coefficient might result in a higher non-dimensional amplitude of excitation in both graphs. Moreover, by raising the coefficient constant, the bigger value of forced amplitude leads to the presence of all three kinds of bifurcation points: supercritical, subcritical, and saddle.

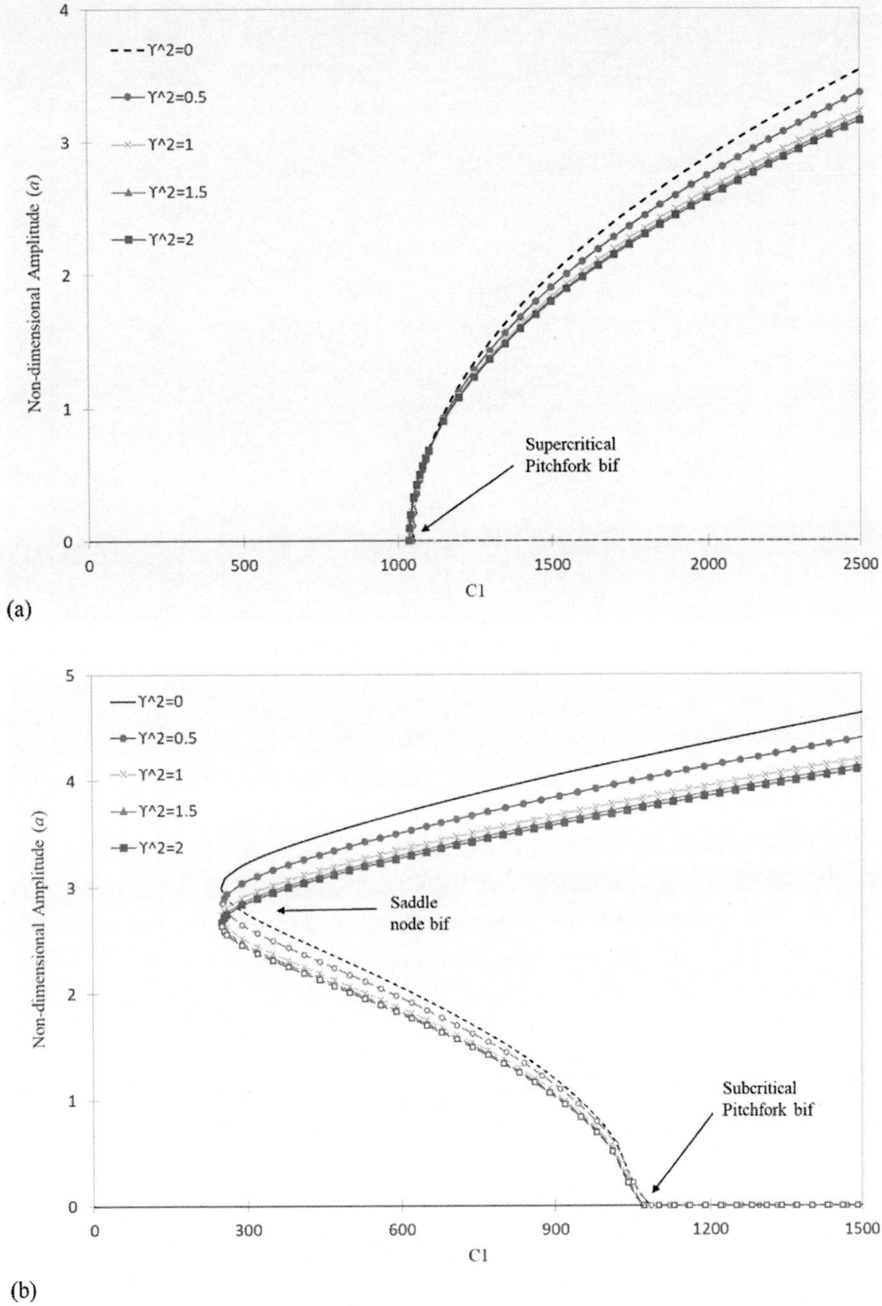

FIGURE 8.27 Non-dimensional amplitude versus force amplitude of parametric excitation curves: effects of nonlocal parameter at (a) $\sigma = -20$ and (b) $\sigma = 20$.

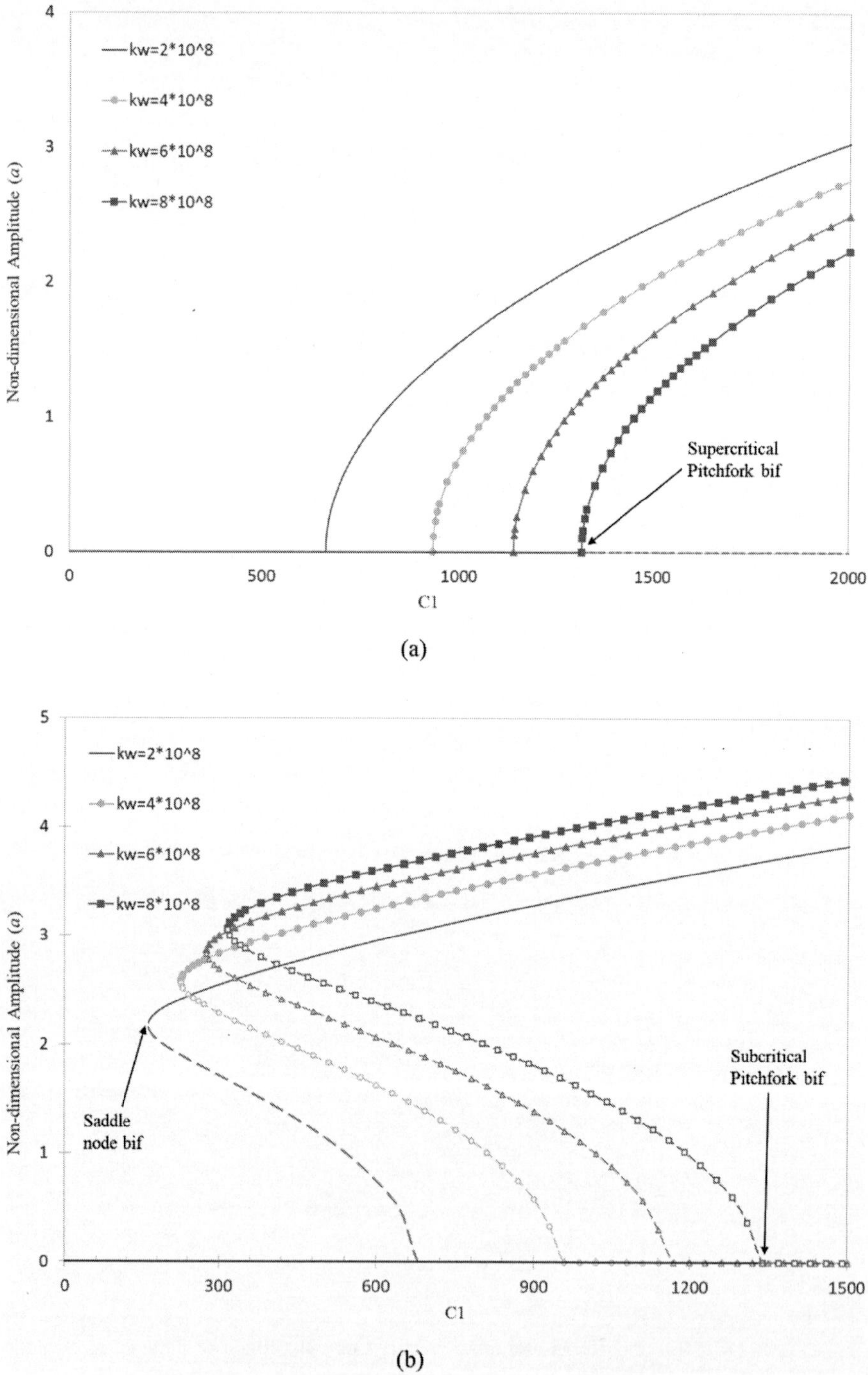

FIGURE 8.28 Non-dimensional amplitude versus force amplitude of parametric excitation curves: effects of Winkler stiffness at (a) $\sigma = -20$ and (b) $\sigma = 20$.

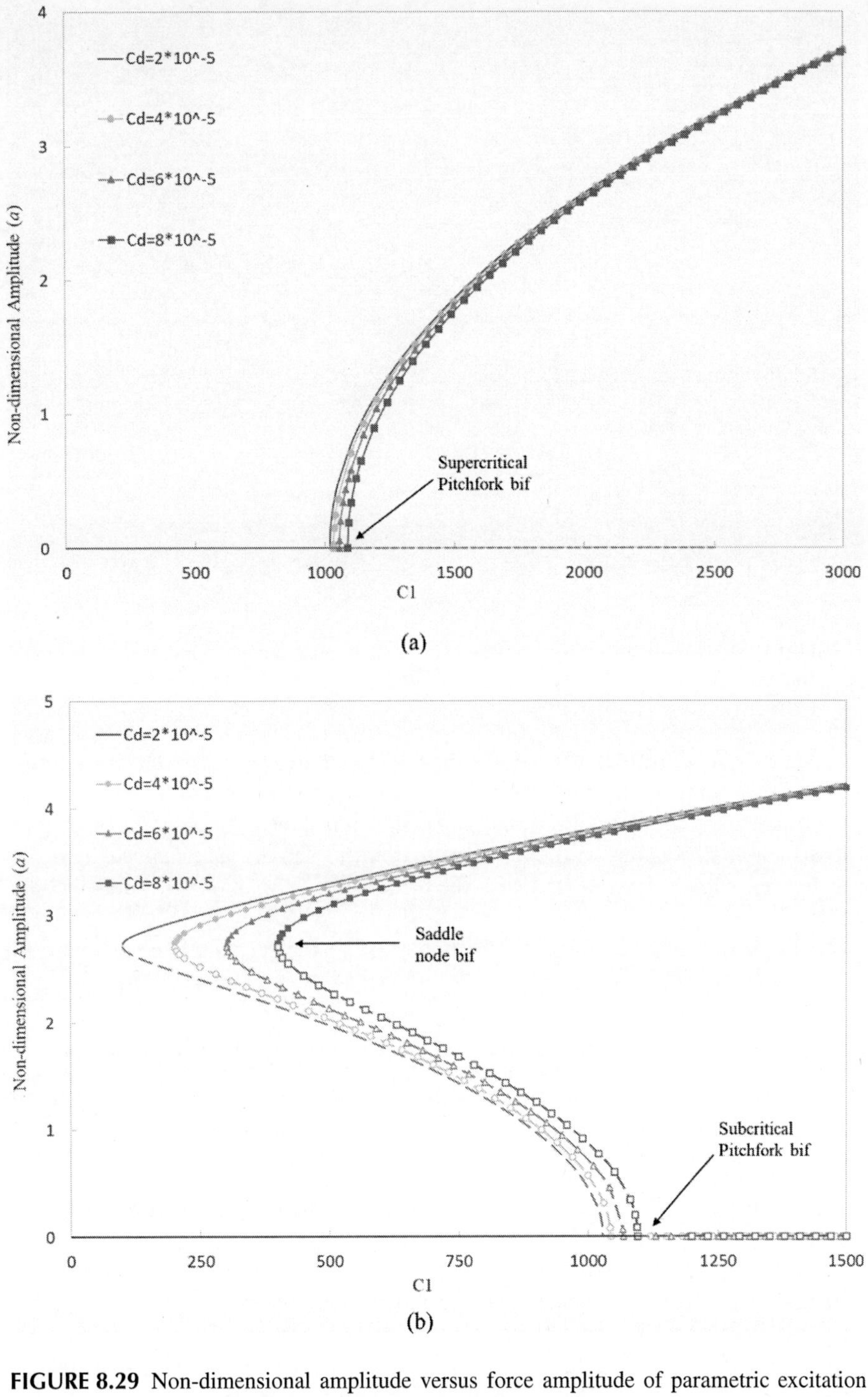

FIGURE 8.29 Non-dimensional amplitude versus force amplitude of parametric excitation curves: effects of viscous damping at (a) $\sigma = -20$ and (b) $\sigma = 20$.

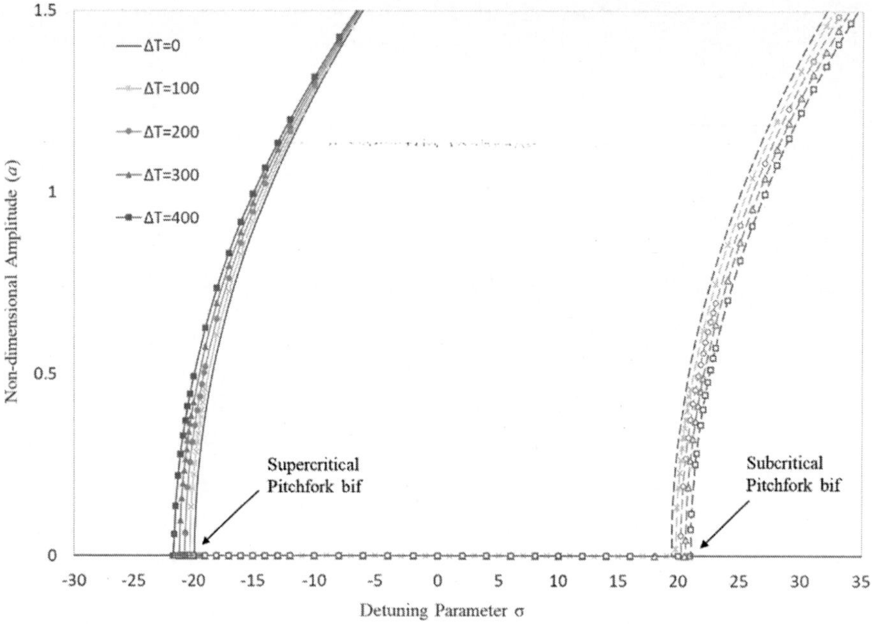

FIGURE 8.30 Non-dimensional amplitude versus detuning parametric curves: effects of temperature variation.

Figure 8.30 illustrates the influence of temperature variations on the frequency-response curve. Clearly, a higher temperature results in a greater non-dimensional amplitude of the solution. Furthermore, it is shown that the position of the bifurcation points varies depending on the temperature values. The distance between two sites of division, which is associated with the unstable basic solution, grows as the temperature rises.

Figures 8.31 and 8.32 illustrate the impact of geometric parameters, namely, the thickness ratio $(\frac{h_c}{h_f})$ and slenderness ratio $(\frac{L}{h})$, on both the trivial and non-trivial solutions. The graphs depict the relationship between the force amplitude of parametric excitation and the amplitude of the nanobeam. Both graphs exhibit two bifurcation points, namely, the saddle node and subcritical pitchfork. By examining the impact of thickness and slenderness ratio on the stiffness of a nanobeam, it can be inferred that an increase in the $(\frac{h_c}{h_f})$ and $(\frac{L}{h})$ ratio would result in a greater non-dimensional amplitude. Furthermore, when the detuning parameter is set to $\sigma = +25$, increasing the slenderness and thickness ratio results in the expansion of the non-trivial unstable branch of the solution.

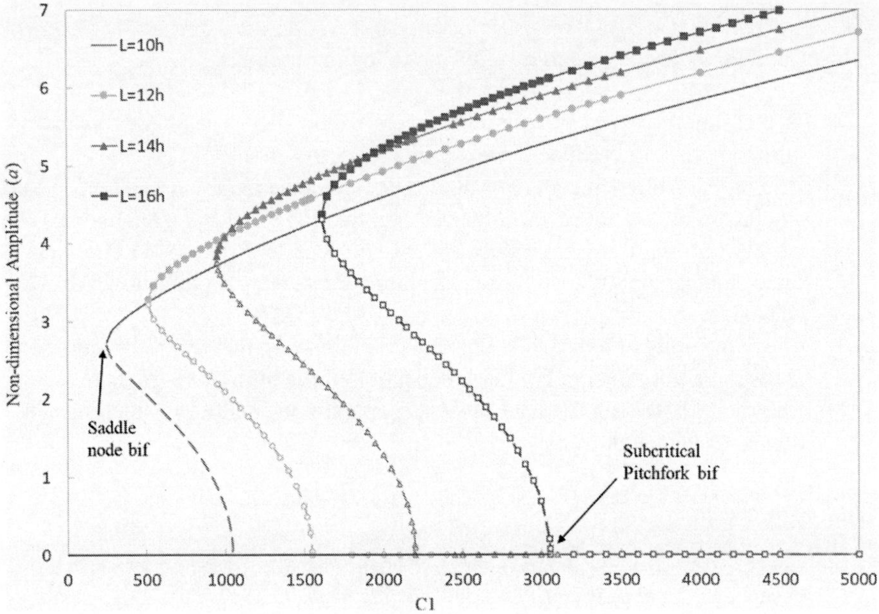

FIGURE 8.31 Non-dimensional amplitude versus force amplitude of parametric excitation curves: effects of slenderness ratios at $\sigma = 20$.

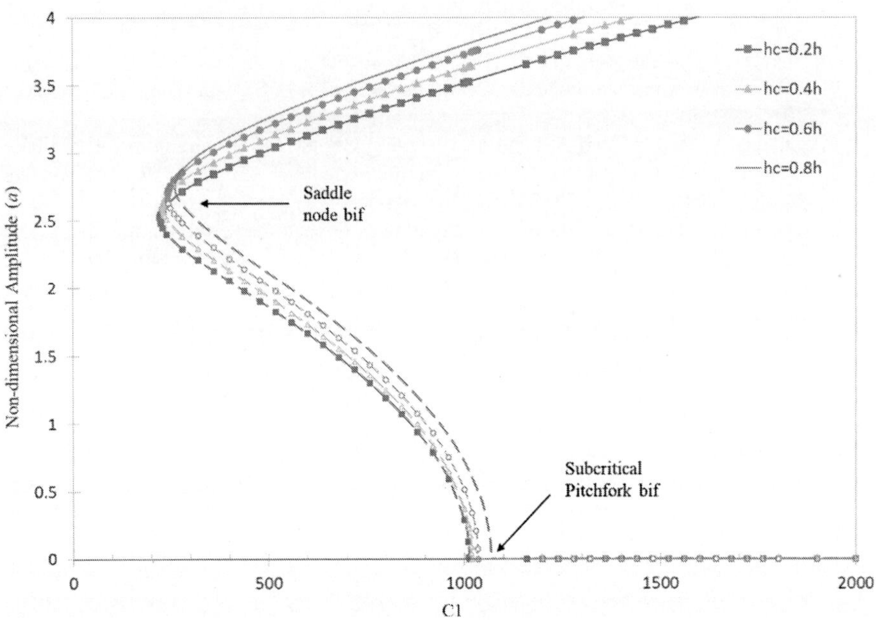

FIGURE 8.32 Non-dimensional amplitude versus force amplitude of parametric excitation curves: effects of thickness ratios at $\sigma = 20$.

8.5.6 Conclusion

The examination of this research yields the following findings.

- Decreasing the values of the porosity coefficient, slenderness, and thickness ratio may result in reduced levels of nonlinear reactions.
- Among the factors that were evaluated, the nonlocal parameter had the least significant effect on the instability and bifurcation state of the nanobeams.
- An elevation in the Winkler coefficient will lead to a decrease in the non-dimensional amplitude and a displacement of bifurcation points towards the right side.
- The dampening of the foundation has little impact on the dynamic response. It has the ability to simply alter the location of the bifurcation points.
- The rising temperatures lead to greater amplitudes and an unstable trivial branch of the nanobeams.

REFERENCES

[1] Rao, S.S., *Vibration of continuous systems*. 2019: John Wiley & Sons.
[2] Amabili, M., *Nonlinear vibrations and stability of shells and plates*. 2008: Cambridge University Press.
[3] Shafiei, N., M. Kazemi, and M. Ghadiri, Nonlinear vibration of axially functionally graded tapered microbeams. *International Journal of Engineering Science*, 2016. **102**: p. 12–26.
[4] Shafiei, N., M. Kazemi, M. Safi, and M. Ghadiri, Nonlinear vibration of axially functionally graded non-uniform nanobeams. *International Journal of Engineering Science*, 2016. **106**: p. 77–94.
[5] Shafiei, N., M. Kazemi, and M. Ghadiri, On size-dependent vibration of rotary axially functionally graded microbeam. *International Journal of Engineering Science*, 2016. **101**: p. 29–44.
[6] Ghadiri, M. and N. Shafiei, Nonlinear bending vibration of a rotating nanobeam based on nonlocal Eringen's theory using differential quadrature method. *Microsystem Technologies*, 2016. **22**: p. 2853–2867.
[7] Shayestenia, F. and M. Ghadiri, Investigation of flexoelectric effect on nonlinear vibration and dynamic instability of piezoelectric sandwich micro/nanobeam using the nonlocal strain gradient theory. *International Journal of Structural Stability and Dynamics*, 2023. **23**(04): p. 2350045.
[8] Nayfeh, A.H. and D.T. Mook, *Nonlinear oscillations*. 2008: John Wiley & Sons.
[9] Ghadiri, M. and S.H.S. Hosseini, Parametric excitation of Euler–Bernoulli nanobeams under thermo-magneto-mechanical loads: Nonlinear vibration and dynamic instability. *Composites Part B: Engineering*, 2019. **173**: p. 106928.
[10] Chen, Y. and Z. Yan, Investigation of pull-in behaviors of a nanoswitch tuned by piezoelectric and flexoelectric effects. *International Journal of Mechanical Sciences*, 2019. **161**: p. 105032.
[11] Zanjanchi Nikoo, M., M. Lezgi, S. Sabouri Ghomi, and M. Ghadiri, Dynamic instability analysis of timoshenko FG sandwich nanobeams subjected to parametric excitation. *Proceedings of the Institution of Mechanical Engineers, Part C: Journal of Mechanical Engineering Science*, 2024. **238**(7): p. 2891–2905.
[12] Lezgi, M., M. Zanjanchi Nikoo, and M. Ghadiri, Nonlinear vibration of Timoshenko FG porous sandwich beams subjected to a harmonic axial load. *Earthquake Engineering and Engineering Vibration*, 2024. **23**(3): p. 649–662.

9 Dynamics of Nonlinear Smart Continuous Structures—Plates

9.1 INTRODUCTION

The oscillation of plates is crucial in the analysis of practical systems, including bridge decks, hydraulic structures, pressure tank covers, highway and airport pavements, ship decks, aircraft, missiles, and machine components. The theory of elastic plates approximates three-dimensional elasticity theory in two dimensions, allowing the deformation of each point in the plate to be described just by the deformation of the midplane. A plate is a solid object confined between two surfaces. The separation between the two surfaces determines the plate's thickness, which is considered negligible relative to the lateral dimensions, including the length and width for a rectangular plate and the diameter for a circular plate. A plate is deemed thin when the ratio of its thickness to the lesser lateral dimension (e.g., width for a rectangular plate and diameter for a circular plate) is below 1/20. In this chapter, the governing equations of forced nonlinear vibrations of rectangular plates are developed utilizing thin plate theory. The oscillation of plates on elastic foundations and those subjected to in-plane loads is also described.

9.2 NONLINEAR EQUATION OF TRANSVERSE VIBRATION OF PLATES

9.2.1 Introduction

This section examines the study of plate transverse vibrations under both free and forced situations. The equations governing the plate's movement are formulated based on the concepts of Kirchhoff theory. Kirchhoff's theory is relevant for the analysis of thin plates, as it neglects the effects of rotational inertia and shear deformation. Mindlin's theory accounts for the influences of rotational inertia and shear deformation. Mindlin's theory is pertinent for analyzing structures with substantial cross-sectional dimensions. The equations that describe the transverse vibration of plates can be formulated as fourth-order partial differential equations. These equations are governed by eight boundary conditions at each extremity [1].

To elucidate the principles governing plate movement, it is essential to develop an appropriate coordinate system. This system comprises the x coordinate, denoting the plate's length; the y coordinate, indicating the plate's breadth; and the z coordinate, representing the plate's thickness or height. In addressing plate issues, the applied loads and geometries are configured such that the displacements (u, v, w) along the coordinates (x, y, z) are contingent upon the x, y, and z coordinates.

DOI: 10.1201/9781003470694-9

9.3 NONLINEAR CLASSICAL PLATE THEORY

9.3.1 Introduction

Plates are planar objects with two dimensions significantly bigger than the third dimension. They have the ability to endure several types of stresses, including tensile, compressive, cross-plate, bending, twisting, and transverse shear forces. For instance, a membrane is a planar structure that can only endure tensile and shear forces acting on its central surface.

In classical plate theory, the middle plate is examined as a benchmark plate, aligning with the Kirchhoff hypothesis. Kirchhoff postulates the following [1, 2]:

1. The magnitude of deformation experienced by the middle plate is less than the thickness of the plate.
2. The middle plate remains unaffected by bending and is considered to be in a neutral state.
3. Avoidance of transverse shear forces is implemented.
4. Transverse tension is neglectable compared to other tension components.

However, laboratory studies indicate that traditional plate theory underestimates deformation and overestimates natural frequencies, particularly at high oscillation frequencies. Furthermore, in the context of composite materials, it is crucial to consider shear stresses, which arise due to their relatively low shear strength. Alternative theories were offered to address the limitations of classical plate theory and Kirchhoff hypotheses. One such theory is the first-order shear deformation theory for homologous plates, which yields six partial differential equations.

The presence of significant elastic deformations and rotations in a structure with varying thickness introduces geometric nonlinearities in the equations that govern its behavior. When there are significant transverse displacements in a plate, the central plate experiences strain and assumptions 1 and 2 become invalid. In order to determine the geometric nonlinearities of thin plates, the concept of van Karman strains was previously introduced and explained.

9.3.2 The Nonlinear Governing Equation of a Rectangular Plates

The equations and boundary conditions that govern the static and nonlinear dynamics of plates (as seen in Figure 9.1) are derived using a combination of classical and nonlinear plate theory of the van Karman type. Classical plate theory, as previously stated, relies on Kirchhoff's assumptions that the transverse sections of the plate are flat and rectangular in relation to the reference plate, both before and after deformation.

The displacement field for the plate is often stated as follows:

$$\begin{aligned} u_1 &= u(x,y,t) - z\frac{\partial w(x,t)}{\partial x} \\ u_2 &= v(x,y,t) - z\frac{\partial w(x,t)}{\partial y} \\ u_3 &= w(x,y,t) \end{aligned} \tag{9.1}$$

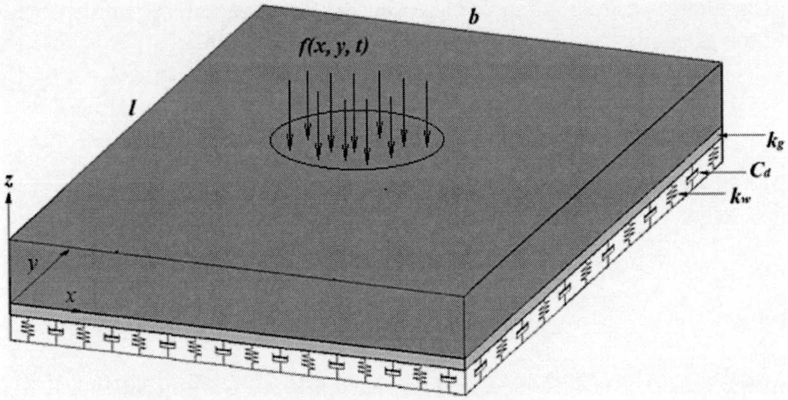

FIGURE 9.1 Schematic of an isotropic rectangular plate, placed on a visco-Pasternak substrate and under a distributed external force.

The displacement components in the x, y, and z directions, denoted as u_1, u_2, and u_3, respectively, are located within the range of Equation (9.1). The displacement field described in Equation (9.1) states that transverse sections, relative to the reference plate, maintain a flat and rectangular shape both before and after deformation. To clarify, both normal transverse stresses and transverse stresses are not considered.

Now, consider the Green's strain-displacement connections in the Lagrangian perspective and take into consideration the assumptions that govern von Karman's theory [3]:

$$\varepsilon_{xx} = \varepsilon_{11} = \frac{1}{2}\left(u_{1,1} + u_{1,1} + u_{3,1}u_{3,1}\right) = u_{1,1} + \frac{1}{2}\left(u_{3,1}\right)^2 = \frac{\partial u_1}{\partial x} + \frac{1}{2}\left(\frac{\partial u_3}{\partial x}\right)^2$$
$$\varepsilon_{xx} = \frac{\partial u}{\partial x} - z\frac{\partial^2 w}{\partial x^2} + \frac{1}{2}\left(\frac{\partial w}{\partial x}\right)^2 \tag{9.2}$$

$$\varepsilon_{yy} = \varepsilon_{22} = \frac{1}{2}\left(u_{2,2} + u_{2,2} + u_{3,2}u_{3,2}\right) \tag{9.3}$$

$$\varepsilon_{zz} = \varepsilon_{33} = \frac{1}{2}\left(u_{3,3} + u_{3,3} + u_{3,3}u_{3,3}\right) = 0 \tag{9.4}$$

$$\varepsilon_{xy} = \varepsilon_{12} = \frac{1}{2}\left(u_{1,2} + u_{2,1} + u_{3,1}u_{3,2}\right) \tag{9.5}$$

$$\varepsilon_{xz} = \varepsilon_{13} = \frac{1}{2}\left(u_{1,3} + u_{3,1} + u_{3,1}u_{3,3}\right) = 0 \tag{9.6}$$

$$\varepsilon_{yz} = \varepsilon_{23} = \frac{1}{2}\left(u_{2,3} + u_{3,2} + u_{3,2}u_{3,3}\right) = 0 \tag{9.7}$$

The Hamilton principle is employed to derive the governing equations. Thus, the Hamilton principle can be expressed in the following manner:

$$\delta \int_{t_1}^{t_2} \left(T - \pi + W_{ext}\right) dt = 0 \tag{9.8}$$

Regarding Equation (9.8), T represents kinetic energy, π represents strain energy, and W_{ext} represents the work done by the external force.

$$\begin{aligned}
\delta\pi &= \int_A \int_{-\frac{h}{2}}^{\frac{h}{2}} \sigma_{ij}\delta\varepsilon_{i,j}\, dzdA = \int_A \int_{-\frac{h}{2}}^{\frac{h}{2}} (\sigma_{xx}\delta\varepsilon_{xx} + \sigma_{yy}\delta\varepsilon_{yy} + 2\sigma_{xy}\delta\varepsilon_{xy})\, dzdA = \\
&\int_A \int_{-\frac{h}{2}}^{\frac{h}{2}} \left[\sigma_{xx}\left(\frac{\partial \delta u}{\partial x} - z\frac{\partial^2 \delta w}{\partial x^2} + \frac{\partial w}{\partial x}\frac{\partial \delta w}{\partial x}\right) + \sigma_{yy}\left(\frac{\partial \delta v}{\partial y} - z\frac{\partial^2 \delta w}{\partial y^2} + \frac{\partial w}{\partial y}\frac{\partial \delta w}{\partial y}\right) + \right. \\
&\left.\sigma_{xx}\left(\frac{\partial \delta u}{\partial y} - 2z\frac{\partial^2 \delta w}{\partial x\partial y} + \frac{\partial w}{\partial x}\frac{\partial \delta w}{\partial y}\right)\right] dzdA = \int_A \left[N_{xx}\left(\frac{\partial \delta u}{\partial x} + \frac{\partial w}{\partial x}\frac{\partial \delta w}{\partial x}\right)\right. \\
&N_{xy}\left(\frac{\partial \delta u}{\partial y} + \frac{\partial \delta v}{\partial x} + \frac{\partial w}{\partial x}\frac{\partial \delta w}{\partial y} + \frac{\partial w}{\partial y}\frac{\partial \delta w}{\partial x}\right) + N_{yy}\left(\frac{\partial \delta v}{\partial y} + \frac{\partial w}{\partial y}\frac{\partial \delta w}{\partial y}\right) - M_{xx}\frac{\partial^2 \delta w}{\partial x^2} \\
&\left. M_{yy}\frac{\partial^2 \delta w}{\partial y^2} - 2M_{xy}\frac{\partial^2 \delta w}{\partial x\partial y}\right] dA = -\int_A \left\{\left(\frac{\partial}{\partial x}N_{xx} + \frac{\partial}{\partial y}N_{xy}\right)\delta u + \left(\frac{\partial}{\partial y}N_{yy} + \right.\right. \\
&\left.\frac{\partial}{\partial x}N_{xy}\right)\delta v + \left[\frac{\partial^2 M_{xx}}{\partial x^2} + \frac{\partial^2 M_{yy}}{\partial y^2} + 2\frac{\partial^2 M_{xy}}{\partial x\partial y} + \frac{\partial}{\partial x}\left(N_{xx}\frac{\partial w}{\partial x}\right) + \frac{\partial}{\partial y}\left(N_{yy}\frac{\partial w}{\partial y}\right) + \right. \\
&\left.\left.\frac{\partial}{\partial y}\left(N_{xy}\frac{\partial w}{\partial x}\right) + \frac{\partial}{\partial x}\left(N_{xy}\frac{\partial w}{\partial y}\right)\right]\delta w\right\} dA + \int_y \left[N_{xx}\delta u + N_{xy}\delta v + \left[\frac{\partial M_{xx}}{\partial x} + 2\frac{\partial M_{xy}}{\partial y} + \right.\right. \\
&\left.\left. N_{xx}\frac{\partial w}{\partial x} + N_{xy}\frac{\partial w}{\partial y}\right)\delta w - M_{xx}\frac{\partial \delta w}{\partial x}\right]_{x=0}^{x=a} dy + \int_x N_{xy}\delta u + N_{yy}\delta v + \\
&\left(\frac{\partial M_{yy}}{\partial y} + 2\frac{\partial M_{xy}}{\partial x} + N_{yy}\frac{\partial w}{\partial y} + N_{xy}\frac{\partial w}{\partial x}\right)\delta w - M_{yy}\frac{\partial \delta w}{\partial y}\Bigg]_{y=0}^{y=b} dx - \\
&2M_{xy}\,\delta w\,\Big|_{(x,y)=(a,0),(0,b)}^{(x,y)=(0,0),(a,b)}
\end{aligned} \tag{9.9}$$

In the context of Equation (9.9), N_{xx} and M_{xx} are defined as follows:

$$\begin{Bmatrix} N_{xx} \\ N_{yy} \\ N_{xy} \end{Bmatrix} = \int_{-\frac{h}{2}}^{\frac{h}{2}} \begin{Bmatrix} \sigma_{xx} \\ \sigma_{yy} \\ \sigma_{xy} \end{Bmatrix} dz$$
$$\begin{Bmatrix} M_{xx} \\ M_{yy} \\ M_{xy} \end{Bmatrix} = \int_{-\frac{h}{2}}^{\frac{h}{2}} \begin{Bmatrix} \sigma_{xx} \\ \sigma_{yy} \\ \sigma_{xy} \end{Bmatrix} zdz \tag{9.10}$$

Regarding the Equation (9.10), N_{xx}, normal force, and M_{xx} represent the values exerted on the A-cross-sectional area. Next, we will utilize T to compute the kinetic energy:

$$\delta T = -\int_A \begin{bmatrix} \left(I_0\ddot{u} - I_1\dfrac{\partial\ddot{w}}{\partial x}\right)\delta u + \left(I_0\ddot{v} - I_1\dfrac{\partial\ddot{w}}{\partial y}\right) \\ \delta v + I_0\ddot{w}\delta w + \left(I_2\dfrac{\partial\ddot{w}}{\partial x} - I_1\ddot{u}\right) \\ \delta\dfrac{\partial w}{\partial x} + \left(I_2\dfrac{\partial\ddot{w}}{\partial y} - I_1\ddot{v}\right)\delta\dfrac{\partial w}{\partial y} \end{bmatrix} dA \tag{9.11}$$

In order to determine the work done by the external force (W_{ext}), we express it in the following manner:

$$\delta\int_{t_1}^{t_2} W_{ext}dt = \int_{t_1}^{t_2}\int_0^l f(x,y,t)\,\delta w dxdydt \tag{9.12}$$

By including the obtained Equations (9.9), (9.11), and (9.12) into the overarching Equation (9.8) and setting the coefficients δu, δv, and δw to zero, we may derive the equations of motion as follows:

$$\begin{aligned} &\delta u: \quad N_{xx,x} + N_{xy,y} = I_0\ddot{u} - I_1\ddot{w}_x \\ &\delta v: \quad N_{xy,x} + N_{yy,y} = I_0\ddot{v} - I_1\ddot{w}_y \\ &\delta w: \quad M_{xx,xx} + 2M_{xy,xy} + M_{yy,yy} + \left(N_{xx}w_{,x}\right)_{,x} + \left(N_{yy}w_{,y}\right)_{,y} + \left(N_{xy}w_{,x}\right)_{,y} + \\ &\left(N_{xy}w_{,y}\right)_{,x} + f(x,y,t) = I_0\ddot{w} + I_2\left\{\ddot{w}_{,xx} + \ddot{w}_{,yy}\right\} + I_1\ddot{u}_x + I_1\ddot{v}_y \end{aligned} \tag{9.13}$$

And boundary conditions as follows:

$$
\begin{aligned}
&u=0,\ \text{Or}\ \left(N_{xx}\right)n_x+\left(N_{xy}\right)n_y=0\\
&v=0,\text{Or}\ \left(N_{xy}\right)n_x+\left(N_{yy}\right)n_y=0\\
&w=0,\ \text{Or}\ \left(M_{xx,x}+M_{xy,y}+N_{xx}w_{,x}+N_{xy}w_{,y}\right)n_x+\left(M_{xy,x}+M_{yy,y}+N_{yy}w_{,y}\right)n_y=0\\
&\frac{\partial w}{\partial x}=0,\ \text{Or}\ \left(M_{xx}\right)n_x+\left(M_{xy}\right)n_y=0\\
&\frac{\partial w}{\partial y}=0,\ \text{Or}\ \left(M_{xy}\right)n_x+\left(M_{yy}\right)n_y=0
\end{aligned}
\tag{9.14}
$$

The governing equations of the plate in terms of displacement expressions are derived in the following manner:

$$
\begin{aligned}
&A_{11}\frac{\partial^2 u}{\partial x^2}+\left(A_{12}+A_{66}\right)\frac{\partial^2 v}{\partial x\partial y}+A_{66}\frac{\partial^2 u}{\partial y^2}+A_{11}\frac{\partial w}{\partial x}\frac{\partial^2 w}{\partial x^2}+A_{12}\frac{\partial w}{\partial y}\frac{\partial^2 w}{\partial x\partial y}+\\
&\qquad A_{66}\left(\frac{\partial w}{\partial y}\frac{\partial^2 w}{\partial x\partial y}+\frac{\partial w}{\partial x}\frac{\partial^2 w}{\partial y^2}\right)=I_0\frac{\partial^2 u}{\partial t^2}-I_1\frac{\partial^3 w}{\partial x\partial t^2}
\end{aligned}
\tag{9.15}
$$

$$
\begin{aligned}
&\left(A_{12}+A_{66}\right)\frac{\partial^2 u}{\partial x\partial y}+A_{66}\frac{\partial^2 v}{\partial x^2}+A_{22}\frac{\partial^2 v}{\partial y^2}+A_{12}\frac{\partial w}{\partial x}\frac{\partial^2 w}{\partial x\partial y}+A_{22}\frac{\partial w}{\partial y}\frac{\partial^2 w}{\partial y^2}+\\
&\qquad A_{66}\left(\frac{\partial w}{\partial y}\frac{\partial^2 w}{\partial x^2}+\frac{\partial w}{\partial x}\frac{\partial^2 w}{\partial x\partial y}\right)=I_0\frac{\partial^2 v}{\partial t^2}-I_1\frac{\partial^3 w}{\partial y\partial t^2}
\end{aligned}
\tag{9.16}
$$

$$
\begin{aligned}
&D_{11}\frac{\partial^4 u}{\partial x^4}+2\left(D_{12}+2D_{66}\right)\frac{\partial^4 w}{\partial x^2\partial y^2}+D_{22}\frac{\partial^4 w}{\partial y^4}+I_0\frac{\partial^2 w}{\partial t^2}=\frac{\partial}{\partial x}\left(N_{xx}\frac{\partial}{\partial x}\right)+\\
&\frac{\partial}{\partial x}\left(N_{yy}\frac{\partial w}{\partial y}\right)+\frac{\partial}{\partial x}\left(N_{xy}\frac{\partial w}{\partial y}\right)+\frac{\partial}{\partial y}\left(N_{xy}\frac{\partial w}{\partial x}\right)+I_2\frac{\partial^2}{\partial t^2}\left(\frac{\partial^2 w}{\partial x^2}+\frac{\partial^2 w}{\partial y^2}\right)+\frac{\partial w}{\partial x}\frac{\partial^2}{\partial t^2}\left(I_0u-\right.\\
&\qquad \left.I_1\frac{\partial w}{\partial x}\right)+\frac{\partial w}{\partial y}\frac{\partial^2}{\partial t^2}\left(I_0v-I_1\frac{\partial w}{\partial y}\right)-I_1\frac{\partial^2}{\partial t^2}\frac{\partial u}{\partial x}-I_1\frac{\partial^2}{\partial t^2}\frac{\partial v}{\partial y}
\end{aligned}
\tag{9.17}
$$

In Equations (9.15), (9.16), and (9.17), the values of A_{ij} and D_{ij} for a homogeneous plate will be as follows:

$$
\left[A_{ij}\right]=\frac{Eh}{1-\nu^2}\begin{bmatrix}1 & \upsilon & 0\\ \nu & 1 & 0\\ 0 & 0 & (1-\nu)/2\end{bmatrix}
\tag{9.18}
$$

$$
\left[D_{ij}\right]=\frac{1}{12}h^2\left[A_{ij}\right]
\tag{9.19}
$$

In Equation (9.18), E represents the modulus of elasticity, and υ represents the Poisson coefficient. Furthermore I_1, I_2, and I_0 are explicitly specified as follows:

$$[I_0, I_1, I_2] = \int_{-\frac{h}{2}}^{\frac{h}{2}} \rho\left[1, z, z^2\right] dz = \rho\left[h, 0, \frac{h^2}{12}\right] \tag{9.20}$$

In Equation (9.20), the presence of symmetry results in I_1 being equal to zero. It is important to mention that the value of I_1 will be zero when the density ρ remains constant and the middle plate is chosen as the reference plate.

In the following, for a symmetrical rectangular plate supported by a visco-Pasternak substrate with isotropic and homogenous material under a distrusted, arbitrary external force, the equations that describe the movement of the plate (9.1) can be succinctly described as follows:

$$\begin{aligned} &\frac{\partial^2 M_{xx}}{\partial x^2} + 2\frac{\partial^2 M_{xy}}{\partial x \partial y} + \frac{\partial^2 M_{yy}}{\partial y^2} + f + \frac{\partial}{\partial x}\left(N_{xx}\frac{\partial w}{\partial x}\right) + \frac{\partial}{\partial y}\left(N_{yy}\frac{\partial w}{\partial y}\right) + \frac{\partial}{\partial x}\left(N_{xy}\frac{\partial w}{\partial y}\right) + \\ &\frac{\partial}{\partial y}\left(N_{xy}\frac{\partial w}{\partial x}\right) = I_0\frac{\partial^2 w}{\partial t^2} - I_2\left(\frac{\partial^4 w}{\partial x^2 \partial t^2} + \frac{\partial^4 w}{\partial y^2 \partial t^2}\right) \end{aligned} \tag{9.21}$$

It is noted that the energy of nonlinear rectangular plate vibrations with significant amplitude is mostly affected by the inertia of the plate, which has a negligible effect on the energy in the middle.

Given the influence of the viscoelastic substrate (consisting of a spring and damper) and the external force f_1 as shown in Figure 9.1, we can express the following:

$$f = k_w w - k_G\left(\frac{\partial^2 w}{\partial x^2} + \frac{\partial^2 w}{\partial y^2}\right) + c_d\frac{\partial w}{\partial t} + f_1 \tag{9.22}$$

The linear coefficient of the Winkler spring in the elastic bed is denoted as k_w, the Pasternak shear coefficient is represented by k_G, the elastic bed damping coefficient is denoted as c_d, and the external transverse force is denoted as f_1.

Additionally, we can express the values for M_{xx}, M_{yy}, and M_{xy}:

$$\begin{aligned} M_{xx} &= -D\left(\frac{\partial^2 w}{\partial x^2} + \nu\frac{\partial^2 w}{\partial y^2}\right) \\ M_{yy} &= -D\left(\frac{\partial^2 w}{\partial y^2} + \nu\frac{\partial^2 w}{\partial x^2}\right) \\ M_{xy} &= -D(1-\nu)\left(\frac{\partial^2 w}{\partial x \partial y}\right) \end{aligned} \tag{9.23}$$

The bending stiffness of the plate, denoted as D, is given by the formula $D = \frac{Eh^3}{12(1-\nu^2)}$. By substituting the aforementioned equations for M_{xx}, M_{yy}, and

M_{xy} into the governing equation, the equation of motion is derived by applying van Karman's nonlinear relationships in the following manner:

$$D\left[\frac{\partial^4 w}{\partial x^4}+2\frac{\partial^4 w}{\partial x^2\partial y^4}+\frac{\partial^4 w}{\partial y^4}\right]+\left[I_0\frac{\partial^2 w}{\partial t^2}-I_2\left(\frac{\partial^4 w}{\partial x^2\partial y^2}+\frac{\partial^4 w}{\partial y^2\partial y^2}\right)\right]-(k_w w-$$
$$k_G\left(\frac{\partial^2 w}{\partial x^2}+\frac{\partial^2 w}{\partial y^2}\right)+c_d\frac{\partial w}{\partial t}\Bigg)-\frac{\partial^2 w}{\partial x^2}\left[\frac{c_{11}}{2}\left(\frac{\partial w}{\partial x}\right)^2+\frac{c_{12}}{2}\left(\frac{\partial w}{\partial y}\right)^2\right]-\frac{\partial w}{\partial x}\left[c_{11}\frac{\partial^2 w}{\partial x^2}\frac{\partial w}{\partial x}+\right.$$
$$\left.c_{12}\frac{\partial^2 w}{\partial x\partial y}\frac{\partial w}{\partial y}\right]-\frac{\partial^2 w}{\partial y^2}\left[\frac{c_{12}}{2}\left(\frac{\partial w}{\partial x}\right)^2+\frac{c_{22}}{2}\left(\frac{\partial w}{\partial y}\right)^2\right]-\frac{\partial w}{\partial y}\left[c_{12}\frac{\partial^2 w}{\partial x\partial y}\frac{\partial w}{\partial x}+c_{22}\frac{\partial^2 w}{\partial y^2}\frac{\partial w}{\partial y}\right]- \tag{9.24}$$
$$c_{66}\frac{\partial^2 w}{\partial x^2}\left(\frac{\partial w}{\partial y}\right)^2-4c_{66}\frac{\partial w}{\partial x}\frac{\partial^2 w}{\partial x\partial y}\frac{\partial w}{\partial y}-c_{66}\frac{\partial^2 w}{\partial y^2}\left(\frac{\partial w}{\partial x}\right)^2=f_1$$

The coefficients c_{11}, c_{22}, c_{12}, and c_{66} are defined as follows:

$$c_{11}=c_{22}=\frac{Eh}{1-\nu^2},\qquad c_{12}=\frac{\nu Eh}{1-\nu^2},\ c_{66}=\frac{Eh}{2(1+\nu)} \tag{9.25}$$

9.4 CASE STUDY 1

9.4.1 Introduction

This study examines the parametric vibration and dynamic instability of a rectangular and symmetric magnetostrictive sandwich composite plate (MSCP) over a visco-Pasternak medium. The MSCP is composed of three layers, with a magnetostrictive layer encompassing the core and composites as its top and bottom surfaces. The study focuses on the MSCP and its response to temperature variations, parametrically stimulating forces, and magnetic loads. The analysis takes into account the geometrical von Karman nonlinearity.

9.4.2 Theory and Formulation

Figure 9.8 displays a sandwich plate composed of three layers atop a visco-Pasternak medium. Figure 9.8 illustrates that the three layers stated are magnetostrictive, forming the core of the sandwich structure. The top and lower faces are believed to be composite layers. The MSCP has dimensions a for length and b for breadth. The core has a thickness of h_c, while the faces have a thickness of h_f [4].

9.4.3 Constitutive Equations

The modelling of sandwich plates in this part utilizes the first-order shear deformation theory (FSDT). FSDT models are widely favored for their ease of analysis and

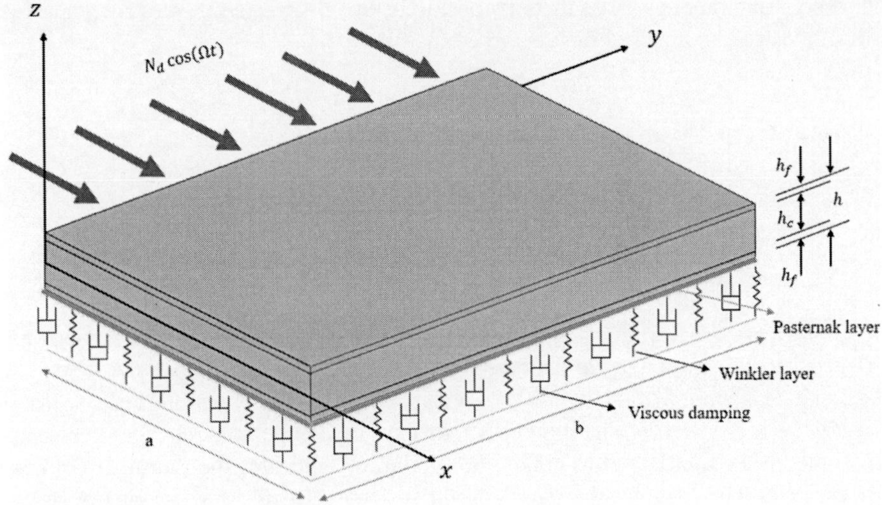

FIGURE 9.2 Schematic of a MSCP subjected to temperature increment, parametrically exciting force, and magnetic field placed on a visco-Pasternak medium.

programming, making them very popular arithmetic models. The displacement fields are provided according to the first-order shear deformation theory of plates.

$$\begin{aligned} \bar{u}(x,y,z,t) &= u(x,y,t) + z\varphi_x(x,y,t) \\ \bar{v}(x,y,z,t) &= v(x,y,t) + z\varphi_y(x,y,t) \\ \bar{w}(x,y,z,t) &= w(x,y,t) \end{aligned} \tag{9.26}$$

The functions $u(x,y,t), v(x,y,t)$, and $w(x,y,t)$ represent the displacement components parallel to the x, y, and z axes, respectively. The functions $\varphi_x(x,y,t)$ and $\varphi_y(x,y,t)$ represent the rotation angles of the normal vector with respect to the y and x axes. The strain field for FSDT, which is not linear, may be mathematically represented using the von Karman hypothesis, as stated in reference.

$$\begin{Bmatrix} \varepsilon_x \\ \varepsilon_y \\ \gamma_{xz} \\ \gamma_{yz} \\ \gamma_{xz} \end{Bmatrix} = \begin{Bmatrix} \varepsilon^0{}_x \\ \varepsilon^0{}_y \\ \gamma^0{}_{xz} \\ \gamma^0{}_{yz} \\ \gamma^0{}_{xz} \end{Bmatrix} + z \begin{Bmatrix} \varepsilon^1{}_x \\ \varepsilon^1{}_y \\ \gamma^1{}_{xz} \\ \gamma^1{}_{yz} \\ \gamma^1{}_{xz} \end{Bmatrix} = \begin{Bmatrix} \dfrac{\partial u}{\partial x} + \dfrac{1}{2}\left(\dfrac{\partial w}{\partial x}\right)^2 \\ \dfrac{\partial v}{\partial y} + \dfrac{1}{2}\left(\dfrac{\partial w}{\partial y}\right)^2 \\ \dfrac{\partial u}{\partial y} + \dfrac{\partial v}{\partial x} + \dfrac{\partial w}{\partial x}\dfrac{\partial w}{\partial y} \\ \dfrac{\partial w}{\partial x} + \varphi_x \\ \dfrac{\partial w}{\partial y} + \varphi_y \end{Bmatrix} + z \begin{Bmatrix} \dfrac{\partial \varphi_x}{\partial x} \\ \dfrac{\partial \varphi_y}{\partial y} \\ \dfrac{\partial \varphi_x}{\partial y} + \dfrac{\partial \varphi_y}{\partial x} \\ 0 \\ 0 \end{Bmatrix} \tag{9.27}$$

The equation that describes the relationship between stress-strain and the impact of magnetostrictive material on it is represented as:

$$\begin{Bmatrix} \sigma_{xx} \\ \sigma_{yy} \\ \sigma_{xy} \\ \sigma_{yz} \\ \sigma_{xz} \end{Bmatrix} = \begin{bmatrix} \bar{Q}^c_{11} & \bar{Q}^c_{12} & \bar{Q}^c_{16} & 0 & 0 \\ \bar{Q}^c_{12} & \bar{Q}^c_{22} & \bar{Q}^c_{26} & 0 & 0 \\ \bar{Q}^c_{16} & \bar{Q}^c_{26} & \bar{Q}^c_{66} & 0 & 0 \\ 0 & 0 & 0 & \bar{Q}^c_{44} & \bar{Q}^c_{45} \\ 0 & 0 & 0 & \bar{Q}^c_{45} & \bar{Q}^c_{55} \end{bmatrix} \left(\begin{Bmatrix} \varepsilon_x \\ \varepsilon_y \\ \gamma_{xy} \\ \gamma_{yz} \\ \gamma_{xz} \end{Bmatrix} - \begin{Bmatrix} \alpha^c_{11} \\ \alpha^c_{22} \\ 0 \\ 0 \\ 0 \end{Bmatrix} \Delta T \right) - \begin{Bmatrix} e_{31} \\ e_{32} \\ e_{36} \\ 0 \\ 0 \end{Bmatrix} H_z \tag{9.28}$$

σ_{ij} represents stress while ε_{ij} represents strain. The model shown above demonstrates that the strain and magnetic field are mutually exclusive. In Appendix A, there is a detailed explanation of the words $\bar{Q}^c{}_{ij}$, which represent the converted elastic stiffness coefficients of the core. The symbol ΔT denotes the temperature changes, which are assumed to be unaffected by time. The symbol α_{ij} indicates the thermal expansion coefficient. The magnetostrictive constant is denoted as e_{ji}, and H_z represents the magnetic field as:

$$H_z = K_c I(x,y,t) = K_c C(t) \frac{\partial w(x,y,z,t)}{\partial t} \tag{9.29}$$

The equation presented involves the constant K_c, which is influenced by several factors including the number of turns and the width of the coil. In this study, the variables $I(t)$, $w(x,y,z,t)$, and $C(t)$, respectively, denote the coil current, transverse displacement, and control gain. It is worth noting that the control gain remains constant and is set to 1. Additionally, the product of K_c and $C(t)$ is referred to as the velocity feedback gain.

The correlation between stress and strain of the face layer is expressed as follows [5]:

$$\sigma^f = c^f \varepsilon^f \tag{9.30}$$

Let f represent the number of layers. According to Hook's rule, we can deduce the following:

$$\begin{Bmatrix} \sigma_{xx} \\ \sigma_{yy} \\ \sigma_{xy} \\ \sigma_{yz} \\ \sigma_{xz} \end{Bmatrix} = \begin{bmatrix} c_{11} & c_{12} & 0 & 0 & 0 \\ c_{12} & c_{22} & 0 & 0 & 0 \\ 0 & 0 & c_{66} & 0 & 0 \\ 0 & 0 & 0 & c_{44} & 0 \\ 0 & 0 & 0 & 0 & c_{55} \end{bmatrix} \left(\begin{Bmatrix} \varepsilon_x \\ \varepsilon_y \\ \gamma_{xy} \\ \gamma_{yz} \\ \gamma_{xz} \end{Bmatrix} - \begin{Bmatrix} \alpha_{11} \\ \alpha_{22} \\ 0 \\ 0 \\ 0 \end{Bmatrix} \Delta T \right) \tag{9.31}$$

where

$$c_{11} = \frac{E_1}{1-\nu_{12}\nu_{21}}, \qquad c_{12} = \frac{\nu_{12}E_2}{1-\nu_{12}\nu_{21}}, \qquad c_{22} = \frac{E_2}{1-\nu_{12}\nu_{21}}$$

$$c_{44} = G_{12}, \qquad c_{55} = G_{13}, \qquad c_{66} = G_{23}$$

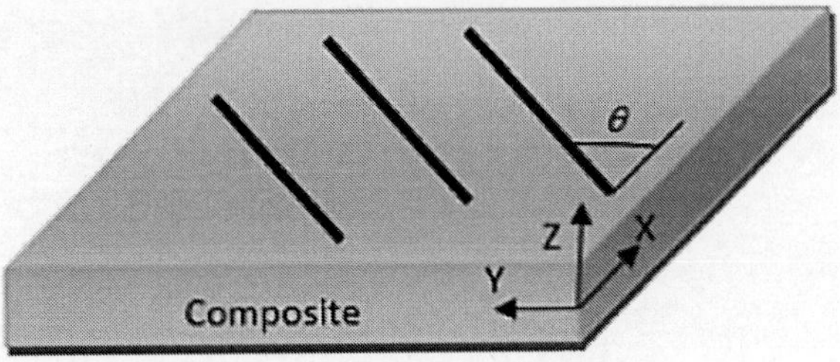

FIGURE 9.3 Schematic of the angle of the fibers in the lamina.

E_1, E_2, ν_{12}, ν_{21}, G_{12}, G_{23}, and G_{13} denote the material characteristics of the composite layers, specifically graphite epoxy (AS/3501) as assumed in this paper. Additionally, Figure 9.3 illustrates that the composite fibers are unidirectional, with θ representing the angle of the fibers relative to the x-axis.

The equation of motion may be derived by using Equation (9.32) using the energy technique and Hamilton's principle.

$$\int_{t_1}^{t_2} \delta\left(U - K - W\right) dt = 0 \tag{9.32}$$

The variables U, K, and W symbolize the energy stored in a system due to deformation, the energy associated with motion, and the energy transferred to or from the system by external forces, respectively.

The equations of motion may be derived by using Hamilton's principle.

$$\begin{aligned}
&\bullet\ \frac{\partial N_{xy}}{\partial x} + \frac{\partial N_{xy}}{\partial y} = I_0 \frac{\partial^2 u}{\partial t^2} + I_1 \frac{\partial^2 \varphi_x}{\partial t^2} \\
&\bullet\ \frac{\partial N_{xy}}{\partial x} + \frac{\partial N_{xy}}{\partial y} = I_0 \frac{\partial^2 v}{\partial t^2} + I_1 \frac{\partial^2 \varphi_y}{\partial t^2} \\
&\bullet\ \frac{\partial}{\partial x}\left(N_x \frac{\partial w}{\partial x} + N_{xy} \frac{\partial w}{\partial y}\right) + \frac{\partial}{\partial y}\left(N_{xy} \frac{\partial w}{\partial x} + N_y \frac{\partial w}{\partial y}\right) + \frac{\partial Q_x}{\partial x} + \frac{\partial Q_y}{\partial y} \\
&\qquad + k_w w - k_p\left(\frac{\partial^2 w}{\partial x^2} + \frac{\partial^2 w}{\partial y^2}\right) + (N_d\, cos(\Omega t)) \frac{\partial^2 w}{\partial x^2} \\
&\qquad + c_d \frac{\partial w}{\partial t} = I_0 \frac{\partial^2 w}{\partial t^2} \\
&\bullet\ \frac{\partial M_{xy}}{\partial x} + \frac{\partial M_y}{\partial y} - Q_y = I_1 \frac{\partial^2 v}{\partial t^2} + I_2 \frac{\partial^2 \varphi_y}{\partial t^2} \\
&\bullet\ \frac{\partial M_{xy}}{\partial x} + \frac{\partial M_y}{\partial y} - Q_y = I_1 \frac{\partial^2 v}{\partial t^2} + I_2 \frac{\partial^2 \varphi_y}{\partial t^2}
\end{aligned} \tag{9.33}$$

Where (I_0, I_1, I_2) represent the mass moments of inertia and are defined as follows:

$$\left(I_0, I_1, I_2\right) = \int_{\frac{h_c}{2}}^{\frac{h_c}{2}+h_t} \rho_t \left(1, z, z^2\right) dz + \int_{\frac{-h_c}{2}}^{\frac{h_c}{2}} \rho_c \left(1, z, z^2\right) dz + \int_{\frac{-h_c}{2}-h_b}^{\frac{-h_c}{2}} \rho_t \left(1, z, z^2\right) dz \tag{9.34}$$

Where ρ_t represents the mass density of the face layer and ρ_c represents the mass density of the core layer. The force resultants Nij and moment resultants. The calculation of M_{ij} may be expressed by using the displacement gradients, magnetic field, and temperature in the following manner:

$$\begin{Bmatrix} N_x \\ N_y \\ N_{xy} \\ M_x \\ M_y \\ M_{xy} \end{Bmatrix} = \begin{bmatrix} A_{11} & A_{12} & A_{16} & B_{11} & B_{12} & B_{16} \\ A_{12} & A_{22} & A_{26} & B_{12} & B_{22} & B_{26} \\ A_{16} & A_{26} & A_{66} & B_{16} & B_{26} & B_{66} \\ B_{11} & B_{12} & B_{16} & D_{11} & D_{12} & D_{16} \\ B_{12} & B_{22} & B_{26} & D_{12} & D_{22} & D_{26} \\ B_{16} & B_{26} & B_{66} & D_{16} & D_{26} & D_{66} \end{bmatrix} \begin{Bmatrix} \varepsilon^0{}_x \\ \varepsilon^0{}_y \\ \gamma^0{}_{xy} \\ \varepsilon^1{}_x \\ \varepsilon^1{}_y \\ \gamma^1{}_{xy} \end{Bmatrix} - \begin{Bmatrix} N^T_x \\ N^T_y \\ N^T_{xy} \\ M^T_x \\ M^T_y \\ M^T_{xy} \end{Bmatrix} - \begin{Bmatrix} N^P_x \\ N^P_y \\ N^P_{xy} \\ M^P_x \\ M^P_y \\ M^P_{xy} \end{Bmatrix} \tag{9.35}$$

$$\begin{Bmatrix} Q_x \\ Q_y \end{Bmatrix} = K \begin{bmatrix} A_{55} & A_{45} \\ A_{45} & A_{44} \end{bmatrix} \begin{Bmatrix} \gamma^0{}_{yz} \\ \gamma^0{}_{xz} \end{Bmatrix}$$

K represents the shear correction factor, which is precisely equal to 5/6.

$$\begin{aligned} \begin{Bmatrix} N^T_x & ,M^T_x \\ N^T_y & ,M^T_y \\ N^T_{xy} & ,M^T_{xy} \end{Bmatrix} &= \int_{\frac{h_c}{2}}^{\frac{h_c}{2}+h_f} \begin{bmatrix} Q_{11} & Q_{12} & 0 \\ Q_{12} & Q_{22} & 0 \\ 0 & 0 & Q_{66} \end{bmatrix} \begin{Bmatrix} \alpha_1 \\ \alpha_2 \\ 0 \end{Bmatrix} (1, z) \Delta T(z)\, dz \\ &+ \int_{-\frac{h_c}{2}}^{\frac{h_c}{2}} \begin{bmatrix} Q_{11} & Q_{12} & 0 \\ Q_{12} & Q_{22} & 0 \\ 0 & 0 & Q_{66} \end{bmatrix} \begin{Bmatrix} \alpha_1 \\ \alpha_2 \\ 0 \end{Bmatrix} (1, z) \Delta T(z)\, dz + \\ &\int_{-\frac{h_c}{2}-h_f}^{-\frac{h_c}{2}} \begin{bmatrix} Q_{11} & Q_{12} & 0 \\ Q_{12} & Q_{22} & 0 \\ 0 & 0 & Q_{66} \end{bmatrix} \begin{Bmatrix} \alpha_1 \\ \alpha_2 \\ 0 \end{Bmatrix} (1, z) \Delta T(z)\, dz \end{aligned} \tag{9.36}$$

$$\begin{Bmatrix} N^P_x & ,M^P_x \\ N^P_y & ,M^P_y \\ N^P_{xy} & ,M^P_{xy} \end{Bmatrix} = \int_{-\frac{h_c}{2}}^{\frac{h_c}{2}} \begin{Bmatrix} e_{13} \\ e_{32} \\ 0 \end{Bmatrix} (1, z) K_c C(t) \frac{\partial w}{\partial t} dz \tag{9.37}$$

A_{ij}, D_{ij}, and B_{ij} represent the stiffness values for extensional, bending, and bending-extensional coupling, respectively. These values are specified in relation to the lamina stiffness Q_{ij}, which is given as follows:

$$\left(A_{ij},B_{ij},D_{ij}\right)=\int_{\frac{h_c}{2}}^{\frac{h_c}{2}+h_f}Q_{ij}^{face}\left(1,z,z^2\right)dz+\int_{-\frac{h_c}{2}}^{\frac{h_c}{2}}Q_{ij}^{core}\left(1,z,z^2\right)dz+ \tag{9.38}$$
$$\int_{-\frac{h_c}{2}-h_f}^{-\frac{h_c}{2}}Q_{ij}^{face}\left(1,z,z^2\right)dz \quad \left(i,j=1,2,6\right)$$

The equations of motion (Equation (9.53)) may be reformulated in terms of displacements $\left(u.v.w\right)$ by substituting the expressions for the force and moment resultants from Equations (9.35) to (9.37).

$$A_{11}\left(\frac{\partial^2 u}{\partial x^2}+\frac{\partial w}{\partial x}\frac{\partial^2 w}{\partial x^2}\right)+A_{12}\left(\frac{\partial^2 v}{\partial y\partial x}+\frac{\partial w}{\partial y}\frac{\partial^2 w}{\partial y\partial x}\right)$$
$$+A_{16}\left(\frac{\partial^2 u}{\partial y\partial x}+\frac{\partial^2 v}{\partial x^2}+\frac{\partial^2 w}{\partial x^2}\frac{\partial w}{\partial y}+\frac{\partial w}{\partial x}\frac{\partial^2 w}{\partial y\partial x}\right)$$
$$+B_{11}\frac{\partial^2\phi_x}{\partial x^2}+B_{12}\frac{\partial^2\phi_y}{\partial y\partial x}+B_{16}\left(\frac{\partial^2\phi_x}{\partial y\partial x}+\frac{\partial^2\phi_y}{\partial x^2}\right)$$
$$+A_{16}\left(\frac{\partial^2 u}{\partial y\partial x}+\frac{\partial w}{\partial x}\frac{\partial^2 w}{\partial y\partial x}\right)+A_{26}\left(\frac{\partial^2 v}{\partial y^2}+\frac{\partial w}{\partial y}\frac{\partial^2 w}{\partial y^2}\right) \tag{9.39}$$
$$A_{66}\left(\frac{\partial^2 u}{\partial y^2}+\frac{\partial^2 v}{\partial y\partial x}+\frac{\partial^2 v}{\partial y\partial x}\frac{\partial w}{\partial y}+\frac{\partial w}{\partial x}\frac{\partial^2 w}{\partial y^2}\right)$$
$$+B_{16}\frac{\partial^2\phi_x}{\partial y\partial x}+B_{12}\frac{\partial^2\phi_y}{\partial y^2}+B_{66}\left(\frac{\partial^2\phi_x}{\partial y^2}+\frac{\partial^2\phi_y}{\partial y\partial x}\right)$$
$$\left(\frac{\partial N_{xx}^T}{\partial x}+\frac{\partial N_{xx}^T}{\partial y}\right)-\left(\frac{\partial N_{xx}^P}{\partial x}+\frac{\partial N_{xx}^P}{\partial y}\right)=I_0\frac{\partial^2 u}{\partial t^2}+I_1\frac{\partial^2\psi_x}{\partial t^2}$$

$$A_{16}\left(\frac{\partial^2 u}{\partial x^2}+\frac{\partial w}{\partial x}\frac{\partial^2 w}{\partial x^2}\right)+A_{26}\left(\frac{\partial^2 v}{\partial y\partial x}+\frac{\partial w}{\partial y}\frac{\partial^2 w}{\partial y\partial x}\right)$$
$$+A_{66}\left(\frac{\partial^2 u}{\partial y\partial x}+\frac{\partial^2 v}{\partial x^2}+\frac{\partial^2 w}{\partial x^2}\frac{\partial w}{\partial y}+\frac{\partial w}{\partial x}\frac{\partial^2 w}{\partial y\partial x}\right)$$
$$+B_{16}\frac{\partial^2\phi_x}{\partial x^2}+B_{26}\frac{\partial^2\phi_y}{\partial y\partial x}+B_{66}\left(\frac{\partial^2\phi_x}{\partial y\partial x}+\frac{\partial^2\phi_y}{\partial x^2}\right)$$
$$+A_{12}\left(\frac{\partial^2 u}{\partial y\partial x}+\frac{\partial w}{\partial x}\frac{\partial^2 w}{\partial y\partial x}\right)+A_{22}\left(\frac{\partial^2 v}{\partial y^2}+\frac{\partial w}{\partial x}\frac{\partial^2 w}{\partial y^2}\right) \tag{9.40}$$
$$A_{66}\left(\frac{\partial^2 u}{\partial y^2}+\frac{\partial^2 v}{\partial y\partial x}+\frac{\partial^2 v}{\partial y\partial x}\frac{\partial w}{\partial y}+\frac{\partial w}{\partial x}\frac{\partial^2 w}{\partial y^2}\right)+B_{12}\frac{\partial^2\phi_x}{\partial y\partial x}+B_{12}\frac{\partial^2\phi_y}{\partial y^2}$$
$$+B_{22}\frac{\partial^2\phi_x}{\partial y^2}+B_{26}\left(\frac{\partial^2\phi_x}{\partial y^2}+\frac{\partial^2\phi_y}{\partial y\partial x}\right)\left(\frac{\partial N_{xy}^T}{\partial x}+\frac{\partial N_{yy}^T}{\partial y}\right)$$
$$-\left(\frac{\partial N_{xy}^P}{\partial x}+\frac{\partial N_{yy}^P}{\partial y}\right)=I_0\frac{\partial^2 v}{\partial t^2}+I_1\frac{\partial^2\psi_y}{\partial t^2}$$

$$KA_{55}\left(\frac{\partial^2 w}{\partial x^2}+\frac{\partial\phi_x}{\partial x}\right)+KA_{45}\left(\frac{\partial^2 w}{\partial y\partial x}+\frac{\partial\phi_y}{\partial x}\right)+KA_{45}\left(\frac{\partial^2 w}{\partial y\partial x}+\frac{\partial\phi_x}{\partial y}\right)+KA_{44}\left(\frac{\partial^2 w}{\partial y^2}+\frac{\partial\phi_y}{\partial y}\right)$$
$$\frac{\partial}{\partial x}\left(N_x\frac{\partial w}{\partial x}+N_{xy}\frac{\partial w}{\partial y}\right)+\frac{\partial}{\partial y}\left(N_{xy}\frac{\partial w}{\partial x}+N_y\frac{\partial w}{\partial y}\right)-\left(\frac{\partial Q_x^P}{\partial x}+\frac{\partial Q_y^P}{\partial y}\right)=I_0\frac{\partial^2 w}{\partial t^2}+$$
$$k_w w-k_p\left(\frac{\partial^2 w}{\partial x^2}+\frac{\partial^2 w}{\partial y^2}\right)+\left(N_d\cos(\Omega t)\right)\frac{\partial^2 w}{\partial x^2}+c_d\frac{\partial w}{\partial t} \tag{9.41}$$

$$B_{11}\left(\frac{\partial^2 u}{\partial x^2}+\frac{\partial w}{\partial x}\frac{\partial^2 w}{\partial x^2}\right)+B_{12}\left(\frac{\partial^2 v}{\partial y\partial x}+\frac{\partial w}{\partial y}\frac{\partial^2 w}{\partial y\partial x}\right)$$
$$+B_{16}\left(\frac{\partial^2 u}{\partial y\partial x}+\frac{\partial^2 v}{\partial x^2}+\frac{\partial^2 w}{\partial x^2}\frac{\partial w}{\partial y}+\frac{\partial w}{\partial x}\frac{\partial^2 w}{\partial y\partial x}\right)+$$
$$D_{11}\frac{\partial^2\phi_x}{\partial x^2}+D_{12}\frac{\partial^2\phi_y}{\partial y\partial x}+D_{16}\left(\frac{\partial^2\phi_x}{\partial y\partial x}+\frac{\partial^2\phi_y}{\partial x^2}\right) \tag{9.42}$$

$$+B_{16}\left(\frac{\partial^2 u}{\partial y\partial x}+\frac{\partial w}{\partial x}\frac{\partial^2 w}{\partial y\partial x}\right)+B_{26}\left(\frac{\partial^2 v}{\partial y^2}+\frac{\partial w}{\partial x}\frac{\partial^2 w}{\partial y^2}\right)$$

$$B_{66}\left(\frac{\partial^2 u}{\partial y^2}+\frac{\partial^2 v}{\partial y\partial x}+\frac{\partial^2 v}{\partial y\partial x}\frac{\partial w}{\partial y}+\frac{\partial w}{\partial x}\frac{\partial^2 w}{\partial y^2}\right)+D_{16}\frac{\partial^2 \phi_x}{\partial y\partial x}$$

$$+D_{26}\frac{\partial^2 \phi_y}{\partial y^2}+D_{66}\left(\frac{\partial^2 \phi_x}{\partial y^2}+\frac{\partial^2 \phi_y}{\partial y\partial x}\right)-$$

$$KA_{55}\left(\frac{\partial w}{\partial x}+\phi_x\right)-KA_{45}\left(\frac{\partial w}{\partial y}+\phi_y\right)-\left(\frac{\partial M_{xx}^T}{\partial x}+\frac{\partial M_{xy}^T}{\partial y}\right)$$

$$-\left(\frac{\partial M_{xx}^P}{\partial x}+\frac{\partial M_{xy}^P}{\partial y}-Q_x^P\right)=I_1\frac{\partial^2 u}{\partial t^2}+I_1\frac{\partial^2 \psi_x}{\partial t^2}$$

$$B_{16}\left(\frac{\partial^2 u}{\partial x^2}+\frac{\partial w}{\partial x}\frac{\partial^2 w}{\partial x^2}\right)+B_{26}\left(\frac{\partial^2 v}{\partial y\partial x}+\frac{\partial w}{\partial y}\frac{\partial^2 w}{\partial y\partial x}\right)$$

$$+B_{16}\left(\frac{\partial^2 u}{\partial y\partial x}+\frac{\partial^2 v}{\partial x^2}+\frac{\partial^2 w}{\partial x^2}\frac{\partial w}{\partial y}+\frac{\partial w}{\partial x}\frac{\partial^2 w}{\partial y\partial x}\right)+$$

$$D_{16}\frac{\partial^2 \phi_x}{\partial x^2}+D_{26}\frac{\partial^2 \phi_y}{\partial y\partial x}+D_{66}\left(\frac{\partial^2 \phi_x}{\partial y\partial x}+\frac{\partial^2 \phi_y}{\partial x^2}\right)+B_{12}\left(\frac{\partial^2 u}{\partial y\partial x}+\frac{\partial w}{\partial x}\frac{\partial^2 w}{\partial y\partial x}\right)$$

$$+B_{26}\left(\frac{\partial^2 v}{\partial y^2}+\frac{\partial w}{\partial x}\frac{\partial^2 w}{\partial y^2}\right)$$

$$B_{26}\left(\frac{\partial^2 u}{\partial y^2}+\frac{\partial^2 v}{\partial y\partial x}+\frac{\partial^2 v}{\partial y\partial x}\frac{\partial w}{\partial y}+\frac{\partial w}{\partial x}\frac{\partial^2 w}{\partial y^2}\right)+D_{12}\frac{\partial^2 \phi_x}{\partial y\partial x}+D_{22}\frac{\partial^2 \phi_y}{\partial y^2} \tag{9.43}$$

$$+D_{26}\left(\frac{\partial^2 \phi_x}{\partial y^2}+\frac{\partial^2 \phi_y}{\partial y\partial x}\right)-$$

$$KA_{45}\left(\frac{\partial w}{\partial x}+\phi_x\right)-KA_{44}\left(\frac{\partial w}{\partial y}+\phi_y\right)-\left(\frac{\partial M_{xx}^T}{\partial x}+\frac{\partial M_{xy}^T}{\partial y}\right)$$

$$-\left(\frac{\partial M_{xy}^P}{\partial x}+\frac{\partial M_{yy}^P}{\partial y}-Q_y^P\right)=I_1\frac{\partial^2 v}{\partial t^2}+$$

$$I_1 \qquad\qquad I_2\frac{\partial^2 \psi_x}{\partial t^2}$$

Where:

$$N_y = A_{12}\left(\frac{\partial u}{\partial x}+\frac{1}{2}(\frac{\partial w}{\partial x})^2\right)+A_{22}\left(\frac{\partial v}{\partial y}+\frac{1}{2}(\frac{\partial w}{\partial x})^2\right)+A_{26}\left(\frac{\partial u}{\partial y}+\frac{\partial w}{\partial x}+\frac{\partial w}{\partial y}\frac{\partial w}{\partial x}\right)$$
$$B_{12}\frac{\partial \phi_x}{\partial x}+B_{22}\frac{\partial \phi_y}{\partial y}+B_{26}\left(\frac{\partial \phi_x}{\partial y}+\frac{\partial \phi_y}{\partial x}\right)-N_{yy}^T-N_{yy}^P$$
$$N_{xy} = A_{16}\left(\frac{\partial u}{\partial x}+\frac{1}{2}(\frac{\partial w}{\partial x})^2\right)+A_{26}\left(\frac{\partial v}{\partial y}+\frac{1}{2}(\frac{\partial w}{\partial x})^2\right)+A_{66}\left(\frac{\partial u}{\partial y}+\frac{\partial w}{\partial x}+\frac{\partial w}{\partial y}\frac{\partial w}{\partial x}\right)$$
$$+B_{26}\frac{\partial \phi_x}{\partial x}+B_{26}\frac{\partial \phi_y}{\partial y}+B_{66}\left(\frac{\partial \phi_x}{\partial y}+\frac{\partial \phi_y}{\partial x}\right) \tag{9.44}$$

9.4.4 Solution Method

The governing equations of the sandwich plate have been computed under the "Theory and Formulation" section. The equations of motion were solved using the Galerkin technique. The Galerkin technique is used to solve the governing equation of motion under simple boundary conditions, assuming the answer to be the following [6]:

$$u(x,y,t)=\sum_n^\infty\sum_m^\infty U[t]\cos[\alpha x]\sin[\beta y]$$
$$v(x,y,t)=\sum_n^\infty\sum_m^\infty V[t]\sin[\alpha x]\cos[\beta y]$$
$$w(x,y,t)=\sum_n^\infty\sum_m^\infty W[t]\sin[\alpha x]\sin[\beta y]$$
$$\phi_x(x,y,t)=\sum_n^\infty\sum_m^\infty X[t]\cos[\alpha x]\sin[\beta y]$$
$$\phi_y(x,y,t)=\sum_n^\infty\sum_m^\infty Y[t]\sin[\alpha x]\cos[\beta y]$$
$$\alpha=\frac{m\pi}{a},\quad \beta=\frac{n\pi}{b} \tag{9.45}$$

By putting Equation (9.45) into Equations (9.39) to (9.43), the resulting equations may be stated as follows:

$$C_{11}W^2+C_{12}U+C_{13}V+C_{14}X+C_{15}Y+C_{16}\dot{W}=0$$
$$C_{21}W^2+C_{22}U+C_{23}V+C_{24}X+C_{25}Y+C_{26}\dot{W}=0$$
$$C_{31}W^3+C_{32}UW+C_{33}VW+C_{34}XW+C_{35}YW+C_{36}X+C_{37}Y$$
$$+C_{38}W+C_{39}\dot{W}W \tag{9.46}$$
$$+C_{310}\dot{W}+C_{311}\ddot{W}+C_{312}\cos(\Omega t)=0$$
$$C_{41}W^2+C_{42}U+C_{43}V+C_{44}X+C_{45}Y+C_{46}W=0$$
$$C_{51}W^2+C_{52}U+C_{53}V+C_{54}X+C_{55}Y+C_{56}W=0$$

The C_{ij} constants are extended in Appendix *B*. The variables (U,V,X,Y) were obtained by computing them in terms of *W*.

$$\begin{aligned} U &= S_{11}W + S_{12}W^2 + S_{13}\dot{W} \\ V &= S_{21}W + S_{22}W^2 + S_{23}\dot{W} \\ X &= S_{31}W + S_{32}W^2 + S_{33}\dot{W} \\ Y &= S_{41}W + S_{42}W^2 + S_{43}\dot{W} \end{aligned} \tag{9.47}$$

The S_{ij} constants are enlarged in Appendix B in terms of C_{ij}.

Next, we replace Equation (9.47) into the third expression of Equation (9.46) to get the governing equation of motion:

$$\ddot{W} + \varepsilon(c_1W + c_2)\dot{W} + \omega^2W + \varepsilon K_1\cos(\Omega t)W + \varepsilon K_2W^2 + \varepsilon^2K_3W^3 = 0 \tag{9.48}$$

Where:

$$c_1 = \frac{(C_{39} + C_{32}S_{13} + C_{33}S_{23} + C_{34}S_{33} + C_{35}S_{43})}{C_{311}},\; c_2 = \frac{C_{310} + C_{36}S_{33}}{C_{311}}$$

$$\omega^2 = \frac{C_{38} + C_{36}S_{31} + C_{37}S_{41}}{C_{311}}, \qquad K_1 = \frac{C_{312}}{C_{311}},$$

$$K_2 = \frac{C_{32}S_{11} + C_{33}S_{21} + C_{34}S_{31} + C_{36}S_{32} + C_{35}S_{41} + C_{37}S_{42}}{C_{311}}$$

$$K_3 = \frac{C_{31} + C_{32}S_{12} + C_{33}S_{22} + C_{34}S_{32} + C_{35}S_{42}}{C_{311}}$$

9.4.4.1 Multiple Scale Method

The approach proposed by Nayfeh and Mook [7], known as the multiple scale method, is used to solve Equation (9.48). According to this approach, the expansion of W is expressed in the form of a second-order approximation:

$$W(t) = W_0(T_0,T_1,T_2) + \varepsilon W_1(T_0,T_1,T_2) + \varepsilon^2W_2(T_0,T_1,T_2) \tag{9.49}$$

T_0 represents rapid time scales, whereas T_1 and T_2 represent slower scales that correspond to variations in amplitude and phase resulting from nonlinearity, damping, and resonances, respectively.

$$T_0 = t, \quad T_1 = \varepsilon t, \quad T_2 = \varepsilon^2 t \tag{9.50}$$

By substituting the Equation (9.49) into Equation (9.48) and equating at the same order of ε, we may get the following partial differential equations.

$$\left(\omega^2 + D_0^{\,2}\right)W_0 = 0 \tag{9.51}$$

$$\left(\omega^2+D_0^{\,2}\right)W_1=-2D_0D_1W_0-K_2W_0^{\,2}-K_1\cos\left(\Omega t\right)W_0-c_1D_0W_0^2-c_2D_0w_0 \quad (9.52)$$

$$-2C_1D_0W_0W_1-C_1D_1W_0^{\,2}-K_3W_0^{\,3}-K_1\cos\left(\Omega t\right)W_1 \quad (9.53)$$

Initially, our objective is to find the solution to a partial differential Equation (9.51) in the specified format.

$$W_0=A\left(T_1,T_2\right)\exp\left[i\omega T_0\right]+\bar{A}\left(T_1,T_2\right)\exp\left[-i\omega T_0\right] \quad (9.54)$$

The function A is unknown and depends on T_1 and T_2. The pairs $(A,\ \bar{A})$ are complex conjugates. By replacing the Equation (9.54) in the right-hand side of Equation (9.52):

$$\begin{aligned}\left(\omega^2+D_0^{\,2}\right)W_1&=-i\omega\left(2D_1A+c_2A\right)e^{i\omega T_0}-K_2A\bar{A}-\left(2ic_1\omega+K_2\right)A^2e^{2i\omega T_0}\\&-\frac{1}{2}K_1\left(Ae^{i(\omega+\Omega)T_0}+\bar{A}e^{i(\Omega-\omega)T_0}\right)+cc\end{aligned} \quad (9.55)$$

The function cc represents the complex conjugate of the previous terms. The resonance conditions derived in this example are primarily the principal resonance $\Omega\cong 2\hat{\omega}$.

$$\Omega=2\omega+\varepsilon\sigma \quad (9.56)$$

σ represents the detuning parameter. By omitting the secular and small divisor components, Equation (9.55) may be simplified when $\Omega\approx 2\omega$.

$$-i\omega\left(2D_1A+c_2A\right)-\frac{1}{2}K_1\bar{A}e^{i\sigma T_1}=0 \quad (9.57)$$

The solution to Equation (9.52) is provided as follows:

$$\begin{aligned}W_1&=\frac{2ic_1e^{2i\omega T_0}A\left[T_2\right]^2}{3\omega}+\frac{K_1e^{i(\Omega+\omega)T_0}A\left[T_2\right]}{2\Omega\left(\Omega+2\omega\right)}\\&+\frac{K_2e^{2i\omega T_0}A\left[T_2\right]^2}{3\omega^2}-3K_2A\left[T_2\right]\bar{A}\left[T_2\right]+cc\end{aligned} \quad (9.58)$$

Similarly, by replacing Equation (9.54) and (9.58) into Equation (9.53) using the second-order approximation:

$$\begin{aligned}\left(\omega^2+D_0^{\,2}\right)W_2&=H_1\left[exp\left(i\omega T_0\right)\right]+H_2\left[exp\left(2i\omega T_0\right)\right]+H_3\left[exp\left(3i\omega T_0\right)\right]\\&+H_4\left[exp\left(i\left(\omega+\Omega\right)T_0\right)\right]+\\H_5&\left[exp\left(i\left(\omega+2\Omega\right)T_0\right)\right]+H_6\left[exp\left(i\left(2\omega+\Omega\right)T_0\right)\right]+H_7\left[exp(i(2\omega-\\&\Omega)T_0)\right]+H_8\left[exp\left(i\Omega T_0\right)\right]\end{aligned} \quad (9.59)$$

The solvability criteria of Equation (9.59) may be expressed in the following manner, after deleting secular terms.

$$H_1\left[exp\left(i\omega T_0\right)\right]=0\rightarrow D_1^{\ 2}A+2i\omega D_2A-\frac{4}{3}c_1^2A^2\bar{A}+c_2D_1A$$
$$+\frac{K_1^2A}{8\omega\Omega+4\Omega^2}-\frac{2ic_1K_2A^2\bar{A}}{\omega}-\qquad(9.60)$$
$$\frac{10K_2^2A^2\bar{A}}{3\omega^2}+3K_3A^2\bar{A}=0$$

By using Equation (9.57) to delete the terms D_1^2A and D_1A from Equation (9.60), we may get the following result:

$$2i\omega D_2A+\left(-\frac{c_2^{\ 2}}{4}+\frac{3K_1^{\ 2}}{32\omega^2}\right)A+\left(-\frac{10K_2^2}{3\omega^2}+3K_3-\frac{4}{3}c_1^2-\frac{2ic_1K_2}{\omega}\right)$$
$$A^2\bar{A}-\frac{K_1\sigma\bar{A}}{4\omega}e^{i\sigma T_1}\qquad(9.61)$$

It can be shown that Equation (9.57) and (9.61) represent the first two components in a multiple-scale analysis of the following:

$$2i\omega\left(\dot{A}+\varepsilon\frac{c_2}{2}A\right)+\varepsilon^2\left[\left(\frac{3K_1^{\ 2}}{32\omega^2}-\frac{c_2^{\ 2}}{4}\right)A+\left(3K_3-\frac{10K_2^2}{3\omega^2}-\frac{4}{3}c_1^2-\frac{2ic_1K_2}{\omega}\right)A^2\bar{A}\right]+$$
$$\frac{1}{2}\varepsilon K_1\left[1-\frac{\varepsilon\sigma}{2\omega}\right]\bar{A}\,exp\left[i\varepsilon\sigma t\right]=0\qquad(9.62)$$

In Equation (9.62), A and $\bar{A}$ are expressed in polar forms.

$$A\left(t\right)=\frac{1}{2}a\left(t\right)exp\left[i\beta\left(t\right)\right],\quad \bar{A}\left(t\right)=\frac{1}{2}a\left(t\right)exp\left[\text{-}i\beta\left(t\right)\right]\qquad(9.63)$$

Which are the amplitudes and phases of steady-state responses referred to as "*a*" and "*β*". By substituting Equation (9.63) into Equation (9.62), the equations regulating the amplitudes and phases may be derived by separating the real and imaginary components.

$$a\beta'=G_1a+G_2a^3+G_3\,a\cos\left[\gamma\right]\qquad(9.64)$$

$$a'=G_4a+G_5a^3-G_3a\sin\left[\gamma\right]\qquad(9.65)$$

Where:

$$G_1=\frac{3\varepsilon^2K_1^2}{64\omega^3}-\frac{\varepsilon^2c_2^2}{8\omega},\quad G_2=\frac{3\varepsilon^2K_3}{8\omega}-\frac{5\varepsilon^2K_2^2}{12\omega^3}-\frac{c_1^2}{6\omega},\quad G_3=\left(\frac{\varepsilon K_1}{4\omega}\right)\left(1-\frac{\varepsilon\sigma}{2\omega}\right),\qquad(9.66)$$
$$G_4=-0.5\varepsilon c_2,\quad G_5=-\varepsilon^2\left(\frac{c_1K_2}{4\omega^2}\right),\quad \gamma=\varepsilon\sigma t-2\beta$$

To determine the steady-state response, we assume that $a'=0$ and $\gamma'=0$ $(\gamma'=\varepsilon\sigma-2\beta')$; thus, modulation equations for the principal parametric resonance are expressed as follows:

$$\left(G_1-\frac{\varepsilon\sigma}{2}\right)a+G_2a^3+G_3a\cos[\gamma]=0 \tag{9.67}$$

$$G_4a+G_5a^3-G_3a\sin[\gamma]=0 \tag{9.68}$$

9.4.4.1.1 *Non-trivial Steady-State Response*

To get a non-trivial solution for the system, it is necessary to delete the determinant of the coefficient matrix. In order to analyze the system's steady-state reaction, it is necessary to take into account that a is not equal to zero ($a\neq 0$). Therefore, the following will be obtained [8]:

$$\gamma'=0\rightarrow\left(G_1-\frac{\varepsilon\sigma}{2}\right)a+G_2a^3=-G_3a\cos[\gamma] \tag{9.69}$$

$$a'=0\rightarrow G_4a+G_5a^3=G_3a\sin[\gamma] \tag{9.70}$$

By using the trigonometric identity $\cos^2[\gamma]+\sin^2[\gamma]=1$ in Equations (9.69) and (9.70), the resulting equations are as follows:

$$\left(\left(G_1-\frac{\varepsilon\sigma}{2}\right)a+G_2a^3\right)^2+\left(G_4a+G_5a^3\right)^2=\left(G_3a\right)^2 \tag{9.71}$$

Ultimately, the amplitude a is acquired in the following manner:

$$a=\left[\frac{1}{2\left(G_2^2+G_5^2\right)}\left((\varepsilon\sigma-2G_1)G_2-2G_4G_5\pm\right.\right. \tag{9.72}$$
$$\left.\left.\sqrt{4G_2^2\left(G_3^2-G_4^2\right)-4\left(\varepsilon\sigma-2G_1\right)G_2G_4G_5+(-\varepsilon^2\sigma^2+4\varepsilon\sigma G_1-4G_1^2+4G_3^2)G_5^2}\right)\right]^{\frac{1}{2}}$$

The stability of the fixed points and the steady-state solution is contingent upon the characteristics of the roots' natural portion. If the natural component of each root is positive, then the associated steady-state solution is said to be unstable. Furthermore, if the real component of each root is zero or negative, the associated steady-state solution is stable.

9.4.4.1.2 *Trivial Steady-State Response*

In order to ascertain the stability of the trivial solution, we analyze the characteristics of the linear solutions of Equation (9.62), which are the solutions of the following:

$$2i\omega\left(\dot{A}+\varepsilon\frac{c_2}{2}A\right)+\varepsilon^2\left(\frac{3K_1^2}{32\omega^2}-\frac{c_2^2}{4}\right)A+\frac{1}{2}\varepsilon K_1\left[1-\frac{\varepsilon\sigma}{2\omega}\right]\bar{A}\exp[i\varepsilon\sigma t]=0 \tag{9.73}$$

To begin with, we define the Cartesian form of the solution as $A = \frac{1}{2}(p - iq)exp[i\beta T_1]$. By inserting this expression into Equation (9.73), we may recast the procedure of the solution as follows:

$$\begin{aligned}&-64p\beta\varepsilon\omega^3 + 32q\varepsilon\omega^3 c_2 - 8p\varepsilon^2\omega^2 c_2^2 - 8p\varepsilon^2\sigma\omega K_1 + 16p\varepsilon\omega^2 K_1 + 3p\varepsilon^2 K_1^2 \\ &+ 64q\beta\varepsilon\omega^3 i + \\ &32p\varepsilon\omega^3 c_2 i + 8q\varepsilon^2\omega^2 c_2^2 i - 8q\varepsilon^2\sigma\omega K_1 i + 16q\varepsilon\omega^2 K_1 i - 3q\varepsilon^2 K_1^2 i + 64\omega^3 ip' + \\ &64\omega^3 q' = 0\end{aligned} \tag{9.74}$$

Simultaneously, in order to render the equation self-sufficient, it is necessary for β to equal $\sigma/2$. In order to achieve autonomy, it is necessary to take into account the phase angle in relation to the detuning parameter. By performing a separation of the real and imaginary components in Equation (9.74), the following outcomes are obtained [9]:

$$p' = -\frac{1}{2}\varepsilon\sigma q - \frac{\varepsilon^2 c_2^2 q}{8\omega} + \frac{\varepsilon^2\sigma K_1 q}{8\omega^2} - \frac{\varepsilon K_1 q}{4\omega} + \frac{3\varepsilon^2 K_1^2 q}{64\omega^3} - \frac{\varepsilon c_2}{2} p \tag{9.75}$$

$$q' = \frac{1}{2}\varepsilon\sigma p + \frac{\varepsilon^2 c_2^2 p}{8\omega} + \frac{\varepsilon^2\sigma K_1 p}{8\omega^2} - \frac{\varepsilon K_1 p}{4\omega} - \frac{3\varepsilon^2 K_1^2 p}{64\omega^3} - \frac{\varepsilon c_2}{2} q \tag{9.76}$$

Now, we construct the Jacobian matrix for Equations (9.75) and (9.76) as shown next:

$$J = \begin{bmatrix} -\frac{\varepsilon c_2}{2} & -\frac{1}{2}\varepsilon\sigma - \frac{\varepsilon^2 c_2^2}{8\omega} + \frac{\varepsilon^2\sigma K_1}{8\omega^2} - \frac{\varepsilon K_1}{4\omega} + \frac{3\varepsilon^2 K_1^2}{64\omega^3} \\ \frac{1}{2}\varepsilon\sigma + \frac{\varepsilon^2 c_2^2}{8\omega} + \frac{\varepsilon^2\sigma K_1}{8\omega^2} - \frac{\varepsilon K_1}{4\omega} - \frac{3\varepsilon^2 K_1^2}{64\omega^3} & -\frac{\varepsilon c_2}{2} \end{bmatrix} \tag{9.77}$$

Thus, we can express the determinant and trace of matrix A in the following manner:

$$\Delta = c_2^2\omega^2 + \left(\sigma\omega + \frac{\varepsilon c_2^2}{4} - \frac{3\varepsilon K_1^2}{32\omega^2}\right)^2 - \frac{K_1^2}{4}\left(1 - \frac{\varepsilon\sigma}{2\omega}\right)^2 \tag{9.78}$$

$$\tau = -\varepsilon c_2 \tag{9.79}$$

Ultimately, if the condition $c_2 > 0$ holds true, we can determine the stability of the system.

9.4.5 Results and Discussion

This study uses the first shear deformation theory to derive the equations of motion for an embedded MSCP (multi-span continuous plate) that is supported by a viscous-Winkler-Pasternak foundation. The plate is subjected to uniform parametric excitation and a magnetic field. The study investigates the impact of several factors, including different fiber angles, viscous-Winkler-Pasternak parameters, temperature, thickness ratio, and velocity feedback gain, on the stability and bifurcation areas of the magnetostrictive sandwich composite plate. The characteristics of the materials are outlined in Tables 9.1 and 9.2, while other aspects are taken into account.

9.4.6 Numerical Results

This work introduces a theoretical analysis of the vibrations of a laminated composite plate. The plate is exposed to an in-plane force, heat load, and is supported by a viscoelastic foundation. This section investigates various significant parameters related to the dynamics of the sandwich plate with a magnetostrictive material core. In addition to parametric studies, factors such as velocity feedback gain, the influence of the foundation, temperature change, thickness ratio, and fiber angle are examined. The results are presented in plotted and charted form.

Tables 9.3 and 9.4, as well as Figures 9.4 and 9.5, illustrate the impact of velocity feedback gain (*C(t)Kc*) on both the amplitude response and dynamic instability. By adjusting the velocity feedback gain parameter, one may manipulate the amplitude of the MSCP. The increase in the damping parameter is found to be directly proportional to the increase in the feedback parameter. Figures 9.4 and 9.5 demonstrate that varying feedback gains have no impact on the subcritical and supercritical pitchfork bifurcation points. However, they may lead to alterations in the saddle points.

TABLE 9.1
Elastic properties of Terfenol-D [4].

Properties	E_c	υ_c	ρ_c	$e_{31} = e_{32}$
	30 GPa	0.25	9.25×10^3 kg/m^3	442.55 N/(mA)

TABLE 9.2
Properties of composite layer [4].

	c_{11} (Gpa)	c_{22} (Gpa)	$c_{12} = c_{21}$ (Gpa)	c_{66} (Gpa)	$c_{44} = c_{55}$ (Gpa)
$\theta = 0$	140.90	10.06	3.02	7	0
$\theta = 45$	46.25	46.25	32.25	36.23	7
$\theta = 90$	10.06	140.9	3.02	7	0

$a = 1m$, $b = a$, $h = 0.1a$, $h_c = 0.8h$, $k_w = 5\times10^7$Pa/m, $k_p = 5\times10^6$Pa.m, $c_d = 5\times10^3$ Pa.s, $\varepsilon = 0.01$.

TABLE 9.3
Effect of Foundation Parameters and Velocity Feedback Gain on Amplitude Response

			C(t)Kc			
K_w	K_p	C_d	10^4	2×10^4	3×10^4	4×10^4
10^7	10^6	0	0.012276	0.00607878	0.00404536	0.00303214
		10^3	0.01226	0.00607123	0.00404033	0.00302838
		5x10^3	0.011884	0.00588504	0.00391642	0.0029355
		10x10^3	0.010559	0.00522888	0.00347975	0.0026082
	5x10^6	0	0.011903	0.00589398	0.00392237	0.00293996
		10^3	0.011886	0.00588567	0.00391684	0.00293582
		5x10^3	0.011471	0.00568023	0.00378012	0.00283334
		10x10^3	0.009986	0.00494515	0.00329094	0.00246668
	10x10^6	0	0.011436	0.00566287	0.00376857	0.00282468
		10^3	0.011417	0.00565355	0.00376237	0.00282003
		5x10^3	0.01095	0.00542206	0.00360832	0.00270457
		10x10^3	0.00924	0.00457541	0.00304488	0.00228225
5x10^7	10^6	0	0.012087	0.00598513	0.00398303	0.00298543
		10^3	0.012071	0.0059772	0.00397775	0.00298147
		5x10^3	0.011675	0.00578142	0.00384746	0.00288382
		10x10^3	0.010271	0.0050863	0.00338487	0.00253708
	5x10^6	0	0.011714	0.00580038	0.00386008	0.00289327
		10^3	0.011696	0.00579167	0.00385429	0.00288893
		5x10^3	0.01126	0.00557596	0.00371073	0.00278133
		10x10^3	0.009688	0.00479756	0.00319272	0.00239306
	10x10^6	0	0.011246	0.00556896	0.00370607	0.00277784
		10^3	0.011226	0.0055592	0.00369958	0.00277297
		5x10^3	0.010736	0.00531645	0.00353803	0.00265189
		10x10^3	0.008925	0.00441962	0.00294121	0.00220454
10x10^7	10^6	0	0.01185	0.00586814	0.00390517	0.00292707
		10^3	0.011833	0.00585972	0.00389957	0.00292287
		5x10^3	0.011413	0.00565148	0.00376099	0.002819
		10x10^3	0.009905	0.00490469	0.00326401	0.00244649
	5x10^6	0	0.011477	0.00568326	0.00378214	0.00283485
		10^3	0.011458	0.00567404	0.003776	0.00283025
		5x10^3	0.010996	0.00544494	0.00362354	0.00271598

(Continued)

TABLE 9.3 (*Continued*)
Effect of Foundation Parameters and Velocity Feedback Gain on Amplitude Response

			C(t)Kc			
K_w	K_p	C_d	10^4	2×10^4	3×10^4	4×10^4
		$10\text{x}10^3$	0.009307	0.00460878	0.00306708	0.00229889
	$10\text{x}10^6$	0	0.011008	0.00545121	0.00362771	0.0027191
		10^3	0.010988	0.00544089	0.00362084	0.00271395
		$5\text{x}10^3$	0.010468	0.00518342	0.0034495	0.00258553
		$10\text{x}10^3$	0.00852	0.00421908	0.00280774	0.00210451

TABLE 9.4
Effect of Geometrical Parameters, Temperature, and Velocity Feedback Gain on the Amplitude Response

			C(t)Kc			
ΔT	*a/b*	h_c/h	10^4	2×10^4	3×10^4	4×10^4
0	1	0.7	0.00965563	0.00476404	0.00316834	0.00237424
		0.8	0.00826376	0.00409209	0.00272323	0.00204116
		0.9	0.00724437	0.00359747	0.00239529	0.00179568
	1.5	0.7	0.0178013	0.00867619	0.00575764	0.00431134
		0.8	0.0163826	0.00804405	0.0053451	0.00400423
		0.9	0.0156984	0.00775188	0.00515617	0.00386405
	2	0.7	0.0280242	0.0132632	0.00875818	0.00654701
		0.8	0.0260732	0.0125373	0.00830063	0.00621058
		0.9	0.0255658	0.012438	0.00825143	0.00617802
150	1	0.7	0.00979356	0.00483209	0.0032136	0.00240816
		0.8	0.00839241	0.0041558	0.00276563	0.00207294
		0.9	0.00736044	0.0036551	0.00243367	0.00182445
	1.5	0.7	0.0179629	0.00875496	0.00580991	0.00435048
		0.8	0.0165258	0.00811439	0.00539183	0.00403925
		0.9	0.0158232	0.00781347	0.00519713	0.00389475
	2	0.7	0.028248	0.0133691	0.00882812	0.0065993
		0.8	0.0262684	0.0126312	0.00836279	0.00625708
		0.9	0.0257358	0.0125207	0.00830629	0.0062191

TABLE 9.4 (*Continued*)
Effect of Geometrical Parameters, Temperature, and Velocity Feedback Gain on the Amplitude Response

			C(t)Kc			
ΔT	*a/b*	h_c/h	10^4	2×10^4	3×10^4	4×10^4
300	1	0.7	0.0099305	0.00489966	0.00325853	0.00244183
		0.8	0.00851997	0.00421896	0.00280766	0.00210444
		0.9	0.00747546	0.00371222	0.0024717	0.00185296
	1.5	0.7	0.0181243	0.00883361	0.00586211	0.00438957
		0.8	0.0166689	0.00818464	0.00543852	0.00407422
		0.9	0.0159479	0.00787506	0.0052381	0.00392545
	2	0.7	0.0284718	0.013475	0.00889805	0.00665157
		0.8	0.0264637	0.0127251	0.00842496	0.0063036
		0.9	0.0259061	0.0126036	0.00836126	0.00626026

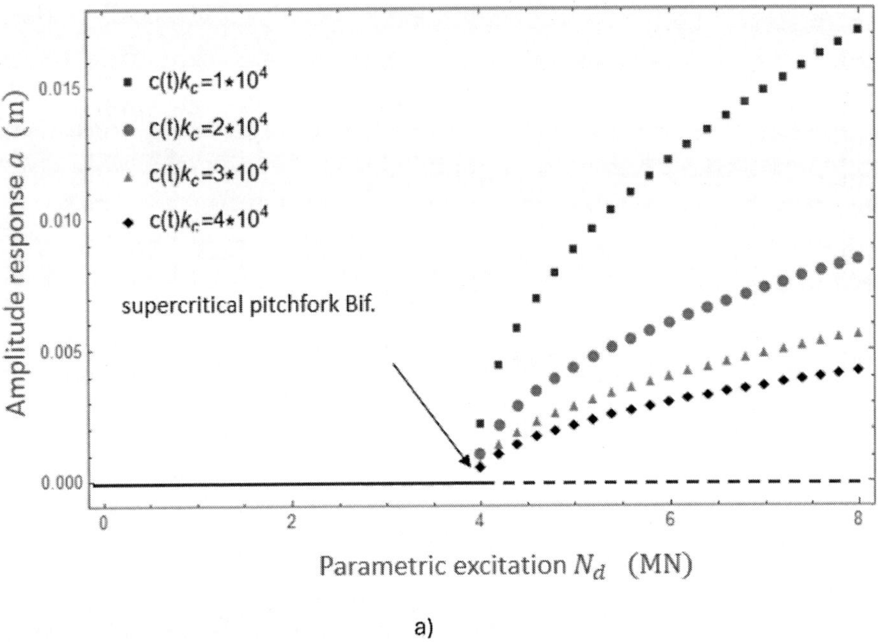

FIGURE 9.4 Amplitude response versus force amplitude of parametric excitation curves: Effect of velocity feedback gain at (a) σ = −20 and (b) σ = 20.

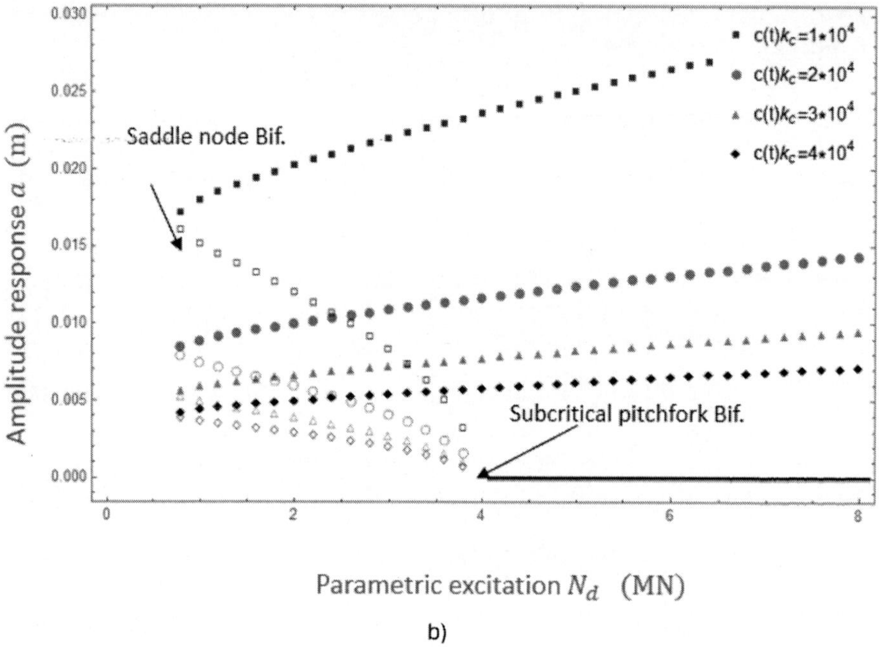

FIGURE 9.4 (Continued)

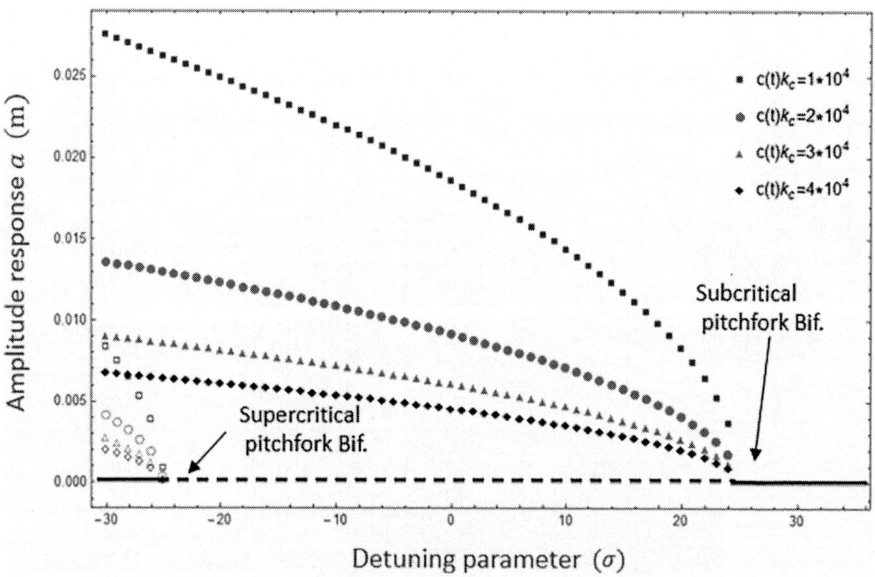

FIGURE 9.5 Amplitude response versus detuning parametric curves: Effect of velocity feedback gain at $N_d = 5\times10^6\,N$.

Figures 9.6 and 9.7 illustrate the impact of foundation characteristics, such as the Winkler and Pasternak coefficient, as well as the influence of viscous damping on the stability of the system. The results indicate that the amplitude response is directly influenced by the foundation coefficients. As the Winkler and Pasternak coefficients increase, the amplitude response decreases, while keeping the excitation constant. In addition, the crucial points of the pitchfork bifurcation shift towards the right. Figure 9.7a clearly demonstrates that the impact of the viscous damping parameter has a

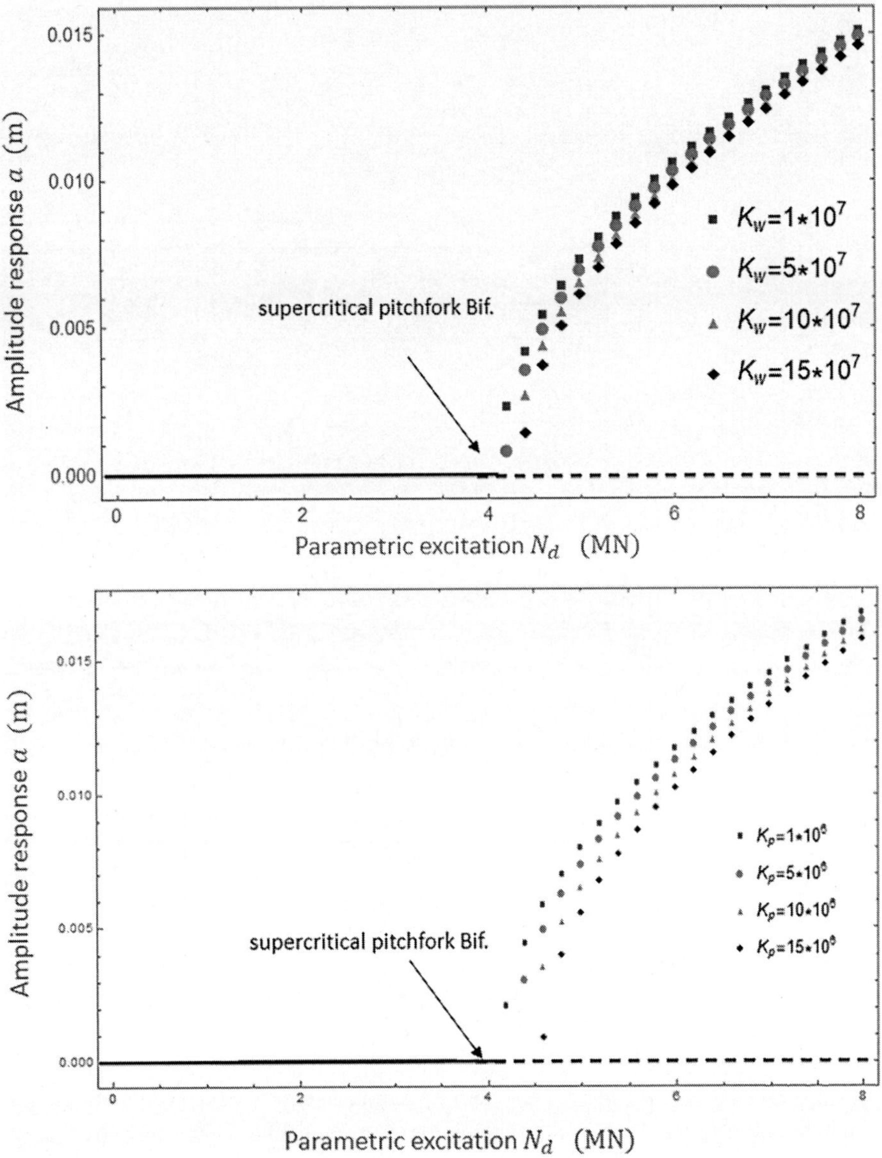

FIGURE 9.6 Amplitude response versus force amplitude of parametric excitation curves: Effect of (a) Winkler stiffness and (b) Pasternak stiffness at $\sigma = 20$.

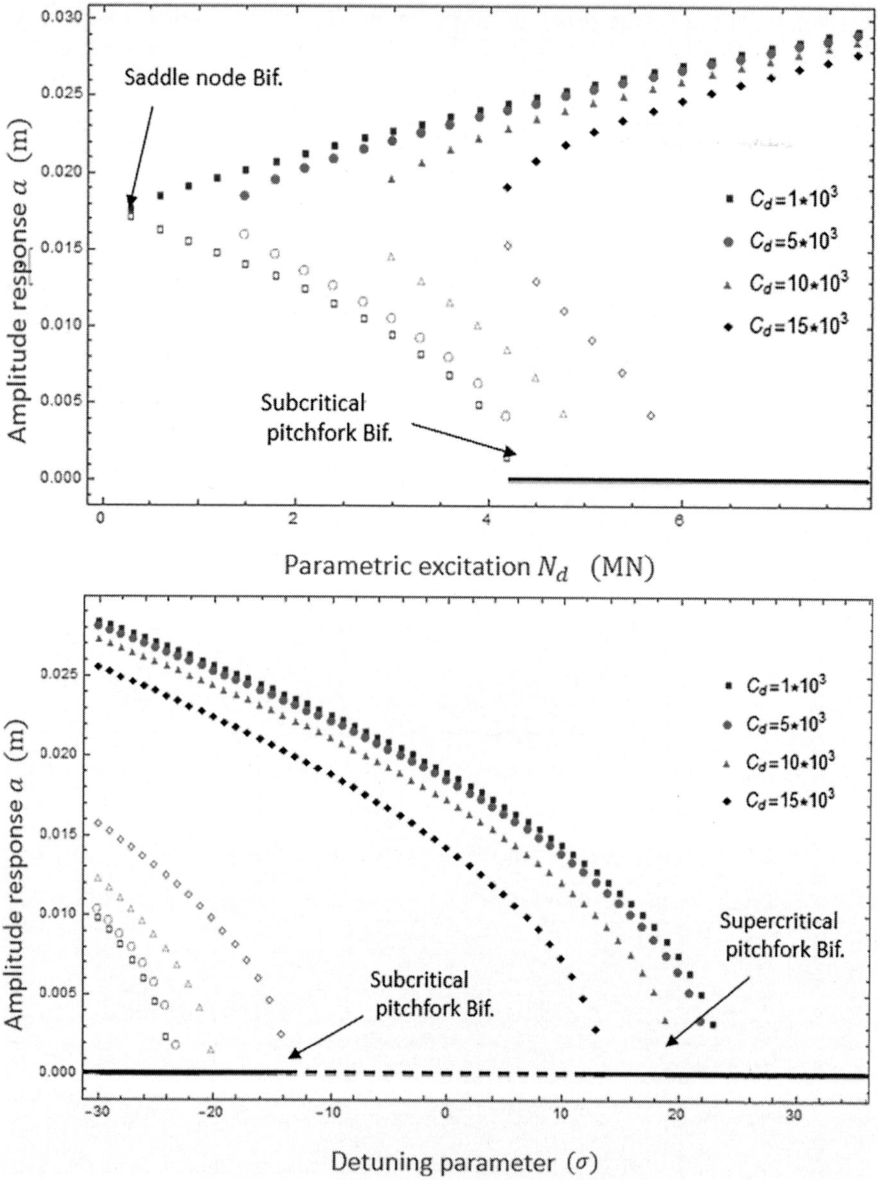

FIGURE 9.7 Effect of viscous damping on (a) amplitude response versus force amplitude of parametric excitation curves at $\sigma = -20$ and (b) amplitude response versus detuning parametric curves.

greater influence on the amplitude response compared to the effect of the foundation parameters, although in a similar manner. Furthermore, it has been shown that reducing the viscous damping parameter causes a displacement of bifurcation points, with a more pronounced effect on saddle points compared to subcritical pitchfork bifurcation sites. Furthermore, it is evident that as the viscous damping parameter increases, while keeping the parametric excitation constant, the unstable region between the supercritical pitchfork and subcritical pitchfork bifurcation points decreases in size, resulting in a larger disparity between the stable and unstable curves.

Figure 9.8(a) and Table 9.4 provide the amplitude response values of the system for different ratios of magnetostrictive layer thickness to total plate thickness

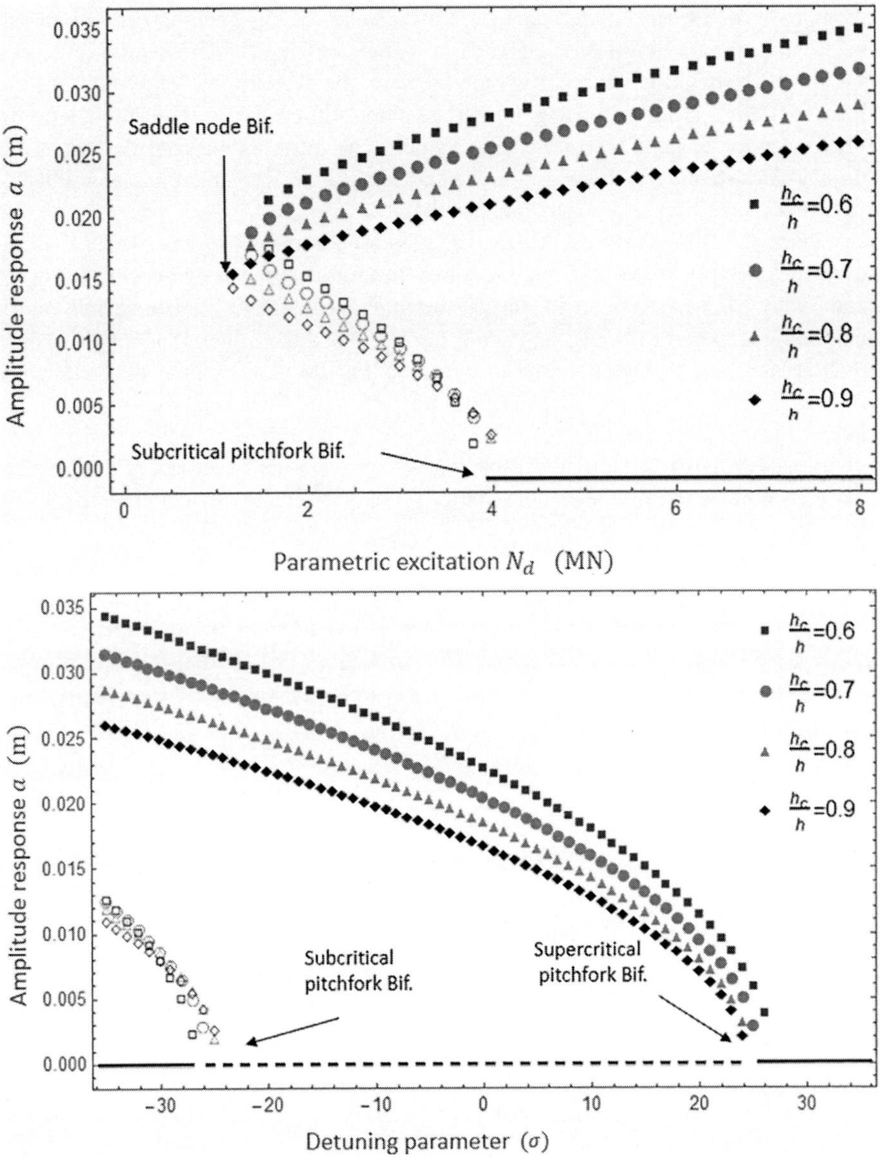

FIGURE 9.8 Effect of thickness ratios on (a) amplitude response versus force amplitude of parametric excitation curves at σ = −20 and (b) amplitude response versus detuning parametric curves.

($\frac{h_c}{h}$) and amplitudes of parametric excitation (N_d). While the location of the subcritical pitchfork bifurcation point remains same, the thickness ratio has a substantial impact on the amplitude response magnitude. It was observed that when the thickness ratio increases, the amplitude response decreases. One potential explanation for this phenomenon might be that when the phrase related to magnetostrictive material is used in the context of damping, an increase in damping leads to a reduction in the magnitude of the amplitude response. Furthermore, it has been shown that saddle bifurcation points manifest at reduced amplitudes as the thickness ratio increases. Figure 9.8(b) examines the correlation between the stability of the system and the thickness ratio. It does this by graphing the amplitude response against the detuning parameter for various thickness ratio values. The findings indicate that when the thickness ratio lowers, there is a noticeable separation between stable and unstable pitchforks, resulting in a wider stable zone.

Figure 9.9 illustrates the impact of the aspect ratio (*a/b*) on the nonlinear dynamic stability of the system. Based on the information provided in Figure 9.9 and Table 9.4, it can be concluded that the impact of changes in aspect ratio on the amplitude response of the system is more pronounced than the impact of changes in thickness ratio that were previously described. Furthermore, it is evident that when the aspect ratio increases, the stable and unstable pitchforks separate, resulting in a larger distance between them.

By analyzing Figure 9.10 and Table 9.4, which examine the impact of temperature variations on the amplitude response of the system, it becomes evident that this factor has a negligible influence on the system's dynamic response and stability.

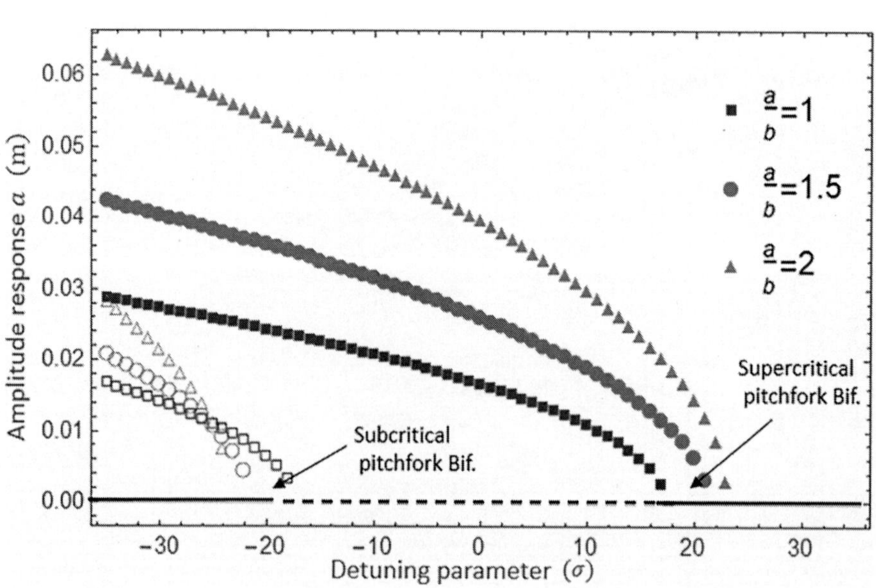

FIGURE 9.9 Amplitude response versus detuning parametric curves: Effect aspect ratio at $N_d = 5\times10^6\,N$.

The impact of the fiber angle is seen in Figure 9.11. This graphic displays the amplitude response versus the parametric excitation parameter at three distinct angles: $\theta = 0, 45$, and 90 degrees. The fiber angle has a noticeable impact on the amplitude response of the composite plate, while the orientation effects the elastic

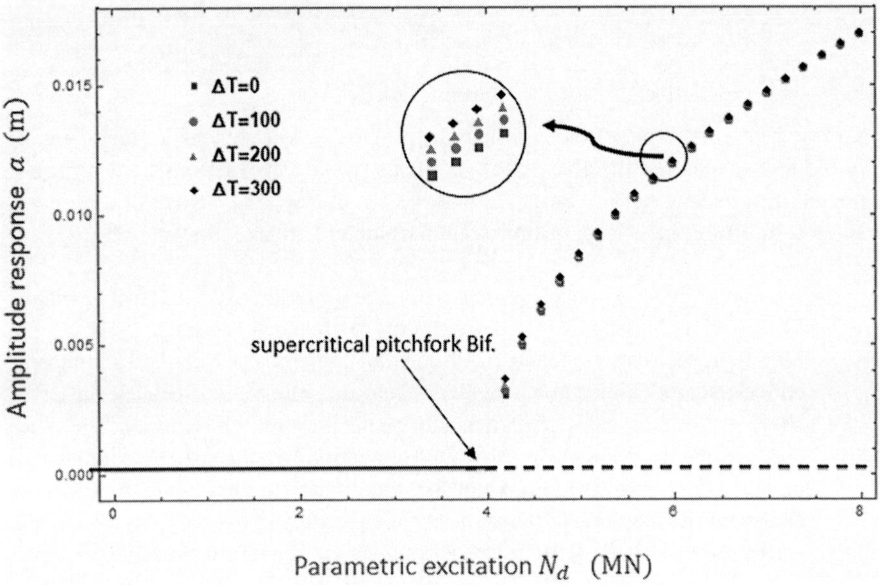

FIGURE 9.10 Amplitude response versus force amplitude of parametric excitation curves: Effects of temperature variation at σ = 20.

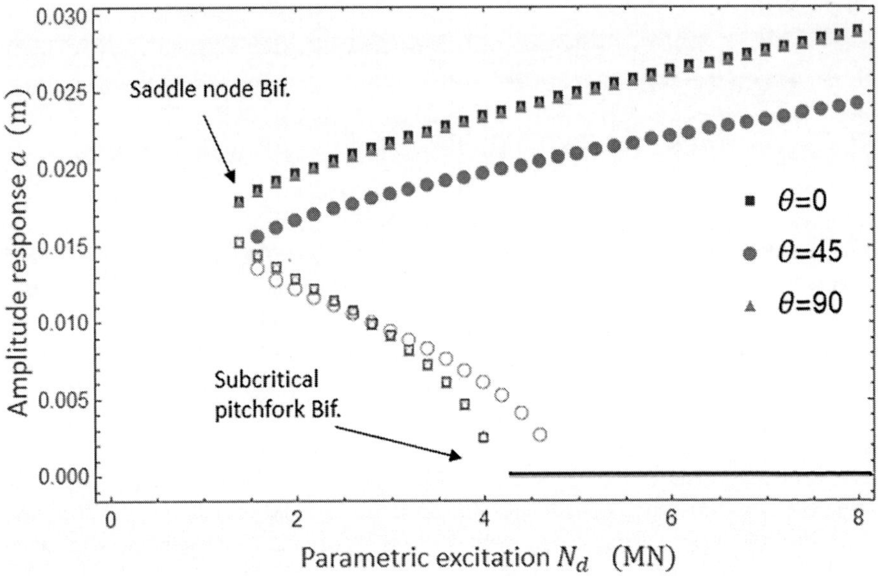

FIGURE 9.11 Amplitude response versus force amplitude of parametric excitation curves: Effects of fiber angle at σ = −20.

characteristics of the layers and the stiffness of the MSCP. This diagram illustrates the impact of the fiber angle on a square plate with a length-to-width ratio of 1. As anticipated, the outcomes for angles of *0* and *90* degrees are comparable for a square plate. The greatest value is obtained for the outcomes of these two angles, whereas the amplitude response for $\theta = 45$ is mostly lower. Additionally, it was determined that the impact of fiber angle on thick plates is greater than on thin plates.

9.4.7 Conclusion

The primary objective of this work is to provide a comprehensive understanding of the use of magnetostrictive materials in sandwich constructions, which have the potential to be utilized in various composite structures. The results of a numerical study examining the impact of important parameters are as follows:

- By adjusting the feedback velocity gain, the vibration of MSCP can be managed, and increasing this parameter leads to greater damping.
- Modifying the velocity feedback gain does not influence the subcritical and supercritical pitchfork bifurcation points, but it may cause changes to saddle points.
- Viscous damping has a greater impact on the amplitude response of the system compared to the stiffness of the foundation. Additionally, its influence on the saddle point is more noticeable than its effect on the pitchfork bifurcation point when the circumstances are subcritical.
- As the thickness ratio falls, the space between the two bifurcation sites widens.
- An increase in aspect ratio causes stable and unstable pitchforks to diverge, resulting in a broader unstable region.
- The temperature does not have a substantial impact on the dynamic behavior and stability of the system.

9.5 CASE STUDY 2

9.5.1 Introduction

The objective of this work is to examine the nonlinear vibration and instability of a sandwich plate. The plate consists of an auxetic honeycomb core and a carbon nanotube reinforced composite (CNTRC) face layer. The plate is placed on a viscous elastic foundation and subjected to parametric excitation. The analytical model incorporates the Hamilton principle and utilizes nonlinear strain-displacement relations derived from the von Karman theory and the first shear deformation plate theory (FSDT).

9.5.2 Mathematical Model

This research examines a sandwich plate composed of three layers: two face sheets reinforced with nanocomposite carbon nanotubes (CNTs) and an auxetic core layer supported by Winkler-Pasternak foundations. Figure 9.12 displays the measurements and proportions of the sandwich plate, which include the length (a), width (b), and the height of both the core (h_c) and the face layers (h_t) [10].

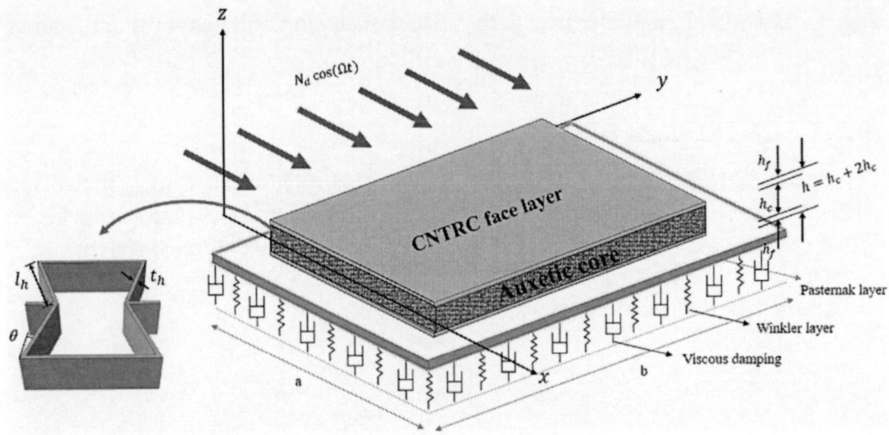

FIGURE 9.12 A schematic diagram of sandwich plate with an auxetic honeycomb core and a carbon nanotube reinforced composite (CNTRC) face layer placed on a visco-Winkler-Pasternak medium.

9.5.3 Auxetic Core Materials

The study presents a plate that has a honeycomb core with an auxetic property, meaning it has a negative Poisson's ratio. The core material consists of unit cells that possess important characteristics that are essential to its properties. The work uses the formulae developed by Zhu et al. to compute the mechanical characteristics of the honeycomb. The material has a distinctive characteristic known as a negative Poisson ratio, which has the potential to provide benefits by lowering stress concentrations and improving resilience [11, 12].

$$E_1^C = E\left(\frac{t_h}{l_h}\right)^3 \frac{\cos\theta}{\left(\frac{h_h}{l_h}+\sin\theta\right)\sin^2\theta}, \qquad E_2^C = E\left(\frac{t_h}{l_h}\right)^3 \frac{\left(\frac{h_h}{l_h}+\sin\theta\right)}{\cos^3\theta}$$

$$G_{12}^C = E\left(\frac{t_h}{l_h}\right)^3 \frac{\left(\frac{h_h}{l_h}+\sin\theta\right)}{\left(\frac{h_h}{l_h}\right)^2\left(1+2\frac{h_h}{l_h}\right)\cos\theta}, \quad G_{13}^C = G\frac{t_h}{l_h}\frac{\cos\theta}{\left(\frac{h_h}{l_h}+\sin\theta\right)}, \quad G_{23}^C = G\frac{t_h}{l_h}\frac{1+2\sin^2\theta}{2\cos\theta\left(\frac{h_h}{l_h}+\sin\theta\right)} \tag{9.80}$$

$$\nu_{12}^C = \frac{\cos^2\theta}{\left(\frac{h_h}{l_h}+\sin\theta\right)\sin\theta}, \qquad \rho^C = \rho\frac{\frac{t_h}{l_h}\left(\frac{h}{l}+2\right)}{2\cos\theta\left(\frac{h_h}{l_h}+\sin\theta\right)}$$

The variables t_h, l_h, and θ represent the thickness, length of the vertical cell, and the angle of the rib, respectively.

9.5.4 CNTRC Face Sheets

The study introduces three variations of sandwich plates, consisting of two face sheets composed of poly(methyl methacrylate) (PMMA) that are reinforced with single-walled carbon nanotubes (SWCNTs). The panels represent three different configurations of carbon nanotube reinforcement (CNTRCs), where the volume proportion of carbon nanotubes may be expressed as follows:

$$\text{FG-UU} \rightarrow V_{CNT} = V_{CNT}^{*}$$

$$\text{FG-OO} \rightarrow V_{CNT} = \begin{cases} 2\left(1-\left|\dfrac{2z-h_c-h_f}{-h_f}\right|\right)V_{CNT}^{*} & \text{(top layer)} \\ 2\left(1-\left|\dfrac{2z+h_c+h_f}{-h_f}\right|\right)V_{CNT}^{*} & \text{(bottom layer)} \end{cases} \tag{9.81}$$

$$\text{FG-XX} \rightarrow V_{CNT} = \begin{cases} 2\left(\left|\dfrac{2z-h_c-h_f}{-h_f}\right|\right)V_{CNT}^{*} & \text{(top layer)} \\ 2\left(\left|\dfrac{2z+h_c+h_f}{-h_f}\right|\right)V_{CNT}^{*} & \text{(bottom layer)} \end{cases}$$

The equation $V_{CNT}^{*} = \dfrac{\omega_{CNT}}{\omega_{CNT} + (\rho_{CNT}/\rho_m) - (\rho_{CNT}/\rho_m)\omega_{CNT}}$ represents the relationship for V_{CNT}^{*}. Here, the subscripts CNT and m refer to carbon nanotubes and the matrix, respectively. The variables ω and ρ denote mass fraction and density, respectively. The effective shear and Young's modulus of a CNTRC face sheet may be expressed as shown next:

$$E_{11}^{f} = \eta_1 V_{CNT} E_{11}^{CNT} + v_m E_m, \quad E_{22}^{f} = \frac{\eta_2}{\dfrac{V_{CNT}}{E_{22}^{CNT}} + \dfrac{v_m}{E_m}}, \quad G_{12}^{f} = \frac{\eta_3}{\dfrac{V_{CNT}}{G_{12}^{CNT}} + \dfrac{v_{mb}}{G_m}}, \tag{9.82}$$

$$G_{13}^{f} = G_{12}^{f}, \qquad G_{23}^{f} = 1.2 G_{12}^{f}$$

V represents the volume fractions in the given context. The values of the correction parameters η_i are shown in Table 9.5. The CNTRC face sheets possess effective Poisson's ratio and thermal expansion coefficients as follows:

$$\nu_{12}^{f} = \nu_{12}^{CNT} V_{CNT}^{*} + v_m V_m$$

$$\alpha_{11}^{f} = \frac{V_m E_m \alpha_m + V_{CNT} E_{11}^{CNT} \alpha_{11}^{CNT}}{V_m E_m + V_{CNT} E_{11}^{CNT}} \tag{9.83}$$

$$\alpha_{22}^{f} = (1+\nu_m) V_m \alpha_m + (1+\nu_{12}^{CNT}) V_{CNT} \alpha_{22}^{CNT} - \nu_{12}\alpha_{11}$$

This work uses the first-order shear deformation theory to establish the governing equations for sandwich plates that have auxetic honeycomb cores and are placed on visco-Pasternak foundations. These plates are subjected to parametric excitation. The strain displacement, as described by the von Karman nonlinear relation, may be expressed as follows:

$$\begin{Bmatrix} \varepsilon_x \\ \varepsilon_y \\ \gamma_{xy} \\ \gamma_{yz} \\ \gamma_{xz} \end{Bmatrix} = \begin{Bmatrix} \varepsilon^0{}_x \\ \varepsilon^0{}_y \\ \gamma^0{}_{xy} \\ \gamma^0{}_{yz} \\ \gamma^0{}_{xz} \end{Bmatrix} + z \begin{Bmatrix} \varepsilon^1{}_x \\ \varepsilon^1{}_y \\ \gamma^1{}_{xy} \\ \gamma^1{}_{yz} \\ \gamma^1{}_{xz} \end{Bmatrix} = \begin{Bmatrix} \dfrac{\partial u}{\partial x} + \dfrac{1}{2}\left(\dfrac{\partial w}{\partial x}\right)^2 \\ \dfrac{\partial v}{\partial y} + \dfrac{1}{2}\left(\dfrac{\partial w}{\partial y}\right)^2 \\ \dfrac{\partial u}{\partial y} + \dfrac{\partial v}{\partial x} + \dfrac{\partial w}{\partial x}\dfrac{\partial w}{\partial y} \\ \dfrac{\partial w}{\partial y} + \varphi_y \\ \dfrac{\partial w}{\partial x} + \varphi_x \end{Bmatrix} + z \begin{Bmatrix} \dfrac{\partial \varphi_x}{\partial x} \\ \dfrac{\partial \varphi_y}{\partial y} \\ \dfrac{\partial \varphi_x}{\partial y} + \dfrac{\partial \varphi_y}{\partial x} \\ 0 \\ 0 \end{Bmatrix} \tag{9.84}$$

The equations provided indicate that u, v, and w represent displacement coefficients for the coordinates (x, y, z), while φ_x and φ_y represent the rotations of the normals at the midpoint relative to the x and y axes, respectively.

The strain-stress link between material characteristics is dependent on temperature.

$$\begin{Bmatrix} \sigma_x \\ \sigma_y \\ \sigma_{xy} \\ \sigma_{yz} \\ \sigma_{xz} \end{Bmatrix} = \begin{bmatrix} Q_{11} & Q_{12} & 0 & 0 & 0 \\ Q_{12} & Q_{22} & 0 & 0 & 0 \\ 0 & 0 & Q_{66} & 0 & 0 \\ 0 & 0 & 0 & Q_{44} & 0 \\ 0 & 0 & 0 & 0 & Q_{55} \end{bmatrix} \begin{Bmatrix} \varepsilon_x \\ \varepsilon_y \\ \gamma_{xy} \\ \gamma_{yz} \\ \gamma_{xz} \end{Bmatrix} \tag{9.85}$$

Q_{ij} represents the lowered stiffnesses of the layer under plane stress, as determined by the material coefficients.

The formula for calculating the moment and force resultants is determined based on displacement gradients.

$$\begin{Bmatrix} N_x \\ N_y \\ N_{xy} \\ M_x \\ M_y \\ M_{xy} \end{Bmatrix} = \begin{bmatrix} A_{11} & A_{12} & A_{16} & B_{11} & B_{12} & B_{16} \\ A_{12} & A_{22} & A_{26} & B_{12} & B_{22} & B_{26} \\ A_{16} & A_{26} & A_{66} & B_{16} & B_{26} & B_{66} \\ B_{11} & B_{12} & B_{16} & D_{11} & D_{12} & D_{16} \\ B_{12} & B_{22} & B_{26} & D_{12} & D_{22} & D_{26} \\ B_{16} & B_{26} & B_{66} & D_{16} & D_{26} & D_{66} \end{bmatrix} \begin{Bmatrix} \varepsilon^0{}_x \\ \varepsilon^0{}_y \\ \gamma^0{}_{xy} \\ \varepsilon^1{}_x \\ \varepsilon^1{}_y \\ \gamma^1{}_{xy} \end{Bmatrix} \tag{9.86}$$

$$\begin{Bmatrix} Q_x \\ Q_y \end{Bmatrix} = K \begin{bmatrix} A_{55} & A_{45} \\ A_{45} & A_{44} \end{bmatrix} \begin{Bmatrix} \gamma^0{}_{yz} \\ \gamma^0{}_{xz} \end{Bmatrix}$$

The shear correction factor is denoted as K, whereas the extensional stiffness is represented by A_{ij}, the bending stiffness by B_{ij}, and the bending-extensional coupling stiffness by D_{ij}. These values are computed using the following formula:

$$\left(A_{ij}, B_{ij}, D_{ij}\right) = \int_{\frac{h_c}{2}}^{\frac{h_c}{2}+h_f} Q_{ij}^{face}\left(1, z, z^2\right) dz + \int_{-\frac{h_c}{2}}^{\frac{h_c}{2}} Q_{ij}^{core}\left(1, z, z^2\right) dz +$$
$$\int_{-\frac{h_c}{2}-h_f}^{-\frac{h_c}{2}} Q_{ij}^{face}\left(1, z, z^2\right) dz \quad (i, j = 1,2,6) \tag{9.87}$$

The nonlinear equilibrium equations of the sandwich plate may be found using Hamilton's principle.

$$\begin{gathered}
\frac{\partial N_x}{\partial x} + \frac{\partial N_{xy}}{\partial y} = I_0 \frac{\partial^2 u}{\partial t^2} + I_1 \frac{\partial^2 \psi_x}{\partial t^2} \\
\frac{\partial N_{xy}}{\partial x} + \frac{\partial N_y}{\partial y} = I_0 \frac{\partial^2 v}{\partial t^2} + I_1 \frac{\partial^2 \psi_y}{\partial t^2} \\
\frac{\partial}{\partial x}\left(N_x \frac{\partial w}{\partial x} + N_{xy} \frac{\partial w}{\partial x}\right) + \frac{\partial}{\partial y}\left(N_{xy} \frac{\partial w}{\partial x} + N_y \frac{\partial w}{\partial x}\right) + \frac{\partial Q_x}{\partial x} + \frac{\partial Q_y}{\partial y} \\
I_0 \frac{\partial^2 v}{\partial t^2} + K_w w - k_p \left(\frac{\partial^2 w}{\partial x^2} + \frac{\partial^2 w}{\partial y^2}\right) \\
\frac{\partial M_x}{\partial x} + \frac{\partial M_y}{\partial y} - Q_x = I_1 \frac{\partial^2 u}{\partial t^2} + I_2 \frac{\partial^2 \psi_x}{\partial t^2} \\
\frac{\partial M_{xy}}{\partial x} + \frac{\partial M_y}{\partial y} - Q_y = I_1 \frac{\partial^2 v}{\partial t^2} + I_2 \frac{\partial^2 \psi_y}{\partial t^2}
\end{gathered} \tag{9.88}$$

The equations of motion may be expressed as coefficients of (u, v, w) and (ϕ_x, ϕ_y) using the force and moment resultants.

$$\begin{gathered}
A_{11}\left(\frac{\partial^2 u}{\partial x^2} + \frac{\partial w}{\partial x}\frac{\partial^2 w}{\partial x^2}\right) + A_{12}\left(\frac{\partial^2 v}{\partial y \partial x} + \frac{\partial w}{\partial y}\frac{\partial^2 w}{\partial y \partial x}\right) \\
+A_{16}\left(\frac{\partial^2 u}{\partial y \partial x} + \frac{\partial^2 v}{\partial x^2} + \frac{\partial^2 w}{\partial x^2}\frac{\partial w}{\partial y} + \frac{\partial w}{\partial x}\frac{\partial^2 w}{\partial y \partial x}\right) + B_{11}\frac{\partial^2 \phi_x}{\partial x^2} + B_{12}\frac{\partial^2 \phi_y}{\partial y \partial x} \\
+B_{16}\left(\frac{\partial^2 \phi_x}{\partial y \partial x} + \frac{\partial^2 \phi_y}{\partial x^2}\right) + A_{16}\left(\frac{\partial^2 u}{\partial y \partial x} + \frac{\partial w}{\partial x}\frac{\partial^2 w}{\partial y \partial x}\right) \\
+A_{26}\left(\frac{\partial^2 v}{\partial y^2} + \frac{\partial w}{\partial y}\frac{\partial^2 w}{\partial y^2}\right)
\end{gathered} \tag{9.89}$$

$$A_{66}\left(\frac{\partial^2 u}{\partial y^2}+\frac{\partial^2 v}{\partial y\partial x}+\frac{\partial^2 v}{\partial y\partial x}\frac{\partial w}{\partial y}+\frac{\partial w}{\partial x}\frac{\partial^2 w}{\partial y^2}\right)$$
$$+B_{16}\frac{\partial^2\phi_x}{\partial y\partial x}+B_{12}\frac{\partial^2\phi_y}{\partial y^2}+B_{66}\left(\frac{\partial^2\phi_x}{\partial y^2}+\frac{\partial^2\phi_y}{\partial y\partial x}\right)$$
$$\left(\frac{\partial N_{xx}^T}{\partial x}+\frac{\partial N_{xx}^T}{\partial y}\right)-\left(\frac{\partial N_{xx}^P}{\partial x}+\frac{\partial N_{xx}^P}{\partial y}\right)=I_0\frac{\partial^2 u}{\partial t^2}+I_1\frac{\partial^2\psi_x}{\partial t^2}$$

$$A_{16}\left(\frac{\partial^2 u}{\partial x^2}+\frac{\partial w}{\partial x}\frac{\partial^2 w}{\partial x^2}\right)+A_{26}\left(\frac{\partial^2 v}{\partial y\partial x}+\frac{\partial w}{\partial y}\frac{\partial^2 w}{\partial y\partial x}\right)$$
$$+A_{66}\left(\frac{\partial^2 u}{\partial y\partial x}+\frac{\partial^2 v}{\partial x^2}+\frac{\partial^2 w}{\partial x^2}\frac{\partial w}{\partial y}+\frac{\partial w}{\partial x}\frac{\partial^2 w}{\partial y\partial x}\right)$$
$$+B_{16}\frac{\partial^2\phi_x}{\partial x^2}+B_{26}\frac{\partial^2\phi_y}{\partial y\partial x}+B_{66}\left(\frac{\partial^2\phi_x}{\partial y\partial x}+\frac{\partial^2\phi_y}{\partial x^2}\right)+A_{12}\left(\frac{\partial^2 u}{\partial y\partial x}+\frac{\partial w}{\partial x}\frac{\partial^2 w}{\partial y\partial x}\right)$$
$$+A_{22}\left(\frac{\partial^2 v}{\partial y^2}+\frac{\partial w}{\partial x}\frac{\partial^2 w}{\partial y^2}\right) \tag{9.90}$$
$$A_{66}\left(\frac{\partial^2 u}{\partial y^2}+\frac{\partial^2 v}{\partial y\partial x}+\frac{\partial^2 v}{\partial y\partial x}\frac{\partial w}{\partial y}+\frac{\partial w}{\partial x}\frac{\partial^2 w}{\partial y^2}\right)$$
$$+B_{12}\frac{\partial^2\phi_x}{\partial y\partial x}+B_{12}\frac{\partial^2\phi_y}{\partial y^2}+B_{22}\frac{\partial^2\phi_x}{\partial y^2}+B_{26}\left(\frac{\partial^2\phi_x}{\partial y^2}+\frac{\partial^2\phi_y}{\partial y\partial x}\right)$$
$$\left(\frac{\partial N_{xy}^T}{\partial x}+\frac{\partial N_{yy}^T}{\partial y}\right)-\left(\frac{\partial N_{xy}^P}{\partial x}+\frac{\partial N_{yy}^P}{\partial y}\right)=I_0\frac{\partial^2 v}{\partial t^2}+I_1\frac{\partial^2\psi_y}{\partial t^2}$$

$$KA_{55}\left(\frac{\partial^2 w}{\partial x^2}+\frac{\partial\phi_x}{\partial x}\right)+KA_{45}\left(\frac{\partial^2 w}{\partial y\partial x}+\frac{\partial\phi_y}{\partial x}\right)$$
$$+KA_{45}\left(\frac{\partial^2 w}{\partial y\partial x}+\frac{\partial\phi_x}{\partial y}\right)+KA_{44}\left(\frac{\partial^2 w}{\partial y^2}+\frac{\partial\phi_y}{\partial y}\right)$$
$$\frac{\partial}{\partial x}\left(N_x\frac{\partial w}{\partial x}+N_{xy}\frac{\partial w}{\partial y}\right)+\frac{\partial}{\partial y}\left(N_{xy}\frac{\partial w}{\partial x}+N_y\frac{\partial w}{\partial y}\right)$$
$$=I_0\frac{\partial^2 w}{\partial t^2}+k_w w-k_p\left(\frac{\partial^2 w}{\partial x^2}+\frac{\partial^2 w}{\partial y^2}\right) \tag{9.91}$$
$$+\left(N_d\cos(\Omega t)\right)\frac{\partial^2 w}{\partial x^2}+c_d\frac{\partial w}{\partial t}$$

$$
\begin{aligned}
&B_{11}\left(\frac{\partial^2 u}{\partial x^2}+\frac{\partial w}{\partial x}\frac{\partial^2 w}{\partial x^2}\right)+B_{12}\left(\frac{\partial^2 v}{\partial y\partial x}+\frac{\partial w}{\partial y}\frac{\partial^2 w}{\partial y\partial x}\right)+\\
&B_{16}\left(\frac{\partial^2 u}{\partial y\partial x}+\frac{\partial^2 v}{\partial x^2}+\frac{\partial^2 w}{\partial x^2}\frac{\partial w}{\partial y}+\frac{\partial w}{\partial x}\frac{\partial^2 w}{\partial y\partial x}\right)+\\
&D_{11}\frac{\partial^2 \phi_x}{\partial x^2}+D_{12}\frac{\partial^2 \phi_y}{\partial y\partial x}+D_{16}\left(\frac{\partial^2 \phi_x}{\partial y\partial x}+\frac{\partial^2 \phi_y}{\partial x^2}\right)+\\
&B_{16}\left(\frac{\partial^2 u}{\partial y\partial x}+\frac{\partial w}{\partial x}\frac{\partial^2 w}{\partial y\partial x}\right)+B_{26}\left(\frac{\partial^2 v}{\partial y^2}+\frac{\partial w}{\partial x}\frac{\partial^2 w}{\partial y^2}\right)\\
&B_{66}\left(\frac{\partial^2 u}{\partial y^2}+\frac{\partial^2 v}{\partial y\partial x}+\frac{\partial^2 v}{\partial y\partial x}\frac{\partial w}{\partial y}+\frac{\partial w}{\partial x}\frac{\partial^2 w}{\partial y^2}\right)+D_{16}\frac{\partial^2 \phi_x}{\partial y\partial x}+\\
&D_{26}\frac{\partial^2 \phi_y}{\partial y^2}+D_{66}\left(\frac{\partial^2 \phi_x}{\partial y^2}+\frac{\partial^2 \phi_y}{\partial y\partial x}\right)-\\
&KA_{55}\left(\frac{\partial w}{\partial x}+\phi_x\right)-KA_{45}\left(\frac{\partial w}{\partial y}+\phi_y\right)-\left(\frac{\partial M_{xx}^T}{\partial x}+\frac{\partial M_{xy}^T}{\partial y}\right)-\\
&\left(\frac{\partial M_{xx}^P}{\partial x}+\frac{\partial M_{xy}^P}{\partial y}-Q_x^P\right)=I_1\frac{\partial^2 u}{\partial t^2}+I_1\frac{\partial^2 \psi_x}{\partial t^2}
\end{aligned}
\tag{9.92}
$$

$$
\begin{aligned}
&B_{16}\left(\frac{\partial^2 u}{\partial x^2}+\frac{\partial w}{\partial x}\frac{\partial^2 w}{\partial x^2}\right)+B_{26}\left(\frac{\partial^2 v}{\partial y\partial x}+\frac{\partial w}{\partial y}\frac{\partial^2 w}{\partial y\partial x}\right)\\
&\qquad+B_{16}\left(\frac{\partial^2 u}{\partial y\partial x}+\frac{\partial^2 v}{\partial x^2}+\frac{\partial^2 w}{\partial x^2}\frac{\partial w}{\partial y}+\frac{\partial w}{\partial x}\frac{\partial^2 w}{\partial y\partial x}\right)+\\
&D_{16}\frac{\partial^2 \phi_x}{\partial x^2}+D_{26}\frac{\partial^2 \phi_y}{\partial y\partial x}+D_{66}\left(\frac{\partial^2 \phi_x}{\partial y\partial x}+\frac{\partial^2 \phi_y}{\partial x^2}\right)+B_{12}\left(\frac{\partial^2 u}{\partial y\partial x}+\frac{\partial w}{\partial x}\frac{\partial^2 w}{\partial y\partial x}\right)\\
&\qquad+B_{26}\left(\frac{\partial^2 v}{\partial y^2}+\frac{\partial w}{\partial x}\frac{\partial^2 w}{\partial y^2}\right)\\
&B_{26}\left(\frac{\partial^2 u}{\partial y^2}+\frac{\partial^2 v}{\partial y\partial x}+\frac{\partial^2 v}{\partial y\partial x}\frac{\partial w}{\partial y}+\frac{\partial w}{\partial x}\frac{\partial^2 w}{\partial y^2}\right)+D_{12}\frac{\partial^2 \phi_x}{\partial y\partial x}+D_{22}\frac{\partial^2 \phi_y}{\partial y^2}\\
&\qquad+D_{26}\left(\frac{\partial^2 \phi_x}{\partial y^2}+\frac{\partial^2 \phi_y}{\partial y\partial x}\right)-\\
&KA_{45}\left(\frac{\partial w}{\partial x}+\phi_x\right)-KA_{44}\left(\frac{\partial w}{\partial y}+\phi_y\right)-\left(\frac{\partial M_{xx}^T}{\partial x}+\frac{\partial M_{xy}^T}{\partial y}\right)-\left(\frac{\partial M_{xy}^P}{\partial x}+\frac{\partial M_{yy}^P}{\partial y}-Q_y^P\right)\\
&=I_1\frac{\partial^2 v}{\partial t^2}+I_2\frac{\partial^2 \psi_x}{\partial t^2}
\end{aligned}
\tag{9.93}
$$

9.5.5 Solution Method

The Galerkin technique may be used to solve these equations given straightforward boundary conditions. This approach aids in the examination of the free and forced vibration of sandwich plates with auxetic honeycomb cores and CNTRC face layers [6].

$$\begin{aligned}
u(x,y,t) &= \sum_n^{\infty}\sum_m^{\infty} U[t]\cos[\alpha x]\sin[\beta y] \\
v(x,y,t) &= \sum_n^{\infty}\sum_m^{\infty} V[t]\sin[\alpha x]\cos[\beta y] \\
w(x,y,t) &= \sum_n^{\infty}\sum_m^{\infty} W[t]\sin[\alpha x]\sin[\beta y] \\
\phi_x(x,y,t) &= \sum_n^{\infty}\sum_m^{\infty} X[t]\cos[\alpha x]\sin[\beta y] \\
\phi_y(x,y,t) &= \sum_n^{\infty}\sum_m^{\infty} Y[t]\sin[\alpha x]\cos[\beta y] \\
\alpha &= \frac{m\pi}{a}, \quad \beta = \frac{n\pi}{b}
\end{aligned} \tag{9.94}$$

The following equations will be derived by putting Equation (9.94) into Equations (9.89) through (9.93):

$$\begin{aligned}
&H_{11}W^2 + H_{12}U + H_{13}V + H_{14}X + H_{15}Y = 0 \\
&H_{21}W^2 + H_{22}U + H_{23}V + H_{24}X + H_{25}Y = 0 \\
&H_{31}W^3 + H_{32}UW + H_{33}VW + H_{34}XW + H_{35}YW + H_{36}X \\
&+H_{37}Y + H_{38}W + H_{39}\dot{W} + H_{310}\ddot{W} + H_{311}\cos(\Omega t) = 0 \\
&H_{41}W^2 + H_{42}U + H_{43}V + H_{44}X + H_{45}Y + H_{46}W = 0 \\
&H_{51}W^2 + H_{52}U + H_{53}V + H_{54}X + H_{55}Y + H_{56}W = 0
\end{aligned} \tag{9.95}$$

The coefficients of H_{ij} are given in Appendix C. In dynamics stability, the lateral direction (w) is the most crucial displacement component for various structures. Therefore, the in-plane and rotational inertia factors are disregarded. The variables $U, V, X,$ and Y were computed in relation to W using the following equations:

$$\begin{bmatrix} U \\ V \\ X \\ Y \end{bmatrix} = \begin{bmatrix} L_{11} & L_{12} \\ L_{21} & L_{22} \\ L_{31} & L_{32} \\ L_{41} & L_{42} \end{bmatrix} \begin{Bmatrix} W \\ W^2 \end{Bmatrix} \tag{9.96}$$

Appendix D offers an elucidation of L_{ij} constants in relation to H_{ij}. To get the governing equation of motion, we replace Equation (9.97) into the third formula of Equations (9.96).

$$\ddot{W}+\varepsilon c_1\dot{W}+\omega^2 W+\varepsilon K_1\cos(\Omega t)W+\varepsilon K_2 W^2+\varepsilon^2 K_3 W^3$$
$$=0\delta\int_{t_1}^{t_2}\left(T-\pi+W_{ext}\right)dt=0 \tag{9.97}$$

Where:

$$c_1=\frac{H_{39}}{H_{310}},\quad \omega^2=\frac{H_{38}+H_{36}L_{31}+H_{37}L_{41}}{H_{310}},\quad K_1=\frac{H_{311}}{H_{310}}$$
$$K_2=\frac{H_{32}L_{11}+H_{33}L_{21}+H_{34}L_{31}+H_{36}L_{32}+H_{35}L_{41}+H_{37}L_{42}}{H_{310}},$$
$$K_3=\frac{H_{31}+H_{32}L_{12}+H_{33}L_{22}+H_{34}L_{32}+H_{35}L_{42}}{H_{310}}$$

9.5.5.1 Multiple Scale Method

Nayfeh and Mook suggest a technique of solving Equation (9.98) using several scales. They derive the second-order approximation of W as follows [7]:

$$W(t)=W_0(T_0,T_1,T_2)+\varepsilon W_1(T_0,T_1,T_2)+\varepsilon^2 W_2(T_0,T_1,T_2)$$

Where:

$$T_0=t,\quad T_1=\varepsilon t,\quad T_2=\varepsilon^2 t \tag{9.99}$$

The following partial differential equations may be obtained by substituting Equation (9.99) into Equation (9.98) and assessing them at the same ε scale.

$$\left(\omega^2+D_0{}^2\right)W_0=0 \tag{9.100}$$

$$\left(\omega^2+D_0{}^2\right)W_1=-2D_0D_1W_0-K_2W_0{}^2-K_1\cos(\Omega t)W_0-c_1D_0W_0^2-c_1D_0w_0 \tag{9.101}$$

$$\begin{aligned}\left(\omega^2+D_0{}^2\right)W_2&=-2D_0D_1W_1-C_1D_0W_1-2D_0D_2W_0-D_1{}^2W_0\\&-C_1D_1W_0-2K_2W_0W_1-K_3W_0{}^3-K_1\cos(\Omega t)W_1\end{aligned} \tag{9.102}$$

To begin, the first task is to solve the partial differential Equation (9.101) in the following manner:

$$W_0=A(T_1,T_2)\exp[i\omega T_0]+\bar{A}(T_1,T_2)\exp[-i\omega T_0] \tag{9.103}$$

Where A and $\bar{A}$ are complex conjugates. The Equation (9.102) is expressed in the following format:

$$\left(\omega^2+D_0{}^2\right)W_1=-i\omega\left(2D_1A+c_1A\right)e^{i\omega T_0}-K_2A\bar{A}-K_2A^2e^{2i\omega T_0}$$
$$-\frac{1}{2}K_1\left(Ae^{i(\omega+\Omega)T_0}+\bar{A}e^{i(\Omega-\omega)T_0}\right)+cc \tag{9.104}$$

In Equation (9.105), the word “cc” denotes the complex conjugate. The secular and small denominator factors in Equation (9.105) may be eliminated by introducing the detuning parameter (σ) and assigning it the following value.

$$-i\omega\left(2D_1A+c_1A\right)-\frac{1}{2}K_1\bar{A}e^{i\sigma T_1}=0 \tag{9.105}$$

The answer to Equation (9.102) is as follows:

$$W_1=+\frac{K_1e^{i(\Omega+\hat{\omega})T_0}A}{2\Omega\left(\Omega+2\hat{\omega}\right)}+\frac{K_2e^{2i\hat{\omega}T_0}A^2}{3\hat{\omega}^2}-K_2\frac{A\bar{A}}{\omega^2}+cc \tag{9.106}$$

By using the same approach to eliminate the secular elements in the previous equation of motion, denoted as Equation (9.104), we substitute Equation (9.107) into Equation (9.103), resulting in the following equation:

$$\begin{aligned}&\left(\omega^2+D_0{}^2\right)W_2=J_1\left[\exp\left(i\omega T_0\right)\right]+J_2\left[\exp\left(2i\omega T_0\right)\right]+\\&J_3\left[\exp\left(3i\omega T_0\right)\right]+J_4\left[\exp\left(i\left(\omega+\Omega\right)T_0\right)\right]+\\&J_5\left[\exp\left(i\left(\omega+2\Omega\right)T_0\right)\right]+J_6\left[\exp\left(i\left(2\omega+\Omega\right)T_0\right)\right]+\\&J_7\left[\exp\left(i\left(2\omega-\Omega\right)T_0\right)\right]+J_8\left[\exp\left(i\Omega T_0\right)\right]\end{aligned} \tag{9.107}$$

By eliminating secular components from Equation (9.108), the requirements for its solvability may be stated as follows:

$$\begin{aligned}&J_1\left[\exp\left(i\omega T_0\right)\right]=0\rightarrow D_1{}^2A+2i\omega D_2A+c_1D_1A\\&+\frac{K_1^2A}{8\omega\Omega+4\Omega^2}-\frac{10K_2^2A^2\bar{A}}{3\omega^2}+3K_3A^2\bar{A}=0\end{aligned} \tag{9.108}$$

By use Equation (9.106) to exclude variables x and y from Equation (9.109), one may get the following:

$$\begin{aligned}&2i\omega D_2A+\left(-\frac{c_1{}^2}{4}+\frac{3K_1{}^2}{32\omega^2}\right)A+\left(-\frac{10K_2^2}{3\omega^2}+3K_3-\frac{2ic_1K_2}{\omega}\right)A^2\bar{A}-\\&\frac{K_1\sigma\bar{A}}{4\omega}e^{i\sigma T_1}=0\end{aligned} \tag{9.109}$$

It is easy to show that Equation (9.106) and (9.110) correspond to the first and subsequent components of the multiple-scale analysis of the given statement.

$$\begin{aligned}&2i\omega\left(\dot{A}+\varepsilon\frac{c_1}{2}A\right)+\varepsilon^2\left[\left(\frac{3K_1{}^2}{32\omega^2}-\frac{c_1{}^2}{4}\right)A+\left(3K_3-\frac{10K_2^2}{3\omega^2}\right)A^2\bar{A}\right]+\\&\frac{1}{2}\varepsilon K_1\left[1-\frac{\varepsilon\sigma}{2\omega}\right]\bar{A}\exp\left[i\varepsilon\sigma t\right]=0\end{aligned} \tag{9.110}$$

The polar representations of A and $\bar{A}$ in Equation (9.111) are as follows:

$$A(t)=\frac{1}{2}a(t)\exp[i\beta(t)],\quad \bar{A}(t)=\frac{1}{2}a(t)\exp[-i\beta(t)] \tag{9.111}$$

An amplitude and phase of the steady state response are determined by a and β. To get the governing equations, one may substitute Equation (9.112) into Equation (9.111) and then separate the real and imaginary components.

$$a\beta'=G_1a+G_2a^3+G_3a\cos[\gamma] \tag{9.112}$$

$$a'=G_4a-G_3a\sin[\gamma] \tag{9.113}$$

Where:

$$\begin{aligned}G_1&=\frac{3\varepsilon^2K_1^2}{64\omega^3}-\frac{\varepsilon^2c_1^2}{8\omega},\\ G_2&=\frac{3\varepsilon^2K_3}{8\omega}-\frac{5\varepsilon^2K_2^2}{12\omega^3},\\ G_3&=\left(\frac{\varepsilon K_1}{4\omega}\right)\left(1-\frac{\varepsilon\sigma}{2\omega}\right),\\ G_4&=-0.5\varepsilon c_1,\quad \gamma=\varepsilon\sigma t-2\beta\end{aligned} \tag{9.114}$$

Consequently, the steady-state response may be obtained by assuming the following presumption: $a'=0$. Furthermore, $\gamma'=0$ $(\gamma'=\varepsilon\sigma-2\beta')$. The modulation equations are calculated using the following formulae:

$$\left(G_1-\frac{\varepsilon\sigma}{2}\right)a+G_2a^3+G_3a\,cos[\gamma]=0 \tag{9.115}$$

$$G_4a-G_3a\,sin[\gamma]=0 \tag{9.116}$$

9.5.5.1.1 *Trivial Steady-State Response*

One may assess the stability of the trivial solution by computing the Cartesian form of the solution ($A=\frac{1}{2}(p-iq)\,exp[i\beta T_1]$) in Equation (9.111), where it is used to make the equations autonomous. Thus, the following equations are shown [13]:

$$p'=-\frac{1}{2}\varepsilon\sigma q-\frac{\varepsilon^2c_1^2q}{8\omega}+\frac{\varepsilon^2\sigma K_1q}{8\omega^2}-\frac{\varepsilon K_1q}{4\omega}+\frac{3\varepsilon^2K_1^2q}{64\omega^3}-\frac{\varepsilon c_1}{2}p \tag{9.117}$$

$$q'=\frac{1}{2}\varepsilon\sigma p+\frac{\varepsilon^2c_1^2p}{8\omega}+\frac{\varepsilon^2\sigma K_1p}{8\omega^2}-\frac{\varepsilon K_1p}{4\omega}-\frac{3\varepsilon^2K_1^2p}{64\omega^3}-\frac{\varepsilon c_1}{2}q \tag{9.118}$$

To assess the stability of the system under the condition of a simple steady-state response, we construct the Jacobian matrix for Equations (9.118) and (9.119) as shown next:

$$J = \begin{bmatrix} \dfrac{dp'}{dp} & \dfrac{dp'}{dq} \\ \dfrac{dq'}{dp} & \dfrac{dq'}{dq} \end{bmatrix}_{(p=q=0)}$$

$$J = \begin{bmatrix} -\dfrac{\varepsilon c_1}{2} & -\dfrac{1}{2}\varepsilon\sigma - \dfrac{\varepsilon^2 c_1^2}{8\omega} + \dfrac{\varepsilon^2 \sigma K_1}{8\omega^2} - \dfrac{\varepsilon K_1}{4\omega} + \dfrac{3\varepsilon^2 K_1^2}{64\omega^3} \\ \dfrac{1}{2}\varepsilon\sigma + \dfrac{\varepsilon^2 c_1^2}{8\omega} + \dfrac{\varepsilon^2 \sigma K_1}{8\omega^2} - \dfrac{\varepsilon K_1}{4\omega} - \dfrac{3\varepsilon^2 K_1^2}{64\omega^3} & -\dfrac{\varepsilon c_1}{2} \end{bmatrix} \quad (9.119)$$

Thus, we can express the determinant and trace of matrix A in the following manner:

$$\Delta = c_2^2\omega^2 + \left(\sigma\omega + \frac{\varepsilon c_2^2}{4} - \frac{3\varepsilon K_1^2}{32\omega^2}\right)^2 - \frac{K_1^2}{4}\left(1 - \frac{\varepsilon\sigma}{2\omega}\right)^2 \quad (9.120)$$

$$\tau = -\varepsilon c_2 \quad (9.121)$$

Ultimately, by assuming $c_1 > 0$, we can determine the stability of the system.

9.5.5.1.2 Non-Trivial Steady-State Response

The non-trivial steady-state response can only be evaluated when the value of "a" is not equal to zero ($a \neq 0$) in Equations (9.116) and (9.117). The outcome will be as follows:

$$\gamma' = 0 \rightarrow \left(G_1 - \frac{\varepsilon\sigma}{2}\right)a + G_2 a^3 = -G_3 a\cos[\gamma] \quad (9.122)$$

$$a' = 0 \rightarrow G_4 a = G_3 a\sin[\gamma] \quad (9.123)$$

The frequency-response equation is obtained by analyzing the phrase $(\cos^2[\gamma] + \sin^2[\gamma] = 1)$.

$$a = \left[\frac{1}{G_2}\left(\frac{\varepsilon\sigma}{2} - G_1 \pm \sqrt{G_3^2 - G_4^2}\right)\right]^{\frac{1}{2}} \quad (9.124)$$

Equation (9.125) may be used to calculate the instability and bifurcation diagrams of auxetic honeycomb cores with CNTRC face sheets in sandwich plates.

9.5.6 Results and Discussion

The sandwich plate used in this research included a honeycomb core, an auxetic conductive layer, and a carbon nanotube reinforced composite layer (CNTRC) on both the top and bottom surfaces of the honeycomb core. The material qualities used in a sandwich plate are shown in Table 9.6 and Table 9.7. The sandwich plate has certain dimensions. The interior auxetic core is characterized by $\theta^{\circ} = -45^{\circ}$, $\frac{h_h}{l_h} = 1$ and $\frac{t_h}{l_h} =$ 0.01385. Furthermore, the parameters considered for the elastic foundation and temperature condition are as follows:

$$k_w = 10^6\,Pa,\ \mathrm{k}_w = 10^5\,Pa,\ \mathrm{c}_d\text{=1200Pa.s},\ \mathrm{T} = 300\,^{o}K,\ \varepsilon\text{=0.01}$$

The influence of the geometrical parameters of the core material (namely, the cell angle and length ratio) on the amplitude of parametric excitation of sandwich plates with auxetic cores can be seen in Figure 9.13 and Figure 9.14. Unlike the cell angle,

TABLE 9.5
The Corrective Parameters for CNTs-Polymer Composites

V^*_{CNT}	η_1	η_2	η_3
0.12	**0.137**	**1.022**	**0.715**
0.17	**0.142**	**1.626**	**1.138**
0.28	**0.141**	**1.585**	**1.110**

TABLE 9.6
The Material Properties of the Core and Faces Layers

Material	E (GPa)	υ	α (×10⁻⁶/K)	α (kg/m³)
Honeycomb material	69(1–0.0053 ΔT)	0.33	23(1+0.00072 ΔT)	2700
PMMA	(3.52–0.0034 ΔT)	0.34	45(1+0.005 ΔT)	1150
CNTs	See Table 9.7	0.175	See Table 9.7	1400

TABLE 9.7
The Young's Modulus, Shear Modulus, and Thermal Expansion Coefficients of SWCNTs

T (K)	E_{11} (TPa)	E_{22} (TPa)	G_{12} (TPa)	α_{11} ($\times 10^{-6}$/K)	α_{12} ($\times 10^{-6}$/K)
300	5.6466	7.0800	1.9445	3.4584	5.1682
500	5.5308	6.9348	1.9643	4.5361	5.0189
700	5.4744	6.8641	1.9644	4.6677	4.8943

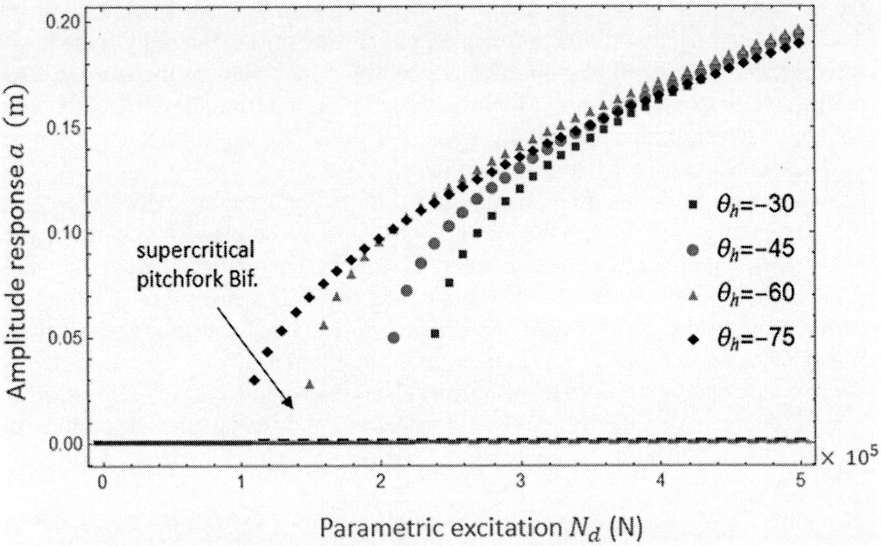

FIGURE 9.13 Influence of cell angle auxetic core on amplitude vibration versus parametric excitation amplitude curves at $\sigma = 30$.

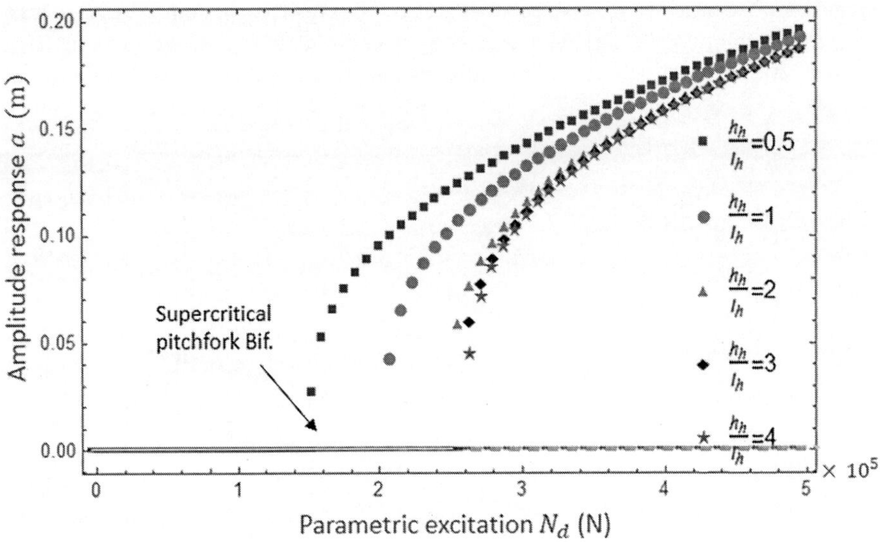

FIGURE 9.14 Influence of $\frac{h_h}{l_h}$ auxetic core on amplitude vibration versus parametric excitation amplitude curves at $\sigma = 30$

the location of the bifurcation point moves towards the left as the cell angle rises. The subcritical pitchfork bifurcation point placements shift to the right as the length ratio increases. Dashed lines represent complex unstable solutions that are not easily predictable or straightforward. Moreover, it was deduced from Figure 9.14 that augmenting the length ratio beyond the value of 2 would not have a substantial impact on the nonlinear characteristics of the sandwich plate.

Figure 9.15 illustrates three distinct forms of carbon nanotube (CNT) reinforcement on the face layer XX, OO, and UU of the sandwich plate. The purpose is to compare the nonlinear dynamic amplitude. The influence of the carbon nanotube (CNT) reinforcement on nonlinear behavior is evident. Moreover, FG-OO amplifies both the magnitude and the distance between supercritical and subcritical bifurcation points, hence, expanding the region of instability. This figure demonstrates that the selection between the other two forms of carbon nanotube (CNT) reinforcement has a little impact on the responses and extent of the unstable zone. Therefore, the reinforcement type used for the nonlinear behavior of sandwich plates in this study is FG-UU.

Figure 9.16 depicts the correlation between the amplitude response (a) and the detuning parameter (b) for different values of CNT volume percent. Figure 9.16 demonstrates that higher CNT volume percentages lead to a decrease in both the amplitude and useable area of the sandwich plate. The analysis of Figure 9.15 and Figure 9.16 has led to the conclusion that carbon nanotubes have a pivotal impact on enhancing the rigidity of a sandwich plate.

Figure 9.17 demonstrates the impact of foundation damping coefficients on both dynamic instability and bifurcation sites. Both figures clearly demonstrate that the reaction amplitude of a sandwich plate may be reduced by raising the damping coefficient. By augmenting the coefficient constant, bigger values of forced amplitude

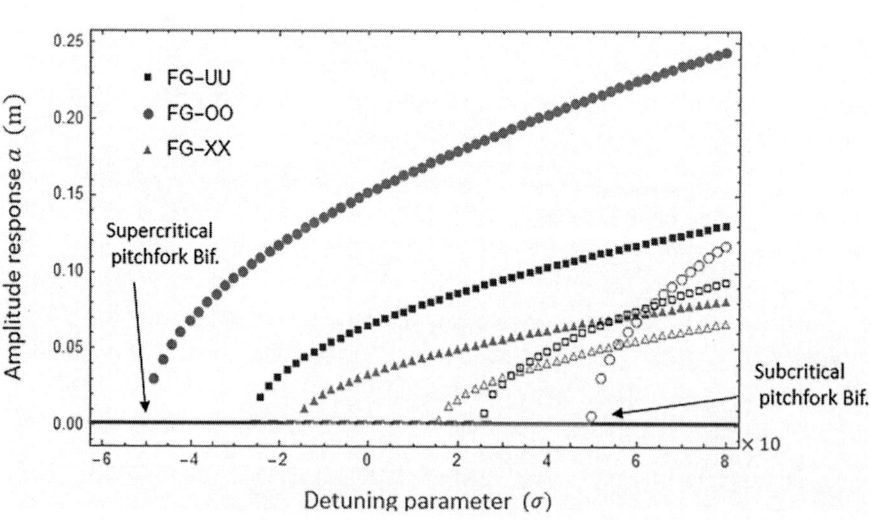

FIGURE 9.15 Effect of different types of CNT reinforcements on amplitude vibration versus detuning parameter curves at $N_d = 10^5$ (N).

will result in the appearance of all three kinds of bifurcation points, namely, supercritical, subcritical, and saddle.

The instability of the sandwich plate is influenced by the Winkler-Pasternak foundation parameters, as seen in Figure 9.18 and Figure 9.19. Therefore, an increase in the

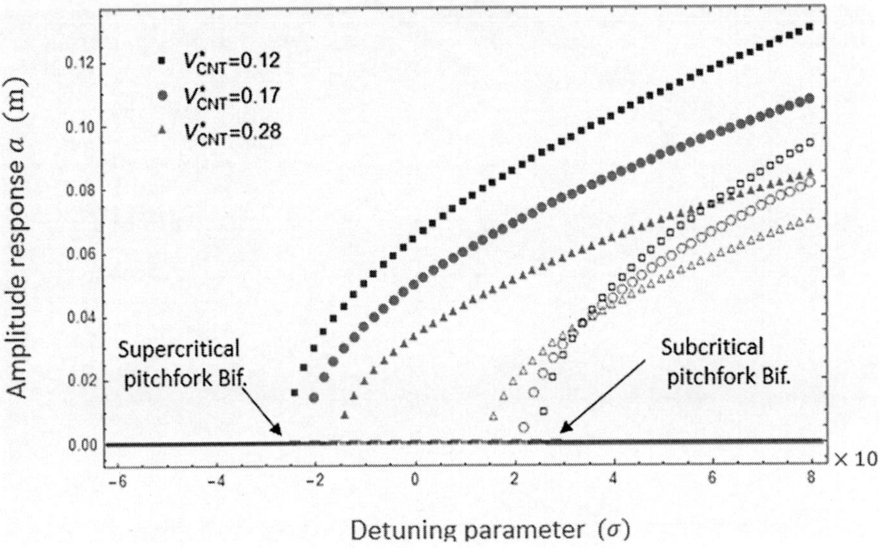

FIGURE 9.16 Effect of different values of CNT volume on amplitude vibration versus detuning parameter curves at $N_d = 10^5$ (N).

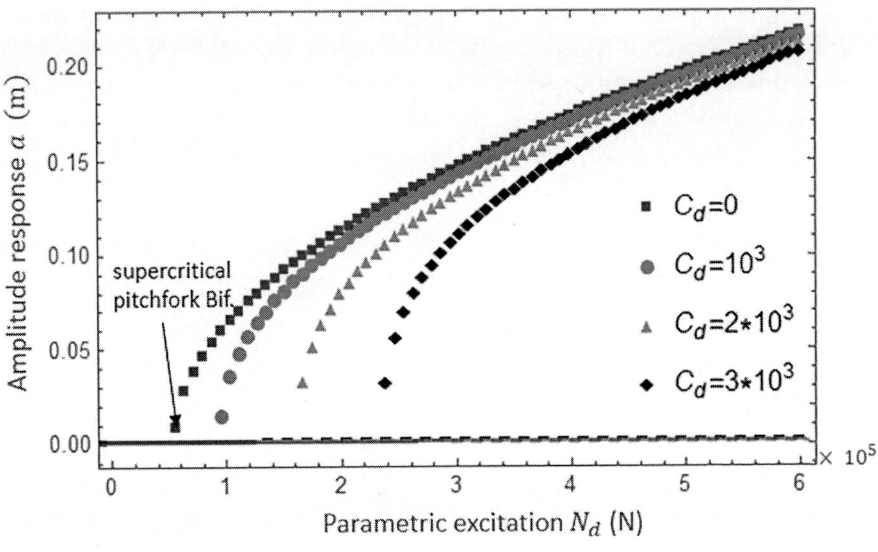

FIGURE 9.17 Influence of viscous damping on bifurcations in the nonlinear instability of a sandwich plate at (a) σ = −30 and (b) σ = 30.

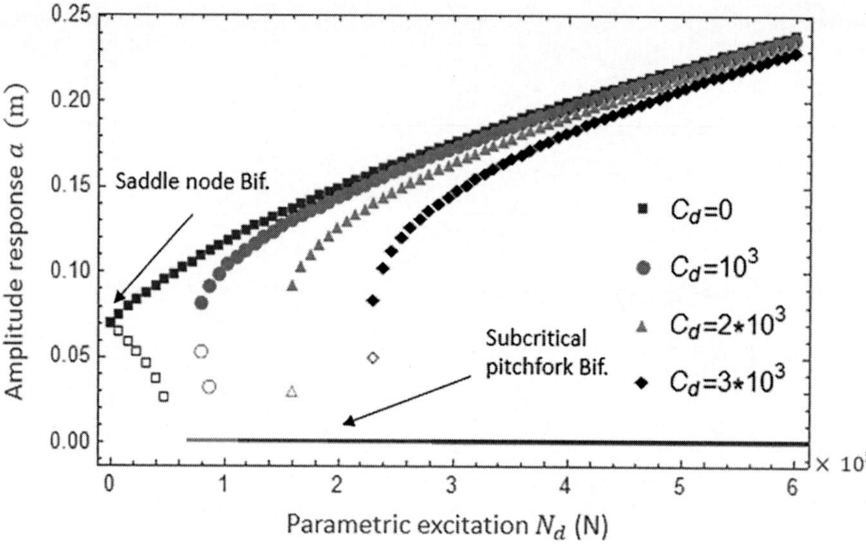

FIGURE 9.17 (Continued)

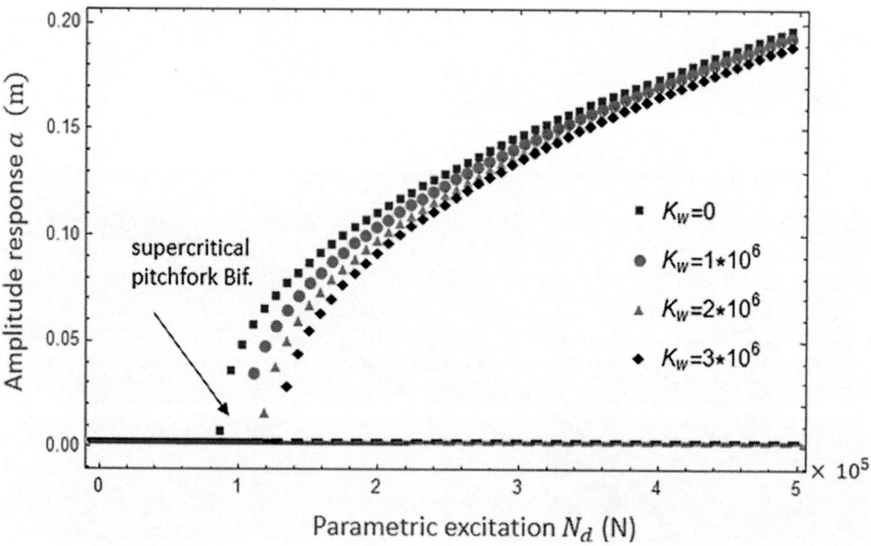

FIGURE 9.18 Influence of Winkler stiffness on bifurcations in the nonlinear instability of a sandwich plate at (a) $\sigma = -30$ and (b) $\sigma = 30$.

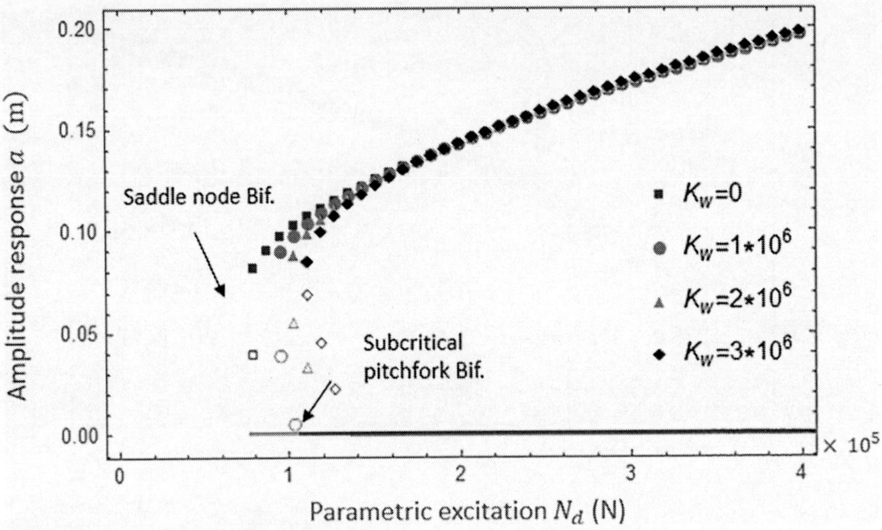

FIGURE 9.18 (Continued)

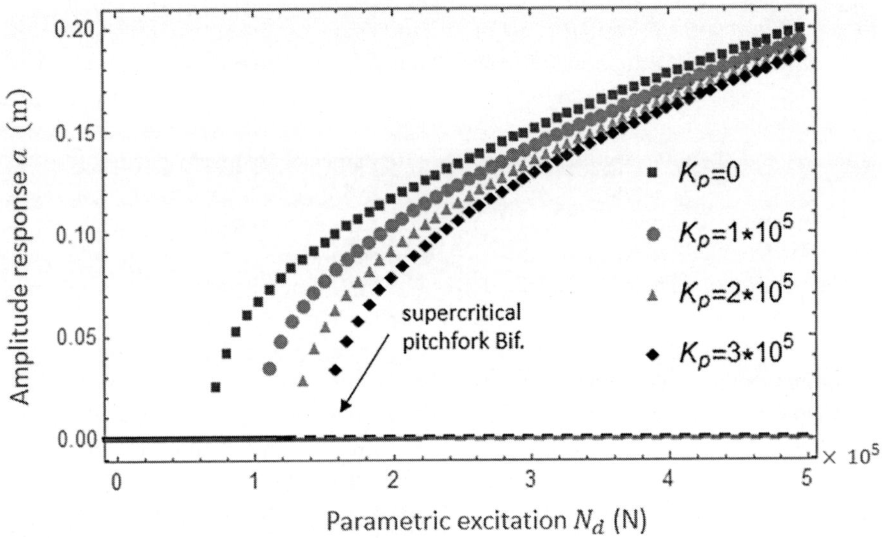

FIGURE 9.19 Influence of Pasternak stiffness on bifurcations in the nonlinear instability of a sandwich plate at (a) σ = –30 and (b) σ = 30.

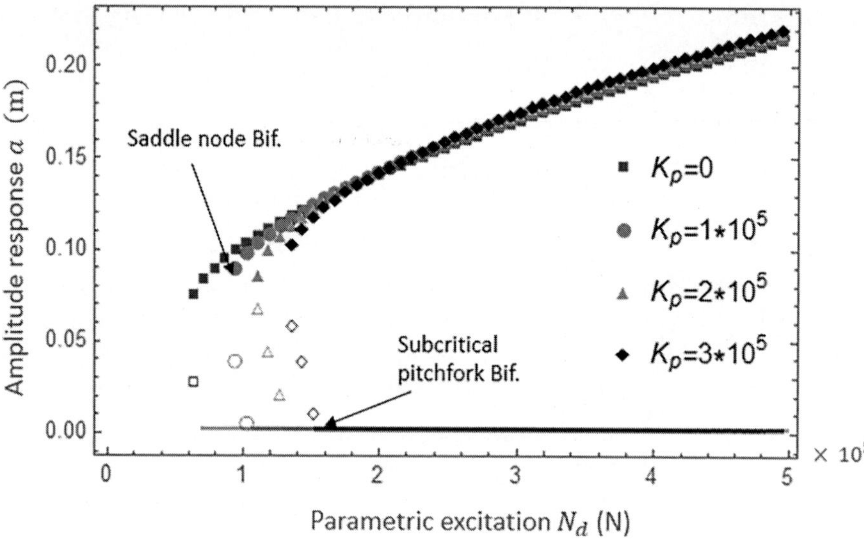

FIGURE 9.19 (Continued)

foundation parameters will lead to a decrease in the response amplitude of the sandwich plate. The placement of the bifurcation points may be influenced by modifying the stiffness characteristics of the foundation, as shown in Figure 9.18 and Figure 9.19. The crucial aspect is that the Pasternak foundation has a greater influence on the system compared to the Winkler foundation. This discrepancy may arise from the nonlinearity of the Pasternak foundation and displacement, in contrast to the Winkler foundation. The foundation serves as a dampening mechanism. Therefore, it is evident that as the foundation coefficients grow, the fundamental frequency decreases.

Figures 9.20, 9.21, and 9.22 depict the relationship between a detuning parameter and the response amplitude. These figures demonstrate how the geometric properties of sandwich plates, such as the ratio of core thickness to width, length-to-width, and slenderness ratio, affect both trivial and non-trivial solutions. All diagrams have two bifurcation points, namely, a supercritical pitchfork and a subcritical pitchfork. Figure 9.20 demonstrates that a greater ratio of (h_c / h) leads to an increased response amplitude and a wider gap between two bifurcation points, indicating an expansion of the unstable region. Similarly, Figure 9.21 demonstrates the relationship between the length-to-width ratio and the properties of the system, and this connection can be understood using the same reasoning. From the analysis of Figure 9.22, it is clear that an augmentation in the thickness of the sandwich plate will result in a reduction in the amplitude level. This occurs due to the enhanced rigidity of the sandwich plate.

The dynamic response of sandwich plates is analyzed by considering a temperature rise with three different values, as seen in Figure 9.23. An increasing temperature has a negative impact on the nonlinear dynamic response of the sandwich plate, causing an increase in its amplitude. The decrease in the rigidity of the sandwich plate may be attributed to the heat impact. Furthermore, changes in the temperature increment values do not seem to have a substantial impact on the location of the bifurcation points.

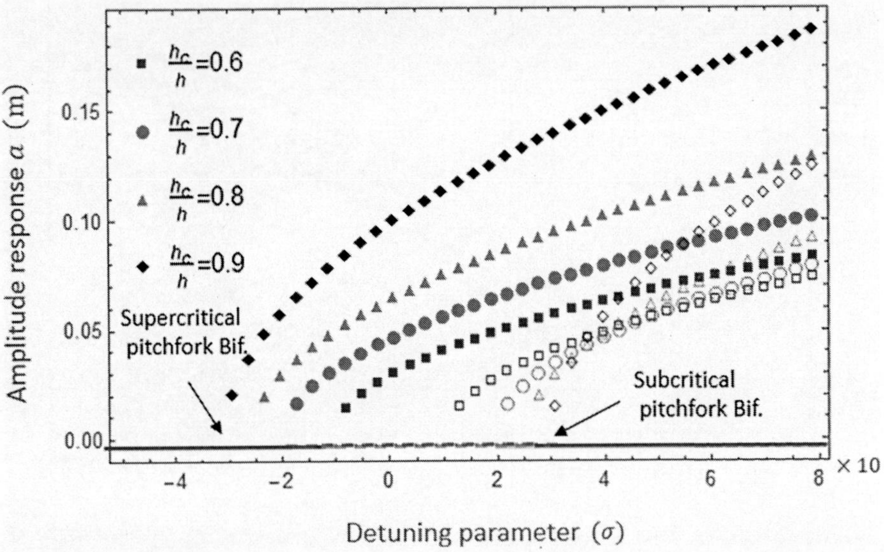

FIGURE 9.20 A comparison of different values of $\frac{h_c}{h}$ on amplitude vibration against detuning parameter diagrams at $N_d = 10^5$ (N).

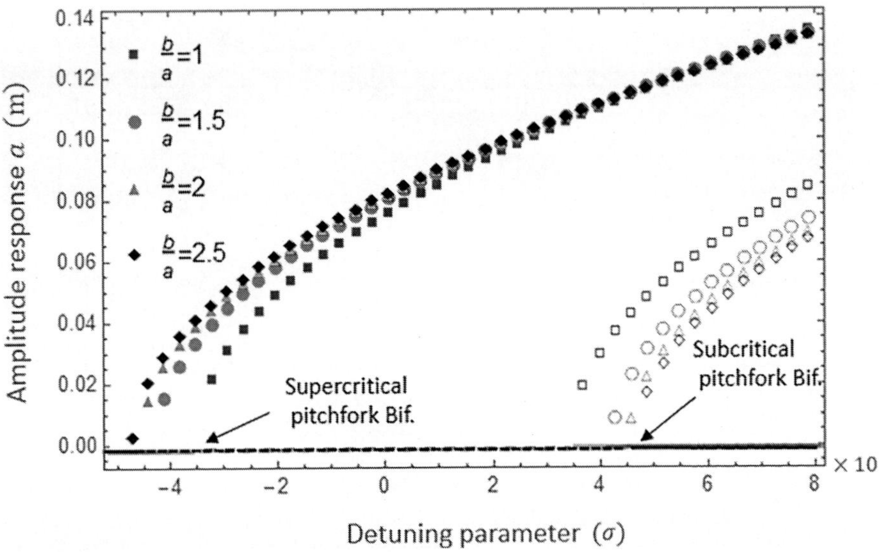

FIGURE 9.21 A comparison of different values of $\frac{b}{a}$ on amplitude vibration against detuning parameter diagrams at $N_d = 10^5$ (N).

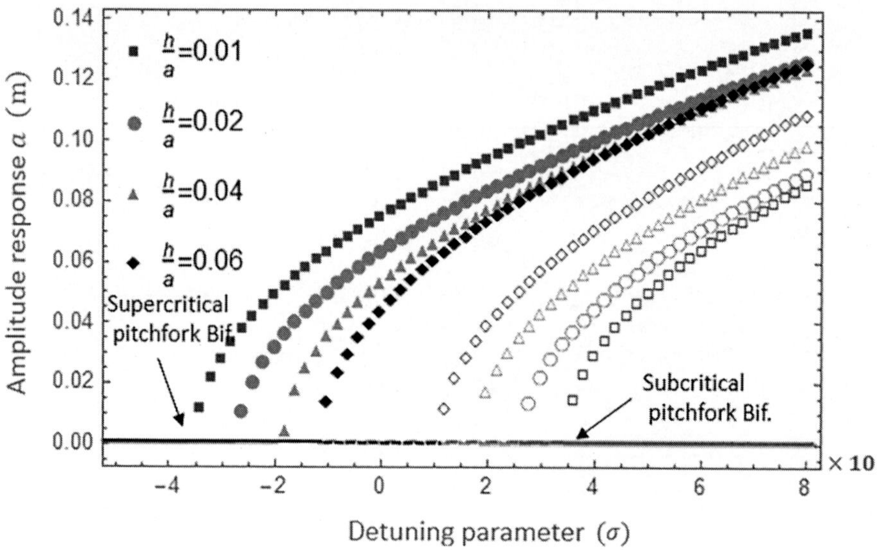

FIGURE 9.22 A comparison of different values of $\frac{h}{a}$ on amplitude vibration against detuning parameter diagrams at $N_d = 10^5$ (N).

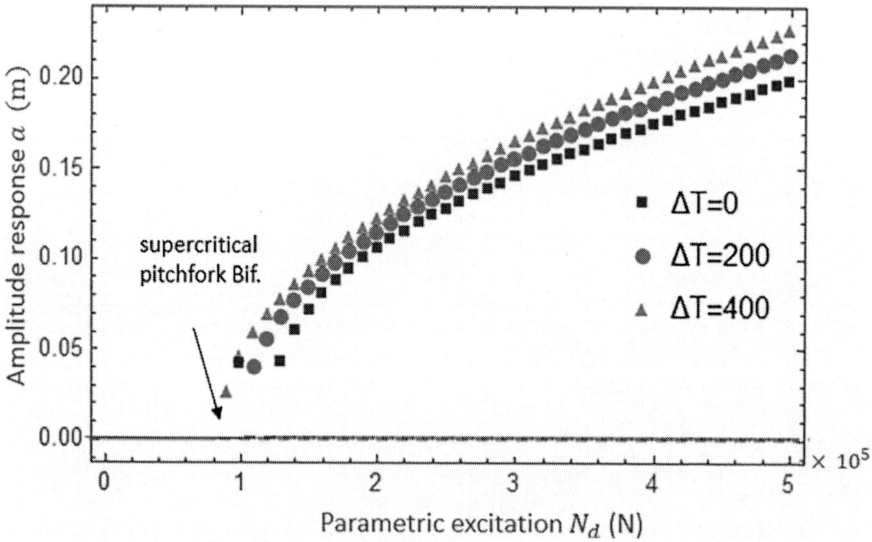

FIGURE 9.23 Influence of temperature increment on bifurcations in the nonlinear instability of a sandwich plate at (a) σ = −30 and (b) σ = 30.

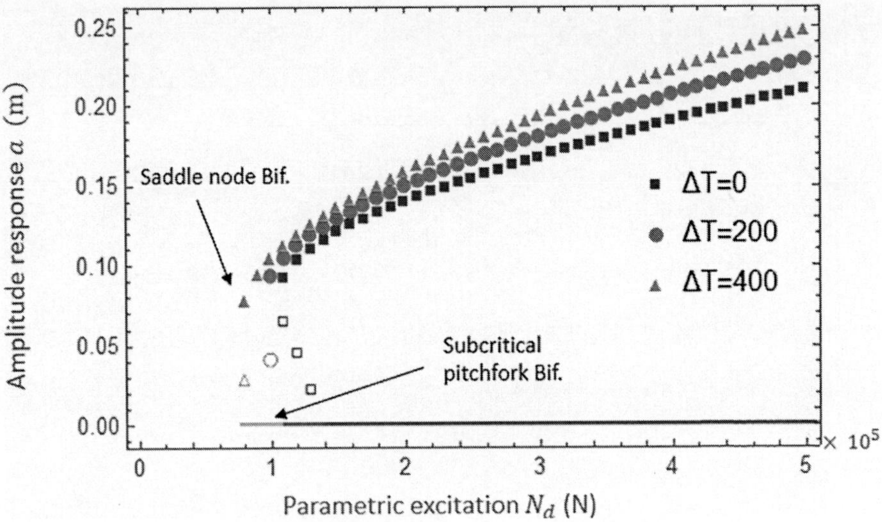

FIGURE 9.23 (Continued)

9.5.7 Conclusion

The study concludes with several key findings.

- The amplitude response curve indicates that increasing the cell angle and decreasing the length ratio of the auxetic core lead to the earlier appearance of supercritical bifurcation points.
- The type of CNT reinforcement has a significant impact on the location of both supercritical and subcritical pitchfork bifurcation, as well as the stability of the system. FG-XX sandwich plates have the most minimal unstable area.
- By increasing the volume fraction of carbon nanotubes (CNTs), the stability of the system is enhanced since it reduces the distance between two bifurcation points.
- Temperature increases have a detrimental impact on the nonlinear vibration of the sandwich plate.
- Sandwich plates can be mechanically reinforced by taking into account the volume fraction of carbon nanotubes (CNTs) and the Viscos-Winkler-Pasternak foundations.
- The unstable area of the plate increases as both the thickness of the auxetic core layer and the height-to-length ratio increase due to the geometric characteristics of the plate.

Appendix A

$$\bar{Q}_{11}^{c}=Q_{11}^{c}\cos^{4}\theta+2\left(Q_{12}^{c}+2Q_{66}^{c}\right)\sin^{2}\theta\cos^{2}\theta+Q_{22}^{c}\sin^{4}\theta$$
$$\bar{Q}_{12}^{c}=\left(Q_{11}^{c}+Q_{22}^{c}-4Q_{66}\right)\sin^{2}\theta\cos^{2}\theta+Q_{12}^{c}\left(\sin^{4}\theta+\cos^{4}\theta\right)$$
$$\bar{Q}_{22}^{c}=Q_{11}^{c}\sin^{4}\theta+2\left(Q_{12}^{c}+2Q_{66}^{c}\right)\sin^{2}\theta\cos^{2}\theta+Q_{22}^{c}\cos^{4}\theta$$
$$\bar{Q}_{16}^{c}=\left(Q_{11}^{c}-Q_{12}^{c}-2Q_{66}^{c}\right)\sin\theta\cos^{3}\theta+\left(Q_{12}^{c}-Q_{22}^{c}+2Q_{66}^{c}\right)\sin^{3}\theta\cos\theta$$
$$\bar{Q}_{26}^{c}=\left(Q_{11}^{c}-Q_{12}^{c}-2Q_{66}^{c}\right)\sin^{3}\theta\cos\theta+\left(Q_{12}^{c}-Q_{22}^{c}+2Q_{66}^{c}\right)\sin\theta\cos^{3}\theta$$
$$\bar{Q}_{66}^{c}=\left(Q_{11}^{c}+Q_{22}^{c}-2Q_{12}^{c}-2Q_{66}^{c}\right)\sin^{2}\theta\cos^{2}\theta+Q_{66}^{c}\left(\sin^{4}\theta+\cos^{4}\theta\right)$$
$$\bar{Q}_{44}^{c}=Q_{44}^{c}\cos^{2}\theta+Q_{55}^{c}\sin^{2}\theta$$
$$\bar{Q}_{45}^{c}=\left(Q_{55}^{c}-Q_{44}^{c}\right)\sin\theta\cos\theta$$
$$\bar{Q}_{55}^{c}=Q_{44}^{c}\sin^{2}\theta+Q_{55}^{c}\cos^{2}\theta$$

$$Q_{11}^{c}=\frac{E}{1-\nu^{2}}\ ,\ Q_{12}^{c}=\frac{\nu E}{1-\nu^{2}}\ ,\ Q_{22}^{c}=\frac{E}{1-\nu^{2}}\ ,\ Q_{44}^{c}=Q_{55}^{c}=Q_{66}^{c}=\frac{E}{2\left(1+\nu\right)}$$

Appendix B

$$C_{11} = -\frac{\pi(32b^2A_{11} + 16a^2A_{12} + 16a^2A_{66})}{36a^2b}, \quad C_{12} = -\frac{\pi(9ab^2\pi A_{11} + 9a^3\pi A_{66}}{36a^2b}, C_{13}$$

$$\frac{\pi(9a\pi^2 b\pi A_{12} + 9a^2b\pi A_{66}}{36a^2b}$$

$$C_{14} = -\frac{\pi(9ab^2\pi B_{11} + 9a^2\pi B_{66}}{36a^2b}, C_{15} = -\frac{\pi(9ab^2\pi B_{12} + 9a^2b\pi B_{66}}{36a^2b}, C_{16} \doteq -\frac{1}{4}be_{31}\pi h_c$$

$$C_{21} = -\frac{\pi(-16b^2A_{12} + 32a^2A_{22} + 16b^2A_{66})}{36ab^2}, C_{22} = -\frac{\pi(9ab^2\pi A_{12} + 9ab^2\pi A_{66}}{36ab^2}, C3$$

$$= -\frac{\pi(9a^2b\pi A_{22} + 9b^3\pi A_{66}}{36ab^2}$$

$$C_{24} = -\frac{\pi(9ab^2\pi B_{12} + 9ab^2\pi B_{66}}{36ab^2}, C_{25} = -\frac{\pi(9a^2b\pi B_{22} + 9b^3\pi B_{66}}{36ab^2}, \; C_{26} = -\frac{1}{4}ae_{32}\pi h_c$$

$$C_{31} = \frac{\begin{array}{l} -2.283025571109432b^4A_{11} + a^2(-4.5660511422b^2A_{12} \\ -2.283025571132a^2A_{22} + 3.044034094812b^2A_{66} \end{array}}{a^3b^3}$$

$$C_{32} = \frac{\begin{array}{l} 5.585053606381853b^3A_{11} + 5.585053606381854a^2bA_{12} \\ +2.792526803190927a^2bA_{66} \end{array}}{a^2b^2}$$

$$C_{33} = \frac{\begin{array}{l} 5.585053606381854ab^2A_{12} + 5.585053606381853\text{a}^3A_{22} \\ +2.792526803190927ab^2A_{66} \end{array}}{a^2b^2}$$

$$C_{34} = \frac{\begin{array}{l} 5.585053606381853b^3B_{11} + 5.585053606381854a^2bB_{12} \\ +2.792526803190927a^2bB_{66} \end{array}}{a^2b^2}$$

$$C_{35} = \frac{5.585053606381854ab^2B_{12} + 5.585053606381853a^3B_{22} + 2.792526803190927ab^2B_{66}}{a^2b^2}$$

$$C_{36} = -0.7853981633974483\,bK\,A_{55}$$

$$C_{37} = -0.7853981633974483\,aK\,A_{44}$$

$$C_{38} = \frac{1}{a^2b^2}(2.46740110027ab^3NTx + 2.4674011002723a^3bNTy - 2.4674011002795a^3bK\,A_{44} - 2.467401100272ab^3K\,A_{55} - 2.467401100272a^3bK_p - 2.467401100272ab^3K_p - 0.25a^3b^3K_{ww})$$

$$C_{39} = \frac{1.77775ab^3e_{31}h_c + 1.77775a^3be_{32}h_c}{a^2b^2}$$

$$C_{310} = 0.25abC_d$$

$$C_{311} = -0.25abI_0$$

$$C_{312} = 2.46740110027ab^3N_d$$

$$C_{41} = -\frac{\pi 32b^2\pi B_{11} - 16a^2\pi B_{12} + 16a^2\pi B_{66}}{36a^2b}, \quad C_{42} = -\frac{9ab^2\pi B_{11} + 9a^3\pi B_{66}}{36a^2b},$$

$$C_{43} = -\frac{9a^2b\pi B_{12} + 9a^2b\pi^2B_{66}}{36a^2b}$$

$$C_{44} = -\frac{9a^3b^2KA_{55} + 9ab^2\pi D_{11} + 9a^3\pi^2D_{66}}{36a^2b}, \quad C_{45} = -\frac{9a^2b\pi^2D_{12} + 9a^2b\pi^2D_{66}}{36a^2b},$$

$$C_{46} = -\frac{1}{4}bK\pi A_{55}$$

$$C_{51} = -\frac{16b^2\pi B_{12} - 32a^2\pi B_{22} - 9b^2\pi^2B_{66}}{36ab^2}, \quad C_{52} = -\frac{9ab^2\pi^2B_{12} + 9ab^2\pi^2B_{66}}{36a^2b},$$

$$C_{53} = -\frac{9a^2b\pi^2B_{22} + 9b^3\pi^2B_{66}}{36ab^2}$$

$$C_{54}=-\frac{9ab^2\pi^2D_{12}+9ab^2\pi^2D_{66}}{36ab^2},\ C_{55}=-\frac{a^2b^2KA_{44}+\pi^2\left(a^2D_{22}+b^2D_{66}\right)}{4ab},$$

$$C_{56}=-\frac{1}{4}aK\pi A_{44}\text{Appendix C}$$

$$S_{11}=\begin{pmatrix}C_{15}(C_{24}(C_{46}C_{53}-C_{43}C_{56})+C_{23}(-C_{46}C_{54}+C_{44}C_{56}))+\\C_{14}(C_{25}(-C_{46}C_{53}+C_{43}C_{56})+C_{23}(C_{46}C_{55}+C_{45}C_{56}))+\\C_{13}(C_{25}(C_{46}C_{54}-C_{44}C_{56})+C_{24}(-C_{46}C_{55}+C_{45}C_{56}))\end{pmatrix}/S_1^*$$

$$\begin{aligned}S_{12}=(&-C_{13}C_{25}C_{44}C_{51}+C_{13}C_{24}C_{45}C_{51}+C_{11}C_{25}C_{44}C_{53}+C_{13}C_{25}C_{41}C_{54}-\\&C_{11}C_{25}C_{43}C_{54}-C_{13}C_{21}C_{45}C_{54}+C_{11}C_{23}C_{45}C_{54}-C_{13}C_{24}C_{41}C_{55}+\\&C_{11}C_{24}C_{43}C_{55}+C_{13}C_{21}C_{44}C_{55}+C_{15}(C_{24}(-C_{43}C_{51}+C_{23}(C_{44}C_{51}-\\&C_{41}C_{54})+C_{21}(-C_{44}C_{53}+C_{43}C_{54}))-C_{11}C_{23}C_{44}C_{55}\\&\qquad+C_{14}(C_{25}(C_{43}C_{51}-C_{41}C_{53})+C_{23}(-C_{45}C_{51}+C_{41}C_{55})\\&\qquad+C_{21}(C_{45}C_{53}-C_{43}C_{55})))/S_1^*\end{aligned}$$

$$S_{13}=\begin{pmatrix}C_{16}(C_{25}(C_{44}C_{53}-C_{43}C_{54})+C_{24}(-C_{45}C_{53}+C_{23}(C_{45}C_{54}-C_{44}C_{55}))+\\C_{26}(C_{15}(-C_{44}C_{53}+C_{43}C_{54})+C_{14}(C_{45}C_{53}-C_{43}C_{55})+C_{13}(-C_{45}C_{54}+C_{44}C_{55}))\end{pmatrix}/S_1^*$$

$$\begin{aligned}S_1^*=(&((C_{15}C_{24}-C_{14}C_{25})(C_{15}C_{43}-C_{13}C_{45})-(C_{15}C_{23}-C_{13}C_{25})(C_{15}C_{44}-C_{14}C_{45}))\\&\times((C_{15}C_{24}-C_{14}C_{25})(C_{15}C_{52}-C_{12}C_{55})-(C_{15}C_{22}-C_{12}C_{25})(C_{15}C_{54}-C_{14}C_{55}))-\\&((C_{15}C_{24}-C_{14}C_{25})(C_{15}C_{42}-C_{12}C_{45})-(C_{15}C_{22}-C_{12}C_{25})(C_{15}C_{44}-C_{14}C_{45}))\\&\times((C_{15}C_{24}-C_{14}C_{25})(C_{15}C_{53}-C_{13}C_{55})-(C_{15}C_{23}-C_{13}C_{25})(C_{15}C_{54}-C_{14}C_{55})))\end{aligned}$$

$$S_{21}=\begin{pmatrix}C_{15}(C_{24}(-C_{46}C_{52}+C_{42}C_{56})+C_{22}(C_{46}C_{54}-C_{44}C_{56}))\\+C_{12}(C_{25}(-C_{46}C_{54}+C_{44}C_{56})+C_{24}(C_{46}C_{55}-C_{45}C_{56}))\\+C_{14}(C_{25}(C_{46}C_{52}-C_{42}C_{56})+C_{22}(-C_{46}-C_{55}+C_{45}C_{56}))\end{pmatrix}/S_2^*$$

$$S_{22} = (C_{12}C_{25}C_{44}C_{51} - C_{12}C_{24}C_{45}C_{51} - C_{11}C_{25}C_{44}C_{52} + C_{11}C_{24}C_{45}C_{52} \\ -C_{12}C_{25}C_{41}C_{54} + C_{11}C_{25}C_{43}C_{54} + C_{12}C_{21}C_{45}C_{54} + C_{15}(C_{24}(C_{42}C_{51} \\ +C_{41}C_{52}) + C_{22}(-C_{44}C_{51} + C_{41}C_{54}) + C_{21}(C_{44}C_{52} - C_{42}C_{54})) \\ -C_{12}C_{21}C_{45}C_{54} + C_{12}C_{24}C_{41}C_{55} - C_{11}C_{24}C_{42}C_{55} - C_{12}C_{21}C_{44}C_{55} \\ +C_{11}C_{22}C_{44}C_{55} - C_{11}C_{22}C_{45}C_{54} + C_{14}(C_{25}(-C_{42}C_{51} + C_{41}C_{52} \\ +C_{41}C_{52}) + C_{22}(C_{45}C_{51} - C_{41}C_{55}) + C_{21}(-C_{45}C_{52} + C_{42}C_{55}))) / S_2^*$$

$$S_{23} = \begin{pmatrix} C_{26}(C_{15}(C_{44}C_{52} - C_{42}C_{52}) + C_{14}(-C_{45}C_{52} + C_{42}C_{55}) + C_{12}(C_{45}C_{54} - C_{44}C_{55})) + \\ C_{16}(C_{25}(-C_{44}C_{52} + C_{42}C_{54}) + C_{24}(C_{45}C_{52} - C_{42}C_{55}) + C_{22}(-C_{45}C_{54} + C_{44}C_{55})) \end{pmatrix} / S_2^*$$

$$S_2^* = (C_{15}C_{24}C_{43}C_{52} - C_{14}C_{25}C_{43}C_{52} - C_{15}C_{23}C_{44}C_{52} + C_{13}C_{25}C_{44}C_{52} \\ +C_{14}C_{23}C_{45}C_{52} - C_{13}C_{24}C_{44}C_{52} - C_{15}C_{24}C_{42}C_{53} \\ -C_{14}C_{25}C_{42}C_{53} + C_{15}C_{52}C_{44}C_{53} - C_{12}C_{25}C_{44}C_{53} - C_{14}C_{22}C_{45}C_{53} \\ +C_{12}C_{24}C_{24}C_{45}C_{53} + C_{15}C_{23}C_{42}C_{53} - C_{13}C_{25}C_{42}C_{53} \\ -C_{15}C_{22}C_{43}C_{54} + C_{12}C_{25}C_{43}C_{54} + C_{13}C_{22}C_{45}C_{53} + C_{12}C_{23}C_{45}C_{54} \\ -C_{14}C_{23}C_{42}C_{55} + C_{13}C_{24}C_{42}C_{55} + C_{14}C_{22}C_{43}C_{55} - C_{12}C_{24}C_{43}C_{55} \\ +C_{13}C_{22}C_{44}C_{55} + C_{12}C_{23}C_{44}C_{55}))$$

$$S_{41} = \begin{pmatrix} C_{14}(C_{23}(-C_{46}C_{52} + C_{42}C_{56}) + C_{22}(C_{46}C_{53} - C_{43}C_{56})) \\ +C_{12}(C_{24}(-C_{46}C_{53} + C_{43}C_{56}) + C_{23}(C_{46}C_{54} - C_{44}C_{56})) \\ +C_{13}(C_{24}(C_{46}C_{52} - C_{42}C_{56}) + C_{22}(-C_{46} - C_{54} + C_{44}C_{56})) \end{pmatrix} / S_4^*$$

$$S_{42} = (C_{12}C_{24}C_{43}C_{51} - C_{12}C_{23}C_{44}C_{51} - C_{11}C_{24}C_{43}C_{52} + C_{11}C_{23}C_{44}C_{52} - C_{12}C_{24}C_{41}C_{53} \\ +C_{11}C_{24}C_{42}C_{53} + C_{12}C_{21}C_{44}C_{53} + C_{14}(C_{23}(C_{42}C_{51} + C_{41}C_{52}) + C_{22}(-C_{43}C_{51} \\ +C_{41}C_{53}) + C_{21}(C_{43}C_{52} - C_{42}C_{53})) - C_{12}C_{23}C_{41}C_{54} - C_{11}C_{23}C_{42}C_{54} \\ -C_{12}C_{21}C_{43}C_{54} + C_{11}C_{22}C_{43}C_{54} - C_{11}C_{22}C_{44}C_{53} + C_{13}(C_{24}(-C_{42}C_{51} \\ +C_{41}C_{52}) + C_{22}(C_{44}C_{51} - C_{41}C_{54}) + C_{21}(-C_{44}C_{52} + C_{42}C_{54}))) / S_4^*$$

$$S_{43} = \begin{pmatrix} C_{26}(C_{14}(C_{43}C_{52} - C_{42}C_{53}) + C_{13}(-C_{44}C_{52} + C_{42}C_{54}) + C_{12}(C_{44}C_{53} - C_{43}C_{54})) \\ +C_{16}(C_{24}(-C_{43}C_{52} + C_{42}C_{53}) + C_{23}(C_{44}C_{52} - C_{42}C_{54}) + C_{22}(-C_{44}C_{53} + C_{43}C_{54})) \end{pmatrix} / S_4^*$$

$$
\begin{aligned}
S_4^* = (&C_{15}C_{24}C_{43}C_{52} - C_{14}C_{25}C_{43}C_{52} - C_{15}C_{23}C_{44}C_{52} + C_{13}C_{25}C_{44}C_{52} \\
&\quad + C_{14}C_{23}C_{45}C_{52} - C_{13}C_{24}C_{45}C_{52} \\
&- C_{15}C_{24}C_{42}C_{53} - C_{14}C_{25}C_{42}C_{53} + C_{15}C_{52}C_{44}C_{53} - C_{12}C_{25}C_{44}C_{53} \\
&\quad - C_{14}C_{22}C_{45}C_{53} + C_{12}C_{24}C_{24}C_{45}C_{53} \\
&+ C_{15}C_{23}C_{42}C_{54} - C_{13}C_{25}C_{42}C_{54} - C_{15}C_{22}C_{43}C_{54} + C_{12}C_{25}C_{43}C_{54} \\
&\quad + C_{13}C_{22}C_{45}C_{54} + C_{12}C_{23}C_{45}C_{54} \\
&- C_{14}C_{23}C_{42}C_{55} + C_{13}C_{24}C_{42}C_{55} + C_{14}C_{22}C_{43}C_{55} - C_{12}C_{24}C_{43}C_{55} \\
&\quad + C_{13}C_{22}C_{44}C_{55} + C_{12}C_{23}C_{44}C_{55}))
\end{aligned}
$$

Appendix C

$$H_{11} = -\frac{\pi(32b^2A_{11} - 16a^2A_{12} + 16a^2A_{66})}{36a^2b}, H_{12} = -\frac{\pi(9ab^2\pi A_{11} - 9a^3\pi A_{66})}{36a^2b},$$

$$H_{13} = -\frac{\pi(9a^2b\pi A_{12} - 9a^2b\pi A_{66})}{36a^2b}$$

$$H_{14} = -\frac{\pi(9ab^2\pi B_{11} - 9a^3\pi B_{66})}{36a^2b}, H_{15} = -\frac{\pi(9a^2b\pi B_{12} - 9a^2b\pi B_{66})}{36a^2b},$$

$$H_{21} = -\frac{\pi(-16b^2A_{12} + 32a^2A_{22} + 16b^2A_{66})}{36ab^2}, H_{22} = -\frac{\pi(9ab^2\pi A_{12} - 9ab^2\pi A_{66})}{36ab^2},$$

$$H_{23} = -\frac{\pi(9a^2b\pi A_{22} + 9b^3\pi A_{66})}{36ab^2}$$

$$H_{24} = -\frac{\pi(9ab^2\pi B_{12} - 9ab^2\pi B_{66})}{36ab^2}, \quad H_{25} = -\frac{\pi(9a^2b\pi B_{22} + 9b^3\pi B_{66})}{36ab^2},$$

$$H_{31} = \frac{\begin{matrix} -2.283025571109432b^4A_{11} + a^2(-4.5660511422b^2A_{12} \\ -2.283025571132a^2A_{22} + 3.044034094812b^2A_{66} \end{matrix}}{a^3b^3}$$

$$H_{33} = \frac{\begin{matrix} 5.585053606381854ab^2A_{12} + 5.585053606381853a^3A_{22} \\ +2.792526803190927ab^2A_{66} \end{matrix}}{a^2b^2}$$

$$H_{34} = \frac{\begin{matrix} 5.585053606381853b^3B_{11} + 5.585053606381854a^2bB_{12} \\ +2.792526803190927a^2bB_{66} \end{matrix}}{a^2b^2}$$

$$H_{35} = \frac{\begin{matrix} 5.585053606381854ab^2B_{12} + 5.585053606381853a^3B_{22} \\ +2.792526803190927ab^2B_{66} \end{matrix}}{a^2b^2}$$

$$H_{36} = -0.7853981633974483\,bK\,A_{55}$$

$$H_{37} = -0.7853981633974483\,aK\,A_{44}$$

$$H_{38} = \frac{1}{a^2b^2}(2.46740110027ab^3NTx + 2.4674011002723a^3bNTy$$
$$-2.4674011002795a^3bK\,A_{44} - 2.467401100272ab^3K\,A_{55}$$
$$-2.467401100272a^3bK_p - 2.467401100272ab^3K_p - 0.25a^3b^3K_{ww})$$

$$H_{39} = 0.25\,ab\,C_d$$
$$H_{310} = -0.25ab\,I_0$$
$$H_{311} = 2.46740110027ab^3N_d$$

$$H_{41} = -\frac{32b^2\pi B_{11} - 16a^2\pi B_{12} + 16a^2\pi B_{66}}{36a^2b},\quad H_{42} = -\frac{9ab^2\pi B_{11} + 9a^3\pi B_{66}}{36a^2b},$$
$$H_{43} - \frac{9a^2b\pi B_{12} + 9a^2b\pi^2 B_{66}}{36a^2b}$$

$$H_{44} = -\frac{9a^3b^2KA_{55} + 9ab^2\pi D_{11} + 9a^3\pi^2 D_{66}}{36a^2b},\ HH_{45} = -\frac{9a^2b\pi^2 D_{12} + 9a^2b\pi^2 D_{66}}{36a^2b},$$
$$H_{46} = -\frac{1}{4}bK\pi A_{55}$$

$$H_{51} = -\frac{16b^2\pi B_{12} - 32a^2\pi B_{22} - 9b^2\pi^2 B_{66}}{36ab^2},\ H_{52} = -\frac{9ab^2\pi^2 B_{12} + 9ab^2\pi^2 B_{66}}{36a^2b},$$
$$H_{53} = -\frac{9a^2b\pi^2 B_{22} + 9b^3\pi^2 B_{66}}{36ab^2}$$

$$H_{54} = -\frac{9ab^2\pi^2 D_{12} + 9ab^2\pi^2 D_{66}}{36ab^2},\qquad H_{55} = -\frac{a^2b^2KA_{44} + \pi^2\left(a^2D_{22} + b^2D_{66}\right)}{4ab},$$
$$H_{56} = -\frac{1}{4}aK\pi A_{44}$$

Appendix D

$$L_{11} = \begin{pmatrix} H_{15}(H_{24}(-H_{46}H_{52} + H_{42}H_{56}) + H_{22}(H_{46}H_{54} - H_{44}H_{56})) \\ +H_{12}(H_{25}(-H_{46}H_{54} + H_{44}H_{56}) + H_{24}(H_{46}H_{55} - H_{45}H_{56})) \\ +H_{14}(H_{25}(H_{46}H_{52} - H_{42}H_{56}) + H_{22}(-H_{46} - H_{55} + H_{45}H_{56})) \end{pmatrix} / L_2^*$$

$$\begin{aligned} L_{12} = (&-H_{13}H_{25}H_{44}H_{51} - H_{13}H_{24}H_{45}H_{51} - H_{11}H_{25}H_{44}H_{53} + H_{11}H_{24}H_{45}H_{53} - \\ &H_{13}H_{25}H_{41}H_{54} - H_{11}H_{25}H_{43}H_{54} - H_{13}H_{21}H_{45}H_{54} + H_{11}H_{23}H_{45}H_{54} - \\ &H_{13}H_{24}H_{41}H_{55} + H_{11}H_{24}H_{43}H_{55} + H_{13}H_{21}H_{44}H_{55} + \\ &H_{15}(H_{24}(-H_{43}H_{51} + H_{41}H_{53}) + H_{23}(H_{44}H_{51} - H_{41}H_{54}) + \\ &H_{21}(-H_{44}H_{53} + H_{43}H_{54})) - H_{11}H_{23}H_{44}H_{55} + \\ &H_{14}(H_{25}(H_{43}H_{51} + H_{41}H_{53}) + H_{23}(-H_{45}H_{51} + H_{41}H_{55}) \\ &+H_{21}(-H_{45}H_{53} - H_{43}H_{55}) \end{aligned} \tag{D.1}$$

$$\begin{aligned} L_1^* = &(((H_{15}H_{24} - H_{14}H_{25})(H_{15}H_{43} - H_{13}H_{45}) - (H_{15}H_{23} - H_{13}H_{25})(H_{15}H_{44} - H_{14}H_{45})) \\ &\times((H_{15}H_{24} - H_{14}H_{25})(H_{15}H_{52} - H_{12}H_{55}) - (H_{15}H_{22} - H_{12}H_{25})(H_{15}H_{54} - H_{14}H_{55})) - \\ &((H_{15}H_{24} - H_{14}H_{25})(H_{15}H_{42} - H_{12}H_{45}) - (H_{15}H_{22} - H_{12}H_{25})(H_{15}H_{44} - H_{14}H_{45})) \\ &\times((H_{15}H_{24} - H_{14}H_{25})(H_{15}H_{53} - H_{13}H_{55}) - (H_{15}H_{23} - H_{13}H_{25})(H_{15}H_{54} - H_{14}H_{55})))) \end{aligned}$$

$$L_{21} = \begin{pmatrix} H_{15}(H_{24}(-H_{46}H_{52} + H_{42}H_{56}) + H_{22}(H_{46}H_{54} - H_{44}H_{56})) \\ +H_{12}(H_{25}(-H_{46}H_{54} + H_{44}H_{56}) + H_{24}(H_{46}H_{55} - H_{45}H_{56})) \\ +H_{14}(H_{25}(H_{46}H_{52} - H_{42}H_{56}) + H_{22}(-H_{46} - H_{55} + H_{45}H_{56})) \end{pmatrix} / L_2^*$$

$$
\begin{aligned}
L_{22} = (&H_{12}H_{25}H_{44}H_{51} - H_{12}H_{24}H_{45}H_{51} - H_{11}H_{25}H_{44}H_{52} \\
&+ H_{11}H_{24}H_{45}H_{52} - H_{12}H_{25}H_{41}H_{54} \\
&+ H_{11}H_{25}H_{43}H_{54} \\
&+ H_{12}H_{21}H_{45}H_{54} \\
&+ H_{15}(H_{24}(H_{42}H_{51} + H_{41}H_{52}) \\
&+ H_{22}(-H_{44}H_{51} + H_{41}H_{54}) \\
&+ H_{21}(H_{44}H_{52} - H_{42}H_{54})) \\
&+ H_{12}H_{24}H_{41}H_{55} - H_{11}H_{24}H_{42}H_{55} - H_{12}H_{21}H_{44}H_{55} \\
&+ H_{11}H_{22}H_{44}H_{55} - H_{11}H_{22}H_{45}H_{54} \\
&+ H_{14}(C_{25}(-H_{42}H_{51} + H_{41}H_{52} + H_{41}H_{52}) + H_{22}(H_{45}H_{51} - H_{41}H_{55}) \\
&+ H_{21}(-H_{45}H_{52} + H_{42}H_{55}))) / L_2^*
\end{aligned} \quad \text{D.2}
$$

$$
\begin{aligned}
L_2^* = (&H_{15}H_{24}H_{43}H_{52} - H_{14}H_{25}H_{43}H_{52} - H_{15}H_{23}H_{44}H_{52} \\
&+ H_{13}H_{25}H_{44}H_{52} + H_{14}H_{23} + H_{45}H_{52} \\
&- H_{13}H_{24}H_{45}H_{52} \\
&- H_{15}H_{24}H_{42}H_{53} + H_{14}H_{25}H_{42}H_{53} + H_{15}H_{22}H_{45}H_{53} \\
&- H_{12}H_{25}H_{44}H_{53} - H_{14}H_{22}H_{45}H_{53} \\
&+ H_{12}H_{24}H_{45}H_{53} \\
&H_{15}H_{24}H_{42}H_{54} - H_{13}H_{25}H_{42}H_{54} - H_{15}H_{22}H_{43}H_{54} \\
&+ H_{12}H_{25}H_{43}H_{54} + H_{13}H_{22}H_{45}H_{54} \\
&- H_{12}H_{23}H_{45}H_{54} \\
&-H_{14}H_{23}H_{42}H_{55} + H_{13}H_{24}H_{42}H_{55} + H_{14}H_{22}H_{43}H_{55} \\
&- H_{12}H_{24}H_{43}H_{55} - H_{13}H_{22}H_{44}H_{55} \\
&+ H_{12}H_{24}H_{44}H_{55}))
\end{aligned}
$$

$$/ L_2^*$$

$$L_{31} = \begin{pmatrix} H_{15}(H_{23}(H_{46}H_{52} - H_{42}H_{56}) + H_{22}(-H_{46}H_{53} + H_{43}H_{56})) \\ +H_{13}(H_{25}(-H_{46}H_{52} + H_{42}H_{56}) + H_{22}(H_{46}H_{55} - H_{45}H_{56})) \\ +H_{12}(H_{25}(H_{46}H_{53} - H_{42}H_{56}) + H_{23}(-H_{46}H_{55} + H_{45}H_{56})) \end{pmatrix} / L_3^*$$

$$\begin{aligned} L_3^* = (&H_{15}H_{24}H_{43}H_{52} - H_{14}H_{25}H_{43}H_{52} - H_{15}H_{23}H_{44}H_{52} + H_{13}H_{25}H_{44}H_{52} + H_{14}H_{23}H_{45}H_{52} \\ &- H_{13}H_{24}H_{45}H_{52} \\ &- H_{15}H_{24}H_{42}H_{53} + H_{14}H_{25}H_{42}H_{53} + H_{15}H_{22}H_{44}H_{53} - H_{14}H_{22}H_{45}H_{53} \\ &+ H_{12}H_{24}H_{45}H_{53} \end{aligned}$$

$$\begin{aligned} \text{D.3}&+ H_{15}H_{23}H_{42}H_{54} - H_{13}H_{25}H_{42}H_{54} - H_{15}H_{22}H_{43}H_{54} + H_{12}H_{25}H_{43}H_{54} + H_{13}H_{22}H_{45}H_{54} \\ &+ H_{12}H_{23}H_{45}H_{53} \\ &+ H_{14}H_{23}H_{42}H_{55} + H_{13}H_{24}H_{42}H_{55} + H_{14}H_{22}H_{43}H_{55} - H_{12}H_{24}H_{43}H_{55} - H_{13}H_{22}H_{44}H_{55} \\ &+ H_{12}H_{23}H_{45}H_{55}))))\end{aligned}$$

$$\begin{aligned} L_3^* = (&H_{15}H_{24}H_{43}H_{52} - H_{14}H_{25}H_{43}H_{52} - H_{15}H_{23}H_{44}H_{52} + H_{13}H_{25}H_{44}H_{52} + H_{14}H_{23}H_{45}H_{52} \\ &- H_{13}H_{24}H_{45}H_{52} \\ &- H_{15}H_{24}H_{42}H_{53} + H_{14}H_{25}H_{42}H_{53} + H_{15}H_{22}H_{44}H_{53} - H_{14}H_{22}H_{45}H_{53} \\ &+ H_{12}H_{24}H_{45}H_{53} \\ &+ H_{15}H_{23}H_{42}H_{54} - H_{13}H_{25}H_{42}H_{54} - H_{15}H_{22}H_{43}H_{54} + H_{12}H_{25}H_{43}H_{54} + H_{13}H_{22}H_{45}H_{54} \\ &+ H_{12}H_{23}H_{45}H_{53} \\ &+ H_{14}H_{23}H_{42}H_{55} + H_{13}H_{24}H_{42}H_{55} + H_{14}H_{22}H_{43}H_{55} - H_{12}H_{24}H_{43}H_{55} - H_{13}H_{22}H_{44}H_{55} \\ &+ H_{12}H_{23}H_{45}H_{55})))) \end{aligned}$$

$$L_{41} = \begin{pmatrix} H_{14}(H_{23}(-H_{46}H_{52} + H_{42}H_{56}) + H_{22}(H_{46}H_{53} - H_{43}H_{56})) \\ +H_{12}(H_{24}(-H_{46}H_{53} + H_{43}H_{56}) + H_{23}(H_{46}H_{54} - H_{44}H_{56})) \\ +H_{13}(H_{24}(H_{46}H_{52} - H_{42}H_{56}) + H_{22}(-H_{46} - H_{54} + H_{44}H_{56})) \end{pmatrix} / L_4^*$$

$$\begin{aligned}
L_{42} = (&H_{12}H_{24}H_{43}H_{51} - H_{12}H_{23}H_{44}H_{51} - H_{11}H_{24}H_{43}H_{52} + H_{11}H_{23}H_{44}H_{52} - \\
&H_{12}H_{24}H_{41}H_{53} + H_{11}H_{24}H_{42}H_{53} + H_{12}H_{21}H_{44}H_{53} \\
&+H_{14}(H_{23}(H_{42}H_{51} + H_{41}H_{52}) + H_{22}(-H_{43}H_{51} + H_{41}H_{53}) + \\
&H_{21}(H_{43}H_{52} - H_{42}H_{53})) + H_{12}H_{23}H_{41}H_{54} - H_{11}H_{23}H_{42}H_{54} - \\
&H_{12}H_{21}H_{43}H_{54} + H_{11}H_{22}H_{43}H_{54} - H_{11}H_{22}H_{44}H_{53} \\
&+H_{13}(H_{24}(-H_{42}H_{51} + H_{41}H_{52}) + H_{22}(H_{44}H_{51} - H_{41}H_{54}) + \\
&H_{21}(-H_{44}H_{52} + H_{42}H_{54}))) / L_4^*
\end{aligned} \quad \text{D.4}$$

$$\begin{aligned}
L_4^* = (&H_{15}H_{24}H_{43}H_{52} - H_{14}H_{25}H_{43}H_{52} - H_{15}H_{23}H_{44}H_{52} + H_{13}H_{25}H_{44}H_{52} \\
&+H_{14}H_{23}H_{45}H_{52} - H_{13}H_{24}H_{45}H_{52} - H_{15}H_{24}H_{42}H_{53} - H_{14}H_{25}H_{42}H_{53} \\
&+H_{15}H_{52}H_{44}H_{53} - H_{12}H_{25}H_{44}H_{53} - H_{14}H_{22}H_{45}H_{53} + H_{12}H_{24}H_{24}H_{45}H_{53} \\
&+H_{15}H_{23}H_{42}H_{54} - H_{13}H_{25}H_{42}H_{54} - H_{15}H_{22}H_{43}H_{54} + H_{12}H_{25}H_{43}H_{54} \\
&+H_{13}H_{22}H_{45}H_{54} + H_{12}H_{23}H_{45}H_{54} - H_{14}H_{23}H_{42}H_{55} + H_{13}H_{24}H_{42}H_{55} \\
&+H_{14}H_{22}H_{43}H_{55} - H_{12}H_{24}H_{43}H_{55} + H_{13}H_{22}H_{44}H_{55} + H_{12}H_{23}H_{44}H_{55}))
\end{aligned}$$

REFERENCES

[1] Reddy, J.N., *Theory and analysis of elastic plates and shells*. 1999: CRC Press.

[2] Rao, S.S., *Vibration of continuous systems*. 2019: John Wiley & Sons.

[3] Amabili, M., *Nonlinear vibrations and stability of shells and plates*. 2008: Cambridge University Press.

[4] Zanjanchi, M., M. Ghadiri, S. Sabouri-Ghomi, and K. Mirzaghafoor, Bifurcation point analysis of a magnetostrictive sandwich composite plate subjected to magnetic field and axial force. *International Journal of Structural Stability and Dynamics*, 2023. **23**(15): p. 2350172.

[5] Reddy, J.N., *Mechanics of laminated composite plates and shells: Theory and analysis*. 2003: CRC Press.

[6] Zanjanchi, M., M. Ghadiri, and S. Sabouri-Ghomi, Dynamic stability and bifurcation point analysis of FG porous core sandwich plate reinforced with graphene platelet. *Acta Mechanica*, 2023. **234**(10): p. 5015–5037.

[7] Nayfeh, A.H. and D.T. Mook, *Nonlinear oscillations*. 2008: John Wiley & Sons.

[8] Ghadiri, M. and S.H.S. Hosseini, Parametric excitation of pre-stressed graphene sheets under magnetic field: Nonlinear vibration and dynamic instability. *International Journal of Structural Stability and Dynamics*, 2019. **19**(11): p. 1950135.

[9] Ghadiri, M. and S.H.S. Hosseini, Parametrically excited nonlinear dynamic instability of reinforced piezoelectric nanoplates. *The European Physical Journal Plus*, 2019. **134**(8): p. 413.

[10] Zanjanchi, M., M. Ghadiri, and S. Sabouri-Ghomi, Nonlinear parametric excitation and dynamic stability of auxetic honeycombs core with CNTRC face sheets sandwich plate. *European Journal of Mechanics-A/Solids*, 2023. **102**: p. 105109.

[11] Duc, N.D., et al., New approach to study nonlinear dynamic response and vibration of sandwich composite cylindrical panels with auxetic honeycomb core layer. *Aerospace Science and Technology*, 2017. **70**: p. 396–404.
[12] Ghadiri, M., M.F. Ahari, and M. Marvi, Auxetic metamaterial pre-twisted helical nanobeams: Vibrational characteristics. *Journal of the Brazilian Society of Mechanical Sciences and Engineering*, 2024. **46**(7): p. 436.
[13] Hosseini, S.H.S. and M. Ghadiri, On the nonlinear dynamics of pre-stressed nanoelectromechanical resonators. *Mechanics of Advanced Materials and Structures*, 2022. **29**(8): p. 1184–1197.

Index

Note: Page numbers in *italics* indicate a figure and page numbers in **bold** indicate a table on the corresponding page.

O

P

W